DATE DUE

NOV 7 1997			

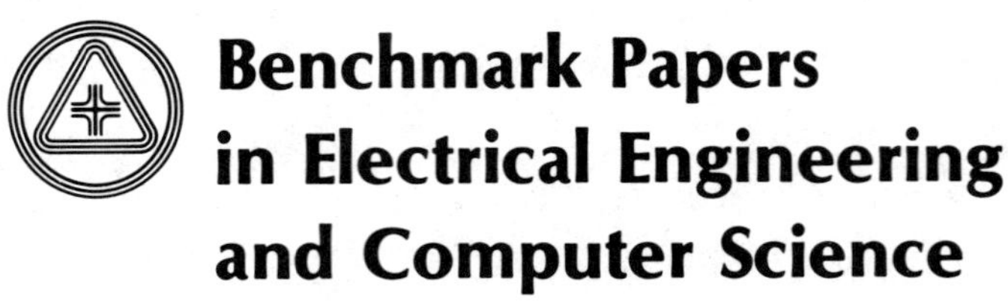

Benchmark Papers
in Electrical Engineering
and Computer Science

Series Editor: John B. Thomas
Princeton University

Volume

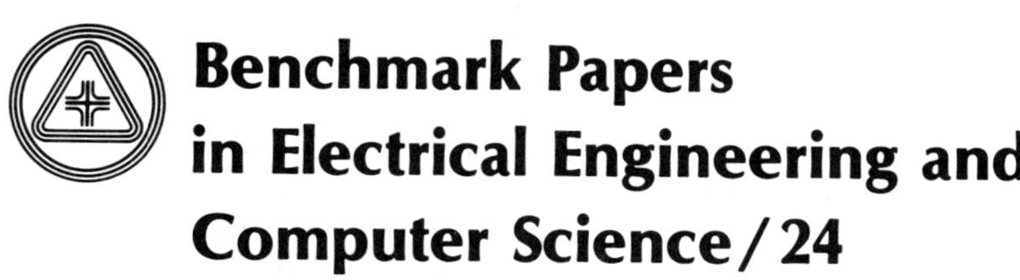

**Benchmark Papers
in Electrical Engineering and
Computer Science / 24**

A BENCHMARK® Books Series

DECONVOLUTION OF
SEISMIC DATA

Edited by

V. K. ARYA

Shell Oil Company
and

J. K. AGGARWAL

The University of Texas at Austin

Hutchinson Ross Publishing Company

Stroudsburg, Pennsylvania

LIBRARY OF CONGRESS CATALOGING IN PUBLICATION DATA
Main entry under title:
Deconvolution of seismic data.
    (Benchmark papers in electrical engineering and computer science; v.24)
    Includes indexes.
    1. Seismology—Data processing—Addresses, essays,
lectures. 2. Seismic reflection method—Deconvolution—
Address, essays, lectures. 3. Time-series analysis—
Addresses, essays, lectures. I. Arya, V. K., 1939–
II. Aggarwal, J. K. (Jagdishkumar Keshoram), 1936–
III. Series.
QE539.D4   551.2′2′02854   81-6311
ISBN 0-87933-406-1   AACR2

Distributed world wide by Academic Press,
a subsidiary of Harcourt Brace Jovanovich,
Publishers.

CONTENTS

Contents

SERIES EDITOR'S FOREWORD

The Benchmark Series in Electrical Engineering and Computer Science sifts, organizes, and makes readily accessible to the reader the vast literature that has accumulated. Although the series is not intended as a complete substitute for a study of this literature, it will serve at least three major critical purposes. First, it provides a practical point of entry into a given area of research. Each volume offers an expert's selection of the critical papers on a given topic as well as his views on its structure, development, and present status. Second, the series provides a convenient and time-saving means for study in areas related to but not contiguous with a reader's principal interests. Last, but by no means least, the series allows the collection, in a particularly compact and convenient form, of the major works on which present research activities and interests are based.

Each volume in the series has been collected, organized, and edited by an authority in the area to which it pertains. In order to present a unified view of the area, the volume editor has prepared an introduction to the subject, has included his comments on each article, and has provided a subject index to facilitate access to the papers.

We believe that this series will provide a manageable working library of the most important technical articles in electrical engineering and computer science. We hope that it will be equally valuable to students, teachers, and researchers.

This volume, *Deconvolution of Seismic Data,* has been edited by V. K. Arya of Shell Oil Company and J. K. Aggarwal of the University of Texas, Austin. It contains seventeen papers on the processing of seismic data, an area of increasing interest and sophistication. The editors are well known for their own contributions to system theory and digital signal processing and have produced a volume that is not only valuable in its own right, but also illustrates the increasingly complex applications of modern digital signal processing techniques.

JOHN B. THOMAS

ix

PREFACE

In contemporary exploration systems, almost all the seismic data are recorded and processed digitally. The volume of the seismic data is astronomical—billions of bits of information are recorded by the seismic crews every day in search for oil and gas accumulations in the subsurface. The success of an exploration venture depends largely on accurate delineation of the "favorable" buried geologic structures and stratigraphic traps. Accurate delineation requires careful and sophisticated processing of digitally recorded seismic data. Deconvolution is one of the major processing steps to resolve the structures and traps. In the last twenty-five years, a considerable amount of literature has evolved dealing with this subject. This volume is a collection of recent contributions describing various important deconvolution techniques as applied to seismic data. A large number of very worthwhile contributions have not been included in this collection due to limitations of space and money. However, every attempt has been made to provide the necessary references to the works that have not been included in this book.

This book is divided into four parts. Each part includes papers that are common in methodology, assumptions, or techniques of deconvolution. Each part contains a summary of the papers describing salient features and important technical points of the contributions. The parts are independent and are self-sufficient in references.

The editors would like to express their thanks to Dr. H. D. Holden of Shell Development and Dr. M. Schoenberger of Exxon Production Research for reviewing the list of selected papers and making numerous constructive suggestions. The editors would also like to express their gratitude to the management of Shell Development Company for the permission to edit this volume.

V. K. ARYA
J. K. AGGARWAL

CONTENTS BY AUTHOR

DECONVOLUTION OF SEISMIC DATA

INTRODUCTION

A large class of systems studied in scientific and engineering research can be represented by a convolution of the input (a band-limited function) and the system impulse response. The system impulse response itself can, in many cases, be represented as a convolution of impulse response functions of many subsystems. *Deconvolution* can be considered as the removal of any or all of the convolutional factors. This book is devoted to the techniques of deconvolution as applied to seismic data. In seismology, a seismogram or a seismic trace is often represented as the convolution of a seismic wavelet (a band-limited function) with the impulse response of the transmission path of the earth's subsurface. This impulse response is a convolution of the earth's reflectivity function, reverberation operator of the water layer (in marine environment), interbed and intrabed multiples within the geologic column, and earth's absorption effects. A deconvolution scheme can be designed to remove any or all of these factors. However, the major goal of seismic deconvolution is to retrieve the earth's reflectivity function. The modern geophysicist wants to know the reflectivity function with as much precision and resolution as possible for a variety of reasons. For example, large reflections may indicate sand layers that are potential hydrocarbon reservoirs. The estimated thickness of a possibly hydrocarbon-bearing sand layer strongly influences the amount of money one can afford to bid for drilling rights on a given tract at a competitive offshore lease sale. Thus the role of deconvolution in seismic data processing can hardly be overemphasized.

Since the early attempts of the Massachusetts Institute of Technology (MIT) Geophysical Analysis Group (Wadsworth et al., 1953; Robinson, 1957) to develop techniques for the deconvolution of seismic data, a large amount of literature has evolved dealing with this subject. This volume is a collection of recent contributions describing various important deconvolution techniques as applied to seismic data. A large number of very worthwhile contributions have not been included in this collection due to lack of space. However, every attempt has been made to provide the necessary references to the works that have not been included in this book.

This book is divided into four parts. Each part includes papers that are common in methodology, assumptions, or techniques of deconvolution. Each part also contains a summary of the papers describing salient features and important technical points of the contributions. The parts are independent and are self-sufficient in references.

Part I contains six papers based on Wiener filtering and predicitve deconvolution. The predictive deconvolution technique assumes that the wavelet is a minimum-phase wavelet and that the reflectivity spectrum is white. Treitel and Robinson (Paper 1) discuss the design of the prediction operator for the autocorrelation of the wavelet using a minimum-phase assumption. They also describe two important parameters, filter lag and filter length, and suggest that one might search for the desired output lag that leads to the minimum error for a selected filter length.

Peacock and Treitel (Paper 2) demonstrate that a deconvolution filter that ideally transforms an unknown signal into an impulse at zero delay is equivalent to a least-squares prediction error filter with unit prediction distance. They also discuss the use of least-squares prediction filters with prediction distance greater than unity. This generalized approach allows one to control the length of the desired output and hence to specify the degree of resolution.

Robinson (Paper 3) describes a method known as "dynamic predicitve deconvolution" that has two characteristic features: (1) the deterministic feature that the primary events are represented by a dynamic reflection series generated by a feedforward system; and (2) the deterministic feature that the reverberations attached to the primary events have the same minimum-phase wavelet generated by a feedback system. The entire trace including all primaries and multiples is used with dynamic predictive deconvolution applied to remove the effect of the feedback components.

Berkhout (Paper 4) discusses the error in the phase spectrum of Wiener filters, even if the wavelet has the minimum-phase property. This error is related to the noise power in the data and may lead to unreliable polarity predictions. He also elaborates on the superiority of two-sided least-squares inverse filtering, which produces zero-phase seismic wavelets of the desired bandwidth.

Unlike the other papers in Part I, in Paper 5, Wood, Heiser, Treitel, and Riley discuss two deterministic deconvolution techniques that require explicit knowledge of the source pulse shape for designing the deconvolution filter. It should be mentioned that all nonrealizable predicitve or Wiener filters require explicit knowledge of the source pulse shape.

The last paper in Part I, by Arya and Holden (Paper 6), also describes the deterministic deconvolution technique. It specifically addresses the effect of shooting geometry and earth's absorption on the useable bandwidth of the seismic data. This paper also gives an overview of the other deconvolution techniques, for example, predicitve deconvolution, homomorphic, and Kalman filtering.

Part II contains three papers. These papers are devoted to Wiener filtering to account for time-varying effects. Seismic data is inherently time-varying due to inelastic absorption of seismic energy in the earth as well as other time-varying phenomena like ghosting, near-surface effects, and receiver-array filtering. Most existing time-varying filtering techniques involve arbitrary division of a seismic trace into a number of gates of given length and assume time stationarity over these gates. Wang (Paper 7) describes a technique to establish optimum gate-lengths directly from the data. This is based on minimizing the mean square error between the true, and a given, approximated time-varying autocorrelation function.

Griffiths, Smolka, and Trembly (Paper 8) introduce an adaptive deconvolution procedure based on the use of a continuously adaptive linear-prediction operator in which the operator coefficients are updated using a steepest-descent method to minimize the mean square error between a given desired response signal and the adaptively filtered signal. Ristow and Kosbahn (Paper 9) also consider time-varying prediction filtering by adaptive updating of the predicitve operator. Their updating technique is essentially similar to that of Griffiths et al. (Paper 8). However, they provide more details on the updating procedure.

Part III deals with homomorphic filtering and consists of five contributions. Homomorphic systems are particularly useful in separating signals that have been combined through convolution. The success of homomorphic filtering depends on the degree of separability of the wavelet and the reflectivity function in the complex cepstrum domain. This method does not require the seismic wavelet to be minimum phase or the reflective series to be random in nature, both of which are essential requirements to Wiener filtering and predictive deconvolution. Ulrych (Paper 10) discusses the use of homomorphic filtering for the seismic deconvolution. He describes the recovery of the wavelet by short-time cepstral gating of the complex cepstrum of seismic trace and illustrates the effects of additive and convolutional noise on the deconvolution results. In particular, Ulrych observes that the additive-noise components complicate the phase spectrum and influence the long-time portion of the complex cepstrum to a much larger degree than the short-time cepstral portion; thus the wavelet can be recovered with good accuracy even in the presence of additive noise. Stoffa, Buhl, and Bryan (Paper 11) describe the application of homomorphic deconvolution to shallow-water marine seismology for wavelet deconvolution and dereverberation. They use an example with a mixed-phase reflector series and non-minimum-phase wavelet to show the potential effectiveness of homomorphic deconvolution where predictive deconvolution will be ineffective. Buttkus (Paper 12) investigates a similar procedure for deep-water reverberations. He suggests simple notch filters in the complex cepstrum domain for the suppression of multiples because of the simplicity and predictability of these components. Buttkus also describes the limitation of homomorphic deconvolution by cepstral gating due to its dependence on overlapping of the cepstral components of the wavelet and reflectivity functions.

Otis and Smith (Paper 13) discuss the improvement in recovering the wavelet by averaging the log spectra of several reflection records. This cepstral stacking will enhance the log spectrum of the wavelet, while averaging out the contributions of the log spectrum of the reflection coefficients.

In Papers 10 through 13, full-band homomorphic systems are used in the analysis. Tribolet (Paper 14) discusses a class of homomorphic systems matched to the bandpass nature of seismic signals and introduces homomorphic bandpass systems. He reports several novel results based on homomorphic bandpass systems and also describes the concept of short-time wavelet estimation that specifically accounts for the time-varying nature of seismic wavelet. Tribolet also discusses a reliable algorithm for phase unwrapping that is crucial in the computations required in homomorphic signal processing.

Part IV contains three papers based on Kalman filtering. The motivation for Kalman filtering is its potential for permitting more flexible modeling assumptions than the Wiener filtering approach. The Wiener filtering method is conventionally applied under the assumptions of a time-invariant system and stationary statistics. The Kalman filtering technique is suitable for handling time-varying models. Ott and Meder (Paper 15) present an application of Kalman filtering for prediction error filtering. Mendel (Paper 16) describes a quantitative evaluation of Ott and Meder's prediction error filter. Mendel and Kormylo (Paper 17) discuss the problem of estimating the reflection coefficients from a seismic trace. One interesting feature of this approach is the reversal of the intuitive role of input and plant, that is, the wavelet is considered as the plant and the reflectivity sequence is considered as the input and is modeled as noise. They observe that the problem is equivalent to one of estimating the random disturbance in a state equation and that the minimum-variance estimates of the reflection coefficient sequences are optimal, smoothed estimates. It should be noted that optimal, filtered estimates of reflection coefficient sequences do not exist.

Besides the techniques discussed in this collection, there are others like L_1 norm (Claerbout and Muir, 1973; Taylor, Banks, and McCoy, 1979) and minimum entropy methods (Wiggins, 1978; Ooe and Ulrych, 1979; Postic, Fourmann, and Claerbout, 1980; Levy and Clowes, 1980) that are currently being reported. The L_1 norm technique has robustness characteristics and results in a limited number of spikes corresponding to major reflection coefficients. The minimum-entropy technique for deconvolution also seeks the smallest number of large spikes that are consistent with the data. It should be noted that the design criterion in minimum-entropy deconvolution refers only to the simplicity of the output. The polarity or delay of the output spikes does not effect the maximization and hence cannot be accurately predicted. The minimum-entropy deconvolution is similar to predictive deconvolution in the sense that operators are determined directly from the data, but predictive deconvolution seeks to maximize entropy or disorder as opposed to the former method, which maximizes order or minimizes entropy.

Each method of deconvolution is based on certain models and assumptions. The success in deconvolution depends on how well those assumptions are met in practice. Wiener filtering (predictive deconvolution), as discussed in Part I, is the most widely used in the seismic data processing industry. It is very effective for the situations in which the wavelet is minimum phase. Generalizations of Wiener filters, as discussed in Part II, Wiener filtering (time-varying case), overcome the deficiency of Wiener filters to accommodate the time-varying nature of seismic data. However, if the wavelet is non-minimum-phase, the techniques of homomorphic filtering (Part III) or Kalman filtering (Part IV) can be applied. In Kalman filtering situations, however, one will have to model the non-minimum-phase behavior of the wavelet, which is generally not available to the geophysicist designing a deconvolution operator. Homomorphic filtering for seismic deconvolution has proven to be quite useful. This method does not require the seismic wavelet to be minimum phase or the reflector series to be random in nature, which are essential requirements for Wiener filtering and predictive deconvolution.

Berkhout and Zaanen (1976) provide an extensive analytical comparison between Wiener filtering, Kalman filtering, and deterministic least-squares estimation. Lines and Ulrych (1977) discuss Wiener filtering, homomorphic filtering, and maximum-entropy method for wavelet estimation and inverse filter design for deconvolution. They note that the autocorrelation function can be used to monitor the nonstationarity of the seismic trace and that the autocorrelation function can be estimated in an optimum fashion by using the maximum entropy method. Differences between minimum-phase Wiener deconvolution and maximum entropy become more pronounced for short data gates. As a result, Lines and Ulrych prefer the maximum-entropy approach for time-adaptive deconvolution. They also discuss the effects of additive noise on the computation of the complex cepstrum and ways to minimize problems in phase unwrapping in homomorphic deconvolution.

Geophysical industry is very competitive in nature; thus a lot of current research in industry is of proprietary nature. Therefore, some of the methods currently used are not available to the general public. However, this collection presents the state of the art in seismic deconvolution as available in literature.

REFERENCES

Berkhout, A. J., and P. R. Zaanen, 1976, A Comparison between Wiener Filtering, Kalman Filtering, and Deterministic Least Squares Estimation, *Geophys. Prospect.* **24**:141–197.

Claerbout, J. F., and F. Muir, 1973, Robust Modelling with Erratic Data, *Geophysics* **38**:826–844.

Levy, S., and R. M. Clowes, 1980, Debubbling: A Generalized Linear Inverse Approach, *Geophys. Prospect.* **28**:840–858.

Lines, L. R., and T. J. Ulrych, 1977, The Old and the New in Seismic Deconvolution and Wavelet Estimation, *Geophys. Prospect.* **25**:512–540.

Ooe, M., and T. J. Ulrych, 1979, Minimum Entropy Deconvolution with an Exponential Transformation, *Geophys. Prospect.* **27**:458–473.

Postic, A., J. Fourmann, and J. Claerbout, 1980, Parsimonious Deconvolution, paper presented at the 50th Annual International Meeting of SEG, Houston.

Robinson, E. A., 1957, Predictive Deconvolution of Seismic Traces, *Geophysics* **22**:767–778.

Taylor, H. L., S. C. Banks, and J. F. McCoy, 1979, Deconvolution with the L_1 Norm. *Geophysics* **44**:39–52.

Wadsworth, G. P., et al., 1953, Detection of Reflections on Seismic Records by Linear Operators, *Geophysics* **18**:539–586.

Wiggins, R. A., 1978, Minimum Entropy Deconvolution, *Geoexploration* **16**:21–35.

ADDITIONAL READINGS

Aftab Alam, M. and C. J. Sicking, 1981, Recursive Removal and Estimation of Minimum-phase Wavelet, *Geophysics* **46**:1379–1391.

Chang, J. Y. and M. M. Backus, 1981, Emergent Angle Dependent Deconvolution, paper presented at the 51st Annual International Meeting of SEG, Los Angeles.

Deeming, T. J. and H. L. Taylor, 1981, Use of an Improved L_1 Norm Minimization for Wavelet Processing of Seismic Sections, paper presented at the 51st Annual International Meeting of SEG, Los Angeles.

Godfrey, R. and F. Rocca, 1981, Zero Memory Non-linear Deconvolution, *Geophys. Prospect.* **29**:189–228.

Hampson, D. and M. Galbraith, 1981, Wavelet Extration by Sonic Log Correlation, paper presented at the 51st Annual International Meeting of SEG, Los Angeles.

Stone, D. G., 1981, Seismic Filter Design from Borehole Data, paper presented at the 51st Annual International Meeting of SEG, Los Angeles.

van Reil, P., A. J. Berkhout, and E. Poggiagliolmi, 1981, Deconvolution by Means of Parameter Estimation, paper presented at the 51st Annual International Meeting of SEG, Los Angeles.

Part I

WIENER FILTERING

Editors' Comments
on Papers 1 Through 6

In seismic exploration of earth's subsurface, an acoustic energy waveform—a seismic wavelet in geophysical terms—is emitted and then pressure- or velocity-sensitive detectors record the reflections to detect density or velocity changes indicative of rock packages at various depths in the subsurface. A seismic trace $[s(n)]$ is assumed to be the convolution of the seismic wavelet $[w(n)]$, with the earth's reflectivity function $[r(n)]$. The goal of a deconvolution scheme is to remove the effect of the wavelet from the seismic trace, thus retrieving the earth's reflectivity function. In 1952 the MIT Geophysical Analysis Group (Wadsworth et al., 1953; Robinson, 1957) began to apply prediction techniques developed by Norbert Wiener (1949) to the deconvolution of seismic data. Basically, the idea is that if a seismic trace is considered to be from a stationary random process, the predictable part of the trace is due to the wavelet whereas the earth's reflectivity is completely unpredictable (that is, a white random process). A Wiener filter is used to derive a prediction operator $a(n)$ that optimally predicts the value of the seismic trace at a later time $m = n + \alpha$

$$\hat{s}(m) = \sum_n s(n)a(m - n)$$

The prediction error, or unpredictable part of the seismic trace $\hat{r}(n)$, is an estimate of the reflectivity function

$$\hat{r}(n) = s(n) - \hat{s}(n)$$

Various authors (Rice, 1962; Wiggins and Robinson, 1965; Robinson and Treitel, 1969; Treitel, 1970) have addressed the problem of design of Wiener filters and their usefulness in seismic deconvolution. Treitel and Robinson (Paper 1) treat the design as follows: the autocorrelation of a seismic trace, under the assumption of a random reflectivity function, becomes the autocorrelation of the wavelet. Using a minimum-phase assumption for the wavelet, one can design a prediction operator from the autocorrelation of the wavelet. The least-squares design criterion leads to a set of normal equations that are usually solved by Levinson's (1947) recursion technique. Further, Treitel and Robinson discuss two important design parameters, filter lag and filter length. One can always improve the performance of the filter by increasing its length. Usually physical and computational considerations prohibit this. Thus they suggest that one might search for that desired output lag that leads to the minimum error for a selected filter length.

Peacock and Treitel (Paper 2) demonstrate that a deconvolution filter that ideally transforms an unknown signal to an impulse at zero delay is equivalent to a least-squares prediction error filter with unit prediction distance. They also discuss the use of least-squares prediction filters with prediction distances greater than unity. This generalized approach allows one to control the length of the desired output and hence to specify the desired degree of resolution. They illustrate the suppression of both short-period and long-period reverberations using the flexibility in prediction distance. Prediction error filters are derived from the autocorrelation of the data. For short data gates the bias and end effects in estimating the autocorrelation function may have adverse effects in the solution of Toeplitz equations using Levinson's recursive scheme. Burg (1975) developed an alternative approach that he refers to as the *maximum entropy technique*. Instead of estimating the autocorrelation directly from the data, he estimates the minimum-phase prediction error filter directly from the data. Various applications of Burg's technique have been reported in the geophysics literature (Ulrych et al., 1973; Lacoss, 1971; Negi and Dimri, 1979). A very good account of Burg's technique, prediction filtering, and least-squares inverse filtering is given by Claerbout (1976). Robinson (1967) gives many useful computer programs for the implementation of prediction error and least-squares inverse filtering. For a good overview of linear prediction, see Makhoul (1975), and for the developments in the theory of least-squares estimation in the last three decades, see Kailath (1974).

The predictive deconvolution technique discussed in Papers 1 and 2 is based on a random reflection model. Robinson (Paper 3) describes an alternate method known as *dynamic predictive deconvolution,* which is based on a dynamic reflection seismic model and is approximately realized by a horizontally layered system subject to normal-incidence wave motion resulting from a unit source spike at the surface. The model has two characteristic features: (1) the deterministic feature that the primary events are approximately represented by a dynamic-reflection series generated by a feedforward system; and (2) the deterministic feature that the reverberations attached to the primary events have the same minimum-phase wavelet generated by a feedback system. It uses the entire trace including all primaries and multiples. It is known that the layered system acts as a pure feedback system in producing the transmitted wave and is thus minimum phase. Robinson also observes that for the seismic trace, that is, reflected wave, the layered system acts with both feedback and feedforward components. Thus one can deconvolve the observed trace by the prediction error operator that contracts the transmitted wave to a spike. The resulting deconvolution is the waveform due to the feedforward components alone representing the wanted dynamic structure. The effects of feedback components, which represent the unwanted reverberation effect of the layered system, is removed by dynamic predictive deconvolution.

Each method of deconvolution is based on certain models and assumptions; the success in deconvolution depends on how well those assumptions are met in practice. In predictive deconvolution, it is assumed that the wavelet is minimum phase. Berkhout (Paper 4) points out that in Wiener filtering, the phase spectra of these filters may be in error even if the seismic wavelet has the minimum-phase property. This error is related to the noise power and may lead to unreliable polarity predictions. He illustrates the comparison of Wiener filtering with wavelet deconvolution. He observes that one-sided, optimum-lag, least-squares inverse filtering actually equals two-sided, least-squares inverse filtering, that is, the two-sided, least-squares inverse filter's being delayed such that it becomes one-sided. Optimizing the lag value actually means optimizing the length of the anticipation part of the two-sided filter with respect to total filter length. He also elaborates on the superiority of two-sided, least-squares inverse filtering, which produces zero-phase seismic wavelets of desirable bandwidth. The properties of zero-phase and minimum-phase wavelets are discussed by Berkhout (1974) and Schoenberger (1974).

In recent years deterministic, rather than statistical, techniques have been used for wavelet deconvolution. Wood et al. (Paper 5) discuss two deterministic deconvolution techniques that require explicit knowledge of the source pulse shape such as the far-field signature. The principal disadvantage of these techniques is that they require the knowledge of source signature. However, zero-phase filtering techniques employing cross-correlation have many significant advantages accruing from the matched filtering (that is, cross-correlation) step. Cross-correlation achieves an improved signal to noise (S/N) ratio prior to designing Wiener inverse filters whereas matched filtering attenuates random noise and improves the S/N ratio by a factor proportional to the ratio of S/N power.

Arya and Holden (Paper 6) also describe a deterministic deconvolution technique using measured far-field signatures. They specifically discuss other physical effects such as earth's absorption, which must be accounted for in a proper deterministic deconvolution scheme. This paper also gives an overview of predictive Wiener filtering, homomorphic filtering, and Kalman filtering. Ziolkowski et al. (1980) also address a deterministic deconvolution scheme. They require a fundamental change in the field technique such that two different seismograms are generated from each source-receiver pair: the source and receiver locations stay the same, but the source used to generate one seismogram is a scaled version of the source used to generate the other. A scaling law is used to provide the relationship between the two source signatures.

REFERENCES

Berkhout, A. J., 1974, Related Properties of Minimum-Phase and Zero-Phase Time Functions, *Geophys. Prospect.* **22**:683–709.
Burg, J. P., 1975, Maximum Entropy Spectral Analysis, Ph.D. thesis, Department of Geophysics, Stanford University, Stanford, Calif.

Claerbout, J. F., 1976, *Fundamentals of Geophysical Data Processing with Applications to Petroleum Prospecting*, McGraw-Hill, New York.

Kailath, T., 1974, A View of Three Decades of Linear Filtering Theory, *IEEE Trans. Inf. Theory* **IT-20**:146–181.

Lacoss, R. T., 1971, Data Adaptive Spectral Analysis Methods, *Geophysics* **36**:661–675.

Levinson, N., 1947, The Wiener RMS (Root Mean Square) Error Criterion in Filter Design and Prediction, *J. Math Phys.* **25**:261–278.

Makhoul, J., 1975, Linear Prediction: A Tutorial Review, *IEEE Proc.* **63**:561–580.

Negi, J. G., and V. P. Dimri, 1979, On Wiener Filter and Maximum Entropy Method for Multichannel Complex Systems, *Geophys. Prospect.* **27**:156–167.

Rice, R. B., 1962, Inverse Convolution Filters, *Geophysics* **27**:4–18.

Robinson, E. A., 1957, Predictive Decomposition of Seismic Traces, *Geophysics* **22**:767–778.

Robinson, E. A., 1967, *Multichannel Time Series Analysis with Digital Computer Programs*, Holden-Day, San Francisco, Calif.

Robinson, E. .A, and S. Treitel, 1969, *Robinson-Treitel Reader*, Siesmograph Sciences Co.

Schoenberger, M., 1974, Resolution Comparison of Minimum-Phase and Zero-Phase Signals, *Geophysics* **39**:826–833.

Treitel, S., 1970, Principles of Digital Multichannel Filtering, *Geophysics* **35**:785–811.

Ulrych, T. J., D. E. Smylie, O. G. Jensen, and G. K. C., Clarke, 1973, Predictive Filtering and Smoothing of Short Records by Using Maximum Entropy, *J. Geophys. Res.* **78**:4959–4964.

Wadsworth, G. P., E. A. Robinson, J. G. Bryan, and P. M. Hurley, 1953, Detection of Reflections on Seismic Records by Linear Operators, *Geophysics* **18**:539–586.

Wiener, N., 1949, *The Extrapolation, Interpolation and Smoothing of Stationary Time Series with Engineering Applications*, M.I.T. Press, Cambridge, Mass.

Wiggins, R. A., and E. A. Robinson, 1965, Recursive Solution to the Multichannel Filtering Problem, *J. Geophys. Res.* **70**:1885–1891.

Ziolkowski, A. M., W. E. Lerwill, D. W. March, and L. G. Peardon, 1980, Wavelet Deconvolution Using a Source Scaling Law, *Geophys. Prospect.* **28**:872–901.

ADDITIONAL READINGS

Deregowski, S. M., 1971, Optimum Digital Filtering and Inverse Filtering in the Frequency Domain, *Geophys. Prospect.* **19**:729–768.

De Voogd, N., 1974, Wavelet Shaping and Noise Reduction, *Goephys. Prospect.* **22**:354–369.

Douze, E. J., 1971, Prediction Error Filters, White Noise and Orthogonal Coordinates, *Geophys. Prospect.* **19**:253–264.

Ekstrom, M. P., 1973, A Spectral Characterization of the Ill Conditioning in Numerical Deconvolution, *IEEE Trans. Audio Electroacoust.* **AU-21**:344–348.

Fourmann, J. M., 1980, Signature Processing-Review and Comparison of Some Processing Principles, paper presented at the 42nd Annual EAEG Meeting, Istanbul.

Ford, W. T., 1978, Optimum Mixed Delay Spiking Filters, *Geophysics* **43**: 125–132.

Ford, W. T., and J. H. Hearne, 1966, Least-Squares Inverse Filtering, *Geophysics* **31**:917–926.

Galbraith, J. N., 1971, Prediction Error as a Criterion for Operator Length, *Geophysics* **36**:261–265.

Lackoff, M. R., and L. R. LeBlanc, 1975, Frequency-Domain Seismic Deconvolution Filtering, *Acoust. Soc. Am. J.* **57**:151–159.

Mereu, R. F., 1976, Exact Wave-Shaping with a Time-Domain Digital Filter of Finite Length, *Geophysics* **41**:659–672.

Pratt, W. K., 1972, Generalized Wiener Filtering Computation Techniques *IEEE Trans. Computers* **C-21**:636–641.

Robinson, J. C., 1972, Computer-Designed Wiener Filters for Seismic Data, *Geophysics* **37**:235–259.

Sicking, C. J., 1980, Windowing and Estimation Variance in Deconvolution, paper presented at the 50th Annual International Meeting of SEG, Houston.

Treitel, S., and R. J. Wang, 1976, The Determination of Digital Wiener Filters from an Ill-Conditioned System of Normal Equations, *Geophys. Prospect.* **24**:317–327.

Wang, R. J., and S. Treitel, 1973, The Determination of Digital Wiener Filters by Means of Gradient Methods, *Geophysics* **38**:310–326.

Webster, G. M., 1978, *Deconvolution*, vol. I, Geophysics Reprint Series.

The Design of High-Resolution Digital Filters

S. TREITEL AND E. A. ROBINSON, MEMBER, IEEE

Abstract—Seismic recordings made with standard filters often afford insufficient resolution for overlapping reflected events. In this treatment we apply least squares Wiener theory to the design of high-resolution digital time-domain filters. Under the assumption that estimates of the seismic pulse shape are available, we present techniques that allow one to calculate digital filters which transform this pulse into one which is sufficiently sharp so that it can be distinguished against a background of noise. The two design criteria governing filter performance are filter lag and filter memory function duration. The performance of a Wiener filter is numerically measurable by a quantity which we call the filter performance parameter P, where $0 \leq P \leq 1$. The quality of the filter output improves as P approaches unity. We thus seek that combination of lag and memory function duration that maximizes P. This goal can be accomplished by the study of a two-dimensional display of P vs. lag and memory function duration. The proposed design techniques are illustrated by means of numerical examples.

I. INTRODUCTION

A MAJOR DIFFICULTY in seismic interpretation is the lack of resolution of field-recorded seismic events. As a result, increased attention is being given to the problem of designing high resolution filters that can increase the amount of detail available on a seismogram. These filters always make use of the high-frequency components present in the recorded seismic signal. This means that field techniques must allow one to record these high frequencies with a minimum of distortion and corrupting noise. Such an aim is difficult to achieve in practice with standard field instrumentation. As a result, the past few years have seen considerable activity in the techniques of digital field recording. One obtains in this way a much larger dynamic range than has been possible before. If the high frequencies in the seismic signal are now recordable with sufficient fidelity, the search for filtering techniques that make use of such high-frequency information becomes justified. Viewed in the time-domain, our objective becomes the design of filters which convert broad, overlapping seismic signals into sharp, clearly resolved pulses. Ideally, we would like to compress a seismic reflection wavelet into a Dirac delta function, that is, into a "spike." Yet we know that this desire can never be satisfied completely, since the spectrum of a spike contains uniform energy at all frequencies. A further serious obstacle is that a filter which converts a broad input pulse into a pure spike must in most cases of practical interest have an infinitely long memory function. Nevertheless, if we are willing to accept filter output wavelet shapes that are

approximations to a pure spike, our search is again a meaningful one. In fact, we need not require the output wavelet to be even an approximation to a spike; since our basic objective is increased resolution, any output wavelet shape which is noticeably sharper than the input wavelet will satisfy our needs. Let us call a filter which shapes an input into some desired output a *"shaping" filter*. Then the above spiking filter becomes a special case of a shaping filter, i.e., our desired shape is, in this case, simply a spike.

One could attempt to build such filters by synthesizing them from lumped elements, but the design problems are quite formidable (Ricker [1]). Therefore, many workers have preferred to attack the problem by means, of digital time-domain filters (Robinson [2], Kunetz [3] Simpson et al. [4], Foster et al [5], Rice [6], Schneider et al. [7], and Treitel and Robinson [8]). We shall use this approach in our present treatment. Our main tool is the digital Wiener filter, which has been described in terms of its seismic applications by the above mentioned workers. We shall give a heuristic description of this technique in Section II of this paper and give some of the necessary mathematics in the Appendixes. Our main objective here is to illustrate the power and flexibility of the high-resolution shaping filter by means of a few simple numerical examples. In so doing, we shall examine some of the design criteria needed in order to obtain satisfactory filter performance.

II. THE ELEMENTS OF WIENER FILTERING

The basic elements of time-domain Wiener filtering are summarized in Fig. 1. Given an input wavelet b_t and a desired output wavelet d_t, the problem is to find a filter f_t whose output $c_t = f_t * b_t$ deviates from the desired output d_t in some minimum sense. Here t is the discrete time variable, and the symbol $*$ denotes convolution. The difference between the desired output d_t and the actual output c_t is the error e_t. If we require that the energy of the error signal

$$I = \sum_t e_t^2$$

be a minimum, the filter f_t is calculable in a straightforward way. The derivation is given in Appendix I. This filter is known as the "Wiener" filter; other names for it in the literature are "least squares" filter or "optimum least squares" filter. It is important to realize that the Wiener filter does not guarantee that the error energy be small in any absolute sense, but only that it be as small as possible consistent with the particular situation at hand. The basic theory and computational

Manuscript received July 14, 1965; revised January 13, 1966.
E. A. Robinson is a Consultant for the Pan American Petroleum Corporation.
S. Treitel is with the Pan American Petroleum Corporation, Tulsa, Okla.

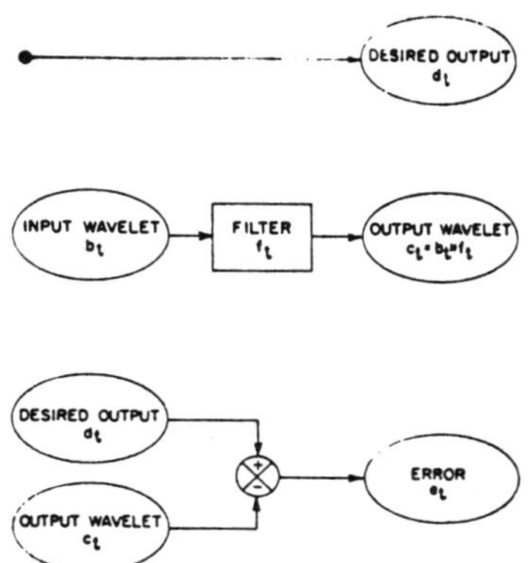

Fig. 1. The elements of Wiener filtering.

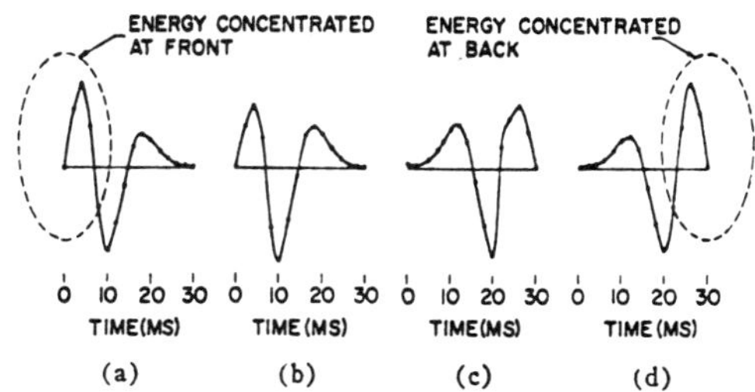

Fig. 2. A suite of four wavelets having the same frequency content. (a) Minimum delay. (b) Mixed delay. (c) Mixed delay. (d) Maximum delay.

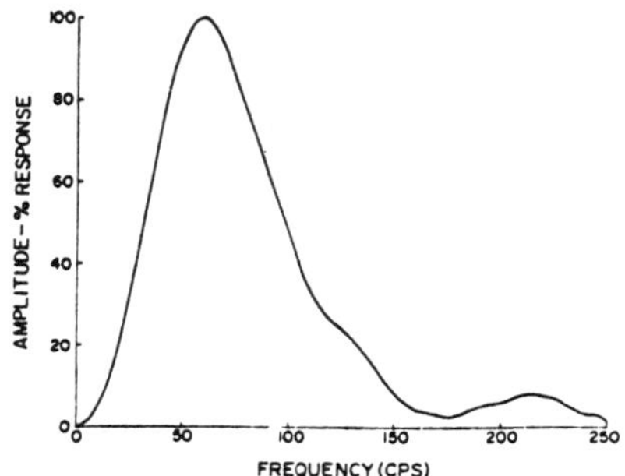

Fig. 3. The common frequency amplitude spectrum of the suite of four wavelets of Fig. 2.

procedures allowing one to obtain the discrete, time-domain Wiener filter are given by Levinson [9]. The computing method used to generate the various shaping filters displayed in the present treatment is summarized in a recent paper by Robinson [10]. A more detailed derivation is given in Appendix III. The method is essentially based on that of Levinson [9] and allows one to calculate time-domain Wiener filters of varying lengths by a fast recursive scheme.

If the desired output d_t is some arbitrary function of time, then f_t is the memory function of a Wiener *shaping* filter. For the special case,

$$d_t = 1 \quad \text{for } t = t_p,$$
$$d_t = 0 \quad \text{for } t \neq t_p,$$

where t_p is a particular value of t, f_t is the memory function of a Wiener *spiking* filter.

III. The Energy Distribution in a Wavelet

Before entering into a discussion of high-resolution-shaping filters, let us consider first how the energy of a seismic wavelet is distributed throughout the time range of its duration. We shall find that the manner in which this energy is distributed provides a basis for understanding how high-resolution filters work. In Fig. 2, four different wavelet shapes are shown. Each of these wavelets has the same time duration and the same frequency content (see Fig. 3) so that the amounts of high frequencies, intermediate frequencies, and low frequencies are precisely the same in each wavelet. But, while these wavelets all have a common frequency amplitude spectrum[1] and a common time duration, they differ in their time distribution of energy.

---

[1] The term "amplitude spectrum," as we use it, refers to the absolute value of the Fourier transform of the time function. That is, by "amplitude spectrum" we mean the magnitude of the complex-valued "amplitude-and-phase spectrum."

Wavelet (a) has its energy concentrated as closely as possible to its front end, that is, the energy is *delayed* in time the smallest possible amount for any wavelet with the amplitude spectrum shown in Fig. 3. *Under the restriction that the frequency content of the wavelet suite of Fig. 2 be fixed*, it is not possible for wavelet (a) to have its energy concentrated still closer to its front end. For this reason, the wavelet (a) is called the *minimum-delay* wavelet of the suite. Of course, if we were allowed to add still higher frequencies to the spectrum of Fig. 3, we could make the wavelet (a) have a still sharper leading edge, but under the restriction that the amplitude spectrum be held constant, the minimum-delay wavelet has a leading edge which is sharper than the leading edge of any other wavelet of the suite. The authors have discussed these matters rigorously elsewhere (Robinson and Treitel [11]).

At the other extreme, we have wavelet (d). This wavelet has its energy concentrated as closely as possible to its back end. In other words, the energy is *delayed* in time the greatest possible amount for *any wavelet* with the same amplitude spectrum and time duration as wavelet (a). It is not possible to make wavelet (d) have its energy concentrated any further to the back under the restriction that the amplitude spectrum and time duration of the wavelet suite of Fig. 2 be as specified. For this reason, wavelet (d) is called a *maximum-delay* wavelet. It is easy to see that wavelet (d) is nothing more than the time reverse of wavelet (a), that is, the maximum-delay wavelet is the time reverse of the minimum-delay wavelet. It also follows that the wavelet (d) has the sharpest possible trailing edge of any wavelet with our given amplitude spectrum and time duration.

Intermediate between the minimum-delay and the maximum-delay wavelets are the mixed-delay wavelets. As we would expect, there are many possible different mixed-delay wavelets as we run through the transition of energy concentration from the front to the back edge of the wavelet. In Fig. 2 we show two of the possible mixed-delay wavelets having the same time duration and amplitude spectrum as wavelets (a) and (d). We see that the energy in wavelet (b) is delayed, with respect to the minimum-delay wavelet (a), but not delayed so much as in the case for wavelet (c).

If our wavelets are described in continuous time, there exists an infinitely large sequence of wavelets ranging from the minimum to the maximum-delay case. All these wavelets have a common amplitude spectrum. If we deal in discrete time, so that the time variable t is replaced by a time index i that assumes integer values corresponding to the time points that are sampled,

$$i = 0, 1, 2, \cdots,$$

it is possible to show that the number of wavelets with the same time duration and having a given amplitude spectrum is finite. Let us assume that a wavelet is sampled at equal increments of time Δt at the points $i = 0, 1, 2, \cdots, n$. We then say that this wavelet, described by $(n+1)$ sampled values, is an $(n\Delta t)$-duration wavelet. In general, there will be at most 2^n different discretely sampled $(n\Delta t)$-duration wavelets having a given amplitude spectrum. There are no other finite duration wavelets with this given amplitude spectrum, although there are an infinite number of infinite-duration wavelets having this given amplitude spectrum. In Fig. 2, the points at which the waveshapes have been sampled are indicated by heavy black dots. We note that all four of these waveshapes are sampled at 2 ms increments at a total of 16 discrete points, i.e., they are 30 ms duration wavelets. We have here chosen to show only four wavelets of the suite having the amplitude spectrum given in Fig. 3. While all members of a suite have a common amplitude spectrum, their phase spectra are distinct. In fact, the minimum-delay wavelet (a) is the "minimum phase-lag" member of the suite, while the maximum-delay wavelet (d) is the "maximum phase-lag" member of the same suite (Robinson and Treitel [11]).

IV. Wavelet Spiking

We wish now to consider the design problem for what we might call a prototype filter, namely, the spiking filter. Let us consider first an isolated event on a seismogram, and let us use for this purpose a reflection wavelet taken from an actual trace. This wavelet is the mixed-

delay pulse (b) of Fig. 2.[2] Our problem is to find a filter which contracts this reflection wavelet to a spike. In theory, this purpose may be achieved *exactly* if we could use a filter whose unit impulse response, or memory function, is allowed to become infinitely long. For *exact* filter performance, we will also need, in general, to delay the desired spike an infinite amount of time relative to the original pulse (Claerbout and Robinson [12]).[3] In practice we must limit ourselves to digital filters whose memory functions have finite duration, and hence, at best, we can achieve our purpose approximately.

Let us suppose that for practical reasons we want first to restrict ourselves to filters which have a memory function that is of the order of the wavelet's duration. Now we are at liberty to place the desired spike in time wherever we please. For example, in Fig. 4 we show six possible positions of the spike. In the first case this desired spike is placed at the very beginning of the wavelet we wish to filter, that is, the desired spike's time *lag* with respect to the onset of our wavelet is zero. The five following cases considered are desired spike locations at lags of 8, 16, 24, 32, and 40 ms, respectively. The duration of our input wavelet is 30 ms, and the sampling increment Δt is 2 ms. Spiking filters were computed for each of the six cases of Fig. 4. The duration of the filter's memory functions was in all cases made equal to 30 ms. The corresponding filter outputs are shown in the first six displays of Fig. 5. We note that the position of the spike is an important factor governing the fidelity with which the actual output resembles the desired spike.

A very convenient way to measure the performance of a Wiener filter is to consider the value of its normalized minimum error energy E (Levinson [9]). This quantity is the value of the average squared error divided by the zeroth lag of the desired output's autocorrelation function. When the filter performs perfectly, $E = 0$, which means that the desired and actual filter outputs agree for all values of t. On the other hand, the case $E = 1$ corresponds to the worst possible case, that is, there is no agreement at all between desired and actual outputs. Instead of the quantity E, it is desirable for a number of reasons to consider the one's complement of E, which we shall call the filter performance parameter, P (see Appendix I),

$$P = 1 - E.$$

Perfect filter performance then occurs when $P = 1$, while the worst possible situation arises when $P = 0$.[4]

The filter outputs, shown in Fig. 5, indicate how the

---

[2] The remaining wavelets (a), (c), and (d) of Fig. 2 were calculated by passing the original seismic wavelet (b) through a series of appropriate digital *phase-shift* filters. The method for doing this has been given in Robinson and Treitel [11].

[3] The exception to this rule occurs when the input wavelet is minimum-delay. Then the desired spike does not need to be delayed for *exact* filter performance, although the filter's memory function must still have infinite duration (see Fig. 11).

[4] Levinson's notation has here been modified. Our E is his V, while our P is his E_M.

value of P varies as the spike is progressively lagged in time. We recall that the filters whose outputs are shown here all have 30 ms duration memory functions. *For a constant filter duration*, then, we might suppose that there must exist at least one value of lag at which P is as large as possible. The display of Fig. 5 shows that the lag of the desired output is of crucial importance in determining filter performance. Our aim must then be to find the *"optimum-lag"* filter at a given constant (and finite) filter duration.

In Fig. 6 we show a plot of P vs. the lag of the desired output spike for a family of 30 ms duration filters. We observe that this curve exhibits several maxima. The highest point of the curve occurs at a lag of 30 ms, and thus, the choice of this lag for the desired spike leads to the "optimum-lag" 30 ms duration filter. The actual output of this filter is shown in the last picture of Fig. 5. We note that the performance of the filter, at lag $= 32$ ms, is not much worse than that of the optimum-lag filter; this is so because the P vs. lag curve of Fig. 6 is nearly flat around its highest point.

The memory function of the optimum-lag 30 ms duration spiking filter is shown in Fig. 7(a). Figure 7(b) gives the amplitude spectrum of this filter. As we expect, this spectrum is richer in higher frequencies than that of the original pulse, whose spectrum we also show for comparison.

Let us next see what happens as we increase the filter memory duration at constant lag. Figure 8 shows a plot of P vs. filter length for a desired spike lag of 16 ms. We now observe that this curve *is* monotonic, and that it asymptotically approaches $P = 0.86$ as the filter length becomes larger and larger.

Now, let us examine the performance of some further "optimum-lag" filters. In Fig. 9 we show the actual outputs of the 40, 60, and 100 ms duration "optimum-lag" spiking filters. In particular, we note that the P-value of 0.98 for the last of these three filters leads to excellent agreement between desired and actual outputs. However, a filter response function which is more than three times as long as the input pulse may be highly undesirable from a physical viewpoint. Thus, the output of such a filter at a given time t would depend too much on seismic trace events far away from the particular event that we wish to spike.

The two important design criteria that we have been discussing here are filter lag and filter memory duration. We can *always* improve performance by increasing the memory function duration, but physical considerations prevent us from making this duration indefinitely long. On the other hand, we may search for that desired output lag which leads to the highest P-value for a given selected filter duration. This lag in filter output harms us in no way and, as we have seen, can improve filter performance drastically.

The filter performance parameter P is a function of

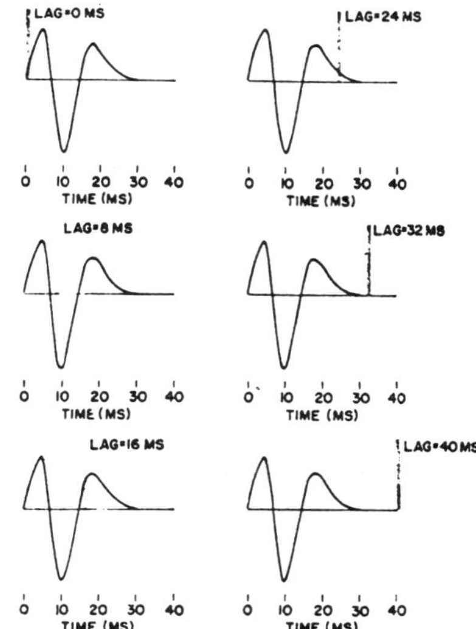

Fig. 4.   Six possible choices of lag for the desired spike.

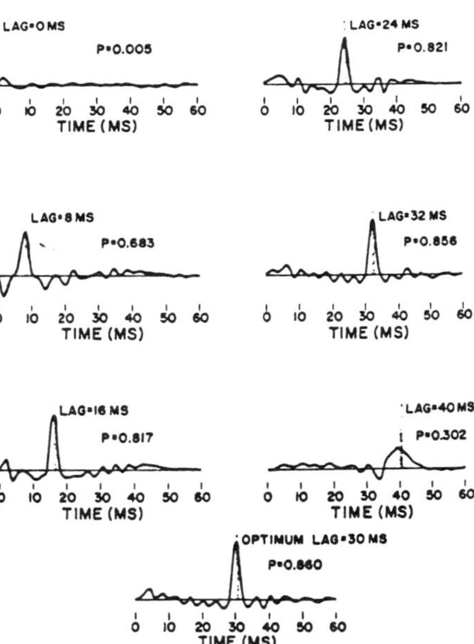

Fig. 5.   The actual outputs of the spiking filters computed for the cases shown in Fig. 4. Also shown is the actual output of the optimum-lag 30 ms duration spiking filter of Fig. 7.

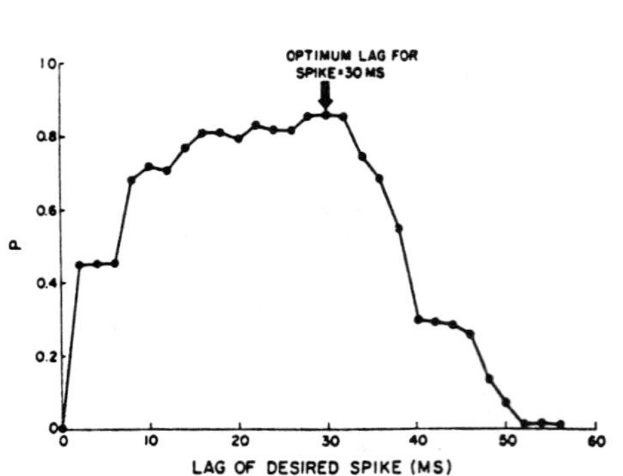

Fig. 6. The filter performance parameter P as a function of desired spike lag for the 30 ms duration spiking filters whose outputs are shown in Fig. 5.

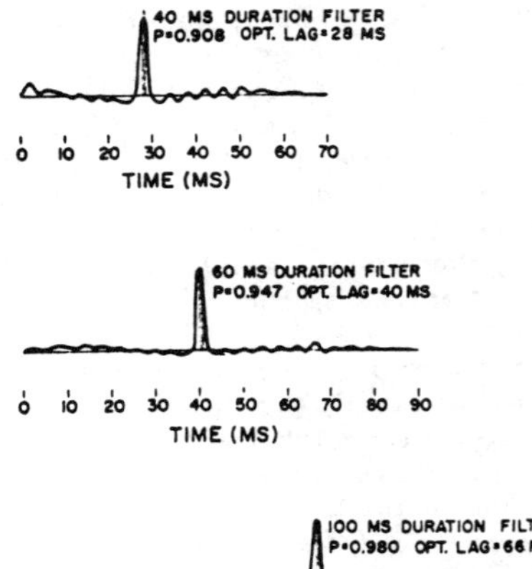

Fig. 9. A comparison between the outputs of various optimum-lag spiking filters for the pulse of Fig. 4.

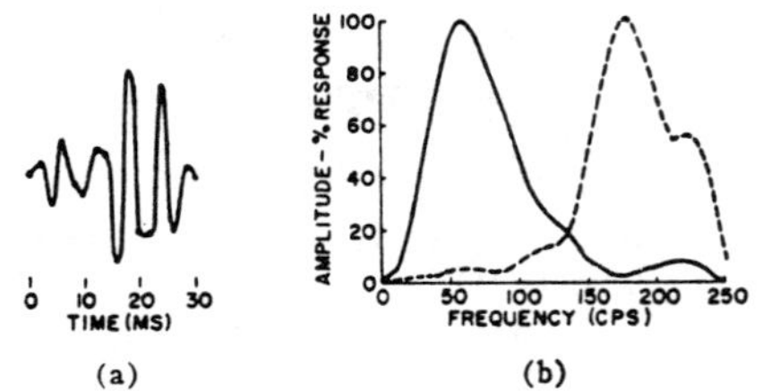

(a)   (b)

Fig. 7. Optimum-lag spiking of the pulse of Fig. 4. (a) The memory function of the optimum-lag 30-ms duration filter. (b) ——— Amplitude spectrum of the pulse of Fig. 4; – – – – Amplitude spectrum of the above optimum-lag, 30 ms duration filter.

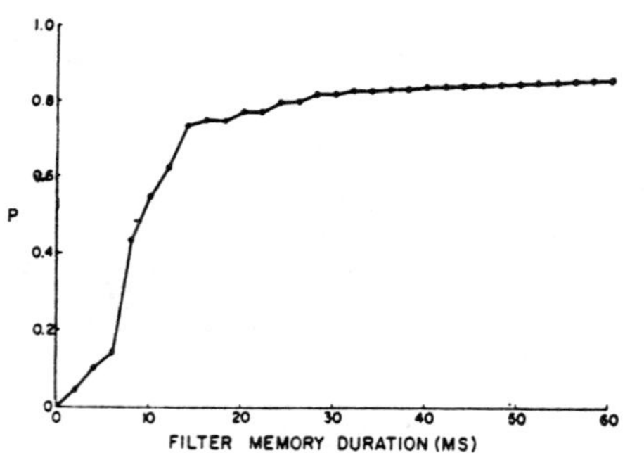

Fig. 8. The filter performance parameter P as a function of filter duration for a desired spike lag of 16 ms. The pulse to be spiked is the one shown in Fig. 4.

lag and duration. Plots of P vs. lag at constant duration (Fig. 6), or of P vs. duration at constant lag (Fig. 8) are helpful, but do not tell us the whole story. Ideally, we would like to investigate the dependence of P on lag and duration for all physically reasonable values of these variables. One way this can be done is to plot P on a two-dimensional grid, with filter lag as the ordinate and filter duration as the abscissa. The array of P values can then be contoured so that we may see at a glance which combinations of lag and duration yield optimum filter performance. Such a "contour map" is given by Fig. 10. The input wavelet is again the one shown in Fig. 4, and our objective is to find a good spiking filter for this wavelet. The section VV' through the contour map then corresponds to the curve of Fig. 6, while the section HH' corresponds to the curve of Fig. 8. The map shows only the contours for $P = 0.85$, 0.90, and 0.95. Obviously, we are most interested in the larger P values, for it is there that best filter performance is obtained. This display enables one to select the best combination of filter lag and duration by inspection.

If we convolve an input wavelet of duration $n\Delta t$ with a filter memory function of duration $m\Delta t$, the resulting output transient is of duration,

(input wavelet duration) + (filter memory duration)

$$= (n\Delta t) + (m\Delta t)$$
$$= (n + m)\Delta t.$$

Now, if this output transient is *shorter* than some assumed lag for the desired spike, the best that the filter can do is to produce an actual output consisting only of zeros. In other words, the output transient never "reaches" the desired spike, which means that the actual output must approximate the desired spike by a zero; therefore, the performance parameter P is also zero. This is the reason for the zero values in the upper left-hand portions of the contour maps of Figs. 10 to 12. Thus, the *meaningful* lag values for a spiking filter lie in the range that can be reached by the actual output; that is, they lie in the range from zero to $(n+m)\Delta t$. For lags outside this range, the performance parameter P must necessarily be zero.

Let us next see what happens when we compute spiking filters for the minimum-delay and maximum-delay pulses of Fig. 2. The contour maps we obtain in this manner are shown in Figs. 11 and 12, respectively. As before, we show only the contours for the higher values of P. The map for the minimum-delay wavelet spiking filter (Fig. 11) peaks for the minimum lag value, which is zero, while the corresponding map for the maximum-delay spiking filter (Fig. 12) peaks for the maximum *meaningful* lag values. Of course, since Fig. 12 and others of its kind depict but a portion of the "duration-lag" plane, the meaningful lag values for the longer filters are not shown there. Finally, the spiking filter display of Fig. 10 represents, as we might expect, a case intermediate between the minimum-delay and maximum-delay extremes.

We have previously discussed (Section III) the problem of energy distribution in a wavelet. From Fig. 11 we see that the best filter performance for the minimum-delay case occurs for *zero-lag value*. We are here attempting to spike a minimum-delay wavelet, namely one whose energy is concentrated near its front end. It is therefore reasonable to expect that a desired spike output at zero-lag value gives better results there than if this spike be more delayed in time. In other words, since the minimum-delay wavelet's energy is concentrated as much as possible near its front end, it makes little sense to lag the desired spike relative to this front end. Similarly, we observe from Fig. 12 that best performance for the maximum-delay case occurs for the *largest meaningful lag values.*[5] This means that because the energy of the maximum-delay wavelet is concentrated as much as possible near its back end, a spiking filter can do the best job if the desired spike is positioned so that it occurs at the larger values of lag. The mixed-delay case of Fig. 10 then represents an intermediate situation. Inspection of Fig. 4 tells us that the energy of this mixed-delay pulse is concentrated roughly at the center of the wavelet's range of duration.

[5] Strictly speaking, these rules hold only for spiking filters that are sufficiently long in relation to the length of the input wavelet. That is, for sufficiently long spiking filters, the rules are that the best spike position occurs at zero for a minimum-delay input wavelet and at $(n+m)\Delta t$ for a maximum-delay input wavelet. For spiking filters that are short in relation to the length of the input wavelet, there may be deviations from these rules for the best spike position; however, as the length of the spiking filter increases, the best spike position settles down to that position given by these rules.

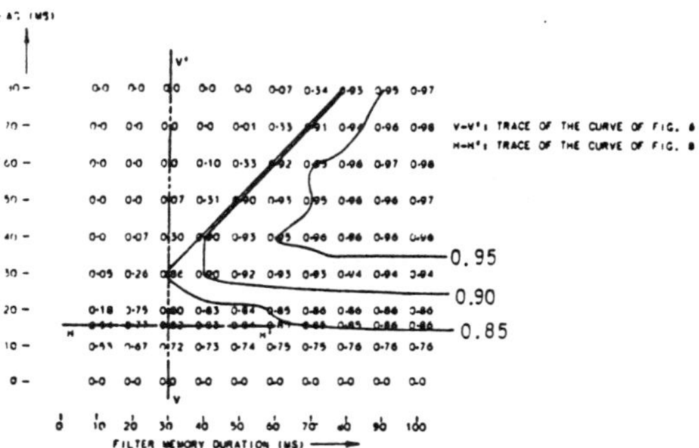

Fig. 10.   *P*-contour map for spike filtering the mixed-delay pulse (b) of Fig. 2. The gridded numbers correspond to values of the filter performance parameter *P*.

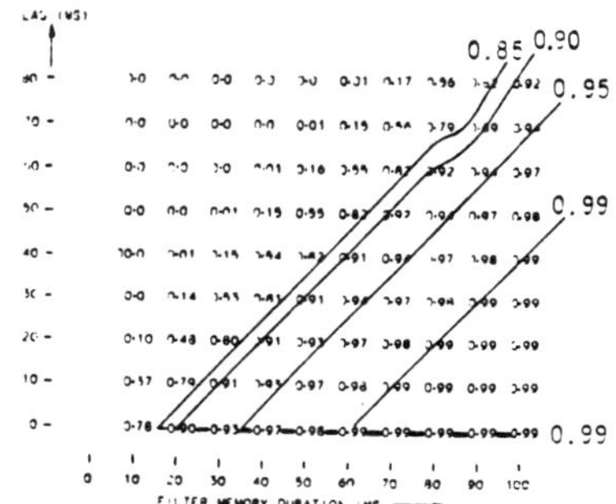

Fig. 11.   *P*-contour map for spike filtering the minimum-delay pulse (a) of Fig. 2. The gridded numbers correspond to values of the filter performance parameter *P*.

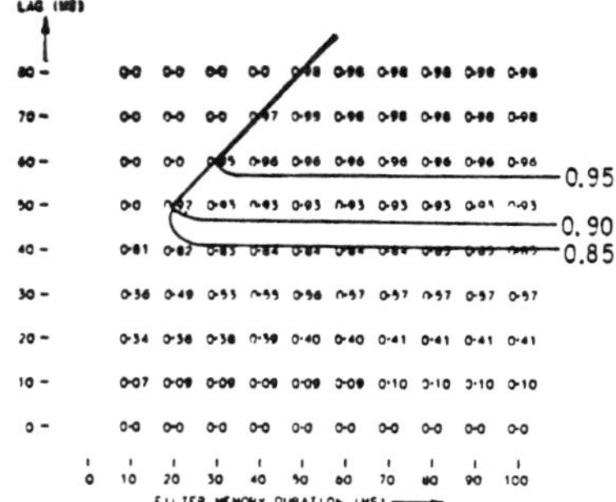

Fig. 12.   *P*-contour map for spike filtering the maximum-delay pulse (d) of Fig. 2. The gridded numbers correspond to values of the filter performance parameter *P*.

We recall that the convolution of a time function of duration $m\Delta t$, with another time function of duration $n\Delta t$, leads to a function of duration $(m+n)\Delta t$. The spiking filter used to obtain the outputs of Fig. 5 has a 30 ms duration, while the wavelet to be spiked (Fig. 4) is also 30 ms long. The filter outputs of Fig. 5 are thus all transients of 60 ms duration. Consider now the output of the optimum-lag spiking filter of Fig. 5. We observe that the best desired spike position for the particular filter duration chosen occurs at a lag of 30 ms. Since the total duration of the output is 60 ms, this "best" spike position occurs in the middle of the filter output transient. A visual study of the pulse of Fig. 4 suggests that its energy is concentrated roughly at the center of its duration range. What happens, then, is that the spiking filter will have the easiest job if the desired spike occupies the same relative position in the filter output function as is taken up by most of the pulse energy in the filter input. This matter is also evident from the contour map of Fig. 10. Let us consider here vertical lines at constant filter duration. For example, at a filter duration of 50 ms, we expect best performance if we position the desired spike at a lag of roughly $\frac{1}{2}(50+30)=40$ ms. Inspection of the contour map shows that the highest P value at a filter length of 50 ms indeed occurs at a lag of 40 ms, and is in fact given by $P=0.93$. If we chose a filter duration of 90 ms, we expect best performance if the desired spike is placed at a lag of $\frac{1}{2}(90+30)=60$ ms. This guess is again confirmed by inspection of the contour map, where we see that the peak P value for a 90 ms filter duration occurs at a lag of 60 ms ($P=0.97$). This rule of thumb is found to be quite useful in practice.

It is evident that contour maps such as the ones shown in Figs. 10 through 12 are powerful tools to display the energy structure of a wavelet. They thus provide a convenient aid for the design of optimum-lag filters.

V. Wavelet Shaping

Having examined the design problems of the spiking filter in some detail, let us next consider the analogous case of shaping-filter design.

The pulse we propose to shape into some desired output wave form is pictured in Fig. 13(a). It is a symmetric Ricker wavelet of 20 ms breadth. The choice of the desired output shape is quite arbitrary, but, since we wish to increase seismic resolution, this desired shape should be considerably sharper than the input wavelet. In Fig. 13 (b) and (c) and in Fig. 14, we show the actual outputs of a series of 50 ms memory duration *optimum-lag* shaping filters that were computed for the input pulse of Fig. 13(a). The desired output shapes are also shown in these figures so that these may be compared directly with the actual outputs. In every case we indicate the value of the filter performance parameter P and of the optimum lag at which this peak P-value is ob-

tained. The 50 ms duration filters we have computed here are slightly longer than the input pulse, which is a 44 ms duration wavelet. The resulting filter outputs are therefore 94 ms duration transients.

The desired shape for the output shown in Fig. 13(b) is a spike. We observe that, although the actual output does peak sharply at the optimum lag of 47 ms, there are sharp events of lesser amplitude at the beginning and end of the output. These spurious signals may often be undesirable, as we shall show below. The comparatively low value of P for this case suggests that the spiking filter here could do better if the filter duration were increased. This is indeed confirmed by a study of the appropriate P-contour map shown in Fig. 15.

Figure 13(c) gives the result for the choice of a saw-tooth desired output. We note that the value of P has increased, and that only one spurious signal occurs near the end of the output transient. The breadth of this spurious signal is, however, very different from the actually obtained approximation to the desired shape. We therefore expect that this shaping filter will be a better one to use than the spiking filter of Fig. 13(a).

Figure 14 shows three more cases that we have investigated. Of the set of five desired shapes treated here and in Fig. 13, we see that the Gaussian desired waveshape assumed in Fig. 14(c) yields the highest P-value for our chosen filter duration of 50 ms. In Fig. 16 we give the P-contour map for the saw-tooth waveshape of Fig. 13(c), while in Fig. 17, we show the P-contour map corresponding to the desired Gaussian waveshape of Fig. 14(c). A comparison of the contour maps of Figs. 15, 16, and 17 shows that these differ significantly among themselves. This is, of course, to be expected, since each contour map corresponds to a different desired output waveshape. The convenience of using such maps in optimum filter design again becomes evident.

It might be asked why the shaping filters with actual outputs shown in Fig. 13(c) and in Fig. 14(a), (b), and (c) perform better than the spiking filter of Fig. 13(b). One reason is that, while the spectrum of a spike is of uniform power at all frequencies, the spectra of all the other desired shapes treated here are not as rich in the higher frequencies. Since the input pulse is also not rich in these higher frequencies, one intuitively expects that other shaping filters will lead to higher P-values for a given filter duration than is the case for the spiking filter of that same duration.

Finally, it is of interest to see how such shaping filters would perform in practice. We treat here an idealized case, which is shown in Fig. 18. Trace 1 (T1) of this illustration gives a hypothetical layered-earth unit impulse response function. Each of the spikes, depicted by a heavy vertical line, represents the arrival of an event at the seismometer. Following standard practice, we assume that the filtering action of the ground and instrumentation are representable in the time domain

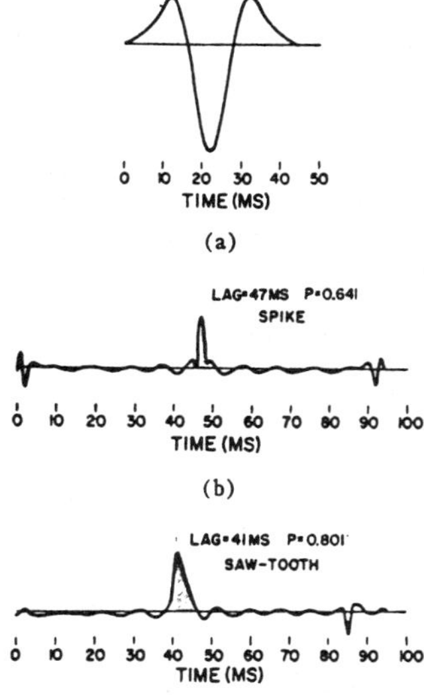

Fig. 13. (a) Input pulse for shape filtering. (b) Actual output for optimum-lag 50 ms duration-spiking filter. (c) Actual output for optimum-lag 50 ms duration-shaping filter. The desired output shapes are as shown.

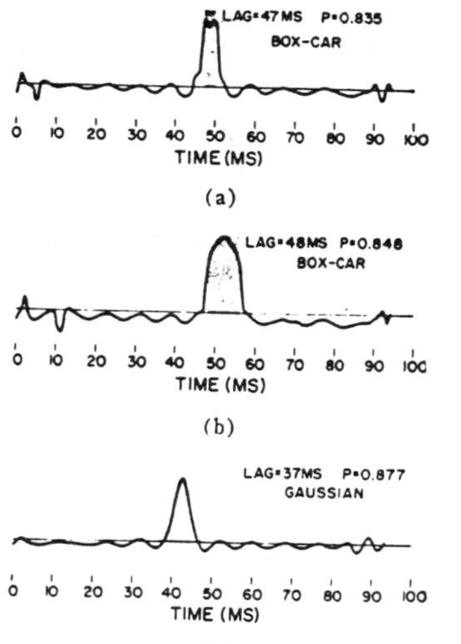

Fig. 14. Actual outputs for three different 50-ms duration-shaping filters for the pulse of Fig. 13(a). The desired output shapes are as shown.

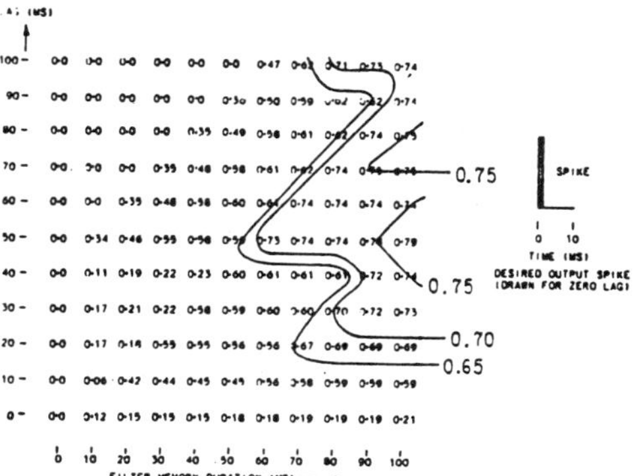

Fig. 15. P-contour map for spiking the pulse of Fig. 13(a). The gridded numbers correspond to values of the filter performance parameter P.

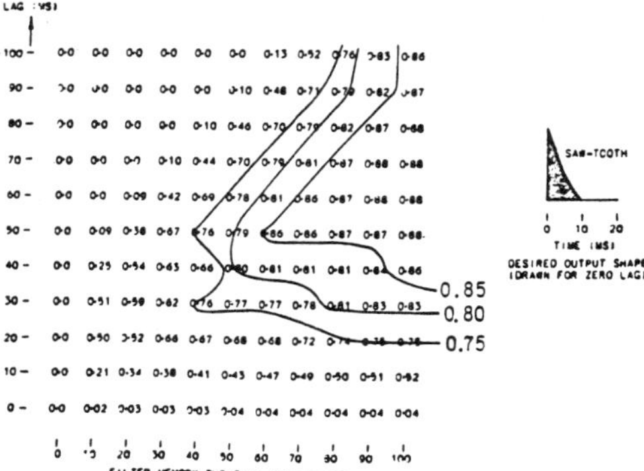

Fig. 16. P-contour map for shaping the pulse of Fig. 13(a) into a desired output as shown. The gridded numbers correspond to values of the filter performance parameter P.

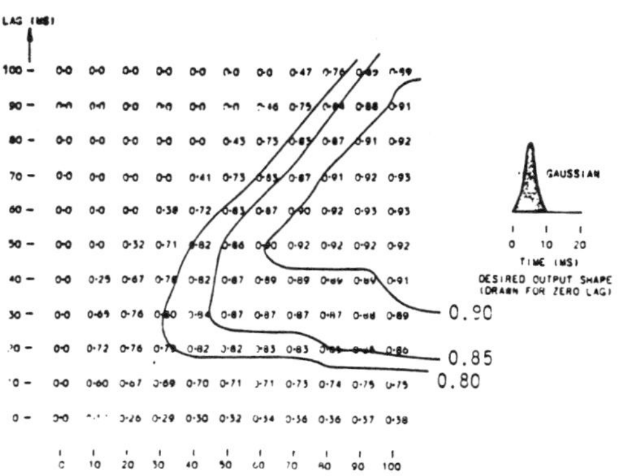

Fig. 17. P-contour map for shaping the pulse of Fig. 13(a) into a desired output as shown. The gridded numbers correspond to values of the filter performance parameter P.

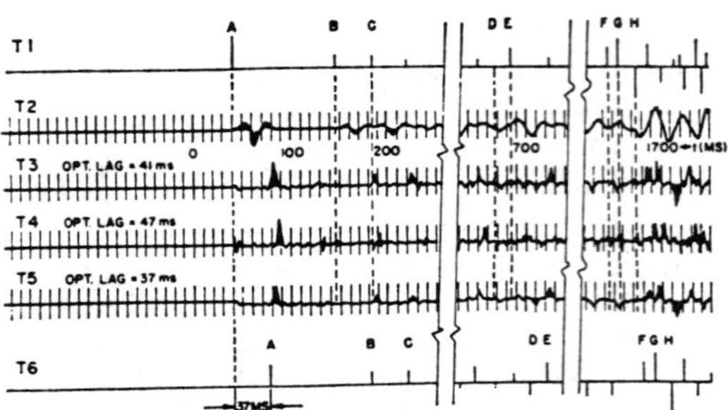

T1: Layered earth unit impulse response.
T2: (T1) * (pulse of Fig. 13(a).
T3: Output from shaping filter; desired output signal: saw-tooth
waveform of Fig. 13(c).
T4: Output from spiking filter.
T5: Output from shaping filter; desired output signal: Gaussian
waveform of Fig. 14(c).
T6: T1 lagged by 37 ms.

Fig. 18. A hypothetical application of shaping filters.

by the pulse shown in Fig. 13(a). The convolution of this pulse with T1 then yields the usual synthetic trace T2. Ideally, we wish to find a filter which, with T2 as input, yields an output closely resembling T1. The output of the optimum-lag 50 ms duration spiking filter is given by T4. We observe that many of the spikes of T1 have been recovered, but that the noise level is very high. This behavior is to be expected because the spiking filter used here generates strong spurious signals [see Fig. 13(b)]. The output of the optimum-lag 50 ms duration shaping filter with a saw-tooth waveform as desired output is shown in T3, while T5 gives the output for such a shaping filter with a Gaussian waveform as desired output (see also Figs. 13 and 14). These latter two traces are not as noisy as T4. Moreover, the choice of desired output shapes other than simple spikes leads to results in which the signal has more diagnostic features that can aid the interpreter in its recognition in a noise background. In order to illustrate this point, let us consider the events labelled "A" through "H" in Fig. 18. In order to locate events on T2 through T5 that correspond to spikes on T1, it is necessary to add the optimum lag to the arrival time of a given spike on T1. The respective optimum lags are indicated at the beginning of each trace. Event A, since it occurs some 150 ms from its nearest neighbor, yields output on all traces uncorrupted by adjacent signals, and the shapes here obtained are those also shown in Figs. 13(b) and (c) and 14(c). Events B and C are closer together, but are still distinguishable as separate signals on the synthetic trace T2. The corresponding shaped signals on T3, T4, and T5 are indicated by appropriate shading. We observe that good resolution has been obtained. On the other hand, events D and E are sufficiently close together so that they are no longer distinguishable as separate signals on the synthetic trace T2. However, our shaping filters do a good job in resolving this pair of pulses on traces 3 and 5; the spiking filter (T4) does not perform well in

this case, as one might expect from what has been said before. A similar situation occurs for events F, G, and H, which are so close together that they are not distinguishable as separate pulses on T2. These events become clearly resolved on T3 and T5; the corresponding result on the spiking filter output of T4 is again of much poorer quality. Many other such examples are observable on the traces, but these have not been explicitly indicated for the sake of brevity. In order to help the reader judge the quality of the results obtained on T5, the ideal unit impulse response T1 is shown again in T6 with a 37 ms lag. In this way dominant events on both traces are brought into vertical alignment.

The above example is admittedly idealized; we know, for example, that the choice of a single seismic waveshape for the entire length of the trace is unrealistic. Even so, the synthetic traces given here do show roughly what one might expect to see in actual seismic applications.

VI. Concluding Remarks

Digital Wiener filters constitute a powerful tool to increase the resolution of overlapping pulses. They are relatively simple to compute and are applicable in a wide variety of ways. Two of the crucial design criteria, filter memory duration and desired output lag, have here been treated in considerable detail. We have shown how one can select the best combination of lag and memory duration for a given problem by a study of an appropriate P-contour map. We have stated that the choice of the desired output shape is arbitrary, although, of course, one wishes to see signals in the actual output which are sharper than their counterparts in the input. Nevertheless, given the input seismic pulse, it might perhaps be possible to arrive by analytical means at desired output pulse shapes that are themselves optimum. In other words, we wish to find that desired signal shape, given an input pulse, which is not only "best" distinguishable in a noise background, but which is achievable also by a short Wiener filter with a high value of the performance parameter P. Such an investigation is perhaps logically the next step to take in this field.

Appendix I
Derivation of the Equations for the Wiener Shaping Filter

Consider the problem of designing an $(m+1)$-length filter $f_t = (f_0, f_1, f_2, \cdots, f_m)$ which converts in the least-error energy sense the $(n+1)$-length input $b_t = (b_0, b_1, b_2, \cdots, b_n)$ into an $(m+n+1)$-length desired output $d_t = (d_0, d_1, d_2, \cdots, d_{m+n})$, where m and n are integers. This problem is illustrated schematically in Fig. 1. The desired output d_t is of arbitrary shape, and hence, the filter f_t is called the least-error energy shaping filter, or simply, the Wiener shaping filter. The actual output c_t is the convolution of the filter f_t with the input b_t, that is,

$$c_t = f_t * b_t$$
$$= (f_0, f_1, f_2, \cdots, f_m) * (b_0, b_1, b_2, \cdots, b_n)$$
$$= (c_0, c_1, c_2, \cdots, c_{m+n}).$$

We must thus determine the filter weighting coefficients f_t, such that the actual output c_t is as close as possible in a least-error energy sense to the desired output d_t. Therefore, we wish to minimize the energy of the error signal $e_t = (e_0, e_1, e_2, \cdots, e_{m+n})$, which is given by the quantity I,

$$I = e_0{}^2 + e_1{}^2 + e_2{}^2 + \cdots + e_{m+n}{}^2$$
$$= (d_0 - c_0)^2 + (d_1 - c_1)^2 + (d_2 - c_2)^2 + \cdots$$
$$+ (d_{m+n} - c_{m+n})^2.$$

Using summation notation, we have

$$I = \sum_{t=0}^{m+n} (d_t - c_t)^2.$$

Inserting into the above expression for I the convolution

$$c_t = \sum_{s=0}^{m} f_s b_{t-s},$$

we obtain

$$I = \sum_{t=0}^{m+n} \left(d_t - \sum_{s=0}^{m} f_s b_{t-s} \right)^2. \qquad (1)$$

The error energy I is minimized if its partial derivatives with respect to each of the filter weighting coefficients $f_0, f_1, f_2, \cdots, f_m$ equal zero. Differentiating I with respect to f_j, where $j = 0, 1, 2, \cdots, m$, and setting each partial derivative equal to zero, we have

$$\frac{\partial I}{\partial f_j} = \sum_{t=0}^{m+n} 2 \left(d_t - \sum_{s=0}^{m} f_s b_{t-s} \right)(-b_{t-j}) = 0,$$

which yields

$$- \sum_{t=0}^{m+n} d_t b_{t-j} + \sum_{t=0}^{m+n} \left(\sum_{s=0}^{m} f_s b_{t-s} \right) b_{t-j} = 0,$$

or

$$\sum_{s=0}^{m} f_s \sum_{t=0}^{m+n} b_{t-s} b_{t-j} = \sum_{t=0}^{m+n} d_t b_{t-j}, \qquad (2)$$

where $j = 0, 1, 2, \cdots, m$. In this set of $(m+1)$ equations we wish to make the substitution

$$\sum_{t=0}^{m+n} b_{t-s} b_{t-j} = r_{j-s}, \qquad (3)$$

where r_{j-s} is the autocorrelation (for index $t = j - s$) of the input b_t. We also define g_j to be

$$g_j = \sum_{t=0}^{m+n} d_t b_{t-j} \qquad (j = 0, 1, 2, \cdots, m), \qquad (4)$$

which is the *cross-product*, (or *cross-correlation*) of the desired output d_t with the input b_t. Substituting (3) and (4) into (2), we obtain

$$\sum_{s=0}^{m} f_s r_{j-s} = g_j \qquad (j = 0, 1, 2, \cdots, m), \qquad (5)$$

where, we repeat:

1) The *unknown* f_s are the weighting coefficients of the Wiener shaping filter.
2) The *known* r_{j-s} are the autocorrelation coefficients of the input b_t.
3) The *known* g_j are the cross-product coefficients between desired output d_t and input b_t.

Let us now write out the set of $(m+1)$ equations in (5). We then have

$$f_0 r_0 + f_1 r_{-1} + f_2 r_{-2} + \cdots + f_m r_{-m} = g_0,$$
$$f_0 r_1 + f_1 r_0 + f_2 r_{-1} + \cdots + f_m r_{1-m} = g_1,$$
$$f_0 r_2 + f_1 r_1 + f_2 r_0 + \cdots + f_m r_{2-m} = g_2,$$
$$\cdots \cdots \cdots \cdots \cdots \cdots \cdots \cdots \cdots$$
$$f_0 r_m + f_1 r_{m-1} + f_2 r_{m-2} + \cdots + f_m r_0 = g_m. \qquad (6)$$

This set of relations is known as the system of *normal* equations. In the case of real-valued scalar functions which we treat here, we can further simplify (6) by recalling that the autocorrelation is then an even function; that is,

$$r_{-t} = r_t.$$

Making use of this property, we finally rewrite (6) in the more compact matrix form,

$$\begin{bmatrix} r_0 & r_1 & r_2 & \cdots & r_m \\ r_1 & r_0 & r_1 & \cdots & r_{m-1} \\ r_2 & r_1 & r_0 & \cdots & r_{m-2} \\ \cdot & \cdot & \cdot & & \cdot \\ r_m & r_{m-1} & r_{m-2} & \cdots & r_0 \end{bmatrix} \begin{bmatrix} f_0 \\ f_1 \\ f_2 \\ \cdots \\ f_m \end{bmatrix} = \begin{bmatrix} g_0 \\ g_1 \\ g_2 \\ \cdots \\ g_m \end{bmatrix}. \qquad (7)$$

Although at first sight it may seem that we are faced with the familiar task of solving a system of $(m+1)$ equations in the $(m+1)$ unknowns $(f_0, f_1, f_2, \cdots, f_m)$, it turns out that the computational labor involved is much less than might be feared. This simplification results from the fact that all elements on any given diagonal of the matrix of autocorrelation coefficients r_t are equal to each other. A recursive scheme to solve the system (7) was first given by Levinson [9] and has been further discussed by Robinson [10] and by Wiggins and Robinson [13]. This latter paper also contains an extremely useful algorithm, due to S. M. Simpson [4], [13], which enables one to compute the Wiener shaping filter for an arbitrarily lagged desired output directly from the Wiener shaping filter calculated for the use of no lag in the desired output. The savings in computer time and storage afforded by the Levinson-Simpson technique are considerable. A simplified derivation of the recursive method, which makes use of z-transform theory, is given in Appendix III. The calculations discussed in this paper have been carried out by means of computer programs based on Levinson's and Simpson's recursive methods.

We finally proceed to derive the expression for the filter performance parameter P. Expanding the expression for the error energy I, given by (1), we obtain

$$I = \sum_{t=0}^{m+n} \left(d_t - \sum_{s=0}^{m} f_s b_{t-s} \right)^2$$

$$= \sum_{t=0}^{m+n} \left(d_t^2 - 2 d_t \sum_{s=0}^{m} f_s b_{t-s} \right.$$

$$\left. + \sum_{s=0}^{m} f_s b_{t-s} \sum_{\sigma=0}^{m} f_\sigma b_{t-\sigma} \right)$$

$$= \sum_{t=0}^{m+n} d_t^2 - 2 \sum_{s=0}^{m} f_s \sum_{t=0}^{m+n} d_t b_{t-s}$$

$$+ \sum_{s=0}^{m} f_s \sum_{\sigma=0}^{m} f_\sigma \sum_{t=0}^{m+n} b_{t-s} b_{t-\sigma},$$

or

$$I = D - 2 \sum_{s=0}^{m} f_s g_s + \sum_{s=0}^{m} f_s \sum_{\sigma=0}^{m} f_\sigma r_{\sigma-s}, \qquad (8)$$

where σ is a dummy summation index, and where we have made use of the expressions for the autocorrelation coefficient r_{j-s} [eq. (3)] and for the cross-product coefficient g_s [eq. (4)]. The quantity D is the energy in the desired output d_t or, in other words, it is the zeroth lag of the autocorrelation of d_t. Now the error energy I will be at a minimum if we substitute into (8) the *normal equations* (5). We then obtain the minimum value for I in the least squares sense, which we call $I_{\min}$,

$$I_{\min} = D - 2 \sum_{s=0}^{m} f_s g_s + \sum_{s=0}^{m} f_s g_s;$$

that is,

$$I_{\min} = D - \sum_{s=0}^{m} f_s g_s. \qquad (9)$$

Here we have made use of the symmetry property of the autocorrelation function, namely that $r_{s-s}=r_{s-\sigma}$. It is convenient to normalize the expression for the minimum error energy $I_{\min}$ by dividing both sides of (9) through by D,

$$\frac{I_{\min}}{D} = 1 - \sum_{s=0}^{m} f_s \frac{g_s}{D}.$$

Letting $I_{\min}/D = E$, and $g_s/D = g_s'$, we obtain

$$E = 1 - \sum_{s=0}^{m} f_s g_s',$$

where E is the normalized minimum error energy. Because E is a sum of squares, it can never be negative;

moreover, E can never be greater than unity, since the value 1 for E can always be obtained by letting the filter f_s be identically zero. The filter performance parameter P is therefore defined as the one's complement of E,

$$P = 1 - E = \sum_{s=0}^{m} f_s g_s', \qquad (10)$$

which constitutes the sought expression for P. It follows directly from the above definition that

$$0 \leq P \leq 1,$$

where the cases $P = 0$ and $P = 1$ correspond, respectively, to no agreement and to perfect agreement of the actual output c_t with the desired output d_t.

APPENDIX II
THE SHAPING FILTER IN THE PRESENCE OF NOISE

In Appendix I we considered the problem of how to design a shaping filter in the noiseless case. Now, we want to extend our analysis to the case where the signal wavelet may be embedded in stationary noise u_t with a known autocorrelation function q_t.

For comparative purposes we shall continue to employ the same notation as we have used in the foregoing discussions. We assume that the received digitized data x_t is the sum of a signal wavelet $(b_0, b_1, \cdots, b_n)$ and stationary noise u_t; that is,

$$x_t = b_t + u_t.$$

We want to design a filter with memory function $(f_0, f_1, \cdots, f_m)$ which will perform two functions, namely,

1) to shape, as well as possible, the signal wavelet $(b_0, b_1, \cdots, b_n)$ into a desired or preferred output wavelet $(d_0, d_1, \cdots, d_{m+n})$, and

2) to produce as little output power as possible when the unwanted stationary noise u_t is its only input.

In the noiseless case, that is, when the unwanted stationary noise u_t is absent, the second of these two functions is no longer pertinent, and hence, this is the case of the shaping filter that we have just treated. On the other hand, if we eliminate the first function above, then an all-stop filter (i.e., one which produces no output) would perform the second function perfectly. In most practical cases, however, we want a filter that performs both of the above functions simultaneously, and hence, we will be faced with the problem of finding some suitable compromise between the two. We do not want to treat the problem of finding a satisfactory compromise here, as it often depends on the particular physical situation and the overall data processing requirements and objectives. Hence, in the filter design we shall merely make use of an arbitrary parameter ν which assigns the relative weighting between the above two functions.

Let us give the mathematical derivation of the shaping filter in the case of autocorrelated noise. In order to select the coefficients of the shaping filter, we will minimize the quantity:

I = (Sum of squared errors between desired output and filtered signal wavelet) + ν(Power of the filtered noise).

Here ν is the preassigned weighting parameter just discussed. In symbols, the quantity to be minimized is

$$I = \sum_{t=0}^{m+n} (d_t - c_t)^2 + \nu E\{v_t^2\},$$

where the notation $E\{\ \}$ indicates the ensemble average, and where

$$v_t = \sum_{s=0}^{m} f_s u_{t-s}$$

represents the filtered noise. Because the received noise u_t is stationary, it follows that the filtered noise v_t is also stationary. Simplifying the expression for I, we obtain

$$I = \sum_{t=0}^{m+n} \left(d_t - \sum_{s=0}^{m} f_s b_{t-s}\right)^2 + \nu \sum_{s=0}^{m} \sum_{t=0}^{m} f_s q_{t-s} f_t.$$

Here,

$$q_{t-s} = E\{u_{\tau-s} u_{\tau-t}\},$$

where τ is a dummy time index, and where q_{t-s} is the autocorrelation of the received stationary noise. Differentiating the expression for I with respect to each of the filter coefficients, and then, setting the derivatives equal to zero, we obtain the set of simultaneous equations

$$\sum_{s=0}^{m} f_s(r_{t-s} + \nu q_{t-s}) = g_t, \qquad \text{for } t = 0, 1, 2, \cdots, m.$$

In this set of normal equations the signal autocorrelation r_{t-s}, the weighting parameter ν, the noise autocorrelation q_{t-s}, and the cross-products g_t, defined as

$$g_t = \sum_{s=0}^{m} d_s b_{s-t}, \qquad \text{for } t = 0, 1, 2, \cdots, m,$$

represent the known quantities, and the filter coefficients f_s represent the unknowns. The solution of the normal equations thereby yields the coefficients of the required shaping filter. Since the matrix of these equations, namely the matrix $[r_{t-s} + \nu q_{t-s}]$, is in the form of an autocorrelation matrix, these equations can be solved efficiently by the recursive method which we will describe in Appendix III.

Appendix III

Recursive Solution of the Simultaneous Equations Involving an Autocorrelation Matrix

We have said that it is possible to take advantage of the special form of an autocorrelation matrix in order to reduce the computational work required for the solution of a set of simultaneous equations involving such a matrix. More specifically, any autocorrelation matrix, say

$$R_n = \begin{bmatrix} r_0 & r_1 & r_2 & \cdots & r_n \\ r_1 & r_0 & r_1 & \cdots & r_{n-1} \\ r_2 & r_1 & r_0 & \cdots & r_{n-2} \\ \cdot & \cdot & \cdot & \cdots & \cdot \\ r_n & r_{n-1} & r_{n-2} & \cdots & r_0 \end{bmatrix},$$

has the property that all the elements on any given diagonal are the same. For example, the zero-lag autocorrelation coefficient r_0 is the element which appears on the main diagonal; the first-lag autocorrelation coefficient r_1 is the element which appears on the first super-diagonal as well as on the first sub-diagonal, etc. Hence, the entire $(n+1) \times (n+1)$ autocorrelation matrix R_n is one which does not involve $(n+1)^2$ distinct elements, but only $(n+1)$ distinct elements, namely r_0, $r_1, r_2, \cdots, r_n$. Let us now see how we can take advantage of this structure (the so-called Toeplitz structure of an autocorrelation matrix).

Suppose that we wish to solve the set of normal equations given by

$$f_0 r_0 + f_1 r_1 + \cdots + f_m r_m = g_0$$
$$f_0 r_1 + f_1 r_0 + \cdots + f_m r_{m-1} = g_1$$
$$\cdots \cdots \cdots \cdots \cdots \cdots \cdots$$
$$f_0 r_m + f_1 r_{m-1} + \cdots + f_m r_0 = g_m, \qquad (11)$$

where the autocorrelation coefficients r_0, $r_1, \cdots, r_m$, as well as the right-hand side coefficients $g_0, g_1, \cdots, g_m$, represent the known quantities, and the filter coefficients $f_0, f_1, \cdots, f_m$ represent the unknown quantities. The structure of the autocorrelation matrix makes possible the following recursive method (Levinson [9]) for the solution of these equations.

The recursive procedure solves the equations in a step-by-step manner. The step $n = 0$ is given as the initial condition, and then the steps $n = 1$, $n = 2$, $\cdots$, $n = m$ are done successively in a recursive manner. The desired filter coefficients are the ones which result on the completion of the final step, namely the step $n = m$. A description of one of these steps will provide the necessary information to enable one to program the method for a digital computer. We will suppose, therefore, that we have completed step n, and we wish to do step $n + 1$.

The completion of step $k = n$ requires that the machine has computed and retained in storage the numerical values of the following quantities:

$$a_{n0}, a_{n1}, \cdots, a_{nn}; \qquad \alpha_n, \beta_n,$$

and

$$f_{n0}, f_{n1}, \cdots, f_{nn}; \qquad \gamma_n.$$

By definition, we suppose that these quantities satisfy the matrix equations:

$$(a_{n0}, a_{n1}, \cdots, a_{nn}, 0)R_{n+1} = (\alpha_n, 0, \cdots, 0, \beta_n), \quad (12)$$

and

$$(f_{n0}, f_{n1}, \cdots, f_{nn}, 0)R_{n+1} = (g_0, g_1, \cdots, g_n, \gamma_n), \quad (13)$$

where the parentheses enclose $1 \times (n+2)$ row vectors, and where R_{n+1} is the $(n+2) \times (n+2)$ autocorrelation matrix.

Because of the Toeplitz structure of the autocorrelation matrix, (12) may be manipulated into the equivalent form given by

$$(0, a_{nn}, \cdots, a_{n1}, a_{n0})R_{n+1} = (\beta_n, 0, \cdots, 0, \alpha_n). \quad (14)$$

Let us now multiply (14) by a constant k_n, as yet undetermined; we then will add the result to (12). We obtain

$$(a_{n0}, a_{n1} + k_n a_{nn}, \cdots, a_{nn} + k_n a_{n1}, k_n a_{n0})R_{n+1}$$
$$= (\alpha_n + k_n \beta_n, 0, \cdots, 0, \beta_n + k_n \alpha_n). \quad (15)$$

Now we want (15) to be identical to

$$(a_{n+1,0}, a_{n+1,1}, \cdots, a_{n+1,n}, a_{n+1,n+1})R_{n+1}$$
$$= (\alpha_{n+1}, 0, \cdots, 0, 0). \quad (16)$$

In order for (15) and (16) to be identical, we must first of all make the requirement that

$$\beta_n + k_n \alpha_n = 0.$$

This equation allows us to determine the constant k_n; that is, we may compute k_n by the formula

$$k_n = -\beta_n/\alpha_n. \quad (17)$$

The identity of (15) and (16), together with our knowledge of k_n, then allows us to compute the quantities given by

$$a_{n+1,0} = a_{n0}$$
$$a_{n+1,1} = a_{n1} + k_n a_{nn}$$
$$\cdots \cdots \cdots$$
$$a_{n+1,n} = a_{nn} + k_n a_{n1}$$
$$a_{n+1,n+1} = k_n a_{n0}, \quad (18)$$

as well as

$$\alpha_{n+1} = \alpha_n + k_n \beta_n. \quad (19)$$

If we define the z-transforms $A_n(z)$ and $A_{n+1}(z)$ as

$$A_n(z) = a_{n0} + a_{n1}z + \cdots + a_{nn}z^n$$

and

$$A_{n+1}(z) = a_{n+1,0} + a_{n+1,1}z + \cdots + a_{n+1,n}z^n$$
$$+ a_{n+1,n+1}z^{n+1},$$

then we see that equations (18) may be encompassed within the confines of one equation, namely,

$$A_{n+1}(z) = A_n(z) + k_n z(a_{nn} + \cdots + a_{n1}z^{n-1} + a_{n0}z^n),$$

which is[6]

---

[6] We recognize $z^n A_n(1/z)$ as the so-called *reverse polynomial* of the polynomial $A_n(z)$.

$$A_{n+1}(z) = A_n(z) + k_n z^{n+1}A_n(1/z). \quad (20)$$

This equation demonstrates the recursion from the known z-transform $A_n(z)$ to the unknown z-transform $A_{n+1}(z)$.

Next, if we compute

$$\beta_{n+1} = a_{n+1,0}r_{n+2} + a_{n+1,1}r_{n+1} + \cdots + a_{n+1,n+1}r_1, \quad (21)$$

then we will have obtained all the quantities in

$$(a_{n+1,0}, a_{n+1,1}, \cdots, a_{n+1,n+1}, 0)R_{n+2}$$
$$= (\alpha_{n+1}, 0, \cdots, 0, \beta_{n+1}), \quad (22)$$

where the parentheses enclose $1 \times (n+3)$ vectors, and where R_{n+2} is the $(n+3) \times (n+3)$ autocorrelation matrix. Equations (12) and (22) are counterparts, where (12) pertains to step n, whereas (22) pertains to step $n+1$.

Because of the Toeplitz structure of the autocorrelation matrix, (16) is the same as

$$(a_{n+1,n+1}, a_{n+1,n}, \cdots, a_{n+1,1}, a_{n+1,0})R_{n+1}$$
$$= (0, 0, \cdots, 0, \alpha_{n+1}). \quad (23)$$

Let us now multiply (23) by a constant q_n, as yet undetermined, and then add the result to (13). We then obtain

$$(f_{n0} + q_n a_{n+1,n+1}, f_{n1} + q_n a_{n+1,n}, \cdots, f_{nn}$$
$$+ q_n a_{n+1,1}, q_n a_{n+1,0})R_{n+1}$$
$$= (g_0, g_1, \cdots, g_n, \gamma_n + q_n \alpha_{n+1}). \quad (24)$$

We want (24) to be identical with

$$(f_{n+1,0}, f_{n+1,1}, \cdots, f_{n+1,n}, f_{n+1,n+1})R_{n+1}$$
$$= (g_0, g_1, \cdots, g_n, g_{n+1}). \quad (25)$$

In order for (24) and (25) to be identical, first of all we must make the requirement that

$$\gamma_n + q_n \alpha_{n+1} = g_{n+1}.$$

This equation allows us to determine the constant q_n; that is, we may compute q_n by the formula

$$q_n = (g_{n+1} - \gamma_n)/\alpha_{n+1}. \quad (26)$$

The identity of equations (24) and (25), together with our knowledge of q_n, allows us to compute the quantities given by

$$f_{n+1,0} = f_{n0} + q_n a_{n+1,n+1}$$
$$f_{n+1,1} = f_{n1} + q_n a_{n+1,n}$$
$$\cdots \cdots \cdots$$
$$f_{n+1,n} = f_{nn} + q_n a_{n+1,1}$$
$$f_{n+1,n+1} = q_n a_{n+1,0}. \quad (27)$$

If we define the z-transforms $F_n(z)$ and $F_{n+1}(z)$ as

$$F_n(z) = f_{n0} + f_{n1}z + \cdots + f_{nn}z^n,$$

and

$$F_{n+1}(z) = f_{n+1,0} + f_{n+1,1}z + \cdots + f_{n+1,n}z^n + f_{n+1,n+1}z^{n+1},$$

then equations (27) may be encompassed within the confines of one equation, namely,

$$F_{n+1}(z) = F_n(z) + q_n z^{n+1} A_{n+1}(1/z). \tag{28}$$

This equation demonstrates the recursion from the known z-transforms $F_n(z)$ and $A_{n+1}(z)$ to the unknown z-transform $F_{n+1}(z)$.

Next, if we compute the quantity

$$\gamma_{n+1} = f_{n+1,0}r_{n+2} + f_{n+1,1}r_{n+1} + \cdots + f_{n+1,n+1}r_1, \tag{29}$$

with the aid of (13), then we will have computed all the quantities in

$$(f_{n+1,0}, f_{n+1,1}, \cdots, f_{n+1,n+1}, 0)R_{n+2}$$
$$= (g_0, g_1, \cdots, g_{n+1}, \gamma_{n+1}), \tag{30}$$

where the parentheses enclose $1 \times (n+3)$ row vectors, and where R_{n+2} is the $(n+3) \times (n+3)$ autocorrelation matrix. Equations (13) and (30) are counterparts, where (13) pertains to step n and (30) pertains to step $n+1$. We have thus completed our task; we have obtained step $n+1$, given that we had step n.

Let us now summarize the computations which one would program on a digital computer. We are given the autocorrelation coefficients

$$r_0, r_1, \cdots, r_m,$$

and the right-hand side coefficients

$$g_0, g_1, \cdots, g_m.$$

At the completion of step n we have the quantities

$$a_{n0}, a_{n1}, \cdots, a_{nn}; \qquad \alpha_n, \beta_n,$$

and

$$f_{n0}, f_{n1}, \cdots, f_{nn}; \qquad \gamma_n$$

in storage. To perform step $n+1$, we make the following computations. We first compute k_n by means of (17). Then, we compute the quantities

$$a_{n+1,0}, a_{n+1,1}, \cdots, a_{n+1,n}, a_{n+1,n+1}; \qquad \alpha_{n+1}, \beta_{n+1}$$

by means of (18), (19), and (21). We next compute q_n by means of (26). Then we compute the quantities

$$f_{n+1,0}, f_{n+1,1}, \cdots, f_{n+1,n}, f_{n+1,n+1}; \qquad \gamma_{n+1}$$

by means of (27) and (29). This completes step $n+1$.

To obtain the final result, namely the solution of the simultaneous equations (11), we must start at step $n=0$, that is, we start with initial quantities which we take as

$$a_{00} = 1, \qquad \alpha_0 = r_0, \qquad \beta_0 = r_1,$$
$$f_{00} = g_0/r_0, \qquad \gamma_0 = f_{00}r_1.$$

The new do the drecursions for $n=1$ to $n=m$; the final T values obtaine for the filter coefficients, namely

$$f_{m,0}, f_{m,1}, \cdots, f_{m,m}$$

represent the desired solution $f_0, f_1, \cdots, f_m$ of the normal equations (11). Instead of fixing the value of m in advance, it is also possible to monitor the recursions according to the value of the error associated with the filter at each step of the recursions. The recursions can be stopped when this error reaches a certain limit.

The machine time required to solve the simultaneous equations (11) for a digital filter with m coefficients is proportional to m^2 for the recursive method, as compared to m^3 for conventional methods of solving simultaneous equations. Another advantage of using this recursive method is that it requires computer storage space proportional to m rather than m^2, as in the case of the conventional methods.

This recursive method has been extended (Robinson [10]) to the matrix-valued case (in which case the autocorrelation matrix has the structure of a block Toeplitz matrix) in order to compute the coefficients of multichannel filters. Also, there is the Simpson sideways recursion (Wiggins and Robinson [13]) which corresponds to shifting the time index of the right-hand side of the simultaneous equations. This sideways recursion is valuable in many applications, for it represents a relatively inexpensive way of determining the filter that incorporates the optimum time lag between the input signal and the desired output signal as described in the text of this paper. As we have seen, the sideways recursion can be used to determine the optimum spike position for a spiking filter and the optimum positioning of the desired output for a shaping filter.

References

[1] N. Ricker, "Wavelet contraction, wavelet expansion, and control of seismic noise," *Geophysics*, vol. 18, pp. 769–792, October 1953.
[2] E. A. Robinson, "Predictive decomposition of seismic traces," *Geophysics*, vol. 22, pp. 767–778, October 1957.
[3] G. Kunetz, "Essai d'analyse de traces sismiques," *Geophysical Prospecting*, vol. 9, pp. 317–341, September 1961.
[4] S. M. Simpson, Jr., et al., Scientific reports 1–11 of contract AF 19 (604) 7378, Advanced Research Projects Agency, Washington, D. C. Project VELA UNIFORM, 1961–1965.
[5] M. R. Foster, W. G. Hicks, and J. T. Nipper, "Optimum inverse filters which shorten the spacing of velocity logs," *Geophysics*, vol. 27, pp. 317–326, June 1962.
[6] R. B. Rice, "Inverse convolution filters," *Geophysics*, vol. 27, pp. 4–18, February 1962.
[7] W. A. Schneider, K. L. Larner, J. P. Burg, and M. M. Backus, "A new data-processing technique for the elimination of ghost arrivals on reflection seismograms," *Geophysics*, vol. 29, pp. 783–805, October 1964.
[8] S. Treitel and E. A. Robinson, "The stability of digital filters," *IEEE Trans. on Geoscience Electronics*, vol. GE-2, pp. 6–18, November 1964.
[9] N. Levinson, "The Wiener RMS (root mean square) error criterion in filter design and prediction," *J. Math. Phys.*, vol. 25, pp. 261–278, January 1947.
[10] E. A. Robinson, "Mathematical development of discrete filters for the detection of nuclear explosions," *J. Geophys. Res.*, vol. 68, pp. 5559–5567, October 1, 1963.
[11] E. A. Robinson and S. Treitel, "Dispersive digital filters," *Reviews of Geophysics*, vol. 3, pp. 433–461, November 1965.
[12] J. C. Claerbout and E. A. Robinson, "The error in least squares inverse filtering," *Geophysics*, vol. 29, pp. 118–120, February 1964.
[13] R. A. Wiggins and E. A. Robinson, "Recursive solution to the multichannel filtering problem," *J. Geophys. Res.*, vol. 70, pp. 1885–1891, April 15, 1965.

PREDICTIVE DECONVOLUTION: THEORY AND PRACTICE†

K. L. PEACOCK* AND SVEN TREITEL*

Least-squares inverse filters have found widespread use in the deconvolution of seismograms. The least-squares prediction filter with unit prediction distance is equivalent within a scale factor to the least-squares, zero-lag inverse filter. The use of least-squares prediction filters with prediction distances greater than unity leads to the method of predictive deconvolution which represents a more generalized approach to this subject.

The predictive technique allows one to control the length of the desired output wavelet, and hence to specify the desired degree of resolution. Events which are periodic within given repetition ranges can be attenuated selectively. The method is thus effective in the suppression of rather complex reverberation patterns.

INTRODUCTION

The Wiener filter is one of the most effective tools for the digital reduction of seismic traces. It constitutes the keystone of many current deconvolution methods. In one realization this filter is used to deconvolve a reverberating pulse train into an approximation of a zero-delay unit impulse. More generally it is possible to arrive at Wiener filters which remove repetitive events having specified periodicities. In this context the Wiener filter is better viewed as a predictor of coherent energy than merely as a spiker of "leggy" wave trains.

The prediction filter used in this treatment gives rise to the method of *predictive deconvolution*. We remark that Robinson's Ph.D. thesis (1954), if written today, would be entitled *"Predictive Deconvolution of Time Series with Applications to Seismic Exploration,"* since the older term *decomposition* has given way to the newer term *deconvolution*. The method of predictive deconvolution has been described in a paper by Robinson (1966), in which the author advocates a prediction distance greater than unity. A discussion of the general properties of the digital Wiener filter has been given by Robinson and Treitel (1967).

BASIC CONCEPTS

The digital filtering process is described by the discrete convolution formula

$$y_\tau = \Delt \sum_t x_t a_{\tau - t},$$

where x_t is the input, a_t is the filter, y_τ is the output, and Δt is the sampling increment. No loss of generality will result if we assume Δt to be unity. In the sequel t and τ are discrete time variables and $\Delta t = 1$ unless otherwise specified.

If a_t is a *prediction* operator with prediction distance α, the output y_τ will be an estimate of the input x_t at some future time $t + \alpha$. We thus write

$$y_\tau = \sum_t x_t a_{\tau - t} = \hat{x}_{t+\alpha}, \tag{1}$$

where $\hat{x}_{t+\alpha}$ is an estimate of $x_{t+\alpha}$.

An error series may be defined as the difference between the true value $x_{t+\alpha}$ and the estimated or predicted value $\hat{x}_{t+\alpha}$,

$$\epsilon_{t \cdot \alpha} = x_{t+\alpha} - \hat{x}_{t+\alpha}. \tag{2}$$

Thus ϵ_t is an output series which represents the nonpredictable part of x_t.

Replacement of the term $\hat{x}_{t+\alpha}$ in equation (2) with its equivalent as defined by equation (1) results in

---

† Presented at the 38th Annual International SEG Meeting in Denver, Colorado, October 1, 1968. Manuscript received by the Editor October 10, 1968.

* Pan American Petroleum Corp., Research Center, Tulsa, Okla.

$$\epsilon_{t+\alpha} = x_{t+\alpha} - \sum_t x_t a_{\tau-t}. \qquad (3)$$

The z-transform of equation (3) is

$$z^{-\alpha} E(z) = z^{-\alpha} X(z) - X(z) A(z). \qquad (4)$$

Multiplication of both sides of equation (4) by z^α yields

$$\begin{aligned} E(z) &= X(z) - z^\alpha X(z) A(z) \\ &= X(z)[1 - z^\alpha A(z)]. \end{aligned} \qquad (5)$$

The quantity $[1 - z^\alpha A(z)]$ is the z-transform of the so-called *prediction error operator*. It is seen to be the difference between the zero-delay unit spike and the prediction operator $A(z)$ delayed by the prediction distance α. Thus one may calculate the error series ϵ_t by computing $\hat{x}_{t+\alpha}$ from equation (1), and follow with a subtraction as defined by equation (2). Alternatively, one may compute the error series in a single step by use of the prediction error operator. Let us assume that a seismic trace is represented by the convolution of an uncorrelated reflection coefficient series with a reverberating pulse train, which by its nature is rich in repetitive energy (see Appendix A). The prediction error filter will then remove the predictable portion of such a trace, which to a good approximation will be given by the repetitive energy in the reverberations. The output of this filtering operation is the error series ϵ_t, which within the framework of this model constitutes the estimate of the reflection coefficient series of the layered subsurface.

Suppose that the prediction operator is given by the n-length series

$$a_t = a_0, a_1, \cdots, a_{n-1}.$$

Then the corresponding prediction error operator with prediction distance α is

$$f_t = 1, \overbrace{0, 0, \cdots, 0}^{\alpha - 1 \text{ zeros}}, -a_0, -a_1, \cdots, -a_{n-1}.$$

We must now deal with the explicit design of the prediction operator, a task which will be accomplished in the next section.

THE LEAST-SQUARES PREDICTIVE FILTERING MODEL

A general least-squares filter model involves the three signals illustrated in Figure 1, namely (1) the input signal x_t, (2) the desired output signal z_t, and (3) the actual output signal y_t. Minimization of the energy existing in the difference between the desired output z_t and the actual output y_t, i.e., minimization of the expression

$$I = \sum_t (z_t - y_t)^2,$$

results in the least-squares, or Wiener filter described by Robinson and Treitel (1967).

The n-length Wiener filter results from the solution of the normal equations with matrix representation,

$$\begin{bmatrix} r_0 & r_1 & \cdots & r_{n-1} \\ r_1 & r_0 & \cdots & r_{n-2} \\ & & \vdots & \\ r_{n-1} & r_{n-2} & \cdots & r_0 \end{bmatrix} \begin{bmatrix} f_0 \\ f_1 \\ \vdots \\ f_{n-1} \end{bmatrix} = \begin{bmatrix} g_0 \\ g_1 \\ \vdots \\ g_{n-1} \end{bmatrix}, \qquad (6)$$

where r_t is the autocorrelation of the input, g_t is the crosscorrelation between the desired output and the input, and f_t is the Wiener filter.

We have seen in the previous section that the prediction operator is that filter which acts on an input trace up to time t and estimates the trace amplitude at some future time $t+\alpha$. Thus it is reasonable to define the desired output for the predictive filter as a time-advanced version of the input x_t. We can now express the prediction filter in terms of a particular Wiener filter, namely the one for which the desired output trace is simply a time-advanced version of the input trace.

In order to solve equation (6) we must know

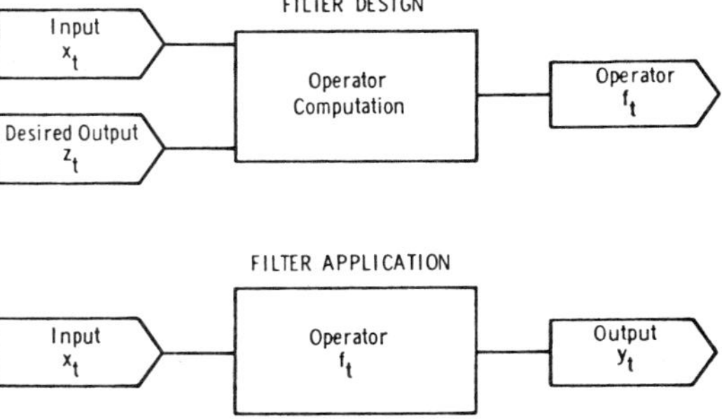

FIG. 1. A general model which illustrates the design and application of the Wiener filter.

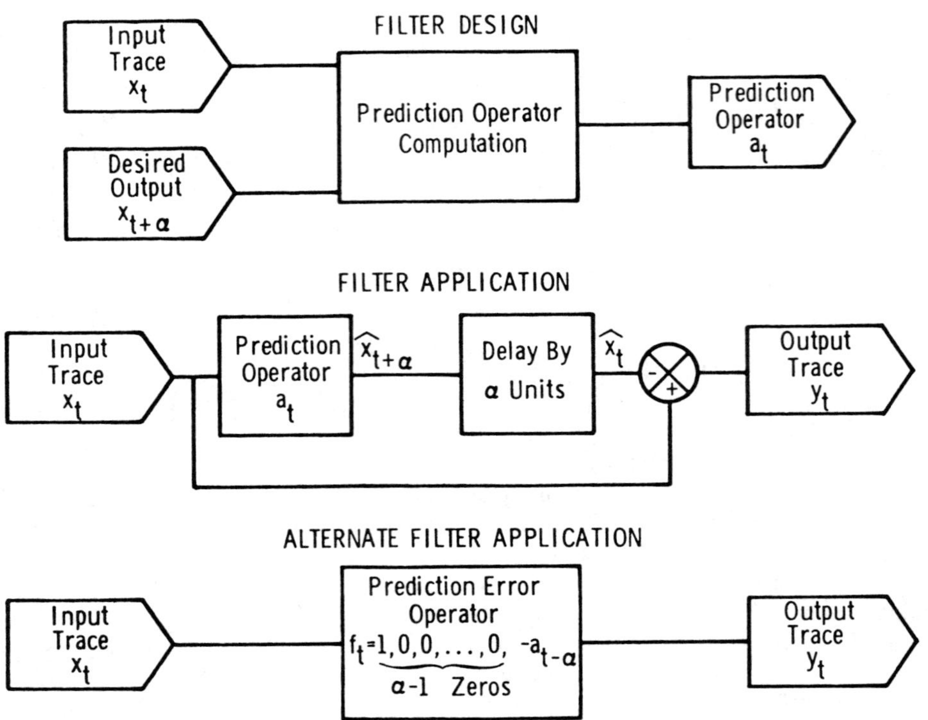

FIG. 2. The predictive filter model which illustrates prediction operator design (upper diagram) and application (center diagram). Alternately, the prediction error operator may be formed and utilized as illustrated in the lower diagram.

the autocorrelation of the input and the positive lag coefficients of the crosscorrelation between the desired output and the input. The autocorrelation of the input trace is given by

$$r_\tau = \sum_t x_t x_{t-\tau},$$

while the crosscorrelation between the desired output and input traces is

$$g_\tau = \sum_t z_t x_{t-\tau}. \tag{7}$$

Since the desired output for the prediction operator is a time-advanced version of the input, i.e., since

$$z_t = x_{t+\alpha},$$

equation (7) becomes

$$g_\tau = \sum_t x_{t+\alpha} x_{t-\tau} = \sum_t x_t x_{t-(\tau+\alpha)} = r_{\tau+\alpha}.$$

Thus, the crosscorrelation between the desired output and the input is by definition equal to the autocorrelation of the input for lags $\geq \alpha$. The normal equations (6) become

$$
\begin{bmatrix}
r_0 & r_1 & \cdots & r_{n-1} \\
r_1 & r_0 & \cdots & r_{n-2} \\
 & & \vdots & \\
r_{n-1} & r_{n-2} & \cdots & r_0
\end{bmatrix}
\begin{bmatrix}
a_0 \\
a_1 \\
\vdots \\
a_{n-1}
\end{bmatrix}
=
\begin{bmatrix}
r_\alpha \\
r_{\alpha+1} \\
\vdots \\
r_{\alpha+n-1}
\end{bmatrix}. \tag{8}
$$

The solution to the above matrix equation yields the prediction operator illustrated in the upper diagram of Figure 2. This prediction operator can be utilized as indicated in the center diagram of Figure 2. Alternately, the corresponding prediction error operator can be formed from the prediction operator and utilized as indicated in the lower diagram of Figure 2.

We have thus shown how one can use the Wiener least squares error criterion to generate the least squares prediction operator, and how the prediction error operator is derived from its corresponding prediction operator. Our aim in the next section is to indicate how the prediction *error* operator can also be expressed in the form of a particular Wiener filter.

Let us modify the above system by first subtracting the coefficient r_i from both sides of the ith row of each equation such that the right-hand side vanishes (the original system is shown within the rectangle),

$$
\begin{array}{rl}
-r_1 + & \boxed{\begin{array}{l} r_0a_0 + r_1a_1 + \cdots + r_{n-1}a_{n-1} = r_1 \\ r_1a_0 + r_0a_1 + \cdots + r_{n-2}a_{n-1} = r_2 \\ \vdots \\ r_{n-1}a_0 + r_{n-2}a_1 + \cdots + r_0a_{n-1} = r_n \end{array}} & \begin{array}{l} -r_1 \\ -r_2 \\ \\ -r_n. \end{array}
\end{array}
$$

Let us next augment the above system in the form,

$$-r_0 + \qquad r_1a_0 + r_2a_1 + \cdots + r_na_{n-1} = -\beta$$

$$
\begin{array}{rl}
-r_1 + & \boxed{\begin{array}{l} r_0a_0 + r_1a_1 + \cdots + r_{n-1}a_{n-1} = r_1 \\ r_1a_0 + r_0a_1 + \cdots + r_{n-2}a_{n-1} = r_2 \\ \vdots \\ r_{n-1}a_0 + r_{n-2}a_1 + \cdots + r_0a_{n-1} = r_n \end{array}} & \begin{array}{l} -r_1 \\ -r_2 \\ \\ -r_n. \end{array}
\end{array}
$$

PREDICTIVE FILTERING AND DECONVOLUTION

We shall demonstrate that the least-squares deconvolution filter which ideally transforms an unknown signal to an impulse at zero delay is equivalent to the prediction *error* filter for which the prediction distance α is unity. The matrix relation for the prediction operator a_t with prediction distance unity ($\alpha = 1$) and length n is obtained by setting $\alpha = 1$ in equation (8),

$$
\begin{bmatrix} r_0 & r_1 & \cdots & r_{n-1} \\ r_1 & r_0 & \cdots & r_{n-2} \\ & & \vdots & \\ r_{n-1} & r_{n-2} & \cdots & r_0 \end{bmatrix} \begin{bmatrix} a_0 \\ a_1 \\ \vdots \\ a_{n-1} \end{bmatrix} = \begin{bmatrix} r_1 \\ r_2 \\ \vdots \\ r_n \end{bmatrix}. \quad (9)
$$

The above system may be written in the form of the n simultaneous linear equations,

$$r_0a_0 + r_1a_1 + \cdots + r_{n-1}a_{n-1} = r_1$$
$$r_1a_0 + r_0a_1 + \cdots + r_{n-2}a_{n-1} = r_2$$
$$\vdots$$
$$r_{n-1}a_0 + r_{n-2}a_1 + \cdots + r_0a_{n-1} = r_n.$$

This system may be written

$$r_0 - r_1a_0 - r_2a_1 - \cdots - r_na_{n-1} = \beta$$
$$r_1 - r_0a_0 - r_1a_1 - \cdots - r_{n-1}a_{n-1} = 0$$
$$\vdots$$
$$r_n - r_{n-1}a_0 - r_{n-2}a_1 - \cdots - r_0a_{n-1} = 0,$$

for which the associated matrix equation is

$$
\begin{bmatrix} r_0 & r_1 & \cdots & r_n \\ r_1 & r_0 & \cdots & r_{n-1} \\ & & \vdots & \\ r_n & r_{n-1} & \cdots & r_0 \end{bmatrix} \begin{bmatrix} 1 \\ -a_0 \\ \vdots \\ -a_{n-1} \end{bmatrix} = \begin{bmatrix} \beta \\ 0 \\ \vdots \\ 0 \end{bmatrix}. \quad (10)
$$

We now see that the Wiener filter of equation (10) can be identified as the unit prediction *error* operator associated with the prediction operator of equation (9). Let us rewrite equation (10) in the form

$$
\begin{bmatrix} r_0 & r_1 & \cdots & r_n \\ r_1 & r_0 & \cdots & r_{n-1} \\ & & \vdots & \\ r_n & r_{n-1} & \cdots & r_0 \end{bmatrix} \begin{bmatrix} b_0 \\ b_1 \\ \vdots \\ b_n \end{bmatrix} = \begin{bmatrix} \beta \\ 0 \\ \vdots \\ 0 \end{bmatrix}, \quad (11)
$$

where

$$ b_0 = 1 $$

$$ b_i = -a_{i-1}, \qquad i = 1, \cdots, n $$

$$ \beta = \sum_{i=0}^{n} b_i r_i. $$

In Appendix B we describe the standard deconvolution method, which is based on the use of the least-squares, zero-delay inverse filter. We note that the system of normal equations for the inverse filter given by equation (B-1) is identical to the system of equations (11), except for a scale factor β. Thus the $(n+1)$-length prediction error operator with prediction distance unity is identical to the zero-delay inverse filter of length $(n+1)$, except for a scale factor.

We shall now show that the predictive filter with prediction distance greater than unity can also serve as a deconvolution operator, and thus it turns out that the predictive filtering technique constitutes a more generalized approach to deconvolution. We remark that under certain assumptions described in Appendix A, the autocorrelation of an input seismic signal can be identified with the autocorrelation of the source wavelet[1].

The inverse filter described in Appendix B shapes the unknown source wavelet to an impulse at zero lag time. We will show here that the predictive filter shapes the unknown source wavelet of length $\alpha+n$ to another unknown wavelet of length α. Thus, by having control of the desired output wavelet length, one may specify the desired degree of resolution.

The predictive filter matrix equation for filter length n and prediction distance α is given by equation (8),

$$
\begin{bmatrix} r_0 & r_1 & \cdots & r_{n-1} \\ r_1 & r_0 & \cdots & r_{n-2} \\ & & \vdots & \\ r_{n-1} & r_{n-2} & \cdots & r_0 \end{bmatrix} \begin{bmatrix} a_0 \\ a_1 \\ \vdots \\ a_{n-1} \end{bmatrix} = \begin{bmatrix} r_\alpha \\ r_{\alpha+1} \\ \vdots \\ r_{\alpha+n-1} \end{bmatrix},
$$

or

$$ r_0 a_0 + r_1 a_1 + \cdots + r_{n-1} a_{n-1} = r_\alpha $$

$$ r_1 a_0 + r_0 a_1 + \cdots + r_{n-2} a_{n-1} = r_{\alpha+1} $$

$$ \vdots \qquad (12) $$

$$ r_{n-1} a_0 + r_{n-2} a_1 + \cdots + r_0 a_{n-1} = r_{\alpha+n-1}. $$

The above system can be augmented in such a way that the prediction operator is converted into its corresponding prediction error operator. This is accomplished by the addition of suitable terms to both sides of the equations (12). Proceeding as in the case of the unit prediction error filter (equations (9) et seq.), one obtains,

$$
\begin{aligned}
-r_0 1 - r_1 0 \quad - \cdots -r_{\alpha-1}0 + r_\alpha a_0 \ + r_{\alpha+1} a_1 + \cdots + r_{\alpha+n-1} a_{n-1} &= -\rho_0 \\
-r_1 1 - r_0 0 \quad - \cdots -r_{\alpha-2}0 + r_{\alpha-1} a_0 + r_\alpha a_1 \ + \cdots + r_{\alpha+n-2} a_{n-1} &= -\rho_1 \\
\vdots \qquad\qquad\qquad & \\
-r_{\alpha-1} 1 - r_{\alpha-2} 0 \quad - \cdots -r_0 0 \ + r_1 a_0 \ + r_2 a_1 + \cdots + \qquad r_n a_{n-1} &= -\rho_{\alpha-1} \\
\vdots \qquad\qquad\qquad & \\
-r_\alpha 1 - r_{\alpha-1} 0 \quad - \cdots -r_1 0 \ + \boxed{\; r_0 a_0 + r_1 a_1 \ + \cdots + r_{n-1} a_{n-1} = r_\alpha \;} & \quad -r_\alpha \\
-r_{\alpha+1} 1 - r_\alpha 0 \quad - \cdots -r_2 0 \ + \boxed{\; r_1 a_0 + r_0 a_1 \ + \cdots + r_{n-2} a_{n-1} = r_{\alpha+1} \;} & \quad -r_{\alpha+1} \\
\vdots \qquad\qquad\qquad & \\
-r_{\alpha+n-1} 1 - r_{\alpha+n-2} 0 - \cdots -r_n 0 \ + \boxed{\; r_{n-1} a_0 + r_{n-2} a_1 + \cdots + r_0 a_{n-1} = r_{\alpha+n-1} \;} & \quad -r_{\alpha+n-1}
\end{aligned}
$$

where the original set (12) is enclosed by the rectangle. The associated matrix equation is

---

[1] The source wavelet is here meant to be the shot pulse modified by near-surface reverberations.

$$
\begin{bmatrix}
r_0 & r_1 & \cdots & r_{\alpha+n-1} \\
r_1 & r_0 & \cdots & r_{\alpha+n-2} \\
 & & \vdots & \\
r_{\alpha-1} & r_{\alpha-2} & \cdots & r_n \\
r_{\alpha} & r_{\alpha-1} & \cdots & r_{n-1} \\
 & & \vdots & \\
r_{\alpha+n-1} & r_{\alpha+n-2} & \cdots & r_0
\end{bmatrix}
\begin{bmatrix}
1 \\ 0 \\ \vdots \\ 0 \\ -a_0 \\ \vdots \\ -a_{n-1}
\end{bmatrix}
=
\begin{bmatrix}
\rho_0 \\ \rho_1 \\ \vdots \\ \rho_{\alpha-1} \\ 0 \\ \vdots \\ 0
\end{bmatrix}, \quad (13)
$$

where

$$\rho_0 = r_0 - (r_{\alpha}a_0 + r_{\alpha+1}a_1 + \cdots + r_{\alpha+n-1}a_{n-1})$$

$$\rho_1 = r_1 - (r_{\alpha-1}a_0 + r_{\alpha}a_1 + \cdots + r_{\alpha+n-2}a_{n-1})$$

$$\vdots$$

$$\rho_{\alpha-1} = r_{\alpha-1} - (r_1 a_0 + r_2 a_1 + \cdots + r_n a_{n-1}).$$

The solution of the above matrix equation yields the prediction error operator with prediction distance α. Let us interpret this equation in terms of the Wiener filter model, where the left-hand matrix is the input autocorrelation matrix, and where the elements of the right-hand column vector constitute the positive lag values of the crosscorrelation between the desired output and the input. Subject to the assumptions given in Appendix A, the autocorrelation function $r_0, r_1, \cdots, r_{\alpha+n-1}$ can be identified with the auto-correlation of a source wavelet of length $\alpha+n$. However, we still require an interpretation of the crosscorrelation,

$$g_\tau = \underbrace{\rho_0, \rho_1, \cdots, \rho_{\alpha-1},}_{\alpha \text{ terms}} \underbrace{0, \cdots, 0.}_{n \text{ zeros}}$$

Although the crosscorrelation function is complicated, we can make one important observation. Since the crosscorrelation vanishes for lags greater than $\alpha-1$, the length of the implied desired output wavelet cannot be greater than α. In other words, the input wavelet is of length $\alpha+n$,

while the implied desired output wavelet is of length α, and hence the prediction error operator shortens an input wavelet of length $\alpha+n$ to an output wavelet of length α. Since α is an independent variable, we are free to select whatever length we choose for the desired output wavelet. We conclude that the predictive filter leads to a more generalized approach to deconvolution, in which one may control the desired degree of resolution or wavelet contraction.

We have shown earlier in this section that the zero-delay least-squares, inverse filter is equal within a scale factor to the prediction error filter with prediction distance unity ($\alpha=1$). Experience has taught us that the output from these filters cannot in general be interpreted with ease. This is due to the presence of high-frequency components in the deconvolved trace, which result from the fact that this kind of deconvolution makes use of inverse, or "spiking" filters. One improves this condition by passing the raw deconvolved trace through suitable low-pass filters, by smoothing the autocorrelation function, or by other related means. We suggest that the use of prediction error filters with arbitrary prediction distance α leads to a deconvolution method in which one has more effective control on the desired degree of resolution. It is also significant to note that the inverse filter deconvolution method requires the insertion of an arbitrary scaling factor into the right side of the normal equations (i.e., the element β of equation (11) is arbitrary). No such scaling factor is needed in the predictive filter model, and hence the trace-to-trace amplitude variation which occurs in the input data can be preserved if so desired. In addition, no time need be spent by the computer in analyzing the output data to determine the scaling factor.

It is of some interest to establish how the predictive filters presented in this section perform on an idealized, noise-free reverberation model. These matters are discussed in Appendix C, where we also show that under appropriate simplifying conditions the prediction error filter becomes identical to the 3-point filter of Backus (1959).

APPLICATIONS OF PREDICTIVE DECONVOLUTION

The concept which permits resolution control by means of the prediction distance parameter has been introduced in the previous section. We have seen that the crosscorrelation between the desired output and the input is zero between lag positions α and $\alpha+n-1$, and we were thus able to deduce that the implied desired output pulse cannot be of length greater than α. We note from the autocorrelation matrix of equation (13) that the predictive filter does not utilize any autocorrelation coefficient beyond lag position $\alpha+n-1$. Our model thus implies that the source wavelet is of length $\alpha+n$.

Since this filter attempts to shape the input into some desired output, we can argue that the autocorrelation of the actual output data will tend to vanish between lag positions α and $\alpha+n-1$. This is because the autocorrelation of the implied desired output wavelet vanishes for lags greater than $\alpha-1$. The predictive filter will thus modify the input in such a way that the autocorrelation of the actual output will tend to vanish between α and $\alpha+n-1$. We cannot expect the autocorrelation to be 0 everywhere beyond lag $=\alpha+n-1$, since the filter computation makes use of no autocorrelation coefficients for lags greater than $\alpha+n-1$.

Anstey (1966) describes how the autocorrelation can be an interpretative aid in the analysis of reverberatory problems. When we consider that the predictive filter is designed only from knowledge of the input autocorrelation and that the magnitude of this input autocorrelation at a particular lag is an indication of the degree of predictability at that lag, we see that the autocorrelogram is a very important entity to gauge the effectiveness of dereverberation by means of predictive deconvolution.

Thus we set our parameters α (the prediction distance) and n (the prediction operator length) such that predictable (i.e., repetitive) energy having periods between α and $\alpha+n-1$ time units will tend to be removed, and hence the autocorrelation of the output will tend to vanish between lags α and $\alpha+n-1$.

Another means to measure the effectiveness of the predictive deconvolution process has been given by Wadsworth et al (1953). These authors point out that the reduction in energy content of the output trace relative to the input trace gives a measure of the predictable energy removed by the filtering operation in the range $t=\alpha$ to $t=\alpha+n-1$.

A given reverberation may be characterized as either "short-period" or "long-period." Long-period reverberations appear on a correlogram as distinct waveforms which are separated by quiet

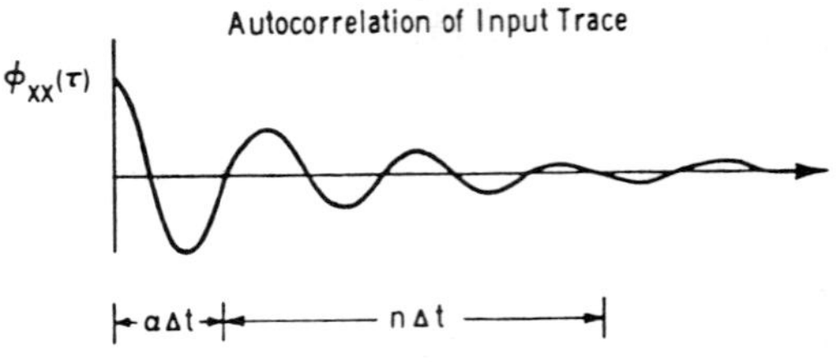

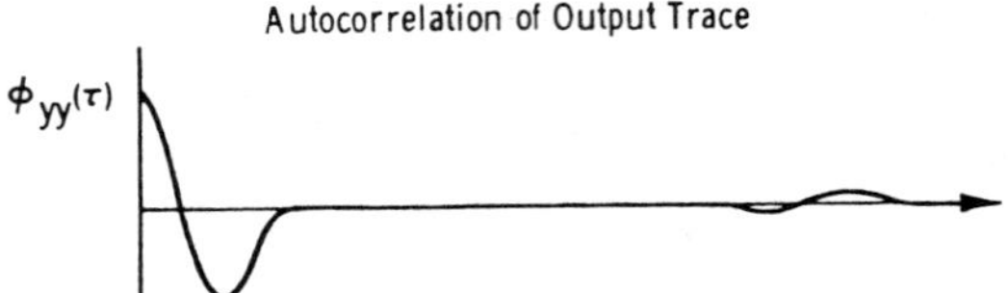

FIG. 3. A typical autocorrelation of an input trace with a moderate amount of short-period reverberation is illustrated by the upper diagram. The lower diagram illustrates the appearance of the output autocorrelation after application of the prediction error operator with prediction distance α and length n, as established in the upper diagram. If the reverberating pattern on the autocorrelation is highly regular, one need set n such that only the first full cycle is spanned. If the pattern is irregular, the significant portion of the reverberation should be spanned.

zones. Short-period reverberations appear on a correlogram in the form of decaying waveforms which are not separated by any noticeable quiet intervals.

The upper diagram of Figure 3 illustrates the autocorrelation of a typical trace which exhibits a moderate degree of short-period reverberation. The prediction distance α is chosen to specify the degree of wavelet contraction desired. As α approaches unity, more contraction and consequently more high-frequency noise is introduced. Thus we choose α so that we may obtain a compromise between wavelet contraction and signal-to-noise ratio in the output trace. Preliminary studies indicate that α should be set roughly equal to the lag that corresponds to the second zero crossing of the autocorrelation function. The lower diagram of Figure 3 illustrates the fact that the autocorrelation of the output signal trace tends to zero between lags α and $\alpha+n-1$.

Figure 4 shows three different predictive deconvolution runs on offshore traces having reverberations with characteristics somewhat inbetween our definitions of the "short-period" and "long-period" types. The product $\alpha\Delta t$ has been given the values 32, 16, and 4 ms. Since this data has a sampling interval of 4 ms, the third run actually corresponds to deconvolution by the

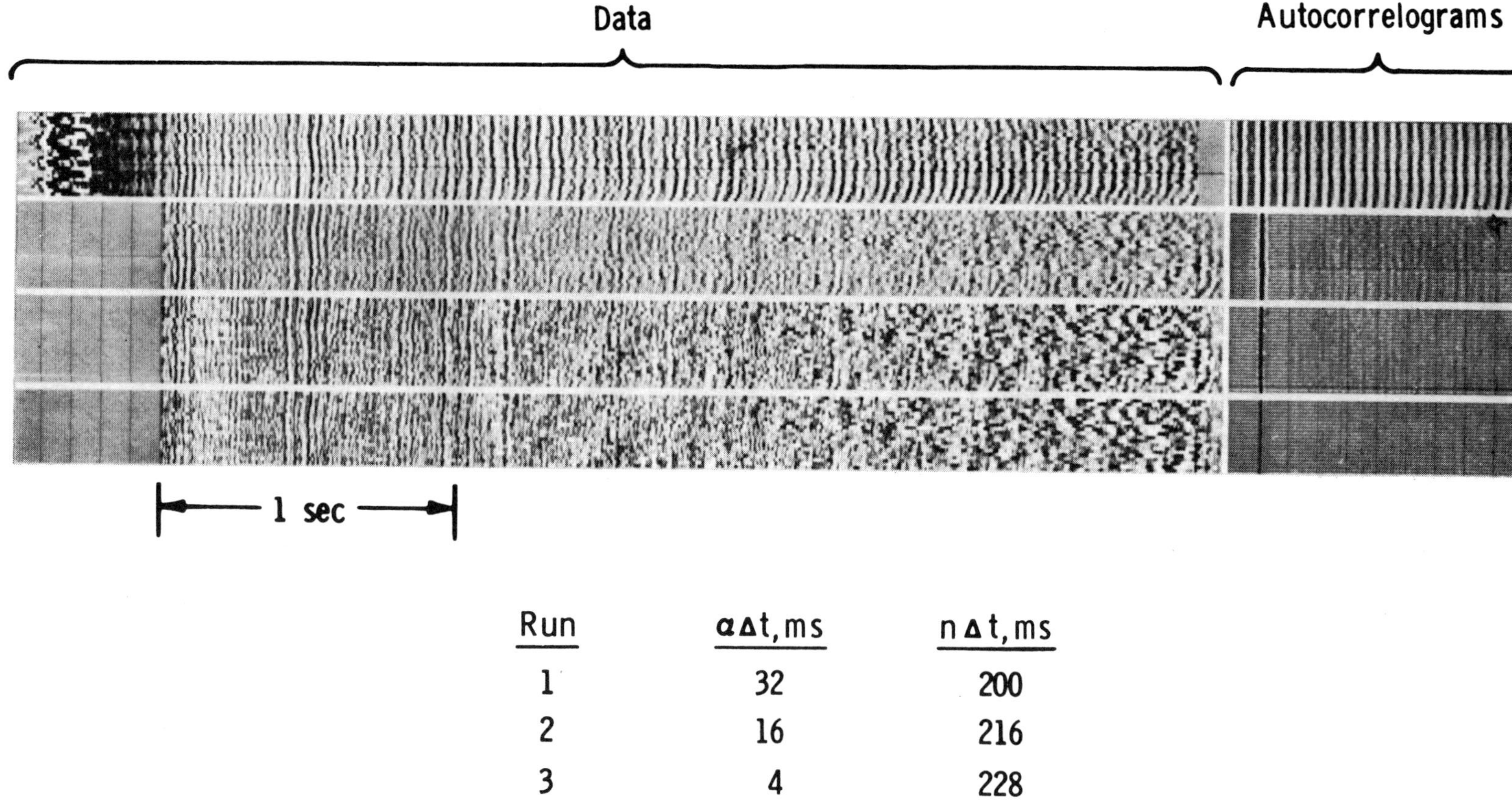

Run	$\alpha\Delta t$, ms	$n\Delta t$, ms
1	32	200
2	16	216
3	4	228

FIG. 4. An illustration of the effect which a variation in α, the prediction distance, has upon the output data. There appears to be some value of α which gives the best compromise between deringing and signal-to-noise ratio. Seismograms—top to bottom. Input; Outputs 1, 2, 3.

zero-delay least-squares inverse filter. We see that for $\alpha\Delta t = 32$ ms ($\alpha = 8$), we obtain a good dereverberation which does not exhibit the noise build-up associated with the smaller prediction distances.

The upper diagram of Figure 5 illustrates the autocorrelation of a typical trace with long-period reverberations. We define the appropriate prediction distance α such that the window to be deleted on the autocorrelogram begins just before the onset of the first multiple indication. Depending upon the nature and period of the reverberation, we define the filter length n such that the window to be deleted spans one, two, or more orders of the multiple pattern. The autocorrelation of the resulting output trace will show very little energy between lags α and $\alpha+n-1$. In addition, further repetitions of the waveform centered at multiples of $\alpha+n/2$ will be attenuated.

We note that the predictive filter enables us to suppress selected waveform portions of the autocorrelation function. This is highly advantageous since some waveforms on the autocorrelation might be due to accidentally strong correlations between certain primary reflections, and in this case we would choose not to suppress them. We may indeed avoid their suppression by the proper selection of the prediction distance α, and we then concentrate on those waveforms associated with reverberations. We remark that we can often successfully remove the long-period multiple indication from the autocorrelation by means of the predictive filter. Even so, the net change on the section itself is not always significant. Perhaps more study will reveal better ways of selecting the parameters such that long-period reverberations will be better attenuated by the predictive filter.

Figure 6 illustrates a predictive deconvolution run on a record with short-period reverberations. The data and associated autocorrelograms show the pulse compression which has been obtained.

Figure 7 depicts a predictive deconvolution run on marine data which exhibits long-period reverberations. We note that the prediction distance in this case is 150 ms and that the filter length is only 60 ms. If one were to deconvolve these traces with the unit prediction error filter, it would be necessary to make the filter length at least equal to 210 ms. This would require much more time to process the data. Kunetz and Fourmann (1968) have reached similar conclusions by a somewhat different line of reasoning. We note

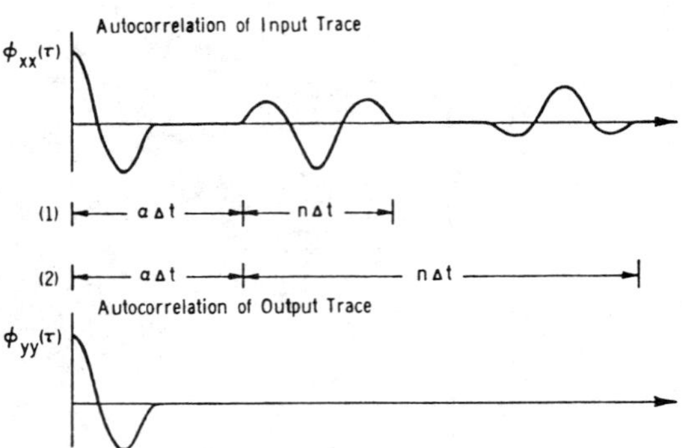

FIG. 5. A typical autocorrelation of an input trace with a moderate amount of long-period reverberation is illustrated by the upper diagram. The lower diagram illustrates the appearance of the output autocorrelation after application of the prediction error operator with prediction distance α and length n as established in the upper diagram. The parameters should be defined as indicated in (1) if the ringing is of a first-order nature, or as indicated in (2) if the ringing is of a second-order nature.

that we have achieved a successful attenuation of the reverberations on the autocorrelograms and a moderately successful dereverberation of the data itself.

CONCLUSIONS

The predictive filter is a very flexible tool for the deconvolution of seismic traces. The ability to specify the prediction distance implies the ability to control output resolution, and this means that a broad range of complex reverberatory problems can be successfully attacked with the present methods. Repetitive waveforms of a particular period can be selectively attenuated, and this is accomplishable without any significant disturbance of waveforms which one may wish to retain. The autocorrelogram is a valuable interpretative device for reverberation analysis and should be used on a routine basis. It would be desirable to have still better criteria for the determination of optimum values of such filter parameters as the prediction distance α and the filter length n. More research and evaluation of the methods presented in this paper are therefore in order.

ACKNOWLEDGMENTS

The authors wish to express their thanks to Mr. C. W. Frasier for the use of some of his unpublished results and to the Pan American Petroleum Corporation for permission to publish this paper.

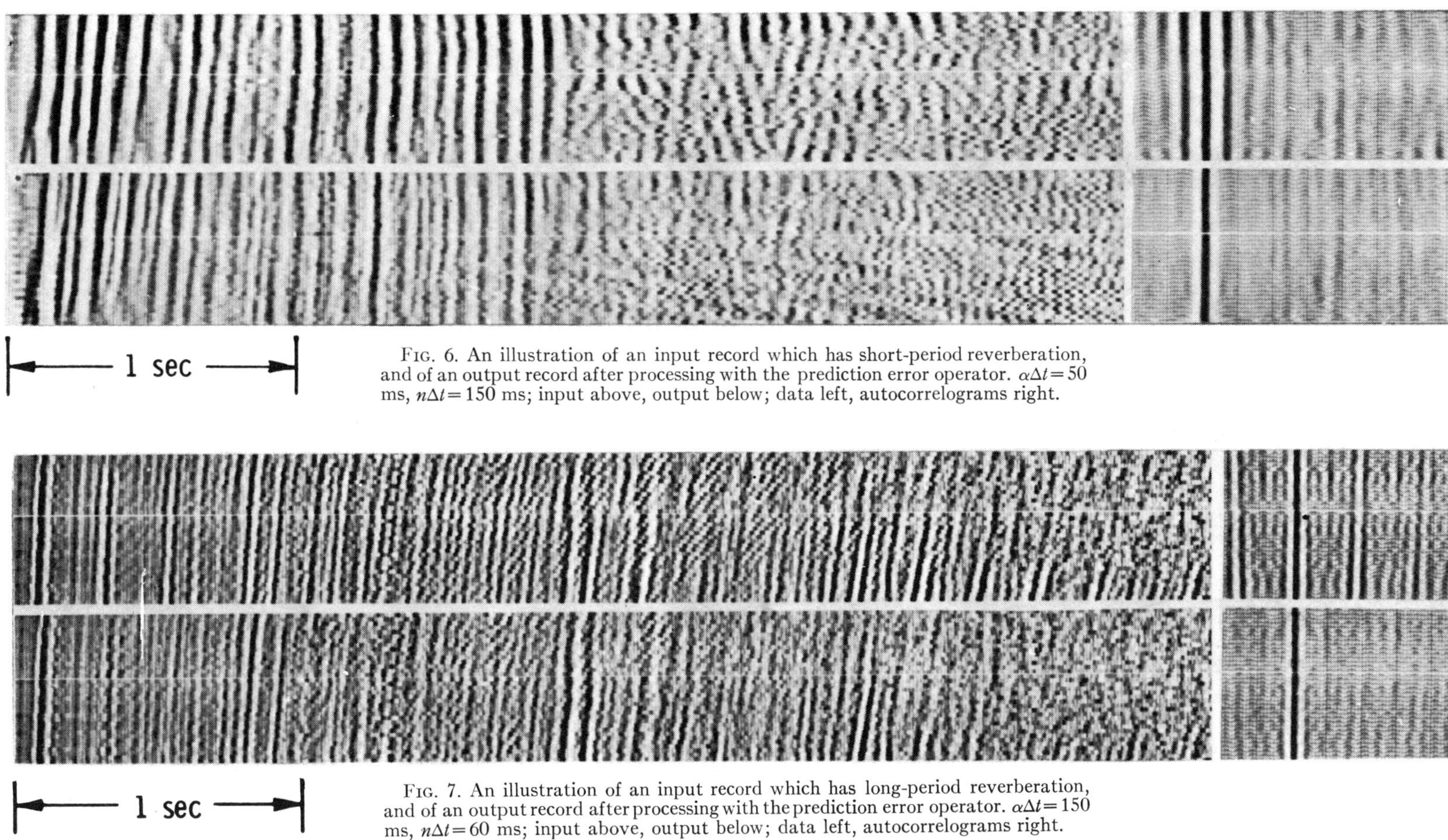

Fig. 6. An illustration of an input record which has short-period reverberation, and of an output record after processing with the prediction error operator. $\alpha\Delta t = 50$ ms, $n\Delta t = 150$ ms; input above, output below; data left, autocorrelograms right.

Fig. 7. An illustration of an input record which has long-period reverberation, and of an output record after processing with the prediction error operator. $\alpha\Delta t = 150$ ms, $n\Delta t = 60$ ms; input above, output below; data left, autocorrelograms right.

REFERENCES

Anstey, N. A., 1966, The sectional autocorrelogram and the sectional retrocorrelogram: Geophys. Prosp., v. 14, p. 389–426.

Backus, M. M., 1959, Water reverberations—Their nature and elimination: Geophysics, v. 24, p. 233–261.

Jenkins, G. M., 1961, General considerations in the analysis of spectra: Technometrics, v. 3, no. 2, p. 133–166.

Kunetz, G., and Fourmann, J. M., 1968, Efficient deconvolution of marine seismic records: Geophysics, v. 33, p. 412–423.

Robinson, E. A., 1954, Predictive decomposition of time series with applications to seismic exploration: Ph.D. thesis, MIT, Cambridge, Mass.

———— 1966, Multichannel z-transforms and minimum-delay: Geophysics, v. 31, p. 482–500.

———— and Treitel, S., 1967, Principles of digital Wiener filtering: Geophys. Prosp., v. 15, p. 311–333.

Wadsworth, G. P., Robinson, E. A., Bryan, J. G., and Hurley, P. M., 1953, Detection of reflections on seismic records by linear operators: Geophysics, v. 18, p. 539–586.

APPENDIX A

THE AUTOCORRELATION OF A SEISMIC TRACE

Under the proper assumptions the autocorrelation of a seismic trace is an estimate of the autocorrelation of the "basic" seismic wavelet.[2] The derivation presented here is similar to one given by Robinson and Treitel (1967).

Suppose we have a signal x_t which results from the convolution of a basic wavelet p_t with an uncorrelated series n_t, where we assume that n_t can be identified with the reflection coefficient series of a layered medium (Robinson, 1954), that is,

$$x_t = p_t * n_t.$$

The z-transform of the autocorrelation of x_t is given by

$$\Phi_{xx}(z) = [P(z)N(z)][P(1/z)N(1/z)].$$

The above equation can be rewritten in the form,

$$\Phi_{xx}(z) = [P(z)P(1/z)][N(z)N(1/z)],$$

which is the z-transform of

$$\phi_{xx}(\tau) = \phi_{pp}(\tau) * \phi_{nn}(\tau). \qquad \text{(A-1)}$$

Therefore the autocorrelation of x_t is equal to the convolution of the autocorrelation of p_t with the autocorrelation of n_t. Since n_t is an uncorrelated series, we obtain

[2] The basic seismic wavelet is assumed to be either the initial shot pulse, or the initial shot pulse modified by near-surface reverberations.

$$\phi_{nn}(\tau) = E_n \qquad \text{for } \tau = 0$$

and

$$\phi_{nn}(\tau) = 0 \qquad \text{for } \tau \neq 0,$$

where E_n is the energy in n_t. Thus equation (A-1) reduces to

$$\phi_{xx}(\tau) = \sum_t \phi_{nn}(t)\phi_{pp}(\tau - t) = E_n\phi_{pp}(\tau),$$

and we see that the autocorrelation of x_t is simply a scaled version of the autocorrelation of p_t. This means that subject to the above assumptions, we can obtain an estimate of ϕ_{pp} even though we do not know p_t itself. The consistency of the autocorrelation estimates can be improved through use of suitable weighting functions. A good discussion of these matters is given by Jenkins (1961).

APPENDIX B

THE INVERSE FILTER MODEL

The Wiener filter model requires that the autocorrelation of the input and the positive lag values of crosscorrelation between the desired output and the input be known. The basic seismic wavelet is generally unknown; however, we can calculate its autocorrelation and the required crosscorrelation if we make the proper assumptions.

Appendix A shows that an estimate of the basic wavelet autocorrelation can be obtained from the input trace. If we assume the desired output to be an impulse at zero lag time, the crosscorrelation between desired output and input also becomes an impulse at zero lag time. In other words, since the crosscorrelation is given by

$$\phi_{dp}(\tau) = \sum_t d_t p_{t-\tau} \quad \text{for } \tau = 0, 1, \cdots, n - 1,$$

where $d_t = 1, 0, 0, \cdots$ is the desired output signal and $p_t = p_0, p_1, p_2, \cdots$ is the basic wavelet or input signal, we see that

$$\phi_{dp}(\tau) = p_0, 0, 0, \cdots ;$$

$$\tau = 0, 1, \cdots, n - 1,$$

which can be scaled in the form,

$$\phi_{dp}(\tau) = 1, 0, 0, \cdots .$$

The matrix equation for the Wiener filter (Robinson and Treitel, 1967) then becomes,

$$
\begin{bmatrix}
r_0 & r_1 & \cdots & r_{n-1} \\
r_1 & r_0 & \cdots & r_{n-2} \\
 & & \vdots & \\
r_{n-1} & r_{n-2} & \cdots & r_0
\end{bmatrix}
\begin{bmatrix}
f_0 \\
f_1 \\
\vdots \\
f_{n-1}
\end{bmatrix}
=
\begin{bmatrix}
1 \\
0 \\
\vdots \\
0
\end{bmatrix}, \quad \text{(B-1)}
$$

where the f_t are the n coefficients which shape the basic wavelet p_t to an approximation of the impulse at zero lag time.

We have thus assumed a model of the form

$$x_t = p_t * n_t,$$

where x_t is the signal trace and n_t is an uncorrelated series which represents the reflection coefficients of the layered subsurface. Since the desired output is an impulse at zero lag time, we see that the model requires a filter f_t such that

$$f_t * p_t \doteq 1$$

or

$$f_t \doteq p_t^{-1},$$

where the symbol $\doteq$ means "approximately equal to." The filter f_t then deconvolves the input trace as follows:

$$y_t \doteq x_t * f_t \doteq p_t * p_t^{-1} * n_t$$

$$y_t \doteq \delta_t * n_t = n_t,$$

where δ_t is the unit impulse function, and in this sense the output trace tends to approximate the subsurface reflection coefficient series.

APPENDIX C

A STUDY OF A TWO-LAYER MARINE REVERBERATION MODEL

Impulse response of first-order component

Figure C-1 represents an idealized noise-free model of an offshore seismic situation. Reflector 1 is the water surface, reflector 2 is the water bottom, reflector 3 is some strong interface beneath the water bottom, and S is the source location just beneath the water surface. The associated normal incidence reflection coefficients are 1, c_1 and c_2, respectively, while the transmission coefficient across reflector 2 is t_1. If c is the downward reflection coefficient, the corresponding upward reflection coefficient is $-c$. From physical considerations, we know that the magnitudes of all re-

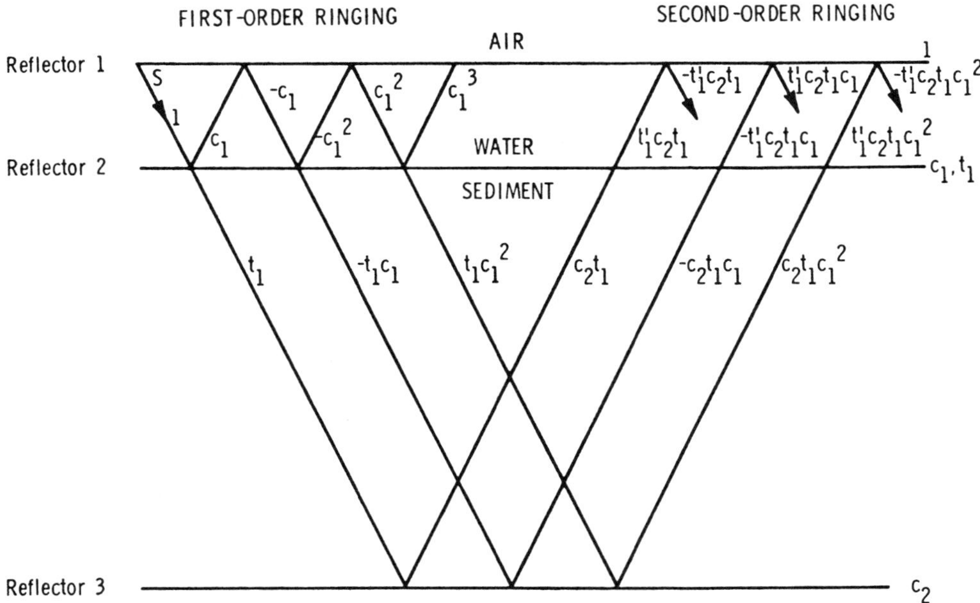

Fig. C-1. First- and second-order ringing in a 2-layer marine model.

flection coefficients are less than unity. The two-way traveltime through the water layer is τ_1.

Let us compute the first-order reverberation portion of the two-layer impulse response as indicated in Figure C-1. This response can be expressed in terms of the following z-transform:

$$R_1(z) = 1 - c_1 z^{\tau_1} + c_1^2 z^{2\tau_1} + \cdots \quad (C\text{-}1)$$

Multiplication of equation (C-1) by $c_1 z^{\tau_1}$ produces

$$c_1 z^{\tau_1} R_1(z) = c_1 z^{\tau_1} - c_1^2 z^{2\tau_1}$$
$$+ c_1^3 z^{3\tau_1} + \cdots , \quad (C\text{-}2)$$

while addition of equations (C-1) and (C-2) yields the expression

$$R_1(z) = \frac{1}{1 + c_1 z^{\tau_1}} . \quad (C\text{-}3)$$

We have thus obtained the z-transform of the first-order reverberation portion of the two-layer impulse response.

(C.1) *Impulse response of second-order component*

Figure C-1 indicates the pattern of all raypaths which contribute to second-order reverberations. A component from the original impulse is reflected from reflector 3 and re-enters the water layer to be reflected from the surface. At this point its amplitude is $-t_1' c_2 t_1$, where t_1' is the upward transmission coefficient across reflector 2. Hence an impulse of amplitude $-t_1' c_2 t_1$ is introduced into the first layer, which again will generate the associated first-order ringing already given by equation (C-3). However, the onset of this ringing occurs with a time delay of $\tau_1 + \tau_2$, where τ_2 is the two-way traveltime in the second layer. The z-transform of this component is

$$R_{2,1}(z) = z^{\tau_1+\tau_2}(-t_1' c_2 t_1)\frac{1}{1 + c_1 z^{\tau_1}} ,$$

where the subscripts of $R(z)$ denote response order and associated components, respectively. Likewise, the next pulse entering the water layer in the above manner generates the second component of the second-order response,

$$R_{2,2}(z) = z^{2\tau_1+\tau_2}(t_1' c_2 t_1 c_1)\frac{1}{1 + c_1 z^{\tau_1}} .$$

The third component of the second-order response

is

$$R_{2,3}(z) = z^{3\tau_1+\tau_2}(-t_1' c_2 t_1 c_1^2)\frac{1}{1 + c_1 z^{\tau_1}} .$$

One can continue this analysis up to any number of additional components. Summation of the above series of equations produces the complete second-order response,

$$R_2(z) = R_{2,1}(z) + R_{2,2}(z) + R_{2,3}(z) + \cdots$$

$$= z^{\tau_1+\tau_2}(-t_1' c_2 t_1)\frac{1}{1 + c_1 z^{\tau_1}}$$

$$\cdot\left[1 - c_1 z^{\tau_1} + c_1^2 z^{2\tau_1} + \cdots\right]$$

$$= z^{\tau_1+\tau_2}(-t_1' c_2 t_1)\frac{1}{(1 + c_1 z^{\tau_1})^2}$$

where we recall that $|c_1| < 1$. Since $z^{\tau_1+\tau_2}$ is simply a delay factor and $-t_1' c_2 t_1$ is a constant, we may shift the time origin and normalize the second-order response. This yields

$$R_2(z) = \frac{1}{(1 + c_1 z^{\tau_1})^2} , \quad (C\text{-}4)$$

an expression which we see to be the square of the first-order response given by equation (C-3).

Removal of first-order ringing

Let us incorporate the first-order impulse response into the predictive deconvolution model. We will assume that the two-way traveltime through the water layer is τ_1 sample units. Thus our impulse response becomes,

$$x_1(t) = 1, \underbrace{0, 0, \cdots, 0,}_{\tau_1 - 1 \text{ zeros}}$$
$$-c_1, \underbrace{0, 0, \cdots, 0,}_{\tau_1 - 1 \text{ zeros}} c_1^2, \cdots .$$

In order to compute the predictive filter, we require the autocorrelation of $x_1(t)$, which is

$$r_\tau = 1 + c_1^2 + c_1^4 + \cdots , \qquad \tau = 0.$$
$$r_\tau = 0, \qquad 0 < \tau < \tau_1.$$
$$r_\tau = -c_1(1 + c_1^2 + c_1^4 + \cdots)$$
$$= -c_1 r_0, \qquad \tau = \tau_1,$$

and so on. Thus the autocorrelation of $x_1(t)$ can

be written

$$r_\tau = E_x, \underbrace{0, 0, \cdots, 0,}_{\tau_1 - 1 \text{ zeros}} -cE_x, \cdots,$$

where E_x is the energy in $x_1(t)$. Let the filter length n be *less* than τ_1, and let the prediction distance be $\alpha = \tau_1$. Then the normal equations become

$$\begin{bmatrix} r_0 & 0 & \cdots & 0 \\ 0 & r_0 & \cdots & 0 \\ & & \cdot & \\ & & \cdot & \\ & & \cdot & \\ 0 & 0 & \cdots & r_0 \end{bmatrix} \begin{bmatrix} a_0 \\ a_1 \\ \cdot \\ \cdot \\ \cdot \\ a_{n-1} \end{bmatrix} = \begin{bmatrix} r_{\tau_1} \\ 0 \\ \cdot \\ \cdot \\ \cdot \\ 0 \end{bmatrix}.$$

The only member of this system whose right side does not vanish is,

$$r_0 a_0 = r_{\tau_1},$$

and thus,

$$a_0 = r_{\tau_1}/r_0 = -c_1 r_0/r_0 = -c_1.$$

The associated prediction error operator is

$$f_2(t) = 1, \underbrace{0, 0, \cdots, 0,}_{\tau_1 - 1 \text{ zeros}} c_1. \qquad \text{(C-5)}$$

In practice it is not necessary to set the prediction distance α exactly to τ_1. The present model permits α to take on any value as long as it is less than or equal to τ_1. Furthermore, the filter length must be such that the inequality $\alpha + n > \tau_1$ holds true.

We note that the z-transform of the prediction error operator of equation (C-5) is $1 + c_1 z^{\tau_1}$ which is the inverse of the first-order impulse response given by equation (C-3). If this first-order impulse response is convolved with the above prediction error filter, the output will be 1; in other words, we will have deconvolved the ringing signal. Figure C-2 illustrates the input signal $x_1(t)$, the prediction error operator $f_1(t)$, and the output signal $y_1(t)$ for this situation.

Removal of second-order ringing

Let us use the predictive deconvolution method on the second-order portion of the two-layer impulse response given by equation (C-4). This response can be written,

$$x_2(t) = 1, \underbrace{0, 0, \cdots, 0,}_{\tau_1 - 1 \text{ zeros}} -2c_1, \underbrace{0, 0, \cdots, 0,}_{\tau_1 - 1 \text{ zeros}}$$
$$\underbrace{3c_1^2, 0, 0, \cdots, 0,}_{\tau_1 - 1 \text{ zeros}} -4c_1^3, \cdots.$$

The autocorrelation of $x_2(t)$ is

$$r_\tau = 1 + 4c_1^2 + 9c_1^4 + 16c_1^9 + \cdots$$
$$= \frac{1 + c_1^2}{(1 - c_1^2)^3}, \qquad \tau = 0.$$
$$r_\tau = 0, \qquad 0 < \tau < \tau_1.$$
$$r_\tau = -2c_1 - 6c_1^3 - 12c_1^5 - 20c_1^7 + \cdots$$
$$= \frac{-2c_1}{(1 - c_1^2)^3}, \qquad \tau = \tau_1.$$
$$r_\tau = 0, \qquad \tau_1 < \tau < 2\tau_1.$$
$$r_\tau = 3c_1^2 + 8c_1^4 + 15c_1^6 + 24c_1^8 + \cdots$$
$$= \frac{-c_1^4 + 3c_1^3}{(1 - c_1^2)^3}, \qquad \tau = 2\tau_1,$$

and so on. Thus the normalized autocorrelation of x_t becomes

$$r_\tau = 1 + c_1^2, \underbrace{0, 0, \cdots, 0,}_{\tau_1 - 1 \text{ zeros}}$$
$$-2c_1, \underbrace{0, 0, \cdots, 0,}_{\tau_1 - 1 \text{ zeros}} 3c_1^2 - c_1^4, \cdots.$$

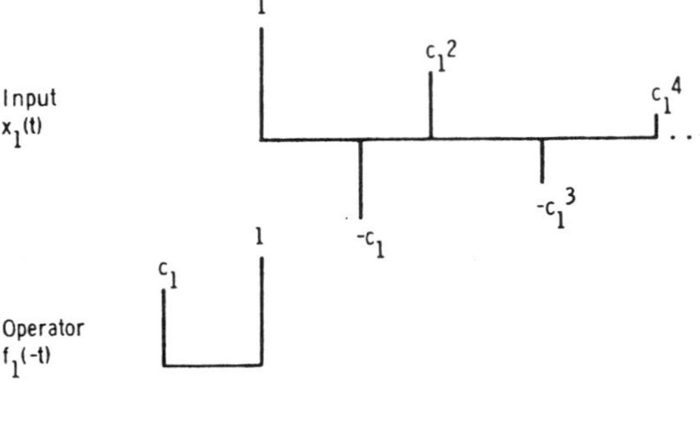
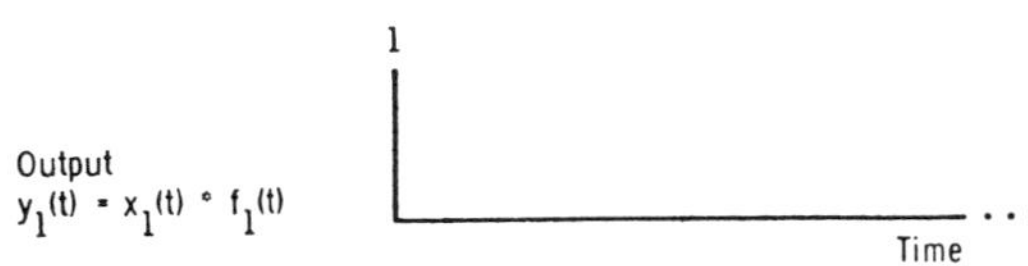

FIG. C-2. Deconvolution of a first-order ringing system. The operator is shown in time-reversed form.

If the filter length is $n = \tau_1 + 1$ and the prediction distance is $\alpha = \tau_1$, the normal equations become,

$$
\begin{array}{c}
\tau_1 - 1 \text{ columns} \\[-2pt]
\tau_1 - 1 \text{ rows} \left\{
\begin{bmatrix}
1 + c_1^2 & 0 & 0 & \cdots & 0 & -2c_1 \\
0 & 0 & 0 & \cdots & 0 & 0 \\
& & & \vdots & & \vdots \\
0 & 0 & 0 & \cdots & 0 & 0 \\
-2c_1 & 0 & 0 & \cdots & 0 & 1 + c_1^2
\end{bmatrix}
\right.
\begin{bmatrix}
a_0 \\ a_1 \\ \vdots \\ a_{\tau_1 - 1} \\ a_{\tau_1}
\end{bmatrix}
=
\begin{bmatrix}
-2c_1 \\ 0 \\ \vdots \\ 0 \\ 3c_1^2 - c_1^4
\end{bmatrix}
\end{array}
$$

The two nonvanishing equations of this system yield the solution

$$a_0 = -2c_1$$

and

$$a_{\tau_1} = -c_1^2.$$

Hence the associated prediction error operator is

$$f_2(t) = 1, \underbrace{0, 0, \cdots, 0,}_{\tau_1 - 1 \text{ zeros}}$$

$$\underbrace{2c_1, 0, 0, \cdots, 0,}_{\tau_1 - 1 \text{ zeros}} c_1^2. \qquad \text{(C-6)}$$

This particular prediction error operator is identical to the three-point filter of Backus (1959). We thus see that in the noise-free case the present predictive deconvolution model yields the classical results obtained on the basis of strictly deterministic considerations. The predictive deconvolution scheme allows a more general attack on the dereverberation problem, as the present treatment has sought to demonstrate.

It is not necessary to set the prediction distance exactly equal to τ_1, nor is it necessary to set the filter length exactly equal to $\tau_1 + 1$. However, these parameters must be set such that $\alpha \leq \tau_1$, $n \geq \tau_1$, and $\alpha + n \geq 2\tau_1$.

The z-transform of equation (C-6) is

$$F_2(z) = 1 + 2c_1 z^{\tau_1} + c_1^2 z^{2\tau_1} = (1 + c_1 z^{\tau_1})^2,$$

which is the inverse of the z-transform of the second-order impulse response given by equation (C-4). Thus, convolution of the second-order impulse response $x_2(t)$ with the prediction error operator $f_2(t)$ produces a zero delay spike (Figure C-3). In other words, the second-order ringing system has been deconvolved by means of the prediction error operator.

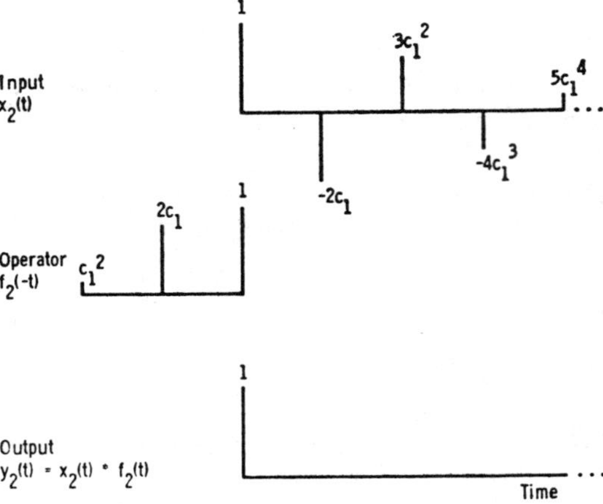

FIG. C-3. Deconvolution of a second-order ringing system. The operator is shown in time-reversed form.

3

DYNAMIC PREDICTIVE DECONVOLUTION*

BY

E. A. ROBINSON**

ABSTRACT

ROBINSON, E. A., Dynamic Predictive Deconvolution, Geophysical Prospecting 23, 780-798.

Dynamic predictive deconvolution makes use of an entire seismic trace including all primary and multiple reflections to yield an approximation to the subsurface structure. We consider plane-wave motion at normal incidence in an horizontally layered system sandwiched between the air and the basement rock. Energy degradation effects are neglected so that the layered system represents a lossless system in which energy is lost only by net transmission downward into the basement or net reflection upward into the air; there is no internal loss of energy by absorption within the layers. The layered system is frequency selective in that the energy from a surface input is divided between that energy which is accepted over time by net transmission downward into the basement and the remaining energy that is rejected over time by net reflection upward into the air. Thus the energy from a downgoing unit spike at the surface as input is divided between the wave transmitted by the layered system into the basement and the wave reflected by the layered system into the air. This reflected wave is the observed seismic trace resulting from the unit spike input. From surface measurements we can compute both the input energy spectrum, which by assumption is unity, and the reflection energy spectrum, which is the energy spectrum of the trace. But, by the conservation of energy, the input energy spectrum is equal to the sum of the reflection energy spectrum and the transmission energy spectrum. Thus we can compute the transmission energy spectrum as the difference of the input energy spectrum and the reflection energy spectrum. Furthermore, we know that the layered system acts as a pure feedback system in producing the transmitted wave, from which it follows that the transmitted wave is minimum-delay. Hence from the computed energy spectrum of the transmitted wave we can compute the prediction-error operator that contracts the transmitted wave to a spike. We also know that the layered system acts as a system with both a feedback component and a feedforward component in producing the reflected wave, that is, the observed seismic trace. Moreover, this feedback component is identical to the pure feedback system that produces the transmitted wave. Thus, we can deconvolve the observed seismic trace by the prediction-error operator computed above; the result of the deconvolution is the waveform due to the feedforward component alone. Now the feedforward component represents the wanted dynamic structure of the layered system whereas the feedback com-

---

* Paper read at the 37th Meeting of the European Association of Exploration Geophysicists, Bergen, June 1975.

** Seismological Institute, Uppsala, Sweden and Texas Geophysical Company, Houston, Texas.

ponent represents the unwanted reverberatory effects of the layered system. Because this deconvolution process yields the wanted dynamic structure and destroys the unwanted reverberatory effects, we call the process dynamic predictive deconvolution. The resulting feedforward waveform in itself represents an approximation to the subsurface structure; a further decomposition yields the reflection coefficients of the interfaces separating the layers. In this work we do not make the assumption as is commonly done that the surface as a perfect reflector; that is, we do not assume that the surface reflection coefficient has magnitude unity.

INTRODUCTION

Each method of deconvolution is based upon a model of the geophysical structure, and the success of the method depends upon how well the conditions imposed by the model are met in practice. The word *convolution* means "folding" and has a well-defined mathematical meaning. The word *deconvolution* means "unfolding" and represents the inverse operation to a convolution operation.

One model leading to a deconvolution operation is the *random-reflection seismic model* (Robinson 1954, 1966). In this model a portion of the seismic trace is given as the convolution of a minimum-delay wavelet with a random reflection series. Such a model is approximately realized by a system of surface layers overlying deep reflection horizons of random spacing and reflection characteristics. These deep horizons give rise to the random reflection series which represents the wanted primary events, whereas the overlying surface layers produce the unwanted reverberatory wavelets attached to each of these primary events. If the system is excited by a unit source spike at the surface, then the reverberatory wavelet results from two-way transmission (that is, down and back up) through the surface layers, and hence is minimum-delay (i.e. minimum-phase). Thus, this model has two characteristic features, namely:

(1) the statistical feature that the primary events are represented by a random-reflection series, and

(2) the deterministic feature that the reverberations attached to the primary events have the same minimum-delay wavelet shape.

The observational data are in the form of the observed seismic trace recorded at the surface. If we compute the autocorrelation of this trace, the contribution from the random-reflection series is averaged out (i.e. destroyed), and we are left with the autocorrelation of the reverberatory wavelet. We can then compute the prediction error operator from this autocorrelation by solving a system of normal equations. Because the reverberatory wavelet is minimum-delay, this prediction error operator is the operator that compresses the reverberatory wavelet to a spike, and hence is the required deconvolution operator. Thus, the method of predictive deconvolution for this model consists of computing the prediction error operator from the autocorrelation of the seismic trace, and then convolving the operator with the trace to yield the random-reflection series as the deconvolved trace. This deconvolved trace gives the required primary events which describe the deep subsurface structure. This deconvolution process may be described as the method of *random predictive deconvolution*, or simply RPD.

In this paper we wish to present an alternative method of deconvolution which we call *dynamic predictive deconvolution*, or simply DPD. This method is based on a *dynamic reflection seismic model*. This model is one in which an entire seismic trace is given as the convolution of a minimum-delay wavelet with a dynamic reflection series. Such a model is approximately realized by a horizontally layered system subject to normal incidence wave motion resulting from a unit source spike at the surface. This model has two characteristic features, namely:

(1) the deterministic feature that the primary events are approximately represented by a dynamic reflection series generated by a feedforward system, and

(2) the deterministic feature that the reverberations attached to the primary events have the same minimum-delay wavelet shape generated by a feedback system.

The observational data consist of the known spike input and the resulting seismic trace recorded at the surface. We compute the autocorrelation of this trace and then subtract this autocorrelation from the autocorrelation of the spike input to obtain a function which is itself an autocorrelation function. We then compute the prediction error operator from this final autocorrelation function by solving a system of normal equations. Because the reverberatory wavelet is minimum-delay, this prediction error operator is the operator that compresses the reverberatory wavelet to a spike, and hence is the required deconvolution operator. Thus, the method of dynamic predictive deconvolution for the given model consists of computing the prediction error operator from the autocorrelation formed as the difference of input and output autocorrelations and then convolving the operator with the trace to yield the dynamic reflection series as the deconvolved trace. This deconvolved trace gives an approximation to the required primary events which describe the subsurface structure. This approximation can be improved upon and the reflection coefficients obtained by a further decomposition.

For reasons of clarity we have tried to make our exposition of this alternative method of predictive deconvolution as simple as possible by treating only its essential mathematical features and so have not included many practical aspects which are required for its operational usage.

The Seismic Model

The seismic model is the familiar horizontally layered elastic medium, each layer being homogeneous and isotropic, subject to plane compressional wave motion at normal incidence. This model has a long pedigree in exploration geophysics, and the reader is referred to the works of Anstey (1960), Baranov and Kunetz (1958, 1960), Berryman, Goupillaud, and Waters (1958), Bois, Chauveau, Grau, and Lavergne (1960), Bortfeld (1960), Goupillaud (1961), Kunetz (1961, 1964), Kunetz and D'Erceville (1962), O'Brien (1961), Seriff (1958), Sherwood and Trorey (1965), Treitel and Robinson (1966), Wuenschel (1960), plus the many subsequent papers on this model that have appeared in *Geophysical Prospecting* and *Geophysics*. For layered-earth models subject to more general wave motion the reader is referred to Båth (1968), and for spectral analysis to Båth (1974).

Whereas most writers number the layers in this model from the top (i.e. the air) downward, we make the opposite convention and number the layers from the bottom (i.e. the basement) upward. Because of this change in convention, let us be quite explicit on how we number the layers and the interfaces (see figure 1).

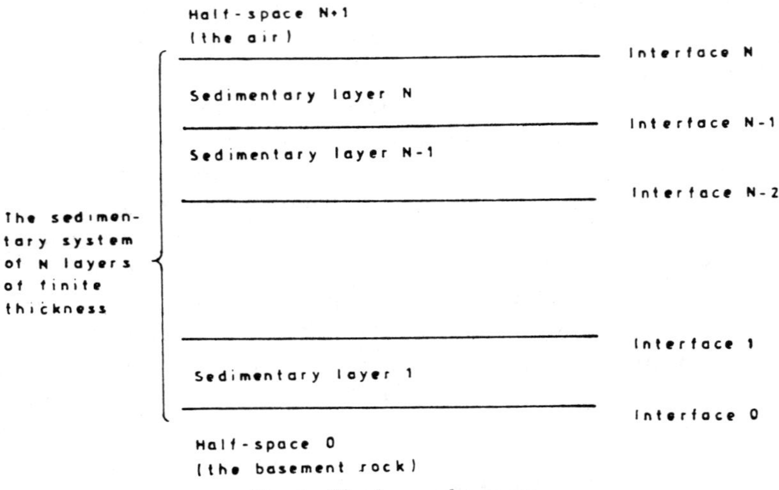

Fig. 1: The layered system

The lowermost layer is actually a half-space which we call the basement and denote by index 0. On top of the basement lie N proper layers of finite thickness which represent the sedimentary column and which we denote by indices running from one at the bottom to N at the top. In other words, layer 1 is the first sedimentary layer in geologic time and is the deepest layer, whereas layer N is the last layer in geologic time and represents the surface layer. Of course, the surface layer would be water in the case of marine exploration. The topmost layer is actually a half-space which we call the air and denote by index $N+1$. Thus, the layered system consists of N sedimentary layers of finite thickness sandwiched in between the basement rock and the air. The term *sedimentary system* will refer to the N layers of finite thickness and will not include the air and basement rock.

There are $N+1$ horizontal interfaces. The lowest interface is denoted by index 0 and represents the top of layer 0, that is, the top of the basement. The highest interface is denoted by index N and represents the top of layer N, that is, the surface of the ground or of the water, as the case may be. Generally, we may say that interface n is the top of layer n, where the integer n runs from 0 through N.

We restrict ourselves to plane compressional wave motion at normal incidence to the horizontal interfaces. To satisfy arbitrary boundary conditions, two plane compressional waves can exist within each layer, one a compressional wave travelling vertically upward and the other a compressional wave travelling vertically downward. The wave motion itself can be measured in terms of any of a number of quantities, such as particle displacement, particle velocity, particle acceleration, stress, or pressure. Whatever quantity is used, reflection and transmission coefficients can be defined for each interface.

If a downgoing unit spike is incident on the top of interface n, then the reflection coefficient r_n is equal to the resulting upgoing spike reflected from the top of interface n, and the transmission coefficient t_n is equal to the resulting downgoing spike transmitted through interface n. If an upgoing unit spike is incident on the bottom of interface n, then the reflection coefficient r_n' is equal to the resulting downgoing spike reflected from the bottom of interface n, and the transmission coefficient t_n' is equal to the resulting upgoing spike transmitted through interface n (see fig. 2).

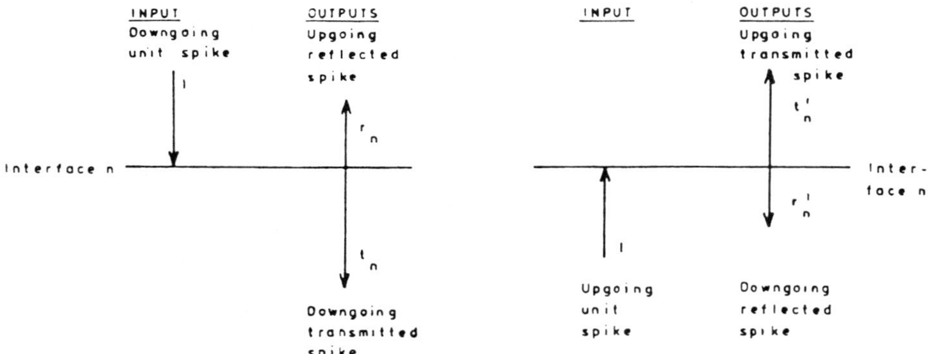

Fig. 2: Schematic diagram illustrating the reflection and transmission coefficients for an interface.

The last three coefficients can be expressed in terms of the first as follows.

$$t_n = 1 + r_n$$
$$r_n' = -r_n$$
$$t_n' = 1 - r_n$$

These coefficients are real numbers, the reflection coefficients always in the range $(-1,1)$ and the transmission coefficient always in the range $(0,2)$. The two-way transmission factor of interface n is equal to

$$t_n t'_n = 1 - r_n^2.$$

We assume that the layered system is lossless; that is, that there are no energy degradation effects such as absorption within the layers. Thus, the only way that the sedimentary system can lose energy is by having a downgoing wave travel into the basement never to return or by having an upgoing wave travel into the air never to return.

In this paper we do not make the assumption that the surface acts as a perfect reflector; that is, we do not assume that the surface reflection coefficient has magnitude unity.

For mathematical simplicity, it is convenient to add hypothetical (i.e. mathematical, not geological) interfaces where necessary, so as to make the two-way travel time in each layer equal to the same quantitiy, which we shall define as one time unit. Of course, the reflection coefficients are zero and the transmission coefficients are one for any such hypothetical interfaces that are added. We let z represent the unit delay operator. Any wave train of spikes a_0, a_1, a_2, ... where a_s denotes the amplitude of the spike at discrete integer time s can be represented by its z-transform.

$$A(z) = a_0 + a_1 z + a_2 z^2 + \cdots$$

Often we shall refer to such a wave train simply as the wave A, where the capital letter A actually denotes the z-transform $A(z)$. The quantity $A(z^{-1})$ is often denoted simply by $\bar{A}$.

HEURISTIC DEVELOPMENT

In order to make the essential ideas clear, we give in this section an heuristic development of the method of dynamic predictive deconvolution. In the remaining we fill in the mathematical and computational details.

For the purposes of this section let us measure the travelling waves in terms of particle velocity. Then the instantaneous energy flow of a travelling wave in a layer is proportional to the product of the characteristic impedance of the layer and the square of the amplitude of the wave. Let Z_0 denote the characteristic impedance of the basement, let Z_{N+1} denote the characteristic impedance of the air. Let the system be at rest and let the input to the sedimentary system be a unit downgoing spike incident on the surface. The output of the sedimentary system is the wave train — denoted by R_N — reflected upward into the air and the wave train — denoted by T_N — transmitted downward into the basement. Because we assume that the sedimentary system is lossless (i.e., that there are no energy degradation effects within the layers), all the input energy must be accounted for by the two outputs R_N and T_N. The energy of the input spike is proportional to Z_{N+1} times unity squared. The energy of the output reflected wave is proportional to $Z_{N+1} R_N \bar{R}_N$ and the energy of the output transmitted wave is proportional to $Z_0 T_N \bar{T}_N$. The law of the conservation of energy states that the energy input to the sedimentary system must equal the energy output from the sedimentary system, that is

$$Z_{N+1} = Z_{N+1} R_N \bar{R}_N + Z_0 T_N \bar{T}_N.$$

The output reflected wave R_N is the observed seismic trace (which includes all primary and multiple reflections) and hence is at our disposal. If we bring this known quantity to the left-hand side of the above equation, we obtain

$$1 - R_N \bar{R}_N = \frac{Z_0}{Z_{N+1}} T_N \bar{T}_N.$$

Let us designate the known left-hand side of this equation by the symbol Φ, which we call the spectral function. That is, the spectral function is defined as

$$\Phi = 1 - R_N \bar{R}_N,$$

and by the law of conservation of energy the spectral function is equal to

$$\Phi = \frac{Z_0}{Z_{N+1}} T_N \bar{T}_N.$$

Let us now find an expression for Z_0 / Z_{N+1}. It is well known that the transmission coefficients t_n and t'_n are given in terms of the characteristic impedances Z_{n+1} and Z_n for the layers $n+1$ and n (where layer $n+1$ is on top of layer n) by

$$t_n = \frac{2Z_{n+1}}{Z_{n+1}+Z_n}, \quad t'_n = \frac{2Z_n}{Z_{n+1}+Z_n}.$$

It follows that

$$\frac{t'_n}{t_n} = \frac{Z_n}{Z_{n+1}}$$

and therefore we have

$$\frac{t'_N \ldots t'_1 t'_0}{t_N \ldots t_1 t_0} = \frac{Z_N}{Z_{N+1}} \ldots \frac{Z_1}{Z_2} \frac{Z_0}{Z_1} = \frac{Z_0}{Z_{N+1}}.$$

That is, the ratio Z_0 / Z_{N+1} is given by the ratio of the transmission factor $t'_0 \, t'_1 \ldots t'_N$ upward through the sedimentary system to the transmission factor $t_N \ldots t_1 t_0$ downward through the sedimentary system.

Next, let us find an expression for T_N. We know that the sedimentary system acts as a pure finite feedback system in producing the transmitted wave, so T_N is proportional to the reciprocal of a polynomial D_N of degree N. If we choose this polynomial so that its leading coefficient is unity, then the proportionality factor is equal to the downward transmission factor $t_N \ldots t_1 t_0$. Thus, the transmitted wave has the form

$$T_N = \frac{t_N \ldots t_1 t_0}{D_N}$$

where we have chosen the time origin of the transmitted wave to be at its first break, where its first break occurs at a delay of $N/2$ time units from the time of the input spike (that is, at a delay of the one-way travel time through the sedimentary layers). Since T_N is the z-transform of a stable one-sided time function, it follows that the denominator polynomial D_N must be minimum-delay. It then follows that T_N itself must be minimum-delay.

Using the results of the above two paragraphs we see that the spectral function is equal to

$$\Phi = \frac{t'_N \ldots t'_1 t'_0}{t_N \ldots t_1 t_0} \; \frac{(t_N \ldots t_1 t_0)^2}{D_N \bar{D}_N} \; .$$

If we define σ_N^2 as the two-way transmission factor of the sedimentary system, that is, if we define

$$\sigma_N^2 = t'_N t_N \ldots t'_1 t_1 t'_0 t_0 = (1 - r_N^2) \ldots (1 - r_1^2)(1 - r_0^2)$$

we see that the spectral function is equal to

$$\Phi = \frac{\sigma_N^2}{D_N \bar{D}_N}$$

Thus, the feedback polynomial D_N and the constant σ_N^2 can be found from the known spectral function Φ by one of the methods of minimum-delay spectral factorization, such as the Fejer-Wold method, or the Kolmogorov method, or the normal-equations method. See Robinson (1954) for a more detailed description of these mathematical concepts.

Because

$$D_N \, T_N = t_N \ldots t_1 t_0 = constant$$

the polynomial D_N is the z-transform of the prediction-error operator that reduces the minimum-delay transmitted wave to a spike.

The sedimentary system acts as a system with both a feedforward component and a feedback component in producing the output reflected wave R_N (i.e. the observed seismic trace). Moreover, the feedback component is identical to the pure feedback system that produces the output transmitted wave. Thus, the z-transform of the seismic trace is given by

$$R_N = \frac{C_N}{D_N}$$

where the polynomial C_N of degree N represents the feedforward component, and the polynomial D_N of degree N represents the feedback component and is the same as the D_N appearing in the expression for T_N. Thus, we can deconvolve the observed seismic trace with the prediction-error operator computed above. In terms of z-transforms this deconvolution is given by the multiplication

$$R_N \, D_N = C_N$$

The result of this deconvolution is the feedforward component C_N. Now the feedforward component represents the wanted dynamic structure of the sedimentary system whereas the feedback component represents the unwanted reverberatory effects of the sedimentary system. In order to see that the feedforward component represents the dynamic structure, let us write down the explicit expression for C_N in terms of the reflection coefficients in the case when $N = 3$. We have

$$C_3(z) = r_3 + (r_2 + r_3 r_2 r_1 + r_3 r_1 r_0)z + (r_1 + r_3 r_2 r_0 + r_2 r_1 r_0)z^2 + r_0 z^3$$

Now the magnitude of any reflection can never exceed unity, and in practice the magnitudes of the reflection coefficients will cluster around zero instead of unity. Hence, generally, the product of three or more reflection coefficients will be of a lower order of magnitude than any single reflection coefficient. Hence, to this approximation, the above feedforward polynomial may be written

$$C_3(z) \approx r_3 + r_2 z + r_1 z^2 + r_0 z^3$$

and, in general, for N sedimentary layers

$$C_N(z) \approx r_N + r_{N-1} z + \ldots + r_1 z^{N-1} + r_0 z^N$$

Since the deconvolution process yields the coefficients of C_N, the process approximately determines the reflection coefficients which represent the wanted dynamic structure of the sedimentary system.

In retrospect we can write the seismic trace as

$$R_N = C_N(1/D_N)$$

which in the time-domain is

*Seismic trace = (Dynamic reflection series) * (Minimum-delay reverberation)* where the asterisk denotes convolution. The deconvolution process is

(*Seismic trace*) * (*Prediction-error operator*) = *Dynamic reflection series*

That is, the deconvolution process yields the wanted dynamic structure and destroys the unwanted reverberatory effect. Thus, we call the process the method of dynamic predictive deconvolution.

If one does not wish to make the above approximation (namely, that the reflection coefficients are approximately equal to the coefficients of the feedforward polynomial), then it is possible to actually decompose the coefficients of the feedforward and feedback polynomials to yield the reflection coefficients exactly.

EINSTEIN ADDITION FORMULA

If one attempts to apply Newtonian mechanical laws to ultra-highspeed charged particles, then an insurmountable contradiction is encountered. That is, the simple addition of velocities does not apply in electrodynamics, and instead one should use the Einstein addition formula for combining velocities which guarantees that the resulting velocity will never exceed the velocity of light (Lorentz, Einstein, Minkowski, and Weyl, 1923).

Likewise in combining reflection coefficients from a system of layers it turns out that the counterpart of the Einstein addition formula has to be used which guarantees that the resulting reflectivity cannot exceed unity in magnitude. In this section we wish to derive this counterpart of the Einstein addition formula.

Because we always let the input be a unit downgoing spike at time zero incident at the topmost interface of a sedimentary system, we can simply refer to the output wave train reflected up into the upper half-space as the *reflection response* and the output wave train transmitted down into the basement at the *transmission response* (see figure 3).

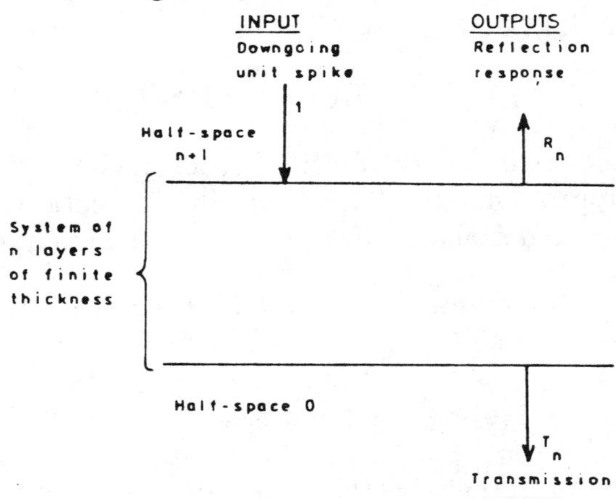

Fig. 3: Schematic diagram illustrating the reflection and transmission responses for a system of *n* layers of finite thickness.

Let us now consider a sedimentary system of n-1 layers with reflection coefficients r_0, r_1, ..., r_{n-1}. Let us also consider another sedimentary system of n layers with the same reflection coefficients r_0, r_1, ..., r_{n-1} plus the additional reflection coefficient r_n. In order for these reflection coefficients to be the same, layer n of the second system must be of the same material as the half-space n of the first system, and all of the layers below must be identical in the two systems (see figure 4).

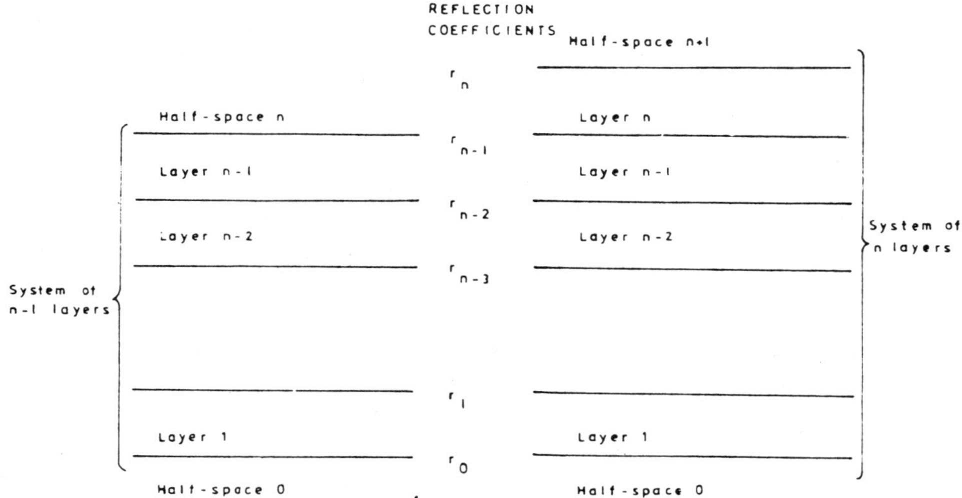

Fig. 4: Two layered systems with the same reflection coefficients r_0, r_1, ..., r_{n-1}. The system of n layers has an additional reflection coefficient r_n.

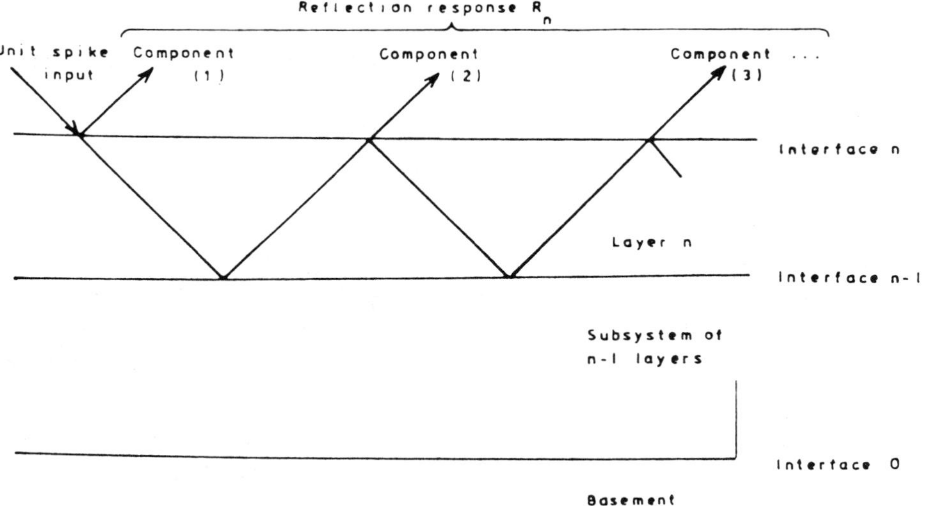

Fig. 5: Schematic make-up of the reflection response R_n.

We now wish to combine the reflection response R_{n-1} of the n-1 layer system with the reflection coefficient r_n in such a way so as to give the reflection response R_n of the n layer system. If we refer to figure 5, we see that the reflection response R_n is made up of an infinite series of components, namely:

(1) the spike r_n resulting from the reflection upward of the source spike from the nth interface,

(2) the spike train $t_n R_{n-1} t'_n$ resulting from the transmission downward of the source spike through the nth interface, reflection upward from the n-1 layer system, and transmission upward through the nth interface,

(3) the spike train $t_n R_{n-1} r'_n R_{n-1} t'_n$ resulting from the transmission downward of the source spike through the nth interface, reflection upward from the n-1 layer system, reflection downward from the nth interface, reflection upward from the n-1 layer system, and transmission upward through the nth interface, and so on. The spike (1) above occurs at the time of the source spike, the spike train (2) above occurs at a delay of one time unit (i.e. at a delay of the two-way travel time through the nth layer), the spike train (3) above occurs at a delay of two time units, and so on. Summing all these contributions, we have

$$R_n = r_n + t_n R_{n-1} t'_n z + t_n R_{n-1} r'_n R_{n-1} t'_n z^2 + \ldots$$

This expression may be factored as

$$R_n = r_n + t_n R_{n-1} t'_n z [1 + r'_n R_{n-1} z + (r'_n R_{n-1} z)^2 + \ldots]$$

which, if we sum the geometric series in brackets, becomes

$$R_n = r_n + \frac{t_n R_{n-1} t'_n z}{1 - r'_n R_{n-1} z} .$$

Using the relationships given earlier between the reflection and transmission coefficients, this expression becomes

$$R_n = \frac{r_n + R_{n-1} z}{1 + r_n R_{n-1} z} .$$

This equation for combining r_n and R_{n-1} to form R_n is of the same mathematical form as the Einstein addition formula in the theory of relativity to combine two velocities to give resulting velocity, and hence represents the required counterpart of the Einstein addition formula in the case of layered media.

In a similar manner the transmission response T_n can be obtained in terms of the reflection coefficient r_n and the transmission coefficient t_n of the nth interface and the reflection response R_{n-1} and the transmission response T_{n-1} of the n-1 layer system. If we refer to figure 6,

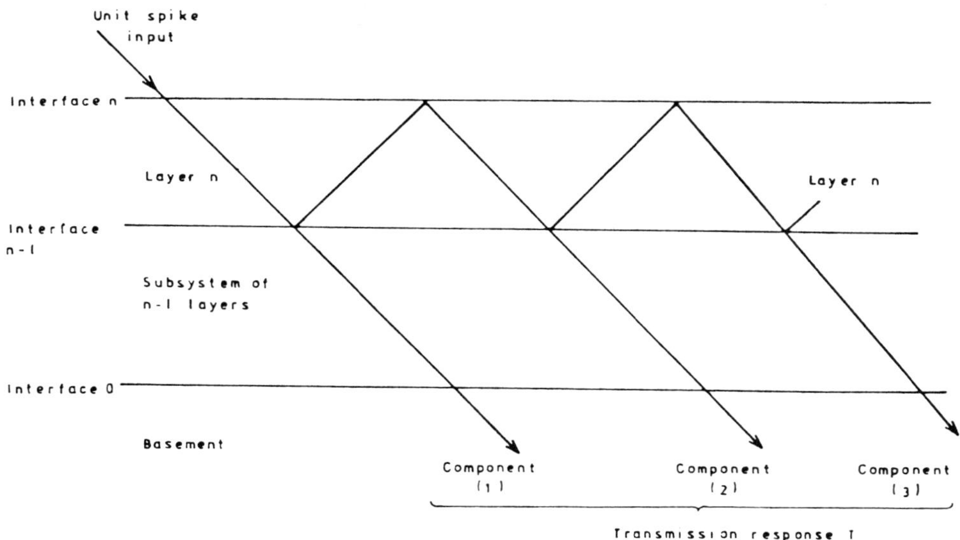

Fig. 6: Schematic make-up of the transmission response T_n.

we see that the transmission response T_n of the n layer system is made up of an infinite series of components, namely

(1) the spike train $t_n\ T_{n-1}$,

(2) the spike train $t_n\ R_{n-1}\ r'_n\ T_{n-1}$,

(3) the spike train $t_n\ R_{n-1}\ r'_n\ R_{n-1}\ r'_n\ T_{n-1}$,

and so on. We choose the time origin of the transmission response as the time of its first break (i.e. at its first non-zero amplitude). Its first break occurs at a delay of $N/2$ (i.e. the one-way travel time) from the time of the surface input spike. With reference to the time origin of the transmission response, the spike train (1) occurs with no delay, spike train (2) occurs with a delay of one time unit, spike train (3) occurs with a delay of two time units, and so on. Summing all these contributions, we have

$$T_n = t_n\ T_{n-1} + t_n\ R_{n-1}\ r'_n\ T_{n-1}\ z + t_n\ R_{n-1}\ r'_n\ R_{n-1}\ r'_n\ z^2 + \ldots$$
$$= t_n\ T_{n-1}\ [1 + R_{n-1}\ r'_n\ z + (R_{n-1}\ r'_n\ z)^2 + \ldots].$$

Summing the geometric series, we have

$$T_n = \frac{t_n T_{n-1}}{1 + r_n R_{n-1}\ z}.$$

We note that the final expressions for R_n and T_n each have the same denominator.

FEEDFORWARD AND FEEDBACK POLYNOMIALS

We now wish to define a sequence of polynomials $C_0, C_1, ..., C_N$ which we call the feedforward polynomials and a sequence of polynomials $D_0, D_1, ..., D_N$ which we call the feedback polynomials. In the case of no layers of finite thickness, the reflection and transmission responses are

$$R_0 = r_0 \quad \text{and} \quad T_0 = t_0.$$

These responses may be expressed in terms of polynomials C_0 and D_0 each of zero degree as

$$R_0 = \frac{C_0}{D_0} \quad \text{and} \quad T_0 = \frac{t_0}{D_0},$$

where the polynomials satisfy

$$C_0 = r_0 \quad \text{and} \quad D_0 = 1.$$

Let us suppose that in the case of n-1 layers of finite thickness the reflection and transmission responses may be expressed in terms of polynomials C_{n-1} and D_{n-1} each of degree n-1 as

$$R_{n-1} = \frac{C_{n-1}}{D_{n-1}} \quad \text{and} \quad T_{n-1} = \frac{t_{n-1} \cdots t_0}{D_{n-1}},$$

where the polynomials satisfy

$$C_{n-1}(0) = r_{n-1} \quad \text{and} \quad D_{n-1}(0) = 1.$$

Using the counterpart of the Einstein addition formula we have

$$R_n = \frac{r_n + \dfrac{C_{n-1}}{D_{n-1}} z}{1 + r_n \dfrac{C_{n-1}}{D_{n-1}} z} = \frac{r_n D_{n-1} + C_{n-1} z}{D_{n-1} + r_n C_{n-1} z}.$$

Let us define the feedforward polynomial C_n of degree n and the feedback polynomial D_n of degree n by the *recursion formulas*

$$C_n = r_n D_{n-1} + C_{n-1} z, \quad D_n = D_{n-1} + r_n C_{n-1} z.$$

From these recursion formulas we see that the polynomials satisfy

$$C_n(0) = r_n \quad \text{and} \quad D_n(0) = 1.$$

The reflection response R_n is given by the ratio

$$R_n = \frac{C_n}{D_n}.$$

Similarly the transmission response is

$$T_n = \frac{t_n \, T_{n-1}}{1 + r_n \, R_{n-1} \, z} = \frac{t_n \, t_{n-1} \cdots t_0 / D_{n-1}}{1 + r_n \dfrac{C_{n-1}}{D_{n-1}} z}$$

$$= \frac{t_n \, t_{n-1} \cdots t_0}{D_{n-1} + r_n \, C_{n-1} \, z} = \frac{t_n \, t_{n-1} \cdots t_0}{D_n} .$$

Because T_n is the z-transform of a stable one-sided time function it follows that the polynomial D_n is minimum-delay. Because the inverse of a minimum-delay function is also minimum-delay, it follows that T_n itself is minimum-delay.

CONSERVATION OF ENERGY

The function $D_n \bar{D}_n$ is the spectral function of the feedback polynomial D_n and the function $C_n \bar{C}_n$ is the spectral function of the feedforward polynomial C_n. Let us now find an expression for the difference of these spectral functions. If we use the recursion formulas given in the above section, we obtain the result

$$D_n \, \bar{D}_n - C_n \, \bar{C}_n = (1 - r_n^2)(D_n \, \bar{D}_{n-1} - C_{n-1} \, \bar{C}_{n-1}).$$

We can now make repeated use of this result to obtain

$$D_n \, \bar{D}_n - C_n \, \bar{C}_n = (1 - r_n^2)(1 - r_{n-1}^2) \cdots (1 - r_0^2).$$

The expression on the right is recognized as the two-way transmission factor σ_n^2 of the n layers of finite thickness. Thus, the difference between the feedback and feedforward spectral functions is equal to the two-way transmission factor, that is

$$D_n \, \bar{D}_n - C_n \, \bar{C}_n = \sigma_n^2 .$$

Let $n = N$ in the above expression, and rewrite it as

$$1 - \frac{C_N \, \bar{C}_N}{D_N \, \bar{D}_N} = \frac{\sigma_N^2}{D_N \, \bar{D}_N} ,$$

which is

$$1 - R_N \, \bar{R}_N = \frac{\sigma_N^2 \, T_N \, \bar{T}_N}{(t_N \cdots t_0)^2} .$$

Since

$$\sigma_N^2 = (t_N \cdots t_0)(t'_N \cdots t'_0),$$

we have

$$1 - R_N \, \bar{R}_N = \frac{t'_N \cdots t'_0}{t_N \cdots t_0} \, T_N \, \bar{T}_N .$$

This equation represents the law of the conservation of energy, namely the input energy minus the output reflected energy is equal to the output transmitted energy. Because the spectral function Φ was defined as $1 - R_N \bar{R}_N$, we have

$$\Phi = \frac{\sigma_N^2}{D_N \bar{D}_N} .$$

We have now established all the mathematical formulas that we made use of in our heuristic development, so now we may turn to the computational formulas required to carry out dynamic predictive deconvolution.

COMPUTATIONAL PROCEDURE

The computational procedure for dynamic predictive deconvolution in terms of the given geophysical model is as follows: Let the observed seismic trace (i.e. the reflection response of the sedimentary system to a unit source spike) be the time series

$$x_0, x_1, x_2, x_3, \ldots$$

(Note: The z-transform of this time series is R_N. Ordinarily, we would use lower case r to denote the coefficients of R_N, but since we have already used r to denote the reflection coefficients, we have used x to denote the coefficients of R_N).

The first computational step is to compute the autocorrelation ψ_s of the seismic trace by the formula

$$\psi_s = \sum_{i=0}^{\infty} x_{i+|s|} \, x_i ,$$

then compute the source autocorrelation, which under the assumption of a unit spike source, is itself a unit spike, and then compute their difference, which is an autocorrelation function φ given by

$$\varphi_0 = 1 - \psi_0$$
$$\varphi_s = -\psi_s \quad \text{for } s \neq 0$$

(Note: The z-transform of the autocorrelation φ_s is seen to be the spectral function Φ).

The second computational step is to compute the prediction error operator (i.e. the deconvolution operator) $d_0 = 1, d_1, d_2, \ldots, d_N$ by solving the *normal equations*

$$\sum_{s=0}^{N} d_s \, \varphi_{t-s} = 0 \quad \text{for } t = 1, 2, \ldots, N$$

and then to compute the two-way transmission factor σ_N^2 by the formula

$$\sigma_N^2 = d_0 \, \psi_0 + d_1 \, \psi_1 + \ldots + d_N \, \psi_N .$$

The coefficients $d_0 = 1, d_1, ..., d_N$ thus found are the coefficients of the feedback polynomial D_N (Note: The normal equations given above are obtained by expressing the equation

$$D_N \, \Phi = \frac{\sigma_N^2}{\bar{D}_N}$$

in terms of powers of z. Because the expansion of $1/\bar{D}_N$ involves only non-positive powers of z, it follows that for positive powers of z the right-hand side of this equation is zero thereby yielding the normal equations). This second computational step is the method of spectral factorization based upon least-squares normal equations. Alternatively, other methods of spectral factorization can be used such as the Kolmogorov spectral factorization or the Fejer-Wold spectral factorization.

The third computational step is to compute the deconvolved seismic trace c_i by the formula

$$c_i = \sum_{s=0}^{N} d_s \, x_{i-s} \qquad \text{for } i = 0, 1, 2, ..., N.$$

The deconvolved trace is the set of coefficients of the feedforward polynomial, and to a first approximation represents the set of reflection coefficients, that is,

$$(c_0, c_1, ..., c_N) \approx (r_N, r_{N-1}, ..., r_0).$$

For the first and last coefficients, an equality sign holds; that is, $c_0 = r_N$ and $c_N = r_0$.

If we wish the remaining reflection coefficients exactly, then a further decomposition is necessary, which is described in the next section.

POLYNOMIAL DECOMPOSITION

At this point in the computation procedure we have calculated the coefficients $(c_0, c_1, ..., c_N)$ of the feedforward polynomial C_N and the coefficients $(d_0, d_1, ..., d_N)$ of the feedback polynomial D_N. In order to decompose these polynomial coefficients into the reflection coefficients, we must invert the recursion formulas given previously. If we solve the recursion formulas for C_{n-1} and D_{n-1} in terms of C_n and D_n we obtain the *inverse recursion formulas*

$$C_{n-1} = (1 - r_n^2)^{-1}(C_n - r_n \, D_n) \, z^{-1}$$
$$D_{n-1} = (1 - r_n^2)^{-1}(D_n - r_n \, C_n).$$

Also, we know that

$$D_n(0) = 1, \qquad\qquad C_n(0) = r_n.$$

After the third computational step we know C_N, D_N, and $r_N = C_N(0)$. We can then use the inverse recursion formulas to obtain C_{N-1}, D_{N-1}, and $r_{N-1} = C_{N-1}(0)$. We can continue using the inverse recursion formulas until we finally obtain all the polynomials and therefore all the reflection coefficients.

For general reference, let us write down the results of the decomposition for $N = 3$.

$$C_3 = r_3 + (r_2 + r_3 r_2 r_1 + r_3 r_1 r_0)z + (r_1 + r_3 r_2 r_0 + r_2 r_1 r_0)z^2 + r_0 z^3$$

$$D_3 = 1 + (r_1 r_0 + r_2 r_1 + r_3 r_2)z + (r_2 r_0 + r_3 r_1 + r_3 r_2 r_1 r_0)z^2 + r_3 r_0 z^3$$

$$C_2 = r_2 + (r_1 + r_2 r_1 r_0)z + r_0 z^2$$

$$D_2 = 1 + (r_1 r_0 + r_2 r_1)z + r_2 r_0 z^2$$

$$C_1 = r_1 + r_0 z$$

$$D_1 = 1 + r_1 r_0 z$$

$$C_0 = r_0$$

$$D_1 = 1.$$

CONCLUSION

The method of dynamic predictive deconvolution or DPD is an alternative deconvolution process to the method of random predictive deconvolution, or RPD. Dynamic predictive deconvolution is based on a dynamic reflection model of the layered earth whereas random predictive deconvolution is based on a random reflection model. Whereas random predictive deconvolution removes the reverberation of the surface layers, dynamic predictive deconvolution removes the reverberation effect of the entire sedimentary column and yields the feedforward dynamic structure of the earth. Both methods of deconvolution make use of the autocorrelation of the observed seismic trace as the starting point in the numerical computations. The use of the autocorrelation of the trace instead of the trace itself makes both methods less sensitive of spurious events and noise that may occur on the trace, and hence both methods have robust statistical characteristics. Because dynamic predictive deconvolution makes use of the entire seismic trace instead of only a portion of the trace as does random predictive deconvolution, and also because the energy characteristics of the source are used in an essential way in dynamic predictive deconvolution, this new method requires careful and thoughtful application.

REFERENCES

ANSTEY, N., 1960, Attacking the problem of the synthetic seismogram: Geophysical Prospecting 8, 242-259.

BARANOV, V., and KUNETZ, G., 1958, Calcul des sismogrammes avec des réflexions multiples: Comptes rendus des séances de l'Académie des Sciences (Paris) 247, 1887-1889.

BARANOV, V., and KUNETZ, G., 1960, Film synthétique avec réflexions multiples, théorie et calcul pratique: Geophysical Prospecting 8, 315-325.

BATH, M., 1968, Mathematical Aspects of Seismology: Elsevier Publishing Company, Amsterdam, London and New York, 415 pages.

BATH, M., 1974, Spectral Analysis in Geophysics: Elsevier Publishing Company, Amsterdam, London and New York, 563 pages.

BERRYMAN, L. H., GOUPILLAUD, P. L., and WATERS, K. H., 1958, Reflections from multiple transition layers: Geophysics 23, 223-252.

BOIS, P., CHAUVEAU, J., GRAU, G., and LAVERGNE, M., 1960, Sismogrammes synthétiques, possibilités techniques de réalisation et limitations: Geophysical Prospecting 8, 260-314.

BORTFELD, R., 1960, Seismic waves in transition layers: Geophysical Prospecting 8, 178-217.

GOUPILLAUD, P. L., 1961, An approach to inverse filtering of near surface layer effects from seismic records: Geophysics 26, 754-760.

KUNETZ, G., 1961, Essai d'analyse de traces sismiques: Geophysical Prospecting 9, 317-341.

KUNETZ, G., 1964, Géneralisation des opérateurs d'antirésonance à un nombre quelconque de réflecteurs: Geophysical Prospecting 12, 283-289.

KUNETZ, G., and D'ERCEVILLE, I., 1962, Sur certaines propriétés d'une onde acoustique plane de compression dans un milieu stratifié: Annales de Géophysique 18, 351-359.

LORENTZ, H. A., EINSTEIN, A., MINKOWSKI, H., and WEYL, H., 1923, The Principles of Relativity, A Collection of Original Memoirs: Methuen and Co., London, Reprinted by Dover Publications, New York, 1958.

O'BRIEN, P. N. S., 1961, A discussion of the nature and magnitude of elastic absorption in seismic prospecting: Geophysical Prospecting 9, 261-275.

ROBINSON, E. A., 1954, Predictive Decomposition of Time Series with Application to Seismic Exploration: Ph. D. Thesis, Massachusetts Institute of Technology, Cambridge, Mass., Reprinted 1967 in Geophysics 32, 418-484.

ROBINSON, E. A., 1966, Multichannel z-transforms and minimum-delay: Geophysics 31, 482-500.

SERIFF, A. J., 1958, Multiple reflections in many layered systems: Paper presented at the 1958 meeting of the SEG, San Antonio, Texas.

SHERWOOD, J. W.C., and TROREY, A. W., 1965, Minimum phase and related properties of the response of a horizontally stratified absorptive earth to plane acoustic waves: Geophysics 30, 191-197.

TREITEL, S., and ROBINSON, E. A., 1966, Seismic wave propagation in layered media in terms of communication theory: Geophysics 31, 17-32.

WUENSCHEL, P. C., 1960, Seismogram synthesis including multiples and transmission coefficients: Geophysics 25, 106-129.

4

LEAST-SQUARES INVERSE FILTERING AND WAVELET DECONVOLUTION

A .   J .   B E R K H O U T *

Detailed comparison between borehole data and seismic data has taught that, in general, conventional seismic inverse filtering is not effective enough to produce desirable deconvolution results, i.e., seismic sections with *broad-band zero-phase* wavelets.

Application of conventional seismic inverse filters has the advantage that very little information is needed from the user. However, as is shown in this paper, the phase spectra of these filters may be seriously in error, even if the seismic wavelet has the *minimum-phase* property.

In wavelet deconvolution the phase spectrum of the filter is correct, provided a good estimate of the seismic wavelet is available. In this paper, wavelet deconvolution is compared with Wiener filtering. The main conclusions are illustrated by examples.

INTRODUCTION

Let us assume that the following model of a seismic trace gate may be used:

$$x(t) = \underbrace{w(t) * u(t)}_{\text{signal}} + \underbrace{w'(t) * u'(t)}_{\text{colored noise}} ,$$

where $w(t)$ is the seismic wavelet, $w'(t)$ the noise wavelet, $u(t)$ the primary reflectogram, and $u'(t)$ white noise,

or

$$x(t) = w(t) * u(t) + n(t), \tag{1}$$

or

$$x(t) = s(t) + n(t),$$

with t being a discrete time parameter,

$$t = \ldots\ldots\ldots -2, -1, 0, +1, +2, \ldots\ldots\ldots,$$

and $*$ denoting convolution. Later in this paper we will set some constraints on the signal part $s(t)$ and the noise part $n(t)$.

The resolution of a seismic trace is determined by the length properties of seismic wavelet $w(t)$. It has been pointed out by Gabor (1946), and later discussed by Berkhout (1974), that a quantitative measure for the effective length of a wavelet $w(t)$ can be given by

$$L^2 = \sum_t t^2 w^2(t) \Big/ \sum_t w^2(t). \tag{2a}$$

Using the Fourier transform of $w(t)$,

$$W(f) = |W(f)|e^{j\phi(f)},$$

the expression for L^2 can be rewritten as

$$L^2 = \int_{-1/2}^{+1/2} \left[\left\{ \frac{d|W(f)|}{df} \right\}^2 \right.$$

$$\left. + |W(f)|^2 \left\{ \frac{d\phi(f)}{df} \right\}^2 \right] df \Big/ \int_{-1/2}^{+1/2} |W(f)|^2 df. \tag{2b}$$

From (2b) it can easily be seen that, for a given amplitude spectrum, L will reach a minimum if $\phi(f) = 0$. Moreover, this minimum will be small if

Manuscript received by the Editor October 6, 1976; revised manuscript received January 15, 1977.
*Delft University of Technology, Delft, Netherlands.

64

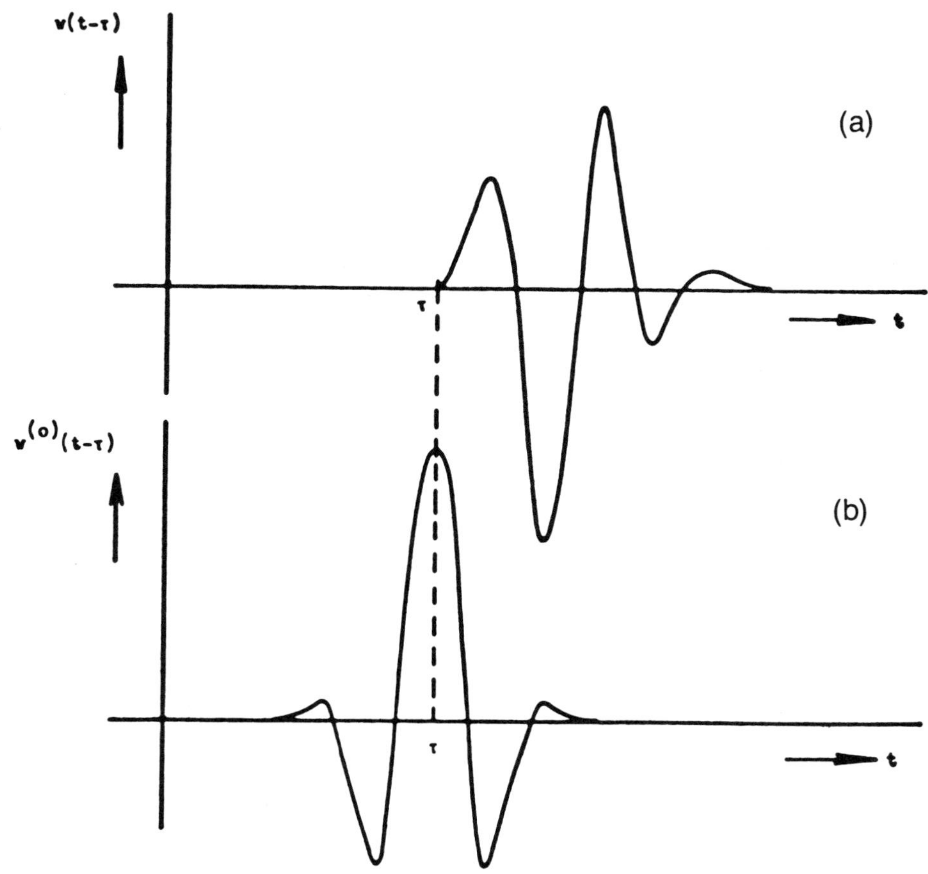

FIG. 1. (a) Minimum-phase reflection $w(t)$ with arrival time τ. (b) Zero-phase correspondent of $w(t)$.

$$\left| \frac{d|W(f)|}{df} \right|$$

is small. Hence, an effective deconvolution procedure should produce seismic wavelets with (a) a *zero-phase* spectrum, and (b) a *smooth* and *broad* amplitude spectrum.

There is an additional important argument for broad-band zero-phase wavelets. If we make use of the inequality

$$\left| \int_{-1/2}^{+1/2} W(f) e^{j2\pi ft} df \right| \le \int_{-1/2}^{+1/2} |W(f)| df, \quad (3)$$

it may be concluded that, for a given amplitude spectrum, the zero-phase wavelet has the largest amplitude. This maximum amplitude will occur at $t = 0$. Hence, if we transform the one-sided reflections (one-sided w.r.t. their arrival times) of a seismic trace into their zero-phase correspondents (zero-phase

w.r.t. their arrival times) then we have maximized the detectability of their arrival times. Figures 1 a and b show a minimum-phase seismic reflection and its zero-phase correspondent. Note the maximum peak amplitude of the zero-phase wavelet at its arrival time. Note also its favorable length property.

Generally, the last step in seismic processing consists of zero-phase band-pass filtering. This means that, if the input consists of *nonzero-phase* reflections, the final processing result will contain *two-sided mixed-phase* reflections. From these reflections it is even more difficult to pick correct arrival times (see Figures 1 c and d).

Bearing in mind that, prior to processing, seismic reflections are one-sided, it may be concluded from Berkhout (1973, 1974) that the most favorable reflection shape before processing is given by a minimum-phase time function. However, after processing seismic reflections may be two-sided and,

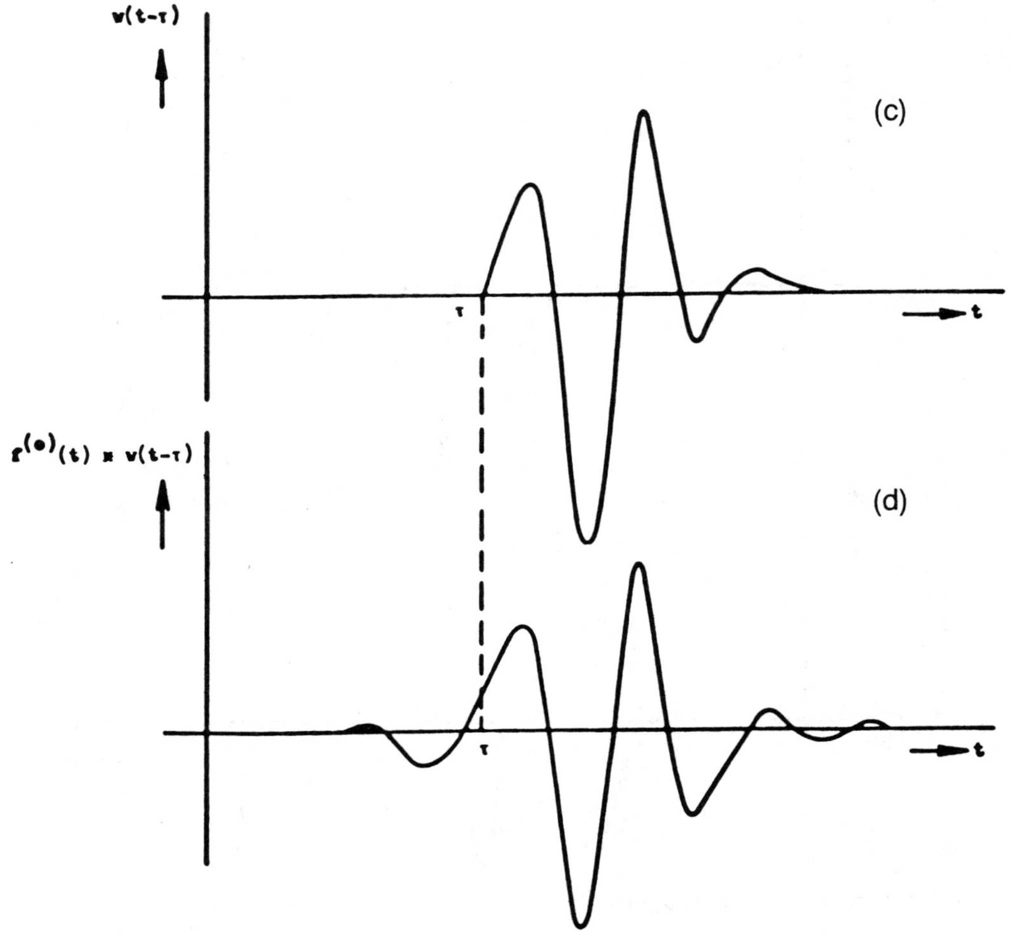

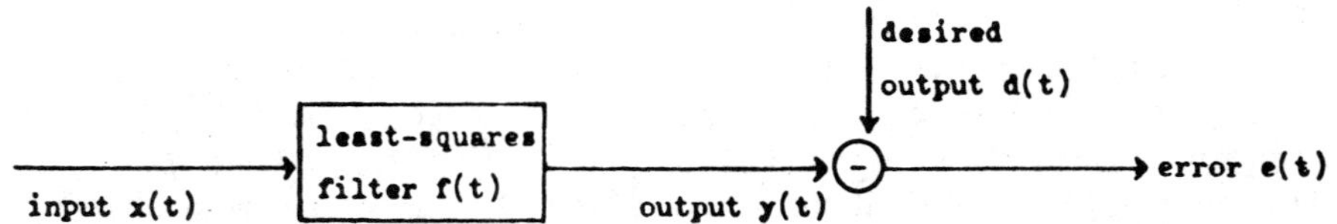

FIG. 1. (c) Minimum-phase reflection $w(t)$ with arrival time τ. (d) Zero-phase band-pass filtered version of $w(t)$.

therefore, after processing the most desirable reflection shape is given by a zero-phase time function.

Finally, it should be mentioned that in Schoenberger's paper on minimum-phase and zero-phase wavelets (1974) very instructive illustrations are shown on the favorable properties of zero-phase wavelets.

It may be worthwhile pointing out that zero-phase and *symmetry* are two different properties. A zero-phase function is always symmetric but a symmetric function need not be zero-phase. An illustrative example on this is given by Berkhout (1974).

In production seismology (i.e., seismology for the evaluation of existing oil and gas fields), the comparison of seismic data with borehole data plays an essential role. For this comparison, high-quality deconvolution results are an absolute necessity. It turns out that, generally, conventional seismic in-

FIG. 2. Principle scheme of least-squares filtering. Least-squares criterion: $\sum_t e^2(t)$ is minimum.

66

verse filters cannot produce sufficiently high-quality data. This is the main reason that, at present, alternative deconvolution techniques are under development based on estimation procedures of the seismic wavelet (wavelet deconvolution).

REVIEW OF LEAST-SQUARES FILTERING

For a given input $x(t)$ and desired output $d(t)$, the least-squares filter $f(t)$ is defined by (Figure 2):

$$E = \sum_t e^2(t) \quad \text{is minimum,} \tag{4a}$$

or

$$E = \sum_t \left[d(t) - y(t) \right]^2 \quad \text{is minimum,} \tag{4b}$$

or

$$E = \sum_t \left[d(t) - \sum_m f(m)x(t - m) \right]^2 \text{is minimum.} \tag{4c}$$

Hence the linear system, represented by the impulse response $f(t)$, transforms the input $x(t)$ as well as possible (according to the least-squares criterion) into the desired output $d(t)$.

E is semidefinite quadratic in the components $f(m)$ so that, if there exists one set of $f(m)$ values for which E is an extreme, this extreme must be a minimum. Differentiation of (4c) by the unknown filter coefficients $f(m)$ yields the equation for the optimum filter:

lates that *input* and minimum error should be orthogonal. From (5d) it follows that the equation for the least-squares filter also formulates that *output* and minimum error are orthogonal:

$$\sum_m f(m)R_{x,e}(m) = 0,$$

or

$$R_{y,e}(0) = 0. \tag{5e}$$

At this stage it should be realized that equation (5a) for the least-squares filter is a rather impractical one, as for the computation of the filter the desired output is required. However, generally, the desired output is unknown as least-squares filtering is called for to estimate the desired output! Hence we have arrived at a vicious circle. The only way to solve this problem is to make use of statistics (Wiener, 1949).

Let us consider the following measurement generating system (Figure 3):

$$x_i(t) = w(t) * u_i(t) + n_i(t).$$

Measurement sequences from this system may be considered as seismic traces recorded with the same source signature. For each measurement sequence $x_i(t)$, we may determine a least-squares filter:

$$\sum_m f(m) \left[\sum_t x(t - n)x(t - m) \right] = \sum_t x(t - n)d(t) \qquad \text{for } n \equiv m, \tag{5a}$$

or

$$\sum_m f(m)R_{x,x}(n - m) = R_{x,d}(n) \qquad \text{for } n \equiv m, \tag{5b}$$

or

$$R_{x,y}(n) = R_{x,d}(n) \qquad \text{for } n \equiv m, \tag{5c}$$

or

$$R_{x,e}(n) = 0 \qquad \text{for } n \equiv m. \tag{5d}$$

Hence the equation for the least-squares filter formu-

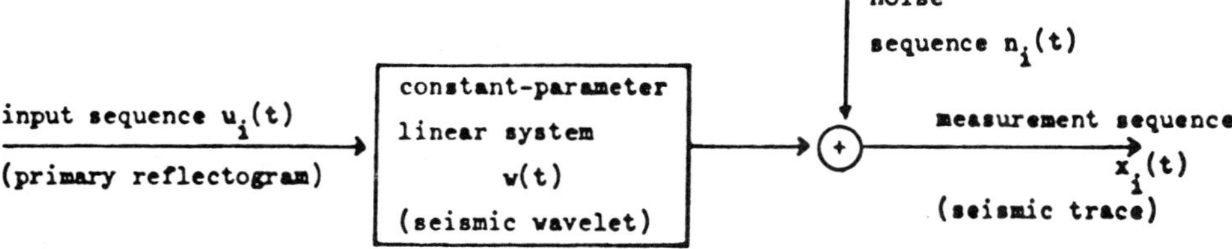

FIG. 3. Measurement-generating system (model of a seismic trace).

$$E_i = \sum_t e_i^2(t)$$

$$= \sum_t \left[d_i(t) - x_i(t) \right]^2$$

$$= \sum_t \left[d_i(t) - \sum_m f_i(m) x_i(t - m) \right]^2, \quad (6a)$$

is minimum, if

$$f_i(t) * R_{x_i, x_i}(t) = R_{x_i, d_i}(t). \quad (6b)$$

Equation (6b) only applies over the time range for which $f_i(t) \neq 0$. If $x_i(t)$ represents a realization of an *ergodic* stochastic process then $f_i(t)$ will approach a limit which is independent of i:

$$f_i(t) \rightarrow f(t) \qquad \text{for increasing observation interval.}$$

However, in practice we do not know whether the measurement process under investigation is ergodic and we also do not measure over infinite time. Generally, we may expect to have available a number of finite-duration measurement sequences. Therefore, let us construct one filter which will minimize the total energy of all available error sequences (Figure 4):

$$E = \frac{1}{N} \sum_{i=1}^{N} E_i$$

$$= \frac{1}{N} \sum_{i=1}^{N} \left[\sum_t e_i^2(t) \right]$$

$$= \frac{1}{N} \sum_{i=1}^{N} \left[\sum_t \left\{ d_i(t) - y_i(t) \right\}^2 \right]$$

$$= \frac{1}{N} \sum_{i=1}^{N} \left[\sum_t \left\{ d_i(t) - \sum_m f(m) x_i(t - m) \right\}^2 \right]$$

$$\text{is minimum.} \quad (7)$$

Differentiation of (7) by the unknown filter coefficients $f(m)$ yields the equation for the optimum filter:

$$\sum_m f(m) R_{x, x}(n - m) = R_{x, d}(n) \qquad \text{for } n \equiv m, \quad (8a)$$

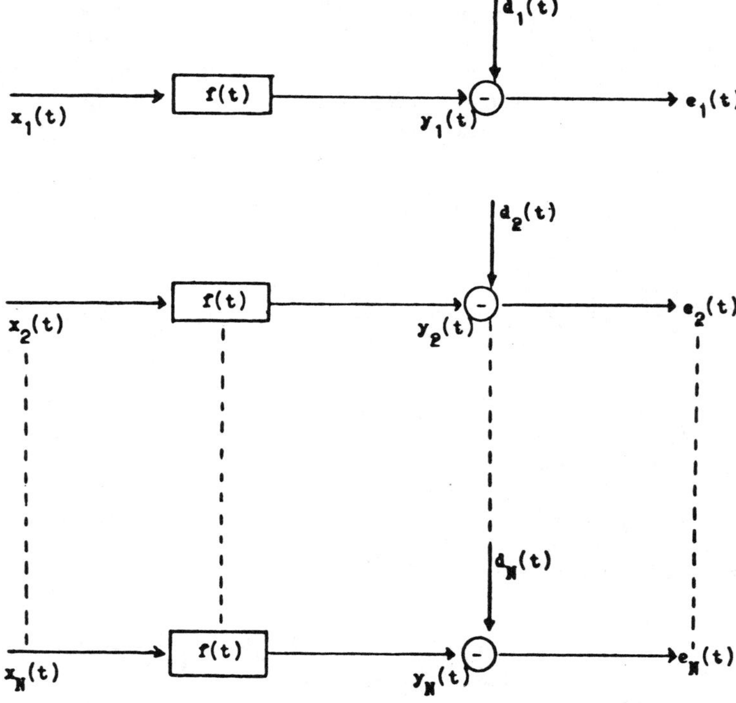

FIG. 4. Least-squares filter that will minimize the average error-energy for N input signals. Least-squares criterion:

$$\frac{1}{N} \sum_{i=1}^{N} \sum_t e_i^2(t) \text{ is minimum.}$$

with

$$R_{x,x}(\tau) = \frac{1}{N} \sum_{i=1}^{N} R_{x_i, x_i}(\tau), \qquad (8b)$$

$$R_{x,d}(\tau) = \frac{1}{N} \sum_{i=1}^{N} R_{x_i, d_i}(\tau). \qquad (8c)$$

If we assume that filter input sequences $x_i(t)$ are independent, then $R_{x,x}(\tau)$ and $R_{x,d}(\tau)$ can be considered as unbiased estimates of ensemble averages $E\{x(t-\tau)x(t)\}$ and $E\{x(t-\tau)d(t)\}$, respectively.

As a consequence, the foregoing can be considered as a practical summary of the statistical least-squares filtering theory. For increasing observation interval and increasing N, the discrete form of Wiener's filter equation is obtained from (8):

$$\sum_{m} f(m)E\{x(t-n)x(t-m)\}$$
$$= E\{x(t-n)d(t)\} \qquad \text{for } n \equiv m. \qquad (9)$$

In the following, we will take (9) as the starting equation for further evaluation. However, it should be realized that in most practical situations ensemble averages $E\{x(t-n)x(t-m)\}$ and $E\{x(t-n)d(t)\}$ are unknown and should be estimated from a *finite* number of realizations of finite duration.

LEAST-SQUARES INVERSE FILTERING

Two-sided least-squares inverse filters

For least-squares inverse filters, the desired signal is given by

$$d(t) = u(t). \qquad (10)$$

Hence, using equation (9), a least-squares inverse filter is defined by

$$\sum_{m} f(m)E\{x(t-n)x(t-m)\}$$
$$= E\{x(t-n)u(t)\} \qquad \text{for } n \equiv m. \qquad (11)$$

Now, we will assign the following properties to the measurement-generating model (trace model):
(a) $u_i(t)$ is a realization of a *white* process $u(t)$,

$$E\{u(t-\tau)u(t)\} = R_{u,u}(\tau) = 1 \qquad (12a)$$

(b) $n_i(t)$ is a realization of a noise process $n(t)$,

$$E\{n(t-\tau)n(t)\} = R_{n,n}(\tau), \qquad (12b)$$

which is uncorrelated with $u(t)$,

$$E\{u(t-\tau)n(t)\} = R_{u,n}(\tau) = 0. \qquad (12c)$$

Note that (12a), (12b), and (12c) only apply if *ensemble* averaging is used. Therefore, if we would only consider one trace gate, then

$$\sum_{t} u_i(t-\tau)u_i(t) \neq \delta_{0,\tau} \quad \text{and .}$$

$$\sum_{t} n_i(t-\tau)n_i(t) \neq R_{n,n}(\tau),$$

and

$$\sum_{t} u_i(t-\tau)n_i(t) \neq 0.$$

Using (12a) and (12c), equation (11) can be rewritten as

$$\sum_{m} f(m)E\{x(t-n)x(t-m)\} = w(-n)$$
$$\text{for } n \equiv m. \qquad (13a)$$

For practical situations, (13a) will become

$$\sum_{m} f(m)R_{x,x}(n-m) = w(-n)$$
$$\text{for } n \equiv m, \qquad (13b)$$

with

$$R_{x,x}(\tau) = \frac{1}{N} \sum_{i=1}^{N} \left[\frac{1}{T} \sum_{t=1}^{T} x_i(t-\tau)x_i(t) \right], \qquad (13c)$$

N being the number of truncated realizations available and T being the observation time.

Before discussing the solution of (13b), let us investigate (13c) in practical situations first. In the frequency domain, the autocorrelation function of one trace gate can be written as

$$\tilde{R}_{x_i, x_i}(f) = |W(f)\tilde{u}_i(f) + \tilde{n}_i(f)|^2$$
$$= |W(f)|^2 |\tilde{u}_i(f)|^2 + |\tilde{n}_i(f)|^2 + \tilde{\epsilon}_i^2(f). \qquad (14)$$

And, after averaging the autocorrelation functions of different trace gates,

$$\tilde{R}_{x,x}(f) = |W(f)|^2 \left\{ \frac{1}{N} \sum_{i=1}^{N} |\tilde{u}_i(f)|^2 \right\}$$
$$+ \frac{1}{N} \sum_{i=1}^{N} |\tilde{n}_i(f)|^2 + \frac{1}{N} \sum_{i=1}^{N} \tilde{\epsilon}_i^2(f). \qquad (15a)$$

If we assume independent trace gates, and we make use of properties (12a), (12b), and (12c), then we may write for increasing N:

$$\frac{1}{N} \sum_{i=1}^{N} |\tilde{u}_i(f)|^2 \rightarrow \tilde{R}_{u,u}(f) = 1, \qquad (15b)$$

$$\frac{1}{N} \sum_{i=1}^{N} |\tilde{n}_i(f)|^2 \rightarrow \tilde{R}_{n,n}(f), \qquad (15c)$$

$$\frac{1}{N} \sum_{i=1}^{N} \tilde{\epsilon}_i^2(f) \ll |W(f)|^2 + \tilde{R}_{n,n}(f), \qquad (15d)$$

and (15a) may be approximated by

$$\tilde{R}_{x,x}(f) = |W(f)|^2 + \tilde{R}_{n,n}(f). \qquad (15e)$$

Unfortunately, it should be mentioned that in practice (15b) does not apply as (1) in many geologic areas $u_i(t)$ cannot be considered as a realization of a white process:

$$\tilde{R}_{u,u}(f) \neq 1, \qquad (16a)$$

(2) neighboring seismic reflectograms cannot be considered to be uncorrelated:

$$\frac{1}{N} \sum_{i=1}^{N} |\tilde{u}_i(f)|^2 \not\rightarrow \tilde{R}_{u,u}(f) \qquad \text{for increasing N.} \qquad (16b)$$

Therefore, we may expect that even after an averaging procedure, the primary reflectogram data will have some influence on the least-squares filter coefficients $f(m)$. This particularly applies to lengthy filters for seismic data of deltaic environments. (See also remark 1 below.)

Note that in conventional least-squares inverse filtering, only one trace gate is used ($N = 1$). From (14) it follows that with single trace deconvolution, the undesirable influence of $\tilde{u}_i(f)$ and $\tilde{\epsilon}_i(f)$ is optimally present. Moreover, this influence is changing from trace to trace.

Equation (13b) can also be written in matrix form. If m ranges from $-M$ to M, then (13b) can be written as

or

$$RF = W. \qquad (17b)$$

Hence, the solution can be formulated as:

$$F = R^{-1}W. \qquad (18)$$

As the matrix R has the Toeplitz structure, R^{-1} can be computed by a very fast algorithm (Levinson scheme).

By computing $f(t)$ with the aid of (18), little insight is obtained into the properties of $f(t)$. Therefore, we will investigate an analytical expression for the solution of (13b) for $M \rightarrow \infty$ and $T \rightarrow \infty$. For this optimum situation, (13b) may be written as

$$F(f)\tilde{R}_{x,x}(f) = W^*(f),$$

or

$$F(f) = \frac{W^*(f)}{\tilde{R}_{x,x}(f)}, \qquad (19a)$$

or, using (12a), (12b), and (12c),

$$F(f) = \frac{W^*(f)}{|W(f)|^2 + \tilde{R}_{n,n}(f)}. \qquad (19b)$$

Using (19b), we may draw the following conclusions for *two-sided* least-squares inverse filters (M, N, and T sufficiently large):

1) Two-sided least-squares inverse filtering produces band-limited zero-phase wavelets:

$$W(f) \rightarrow |W(f)|^2 / [|W(f)|^2 + \tilde{R}_{n,n}(f)]. \qquad (20a)$$

2) Two-sided least-squares inverse filtering approaches wavelet *inverse* filtering for frequency bands with a very high signal-to-noise ratio:

$$F(f) \rightarrow 1/W(f). \qquad (20b)$$

$$\begin{pmatrix} R_{x,x}(0) & R_{x,x}(1) & \cdots & R_{x,x}(2M) \\ R_{x,x}(1) & R_{x,x}(0) & & R_{x,x}(2M-1) \\ \vdots & \vdots & & \vdots \\ R_{x,x}(M-1) & R_{x,x}(M-2) & \cdots & R_{x,x}(M+1) \\ R_{x,x}(M) & R_{x,x}(M-1) & \cdots & R_{x,x}(M) \\ R_{x,x}(M+1) & R_{x,x}(M) & \cdots & R_{x,x}(M-1) \\ \vdots & \vdots & & \vdots \\ R_{x,x}(2M-1) & R_{x,x}(2M-2) & \cdots & R_{x,x}(1) \\ R_{x,x}(2M) & R_{x,x}(2M-1) & \cdots & R_{x,x}(0) \end{pmatrix} \begin{pmatrix} f(-M) \\ f(-M+1) \\ \vdots \\ f(-1) \\ f(0) \\ f(1) \\ \vdots \\ f(M-1) \\ f(M) \end{pmatrix} = \begin{pmatrix} w(M) \\ w(M-1) \\ \vdots \\ w(1) \\ w(0) \\ w(-1) \\ \vdots \\ w(-M+1) \\ w(-M) \end{pmatrix}, \qquad (17a)$$

3) Two-sided least-squares inverse filtering approaches *matched* filtering for frequency bands with a very low signal-to-noise ratio:

$$F(f) \rightarrow W^*(f)/\tilde{R}_{n,n}(f). \qquad (20c)$$

It is interesting to notice that, completely independent from the signal length arguments in the introduction, it follows from two-sided least-squares inverse filtering that a zero-phase time function gives the most desirable wavelet shape after deconvolution. Note also that two-sided least-squares inverse filtering can be subdivided into two steps: (1) A correlation step, the correlation operator being given by the seismic wavelet $w(t)$; and (2) an amplitude-spectrum shaping step, the shaping step being defined by $1/[|W(f)|^2 + \tilde{R}_{n,n}(f)]$. This consideration explains the relationship between vibroseis-type processing and least-squares deconvolution.

Finally, it should be realized that two-sided least-squares inverse filtering should be preceded by a wavelet estimation procedure.

One-sided least-squares inverse filters

In the following we will discuss the situation that wavelet (= impulse response of measurement-generating system) $w(t)$ and filter $f(t)$ are one-sided:

$$w(t) = 0 \qquad \text{for } t < 0,$$

and

$$f(t) = 0 \qquad \text{for } t < 0.$$

For this one-sided situation, equation (13b) for the optimum least-squares inverse filter can be written as:

$$\sum_{m=0}^{M} f(m) R_{x,x}(k-m) = w(0)\, \delta_{0,k}$$

$$\text{for } k = 0, 1, \ldots, M. \qquad (21)$$

Note that one-sided least-squares inverse filtering needs very little information from the user [wavelet $w(t)$ need not be known]. This may explain its immense popularity in the seismic industry.

Like we treated the two-sided least-squares inverse filter, let us investigate the properties of the one-sided least-squares inverse filter by taking $M \rightarrow \infty$ and $T \rightarrow \infty$. Then (21) becomes the famous Wiener-Hopf equation (in discrete form). According to Appendix A, the solution is given by:

$$|F(f)| = \tilde{R}_{x,x}^{-1/2}(f), \qquad (22a)$$

$\psi(f)$ = minimum-phase spectrum that is related to

$$\tilde{R}_{x,x}^{-1/2}(f). \qquad (22b)$$

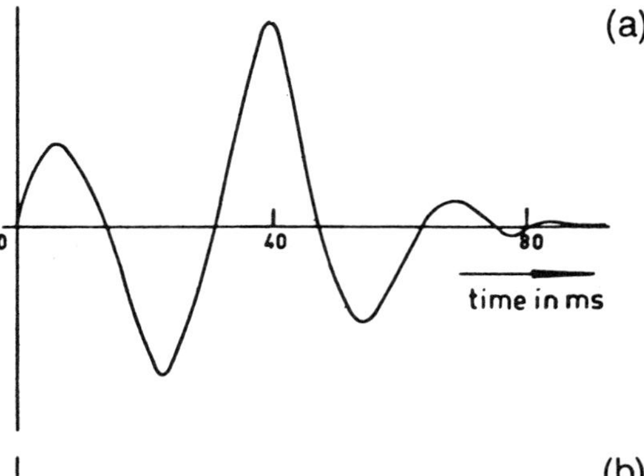

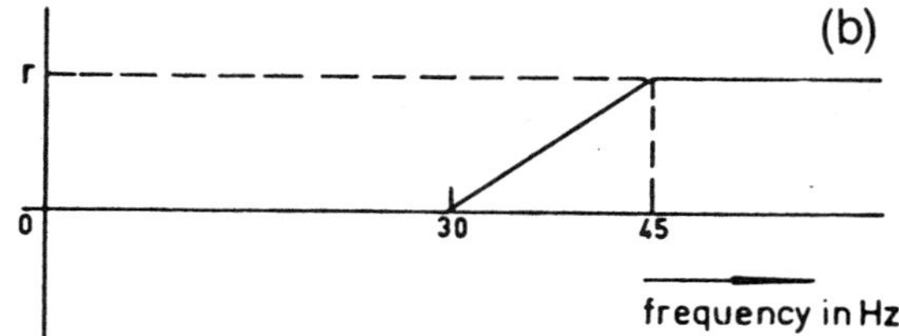

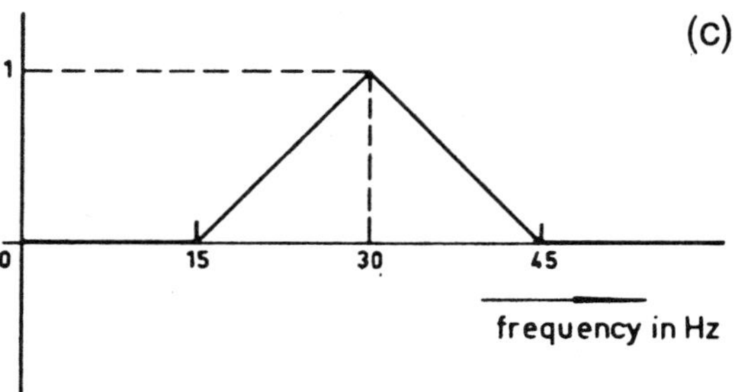

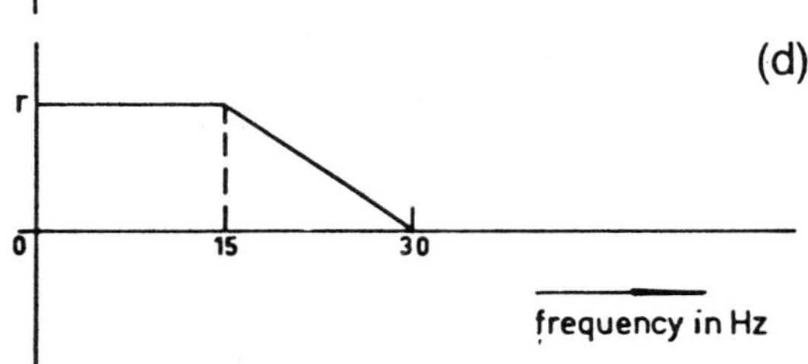

FIG. 5. (a) Minimum-phase seismic wavelet. (b) Energy-density spectrum of seismic wavelet. (c) Power-density spectrum of high-frequency noise. (d) Power-density spectrum of low-frequency noise.

From (22b) it follows that the phase spectrum of the one-sided least-squares inverse filter is correct only if (1) wavelet $w(t)$ has the minimum-phase property, and (2) $\tilde{R}_{x,x}(f) = |W(f)|^2$. Constraint 2 can be met in the noise-free situation, since

$$\tilde{R}_{x,x}(f) = |W(f)|^2 + \tilde{R}_{n,n}(f).$$

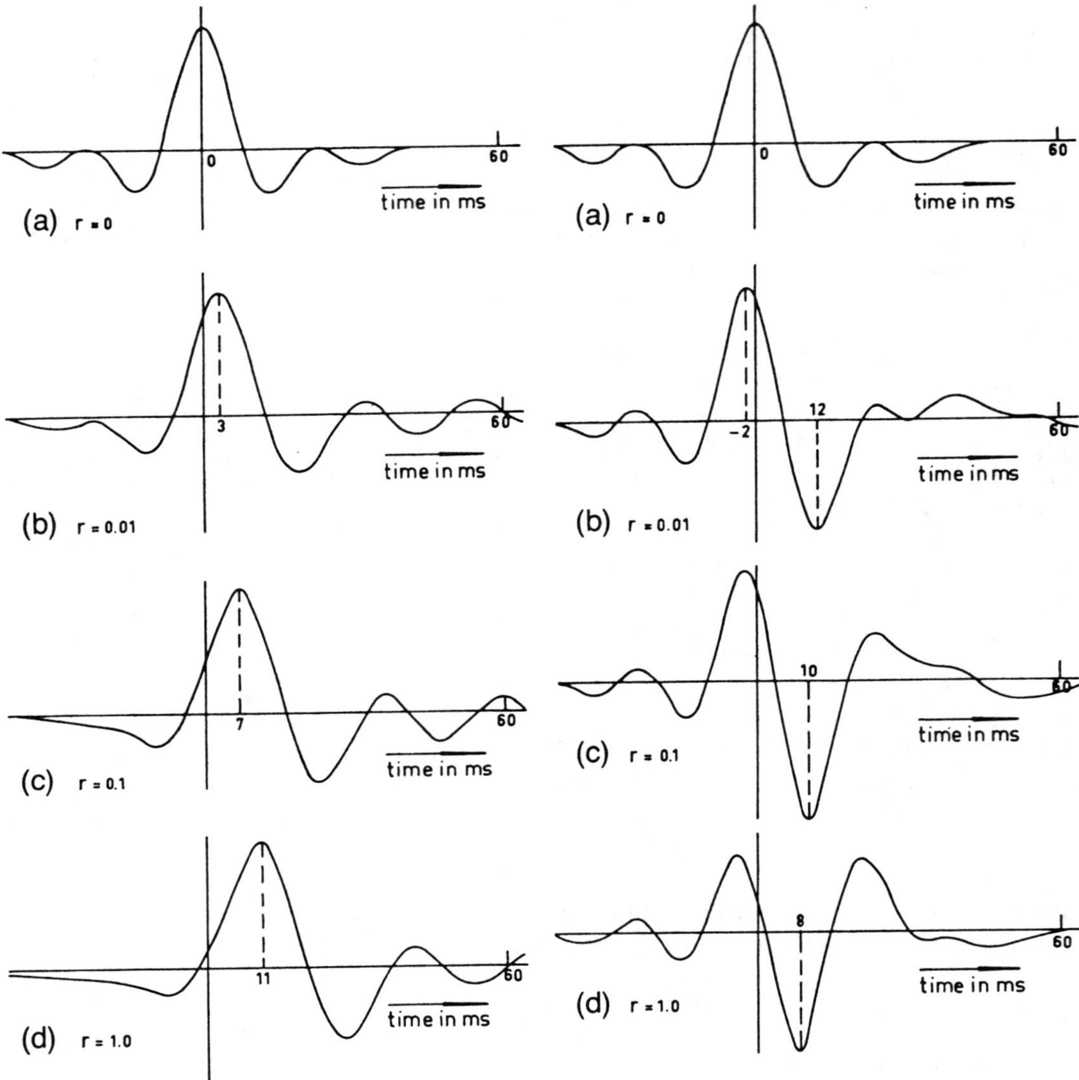

FIG. 6. Shape of wavelet after one-sided least-squares inverse filtering for different noise levels on the seismic trace, the noise being high frequency.

FIG. 7. Shape of wavelet after one-sided least-squares inverse filtering for different noise levels on the seismic trace, the noise being low frequency.

Hence we have derived the important conclusion that, even with minimum-phase wavelets, one-sided least-squares inverse filtering is not very suitable to produce high-resolution results. For the error in the minimum-phase spectrum of the filter $\Delta\psi_n$, we may write $\Delta\psi_n$ = minimum-phase spectrum that is related to

let us consider the minimum-phase wavelet $w(t)$ in Figure 5a. Its energy-density spectrum $|W(f)|^2$ is depicted in Figure 5b. We will compute (using Levinson's scheme) one-sided least-squares inverse filters for this wavelet, assuming different noise levels r on the recording:

$$[1 + \{\bar{R}_{n,n}(f)/|W(f)|^2\}]^{-1/2}. \qquad (23)$$

$$f(t) * [R_{w,w}(t) + rR_{n,n}(t)] = c\delta(t) \qquad \text{for } 0 \le t \le T, \qquad (24a)$$

Examples

To illustrate the effect of $\Delta\psi_n$ on filtered wavelets,

c being chosen such that $f(0) = 1$.

Note that we have represented $R_{x,x}(t)$ in (24a) by

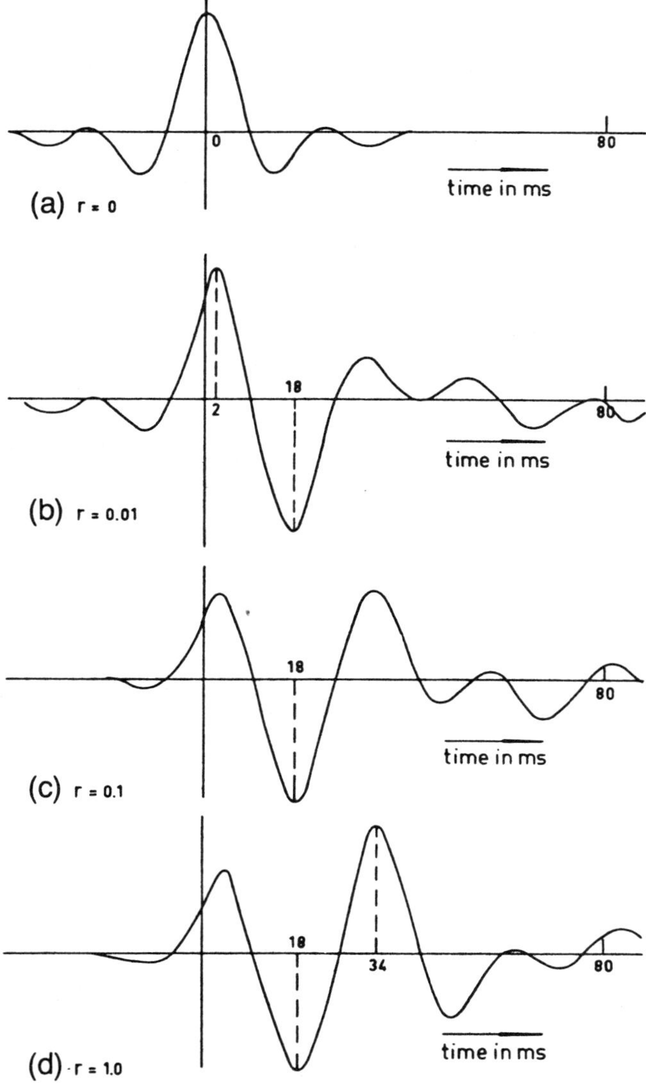

FIG. 8. Shape of wavelet after one-sided least-squares inverse filtering for different noise levels on the seismic trace, the noise being white.

$R_{w,w}(t) + rR_{n,n}(t)$, assuming that the influence of cross terms $\epsilon(t)$ and primary reflectogram information $u(t)$ has been averaged out. In this way, the influence of noise on the optimum filter has been singled out for our investigation (see also remark 1 below). The filter performance will be judged by inspecting the filtered wavelet *without* noise $f(t) * w(t)$. By deleting the output noise $f(t) * n(t)$ in the displays, the deconvolved wavelet $f(t) * w(t)$ can be optimally investigated.

We will consider *high*-frequency noise (Figure 5c), *low*-frequency noise (Figure 5d), and *white* noise.

For r we have chosen 0.01, 0.1 and 1.0. The filter length T will be taken 200 msec.

Finally, in order to facilitate a comparison of all filter results for the same bandwidth, the filtered wavelets $f(t) * w(t)$ were zero-phase band-pass filtered before display, the passband being defined by a trapezoid with the four corner frequencies 10-20-40-60 Hz. By doing this, we also follow a procedure that is standard in the practice of seismic processing (band-pass filtering after deconvolution).

In Figure 6 the filtered wavelet is shown, assuming high-frequency noise. From the results, we see that the presence of high-frequency noise causes a delay in the filter output.

Figure 7 shows the filtered wavelet, assuming low-frequency noise. Note that apart from a delay in the filter output, the presence of low-frequency noise also causes a distortion such that an apparent polarity reversal occurs.

In Figure 8 the filtered wavelet is depicted, assuming white noise. It is interesting to see that already a small level of white noise causes a significant delay and distortion of the filter output. As a consequence, the application of white-noise factors for stabilization of the normal equations should be done with care. Figure 8 suggests that only very small factors should be used. Finally, Figure 9 shows the filtered wavelet, using two-sided least-squares inverse filtering:

$$f(t) * [R_{w,w}(t) + rR_{n,n}(t)] = w(-t), \quad (24b)$$

and white noise (see computational scheme below). Note the superior performance of the two-sided least-squares inverse filters in the presence of noise: zero delay and minimum distortion. Note also that the bandwidth of the filtered wavelet automatically decreases with increasing noise.

If the noise power spectrum is not known, it is difficult to estimate the influence of noise on the filter coefficients. Therefore, in practice, delay and polarity of conventionally deconvolved data are hard to predict.

Finally it is interesting to notice that, apart from the noise free situation, $\Delta\psi_n(f) = 0$ in the rare situation that the power-density spectrum of the noise equals the energy-density spectrum of the wavelet:

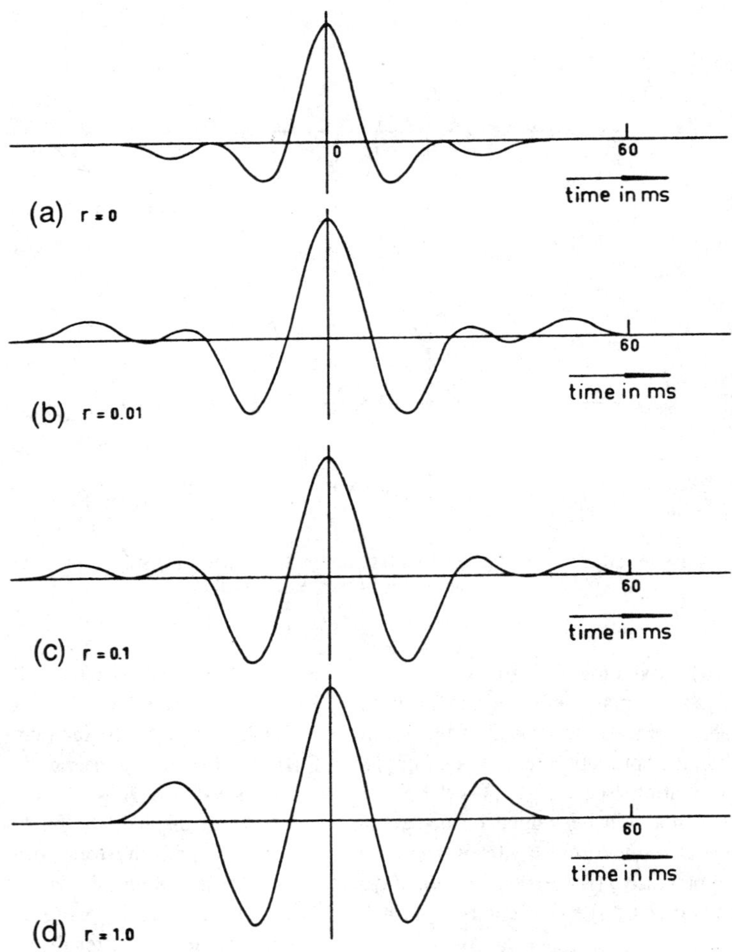

FIG. 9. Shape of wavelet after *two-sided* least-squares inverse filtering for different noise levels on the seismic trace, the noise being white.

$$\frac{\bar{R}_{w,w}(f)}{\bar{R}_{n,n}(f)} \quad \text{is constant.}$$

This property may explain a well-known effect in the practice of deconvolution that, in particularly noisy situations, the results of one-sided least-squares inverse filtering may be improved if zero-phase band-pass filtering is applied first!

Remarks

1) As discussed before, if $|\bar{u}_i(f)|^2 \neq 1$, the primary reflectogram information will have an influence on the filter coefficients as well. For two-sided least-squares inverse filtering, this influence only applies for the filter amplitude spectrum. However, for one-sided least-squares inverse filtering, the phase spectrum of the filter is also influenced: $\Delta\psi(f) = $ minimum-phase spectrum that is related to

$$[|\bar{u}_i(f)|^2 + \{\bar{R}_{n,n}(f)/|W(f)|^2\}]^{-1/2}.$$

2) Let us consider the measurement process once more:

$$x(t) = \underbrace{w(t) * u(t)}_{\text{signal}} + \underbrace{w'(t) * u'(t)}_{\substack{\text{colored} \\ \text{noise}}}.$$

(with the second term labeled white noise above)

It is easy to verify that if we take for one-sided least-squares inverse filtering,

$$d(t) = u'(t),$$

instead of $d(t) = u(t)$, then the remarkable result would be obtained of a fully unchanged filter. Hence the one-sided least-squares inverse filter can not discriminate against noise. This explains the ever existing need of applying band-pass filtering after one-sided least-squares inverse filtering.

3) In exploration seismology, the distortion of one-sided least-squares inverse filtered seismic wavelets, due to the presence of noise, may lead to unreliable polarity predictions ("hard" or "soft"?). In production seismology, the delay and distortion of one-sided least-squares inverse filtered seismic wavelets, due to the presence of noise, may cause false correlations between seismic data and borehole data.

4) One-sided optimum-lag least-squares inverse filtering actually equals two-sided least-squares inverse filtering, the two-sided least-squares inverse filter being delayed such that it becomes one-sided. After this one-sided filter has been applied, the

data should be advanced to obtain the correct arrival times:

$$f(t) * R_{x,x}(t) = w(-t + \tau) \quad \text{for } 0 \leq t \leq T,$$

equals

$$f(t + \tau) * R_{x,x}(t) = w(-t) \quad \text{for } -\tau \leq t \leq T - \tau,$$

τ being the optimum lag. The author prefers to treat these filters as two-sided. Optimizing the value of lag τ actually means optimizing the length of the anticipation part of the two-sided filter with respect to the total filter length.

Computational scheme

For the computation of two-sided least-squares inverse filters, the following practical procedure is proposed. It consists of two simple steps:

1) *Amplitude-spectrum spacing—*. Derive a one-sided least-squares inverse filter with duration T, using the efficient Levinson scheme:

$$h(t) * R_{x,x}(t) = c\delta(t) \quad \text{for } 0 \leq t \leq T,$$

and compute the autocorrelation function of $h(t)$:

$$g(t) = h(-t) * h(t).$$

Note that $g(t)$ represents the zero-phase amplitude-spectrum shaping filter. For large T we may write:

$$G(f) = 1/\bar{R}_{x,x}(f).$$

The value of T determines which amount of detail of $\bar{R}_{x,x}(f)$ will be present in $G(f)$. For small T values, the general shape of $\bar{R}_{x,x}(f)$ will only be inverted.

2) *Correlation step—*. Compute the final filter by correlating $g(t)$ with $w(t)$:

$$f(t) = w(-t) * g(t).$$

Assuming a correct estimate of wavelet $w(t)$, the filter $f(t)$ will always produce zero-phase reflections independent on the choice of T. The value of T has influence on the amplitude spectrum only.

The example in Figure 9 was computed according to this scheme with $T = 100$ msec. It may be pointed out that this computational scheme could be most useful in Vibroseis data deconvolution.

WAVELET DECONVOLUTION

In wavelet deconvolution the filter to be used is a band-limited wavelet-inverse filter derived by a least-squares procedure:

$$E = \sum_t [d(t) - f(t) * w(t)]^2 \quad \text{is minimum,}$$

or

$$E = \sum_t \left[d(t) - \sum_m f(m) w(t - m) \right]^2 \qquad \text{is minimum.}$$
$$(25)$$

Differentiation of (25) by the filter coefficients $f(m)$ yields:

$$\sum_m f(m) R_{w,w}(k - m) = R_{w,d}(k) \qquad \text{for } k \equiv m.$$
$$(26)$$

For the desired wavelet $d(t)$, a zero-phase signal is chosen with some desirable bandwidth. For stabilization purposes a small factor, c^2, is added to $R_{w,w}(0)$:

$$R'_{w,w}(\tau) = \sum_t w(t - \tau) w(t) + c^2 \delta_{0,\tau}.$$

For a two-sided filter of large duration, the stabilized solution of (26) may be written as

$$F(f) = \frac{\tilde{R}_{w,d}(f)}{\tilde{R}'_{w,w}(f)}$$
$$= \frac{W^*(f)}{|W(f)|^2 + c^2} D(f). \qquad (27)$$

From equation (27), it follows that (1) wavelet deconvolution produces zero-phase wavelets, (2) the stabilization factor c simulates the existence of white noise on the input data with power c^2, and (3) $D(f)$ acts as a band-pass filter.

Note that, unlike two-sided least-squares inverse filtering, wavelet deconvolution is *not* adaptive to noise on the input data. The effect of "noise boosting" is (partially) avoided by making use of a stabilization factor and a band-limiter.

CONCLUSIONS

1) Effective inverse filtering (deconvolution) procedures should produce zero-phase wavelets with broad and smooth amplitude spectra. Unfortunately, the latter must depend on the noise content of the input data.

2a) One-sided least-squares inverse filtering is easy to apply as it needs little information from the user. This may explain its immense popularity in the seismic industry. Until now it has been assumed that the main drawback of one-sided least-squares inverse filtering is given by the minimum-phase assumption. In this paper it is demonstrated that, even with minimum-phase wavelets, one-sided least-squares inverse filtering is not very suitable

to produce high-resolution results. It is shown that, in the presence of noise, the sign-reversed minimum-phase spectrum of the filter may differ considerably from the minimum-phase spectrum of the wavelet.

b) Examples indicate that with one-sided least-squares inverse filtering the presence of noise will lead to a *delay* of the filtered wavelet. If low-frequency noise is present an apparent polarity reversal occurs as well.

3a) Two-sided least-squares inverse filtering is greatly superior to one-sided least-squares inverse filtering. It produces zero-phase seismic wavelets of some desirable band width. Two-sided least-squares inverse filtering should be preceded by a wavelet estimation procedure.

b) Two-sided least-squares inverse filtering acts as a wavelet inverse filter for frequency bands with a high signal-to-noise ratio:

$$F(f) = 1/W(f).$$

For frequencies with a low signal-to-noise ratio, two-sided least-squares inverse filtering will approach matched filtering:

$$F(f) = W^*(f)/|N(f)|^2.$$

4) Wavelet deconvolution closely resembles two-sided least-squares inverse filtering, provided the noise on the input data is white.

5) For effective inverse filtering, wavelet (= impulse response of measurement generating system) and noise power spectrum (= additive noise in measurement generating model) should be known. Therefore, future deconvolution research should be devoted to estimation techniques for seismic wavelets and noise power spectra.

REFERENCES

Berkhout, A. J., 1973, On the minimum-length property of one-sided signals: Geophysics, v. 38, p. 657–672.
—— 1974, Related properties of minimum-phase and zero-phase time functions: Geophys. Prosp., v. 22, p. 683–709.
Gabor, D., 1946, Theory of communication: J. IEEE, v. 93, p. 429–441.
Noble, B., 1958, The Wiener Hopf technique: London, Pergamon Press.
Schoenberger, M., 1974, Resolution comparison of minimum-phase and zero-phase signals: Geophysics, v. 39, p. 826–833.
Wiener, N., 1949, Extrapolation, interpolation and smoothing of stationary time series: Cambridge, MIT Press.

APPENDIX

Let us consider the general equation that defines the *one-sided* least-squares filter $f(t)$:

$$f(t) * R_{x,x}(t) = R_{x,d}(t) \qquad \text{for} \qquad t \geq 0, \quad (A\text{--}1)$$

with

$$f(t) = 0 \qquad \text{for} \qquad t < 0,$$

$x(t)$ = input to the filter,

$R_{x,x}(t)$ = (estimate of) the auto-correlation function of $x(t)$,

$d(t)$ = desirable output, and

$R_{x,d}(t)$ = (estimate of) the cross-correlation function of $x(t)$ and $d(t)$.

Equation (A–1) can also be written as

$$f^+(t) * R_{x,x}(t) = p^+(t) + q^-(t) \qquad \text{for all } t, \quad (A\text{–}2)$$

with

$$
\begin{aligned}
f^+(t) &= f(t) & &\text{for all } t, \\
p^+(t) &= R_{x,d}(t) & &\text{for } t \geq 0, \\
p^+(t) &= 0 & &\text{for } t < 0, \\
q^-(t) &= 0 & &\text{for } t \geq 0, \\
q^-(t) &= \text{unknown} & &\text{for } t < 0.
\end{aligned}
$$

Equation (A–2) has to be solved for the two unknowns $f^+(t)$ and $q^-(t)$. This problem is known as the Hilbert problem or the generalized Wiener-Hopf problem (Noble, 1958).

Now let us concentrate on the *discrete* problem. As equation (A–2) holds for all t, application of the z-transform yields:

$$F^+(z)\bar{R}_{x,x}(z) = P^+(z) + Q^-(z)$$
$$\text{at least for } |z| = 1, \quad (A\text{–}3)$$

with

$$F^+(z) = \sum_{n=0}^{\infty} f_n z^n \qquad \text{for } |z| \geq 1,$$

$$\bar{R}_{x,x}(z) = \sum_{n=-\infty}^{+\infty} r_n z^n \qquad \text{for } |z| = 1,$$

$$P^+(z) = \sum_{n=0}^{\infty} p_n z^n \qquad \text{for } |z| \geq 1,$$

$$Q^-(z) = \sum_{n=-\infty}^{-1} q_n z^n \qquad \text{for } |z| \leq 1.$$

To solve (A–3), we begin by factorizing $\bar{R}_{x,x}(z)$:

$$R_{x,x}(z) = S^{\min}(z) S^{\min}(1/z) \qquad \text{for } |z| = 1,$$
$$(A\text{–}4a)$$

with

$$S^{\min}(z) = \sum_{n=0}^{\infty} s_n z^n \qquad \text{for } |z| \geq 1$$

and

$$S^{\min}(1/z) = \sum_{n=-\infty}^{0} s_{-n} z^n \qquad \text{for } |z| \leq 1.$$

$S^{\min}(z)$ has no singularities and no zeros for $|z| < 1$. Hence $S^{\min}(t)$ represents a minimum-phase time function with amplitude spectrum $[\bar{R}_{x,x}(f)]^{1/2}$.

Note that factorization (A–4a) is allowed under a very gentle constraint on $\bar{R}_{x,x}(z)$, the Kolmogorov condition:

$$\oint_{c_1} \ln[1/\bar{R}_{x,x}(z)]\,dz < \infty, \qquad (A\text{–}4b)$$

c_1 representing the unit circle ($|z| = 1$).

Condition (A–4b) is always fulfilled in practice. Actually, in the practice of least-square inverse filtering the use of a stabilization factor (noise factor) assures that a stronger constraint holds:

$$\oint_{c_1} [1/\bar{R}_{x,x}(z)]\,dz < \infty. \qquad (A\text{–}4c)$$

Using the factorization of $\bar{R}_{x,x(z)}$, (A–3) can be rewritten as:

$$F^+(z) S^{\min}(z) S^{\min}(1/z) = P^+(z) + Q^-(z)$$
$$\text{for } |z| = 1,$$

or

$$F^+(z) S^{\min}(z) = \frac{P^+(z)}{S^{\min}(1/z)} + \frac{Q^-(z)}{S^{\min}(1/z)}$$
$$\text{for } |z| = 1. \quad (A\text{–}5)$$

Now, making use of the power series expansion of $S^{\min}$, P^+ and Q^-, it follows that:

$$\frac{P^+(z)}{S^{\min}(1/z)} = \sum_{n=-\infty}^{+\infty} a_n z^n \qquad \text{for } |z| = 1,$$

and

$$\frac{Q^-(z)}{S^{\min}(1/z)} = \sum_{n=-\infty}^{-1} b_n z^n \qquad \text{for } |z| \leq 1.$$

Hence expression (A–5) can be reformulated as

$$F^+(z) S^{\min}(z) = \sum_{n=-\infty}^{+\infty} c_n z^n \qquad \text{for } |z| = 1, \quad (A\text{–}6)$$

with

$$
\begin{aligned}
c_n &= a_n & &\text{for} & n &\geq 0, \\
c_n &= a_n + b_n & &\text{for} & n &< 0.
\end{aligned}
$$

Since $F^+(z) S^{\min}(z)$ represents a "plus function", a solution for (A–6) can only be found if

$$a_n = -b_n \qquad \text{for} \qquad n < 0,$$

or

$$\left[\frac{P^+(z)}{S^{\min}(1/z)}\right]^- = -\frac{Q^-(z)}{S^{\min}(1/z)} \qquad \text{for } |z| \leq 1,$$

and the solution of (A–3) can be formulated as

$$F^+(z) = \frac{1}{S^{\min}(z)} \left[\frac{P^+(z)}{S^{\min}(1/z)} \right]^+ \qquad \text{for } |z| \geq 1.$$

$$(A-7)$$

Let us apply this result to one-sided least-squares inverse filtering. For this type of least-squares filtering, we may write:

$$R_{x,d}(t) = w(-t)$$

$$\text{with} \qquad w(t) = 0 \qquad \text{for } t < 0,$$

or

$$R_{x,d}(z) = \sum_{n=-\infty}^{0} w_{-n} z^n \qquad \text{for } |z| \leq 1.$$

Hence $P^+(z) = w_0$, and

$$F^+(z) = \frac{1}{S^{\min}(z)} \left[\frac{w_0}{S^{\min}(1/z)} \right]^+$$

$$= \frac{w_0/s_0^{\min}}{S^{\min}(z)} \qquad \text{for } |z| \geq 1. \qquad (A-8)$$

Conclusion

The one-sided least-squares inverse filter $f(t)$, of a time function $x(t)$, is given by

$$|F(f)| = c/[\bar{R}_{x,x}(f)]^{1/2},$$

c being some scaling constant and

$$\psi(f) = - \text{ minimum phase spectrum that is related to the amplitude spectrum } [R_{x,x}(f)]^{1/2}.$$

Note that if constraint (A–4c) applies, then

$$\oint_{c_1} |1/S^{\min}(z)|^2 \, dz < \infty,$$

or

$$\oint_{c_1} |F(z)|^2 \, dz < \infty,$$

or

$$\int_{-1/2 \Delta t}^{1/2 \Delta t} |F(f)|^2 \, df < \infty$$

(Δt being the sampling interval) or

$$\sum_{n=0}^{\infty} f_n^2 < \infty.$$

Hence, if constraint (A–4c) is fulfilled, the one-sided least-squares inverse filter is energy bounded.

Finally, it should be mentioned that for *two-sided* least-squares filters, the general equation applies for all t. As a consequence, it can be directly solved by taking the Fourier transform:

$$f(t) * R_{x,x}(t) = R_{x,d}(t) \qquad \text{for all } t.$$

Hence,

$$F(f) = \bar{R}_{x,d}(f)/\bar{R}_{x,x}(f),$$

and, therefore,

$$f(t) = \int_{-1/2 \Delta t}^{1/2 \Delta t} \frac{\bar{R}_{x,d}(f)}{\bar{R}_{x,x}(f)} e^{j2\pi ft} \, df.$$

5

THE DEBUBBLING OF MARINE SOURCE SIGNATURES

L . C . W O O D * , R . C . H E I S E R ‡ , S . T R E I T E L ‡ , AND P . L . R I L E Y ‡

We propose two novel approaches for the removal of bubble oscillations from marine seismograms. Both methods involve filters containing matched as well as least-squares inverse components. The techniques are deterministic rather than statistical because we require explicit knowledge of the source pulse shape such as a far-field signature. The first approach leads to the design of a matched filter in cascade with a two sided Wiener-shaping operator. The second begins with the computation of the zero-delay Wiener filter that is inverse to the source signature and continues with the autocorrelation of the resulting output.

This performance is illustrated with seismic field data. Our techniques are general ones, applicable in principle to any type of source; however, the examples displayed use Maxipulse and Aquapulse seismic records. Acoustic oscillations caused by successive expansions and contractions of gas bubbles initiated by the seismic source can be attentuated effectively with these filters, and the process has been called debubbling by other investigators. Our procedures improve the definition of seismic reflections (resolution), and the debubbling procedure is a first step in data processing prior to applying other deconvolution and pulse compression techniques.

INTRODUCTION

The cancellation of bubble oscillations in marine recording has received considerable attention in recent years. Acoustic oscillations caused by successive expansions and contractions of gas bubbles initiated by the source can be attenuated with a digital filter, and the process has been called debubbling (Mateker, 1971). The purpose of the present contribution is to consider the debubbling problem, and this paper discusses two methods of correcting data for the energy contained in the bubble train caused by seismic sources. Our techniques are general ones, applicable in principle to any type of source; however, the examples displayed in the sequel use Maxipulse™ and Aquapulse™ records acquired in an offshore marine environment. The latter types of filtering fall under the general category of pulse shaping; in other words, a digital filter shapes a narrow-band, reverberating input signal into a broad-band, highly compressed output waveform. Our techniques have been found to be an improvement over a debubbling method described in the literature (Mateker, 1971). This latter method is one of the state-of-the-art methods employed at the present time for debubbling Maxipulse data.

Sharp source pulses are needed to improve the definition of seismic reflections; that is to say, resolution is inversely proportional to pulse time-width. The source signature is assumed to be known throughout our discussion either through field measurements during data acquisition, by wavelet estimation during data processing, or through a priori knowledge. The objective of our methods is to further contract these signatures in data processing prior to applying other deconvolution and pulse compression techniques.

Two distinct implementations of our debubbling procedures are described, and both schemes yield

™ Western Geophysical Co. of America.

Presented at the 47th Annual International SEG Meeting, September 20, 1977 in Calgary. Manuscript received by the Editor March 16, 1977; revised manuscript received November 10, 1977.
*Amoco Production Co., Box 3092, Houston, TX 77001.
‡Amoco Production Co., Box 591, Tulsa, OK 74102.

about equal results. One version consists of designing a matched filter in cascade with a two-sided Wiener-shaping operator. An alternate approach leads to the design of a digital filter containing matched as well as inverse, or "whitening" components. We subsequently show that both implementations yield nearly identical debubbling filters, a fact which we illustrate with field-recorded seismic data. The first part of our paper briefly outlines the pulse compression algorithms for which a detailed development is given in the Appendix. This discussion is then followed with examples of its use on field data recorded with Maxipulse and Aquapulse sources.

THE SEISMIC DEBUBBLING PROBLEM

The signatures produced by marine sources such as dynamite, as well as air-gun and gas-gun source arrays consist typically of a sequence of oscillations whose amplitudes generally diminish as time increases (Giles, 1968). Quite evidently such pulse trains must be compressed in the received seismogram so that acceptable resolution levels result.

Many marine sources create bubble oscillations of varying magnitudes (see Figure 1). An impulsive source creates conditions that cause a rapidly expanding and contracting bubble to form. The bubble continues to expand until the cooling and expanding volume creates a pressure less than hydrostatic. At this point the bubble collapses with increasing speed as water rushes inward. The inward rushing waters create an additional acoustic impulse that may often exceed in amplitude the initial detonation pulse. Compression by the water causes an increase in gas pressure and the bubble again expands to repeat the cycle. Cycles are repeated several times as energy decreases. The seismic energy handbook by Kramer et al (1968) gives an excellent description of this phenomenon. These bubble oscillations contribute to the basic wavelet that combines with the reflection sequence to produce a very complicated seismic record. We would like, of course, for the source wavelet to be a narrow pulse of small time duration.

Maxipulse and Aquapulse have source related oscillations, and significant bubble oscillations occur with both source types. The reader may refer to

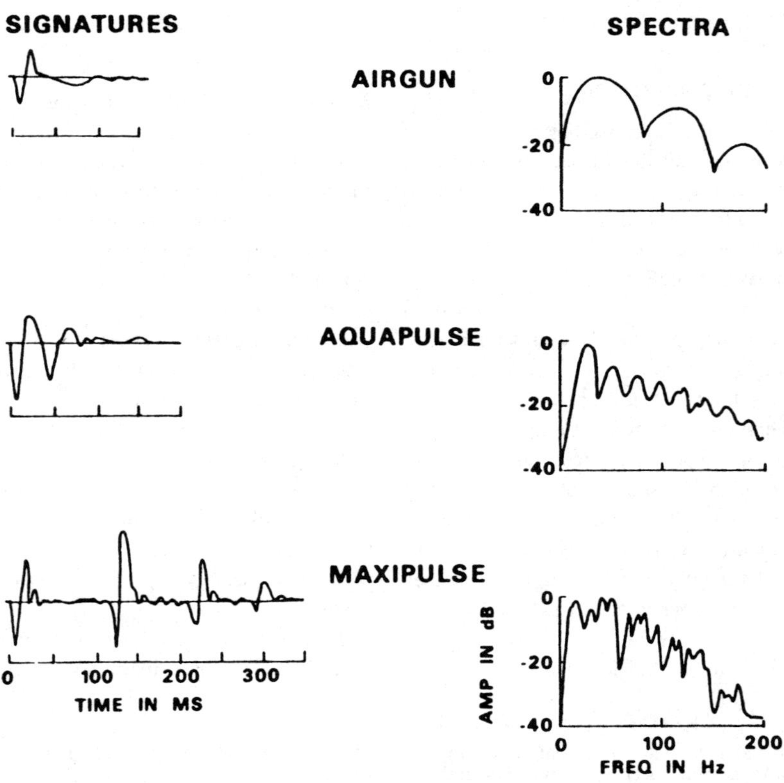

Fig. 1. Comparison of signatures and spectra for Aquapulse, Maxipulse, and air-gun sources.

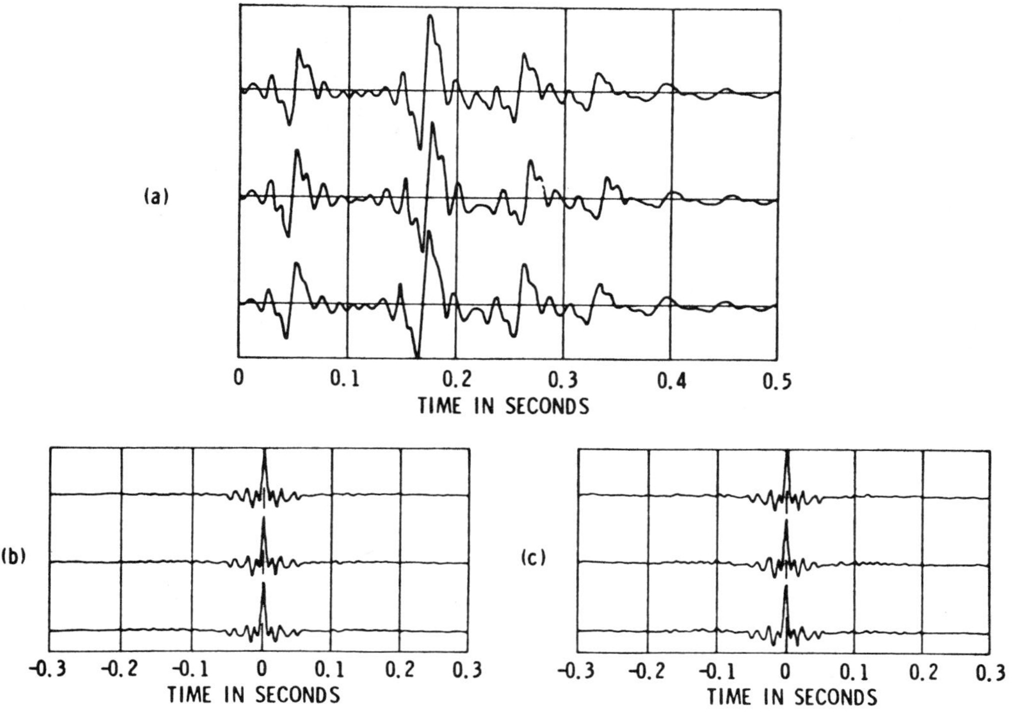

FIG. 2. (a) Typical recorded Maxipulse bubble signatures debubbled with (b) the filter $D(z)$, and (c) the filter $\hat{D}(z)$.

Mateker (1971) and Kramer et al (1968) for excellent discussions of these two types of seismic sources. Air-gun sources also create significant bubble oscillations. For air-gun references, the reader may consult papers by Kologinczak (1974), Kramer et al (1968), Smith (1975), and Wielandt (1975). Maxipulse is a delayed-fuse, dynamite source (Mateker, 1971) producing seismic records whose zero-time references are not synchronized because detonation times and water depths cannot be controlled; in other words, they cannot be synchronized (record-to-record) in the field. Maxipulse bubbles oscillate for some 500–1000 msec depending upon the charge size and depth at which detonation occurs. Aquapulse, on the other hand, is a combustible oxygen-propane gas mixture that expands a rubber sleeve (Kramer et al, 1968, p. 47) producing a reverberating pulse of shorter time duration than Maxipulse. Source reverberations, nevertheless, have significant amplitudes and time duration, and Aquapulse initiations can be synchronized in time. Our experiments show that Aquapulse source wavelets reverberate and should be inverse filtered in a manner similar to Maxipulse.

The bubble oscillations created by the Maxipulse source are recorded routinely and are typical of bubble oscillations produced by an underwater explosion. Debubbling itself is left to subsequent data processing. The Aquapulse source and source-array, however, are usually designed to attenuate the effects of bubble oscillations. Aquapulse signatures do not exhibit typical bubble oscillations, but nevertheless reverberate for a significant length of time. The air-gun signature also reverberates for a considerable length of time.

Figure 2a illustrates Maxipulse source signatures. Our debubbling procedures reduce Aquapulse and Maxipulse source reverberations to well-defined signals of short time duration. The filtering methods described in the next section and in the Appendix have been tested using Maxipulse and Aquapulse source signatures.

DEBUBBLING OF SOURCE SIGNATURES

A marked improvement in the resolution of marine seismograms can be achieved using our debubbling procedures. These techniques reduce significantly the complexity of marine reflection seismograms. We consider the basic wavelet $W(t)$ to be a composite signal obtained by convolving the far-field source signature $x(t)$ with the impulse response $I(t)$ of the recording instruments and an earth effects signature $E(t)$,

$$W(t) = x(t) * I(t) * E(t). \tag{1}$$

The earth effects term $E(t)$ describes such phenomena as the desired reflectivity sequence, near-surface ghosting, small-delay reverberations, absorption, and similar disturbances. Divergence effects, however, may or may not be included in the far-field source signature $x(t)$. If divergence is not included in the source signature, then it is included as part of the earth effects term $E(t)$. We assume throughout this paper that source signature $x(t)$ is known, and the purpose of this paper is to describe an effective method for compressing this pulse. We are not concerned at this initial stage of the debubbling process with the other two parts of the basic wavelet, namely the $I(t)$ and $E(t)$ components. These effects are handled in subsequent deconvolution and whitening processes. It is possible, however, to include a deterministic correction for handling earth effects, such as ghosting, as part of the source signature when they are sufficiently well known; the sole purpose of our algorithm, therefore, is to compress a specified source pulse $x(t)$.

The first method we shall describe consists of treating all signatures in much the same manner as one treats Vibroseis® data. This is to say, the first step consists of crosscorrelating the known signature with the recorded data. Correlation as a general rule increases signal-to-noise ratios. Also, it simplifies phase considerations because the correlated waveform, which should be the autocorrelation of the source signature, has a zero-phase spectrum. After the initial crosscorrelation step, the data should contain the source signature's autocorrelation function as a basic wavelet.

The second step in this filtering technique then is to shape the autocorrelated source wavelet to some highly compressed, desired output signal of very small time duration. This task can be accomplished using Wiener filters (Treitel and Robinson, 1966). In actual practice we have found that it is preferable not to compress the correlated waveform to a Dirac impulse in one step but, rather, to shape the autocorrelation to a zero-phase pulse consisting of the autocorrelation values between the second-zero crossings. The final compression to an approximation of a Dirac impulse is accomplished best at a later stage in processing where other pulse compression methods can be used. We now proceed to describe

®Trademark of Continental Oil Co.

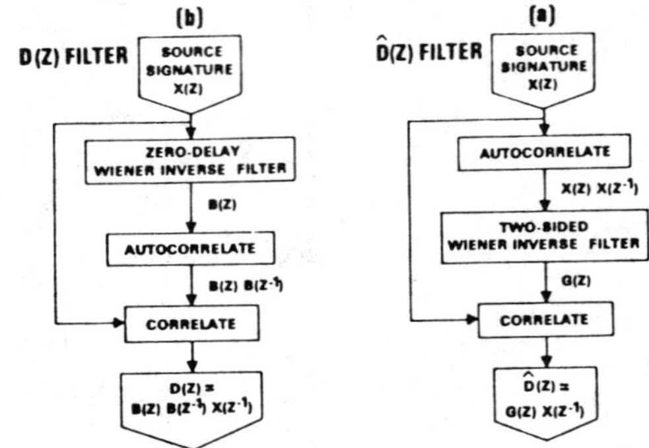

FIG. 3. Block diagram illustrating the calculation of the two debubbling filters $D(z)$ and $\hat{D}(z)$.

the second debubbling technique which yields nearly identical results.

The first step in our second method also consists of matched filtering by crosscorrelating with the known signature. The next step, however, differs considerably from the approach described above in that a *zero-delay* Wiener inverse filter for the signature is calculated. The debubbling procedure consists of three filtering stages in cascade:

1) Matched filtering of the data by crosscorrelating the data with the known source signature.
2) Filtering the data with the zero-delay Wiener inverse filter for the signature.
3) Matched filtering of the data by crosscorrelating the data with the zero-delay Wiener inverse of the source signature (the same filter used in step 2).

At this point, it is still not clear that the two debubbling techniques are nearly identical and that they should be expected to perform equally well as debubbling filters. Let us, therefore, review calculation of the two debubbling filters in greater detail.

Calculation of the filter for the first method (Figure 3a) consists of the following sequence of operations:

1) Given the bubble signature x_t, compute its autocorrelation, whose z-transform is $X_n(z)X_n(z^{-1})$ and $X_n(z)$ is the z-transform of x_t, of length $n + 1$.
2) Given the autocorrelation pulse with z-transform $X_n(z)X_n(z^{-1})$, compute its two-sided Wiener inverse filter $G_p(z)$, where

$$G_p(z) = g_0 + g_1 z + \cdots + g_p z^p \tag{2}$$

is of length $p + 1$. Then for sufficiently large p, we may expect that

$$X_n(z)\dot{X}_n(z^{-1})G_p(z) \sim 1 \tag{3}$$

although in practice this goal can be achieved only as p approaches infinity.

3) Compute the debubbling filter $\hat{D}(z)$ by crosscorrelating the two-sided Wiener inverse filter $G_p(z)$ with the source signature $X_n(z)$. In z-transform notation:

$$\hat{D}(z) = G_p(z)X_n(z^{-1}).$$

Thus we see that the $\hat{D}(z)$ filter is applied in parts; namely, the first part is the $X_n(z^{-1})$ crosscorrelation step, and the second part is the $G_p(z)$ Wiener filter.

Calculation of the debubbling filter $D(z)$ for the second method (Figure 3b), by comparison, is as follows:

1) Compute $B_k(z)$, which is the z-transform of the zero-delay Wiener filter inverse to x_t, of length $k + 1$.

2) Compute $B_k(z^{-1})$, which is the matched filter for the zero-delay Wiener filter, and apply to $B_k(z)$ to obtain the autocorrelation $B_k(z)B_k(z^{-1})$.

3) Compute the debubbling filter by crosscorrelating the autocorrelation $B_k(z)B_k(z^{-1})$ with the source signature $X_n(z)$:

$$D(z) = B_k(z)B_k(z^{-1})X_n(z^{-1}). \tag{4}$$

In practice, we apply the $D(z)$ filter in parts; that is to say, we first crosscorrelate to apply the $X_n(z^{-1})$ term, and then apply the $B_k(z)$ filter followed in cascade by the $B_k(z^{-1})$ filter to achieve the $X_n(z^{-1})B_k(z)B_k(z^{-1})$ debubbling filter.

In the Appendix we show that

$$X_n(z)B_k(z) \sim A_{n-m}(z), \tag{5}$$

where $A_{n-m}(z)$ is a type 1 all-pass, or dispersive filter (Robinson and Treitel, 1965). The numerator and denominator of this all-pass filter are both of degree $n - m$, where $m \le n$. This in turn implies that

$$X_n(z)X_n(z^{-1})B_k(z)B_k(z^{-1})$$

$$\sim A_{n-m}(z)A_{n-m}(z^{-1}) = 1, \tag{6}$$

where we use the fact that the autocorrelation of an all-pass response is the unit spike. We recognize $X_n(z^{-1})$ to be the matched filter for the input x_t (Treitel and Robinson, 1969).

The filter for the first debubbling technique can then be written

$$\hat{D}(z) = X_n(z^{-1})G_p(z), \tag{7}$$

and the second debubbling filter can be written

$$D(z) = X_n(z^{-1})B_k(z)B_k(z^{-1}). \tag{8}$$

Now, since $B_k(z)$ is the zero-delay Wiener inverse filter for the input $X_n(z)$, it follows that $B_k(z^{-1})$ is the zero-delay Wiener inverse filter for the input $X_n(z^{-1})$. But we have just seen that $G_p(z)$ is the generally optimum-lag inverse of $X_n(z)X_n(z^{-1})$, and hence we may expect that for sufficiently large k and p,

$$B_k(z)B_k(z^{-1}) \sim G_p(z). \tag{9}$$

Thus the filters $D(z)$ and $\hat{D}(z)$ will become increasingly similar as the filter lengths k and p become larger. In other words, either $D(z)$ or $\hat{D}(z)$ may be expected to perform comparably well as debubbling filters. Let us note at this point that both $G_p(z)$ and $B_k(z)B_k(z^{-1})$ are zero-phase filters.

The reader is referred to the Appendix for a more detailed discussion of the two debubbling methods. It should be clear at this point in the development, however, that either of the two schemes should produce equivalent results. This latter statement is verified in the next section, where we apply these new filters to the seismic debubbling problem.

The debubbling techniques currently in common use throughout the industry have several disadvantages when compared to either of the two methods described in this paper. One conventional method can be described with ease because it consists of a Wiener shaping filter corresponding to our Wiener inverse filter $B_k(z)$ (Mateker, 1971). Other conventional techniques operate entirely in the frequency domain. The particular implementation of a Wiener shaping filter (Treitel and Robinson, 1966) often provides a solution, but it is not without difficulties. As is well known, the associated normalized squared error (NSE) depends on the delay properties of the input signal. For a given filter length, the smaller values of the NSE will occur for only certain time delays (or lags) of the desired output relative to the input (Treitel and Robinson, loc. cit.). These "optimum-lag" Wiener filters are cumbersome to obtain in practice, and their performance degrades rapidly with decreasing signal-to-noise ratio. Our new filters, however, are an alternate use of the Wiener filter which appears to circumvent these difficulties.

Another disadvantage of the conventional method is that some sources (e.g., Maxipulse) have zero-time references that are not synchronized. Maxipulse recordings are produced by dynamite explosions whose detonation times and water depths cannot be controlled; hence, they cannot be synchronized (record-to-record) in the field. These timing variations can be corrected automatically in the initial crosscorrelation step. The conventional method, however, must determine static-time corrections to adjust for these variations by picking the onset of the Maxipulse bubble signatures. Let us now consider some examples using recorded seismic field data to test and evaluate these various debubbling procedures.

EXAMPLES

Bubble oscillations are most pronounced on Maxipulse recorded seismograms. An offshore Texas line has been recorded with both Aquapulse and Maxipulse in order to compare the effectiveness of each as a marine source. These data will be used also to compare the debubbling effectiveness of the filters $D(z)$ and $\hat{D}(z)$. The first example consists of recorded Maxipulse signatures and their corresponding data traces. Since source signatures vary from shot to shot, it is necessary to record a source signature for each profile. The profile is then debubbled through use of its recorded signature. Figure 2a is a display of three Maxipulse bubble signatures. Figures 2b and c depict the corresponding debubbled signatures, which have been obtained by convolution with the filters $D(z)$ and $\hat{D}(z)$, respectively. Application of the two filters results in apparently identical debubbled signatures.

The two traces of Figure 4a are adjacent six-fold, common-depth-point (CDP) stacked traces. Three gas zones at this CDP location were confirmed by well information. The zones are located at approximate two-way times of 0.510, 0.685, and 0.830 sec,

and are indicated by arrows. The maximum source-to-receiver distance for these traces is 3135 ft. The debubbled traces of Figures 4b and c result from using filters $D(z)$ and $\hat{D}(z)$, respectively, on individual common-source traces before normal moveout correction. We note that traces within a CDP do not share a common source signature; therefore, each trace was debubbled with a filter calculated from its corresponding recorded source signature. Comparison of the traces in Figure 4 indicates that the debubbling filters $D(z)$ and $\hat{D}(z)$ are about equally effective.

We have found that predictive deconvolution (Peacock and Treitel, 1969) is much more effective in removing water-bottom reverberations after our new debubbling filters have been applied as compared with the conventional method (Mateker, 1971). Figure 5 shows an offshore Texas seismic line after debubbling with our first method [the $\hat{D}(z)$ filter] and the same line after debubbling using the conventional state-of-the-art procedure described by Mateker (1971). The only computational variable represented in Figure 5 is the different filtering method used (i.e., debubbling). All other variables are identical. Figure 6 shows the same data of Figure 5 after predictive deconvolution has been applied. The data in Figure 6 obtained using our new method agree with well-log information better than the data debubbled with the state-of-the-art method; the reflections associated with the gas zones are better defined and exhibit more continuity. Figure 7 is an enlargement of part of the data in Figure 6 and shows that predictive deconvolution has not been as effective

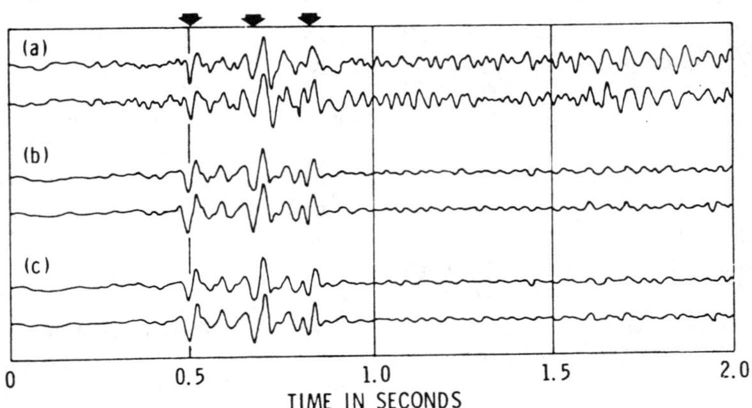

FIG. 4. A comparison of (a) 6-fold stack of traces without debubbling, (b) 6-fold stack of traces debubbled using filter $D(z)$, and (c) 6-fold stack of traces debubbled using filter $\hat{D}(z)$.

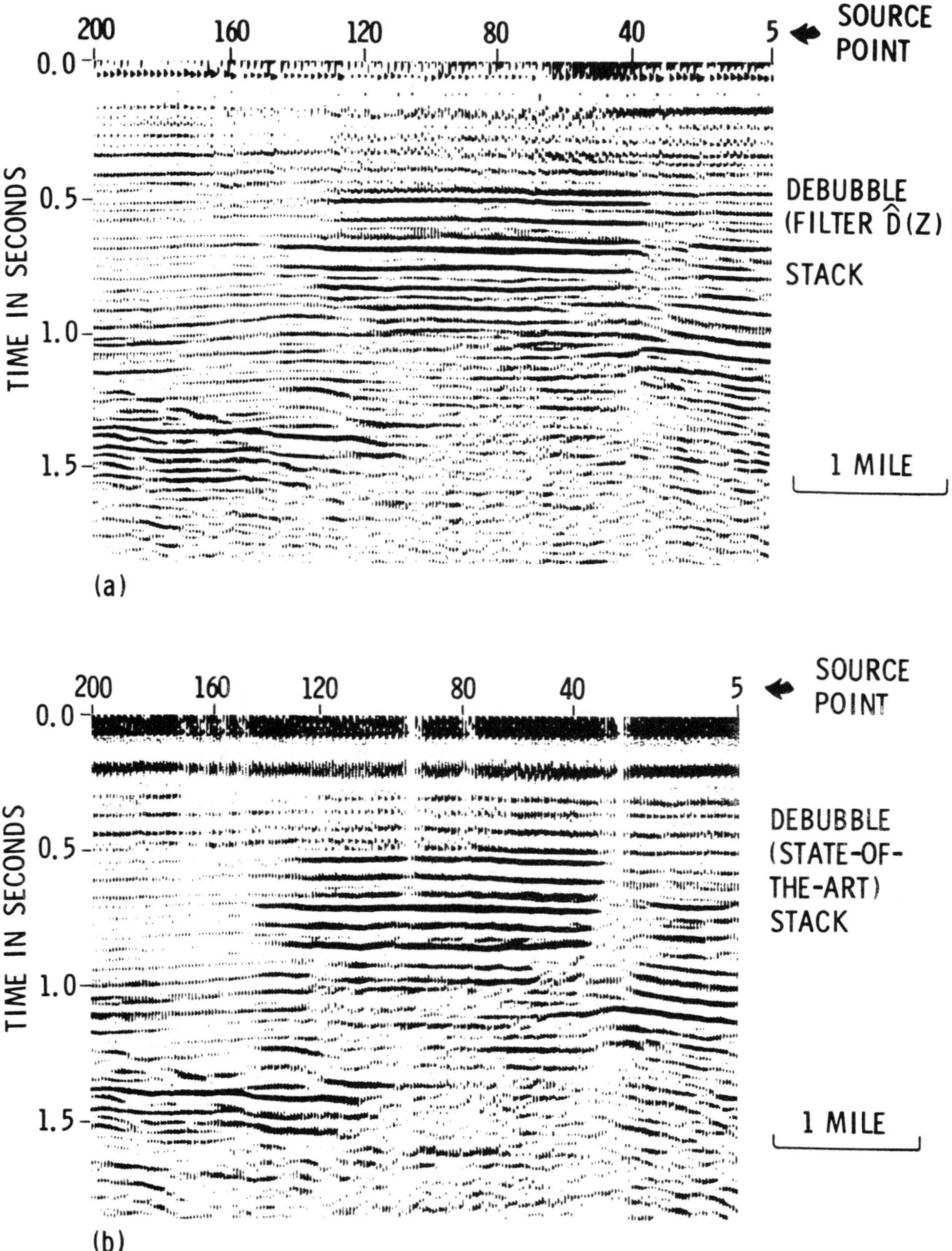

FIG. 5. A comparison of debubbling using (a) filter $\hat{D}(z)$ and (b) the state-of-the-art technique.

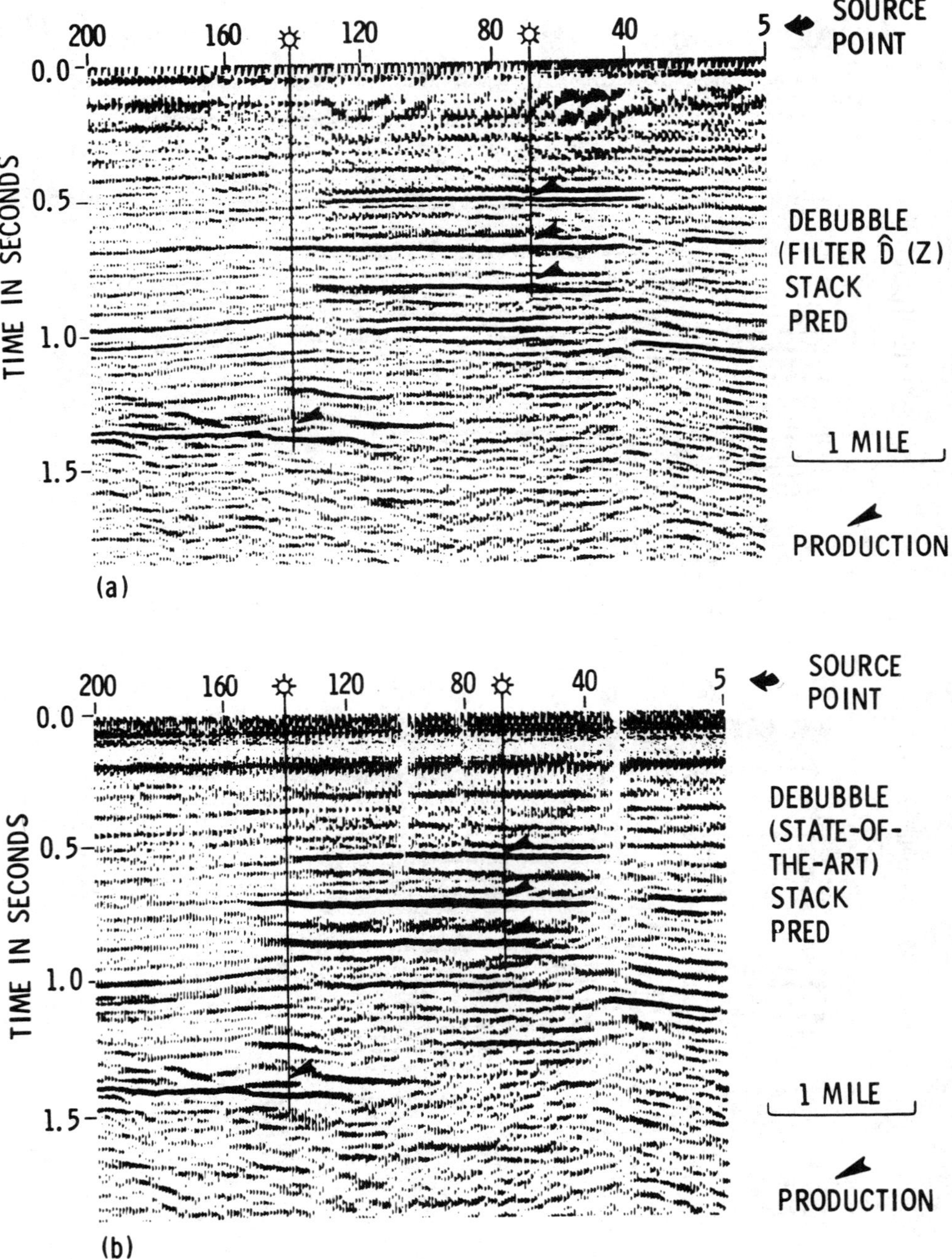

FIG. 6. Predictive deconvolution applied to the data shown in Figure 5.

in removing the water-bottom multiple reflections using the conventional debubbling technique. Note that our debubbling filter $\hat{D}(z)$ results in removal of the water-bottom multiple reflections at about 0.6 and 0.8 sec, whereas the state-of-the-art method does not.

Figure 8 illustrates the same offshore Texas line recorded using an Aquapulse source where the data have been debubbled using the $\hat{D}(z)$ filter based on an average signature for 27 sources, and Figure 9 shows the Aquapulse section with predictive deconvolution applied. The reverberatory character of the Aquapulse signature is evident in Figure 8. We have found that predictive deconvolution is much more effective on the Aquapulse data after applying our debubbling technique. As can be seen in Figure 10,

an enlargement of part of the data in Figure 9, the predictive deconvolution is more effective in removing the water-bottom multiples at 0.62 and 0.83 sec on the data debubbled with the $\hat{D}(z)$ filter.

DISCUSSION

The principal disadvantage of our inverse filtering method is that it requires knowledge of source signatures. Reliable wavelet estimation procedures, however, alleviate this problem. Furthermore, reliable estimates of far-field signatures can be made from near-field measurements [see, for example, the paper by Wielandt (1975)]. Our zero-phase filtering technique employing crosscorrelation has many significant advantages that can be divided in

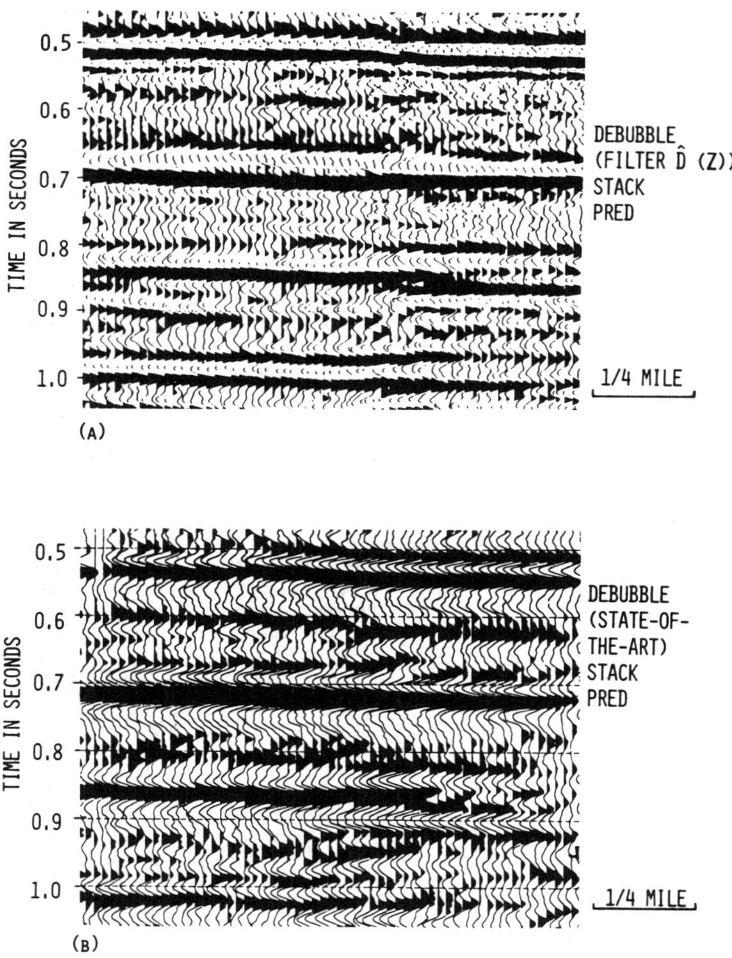

FIG. 7. Enlarged areas of the data shown in Figure 6.

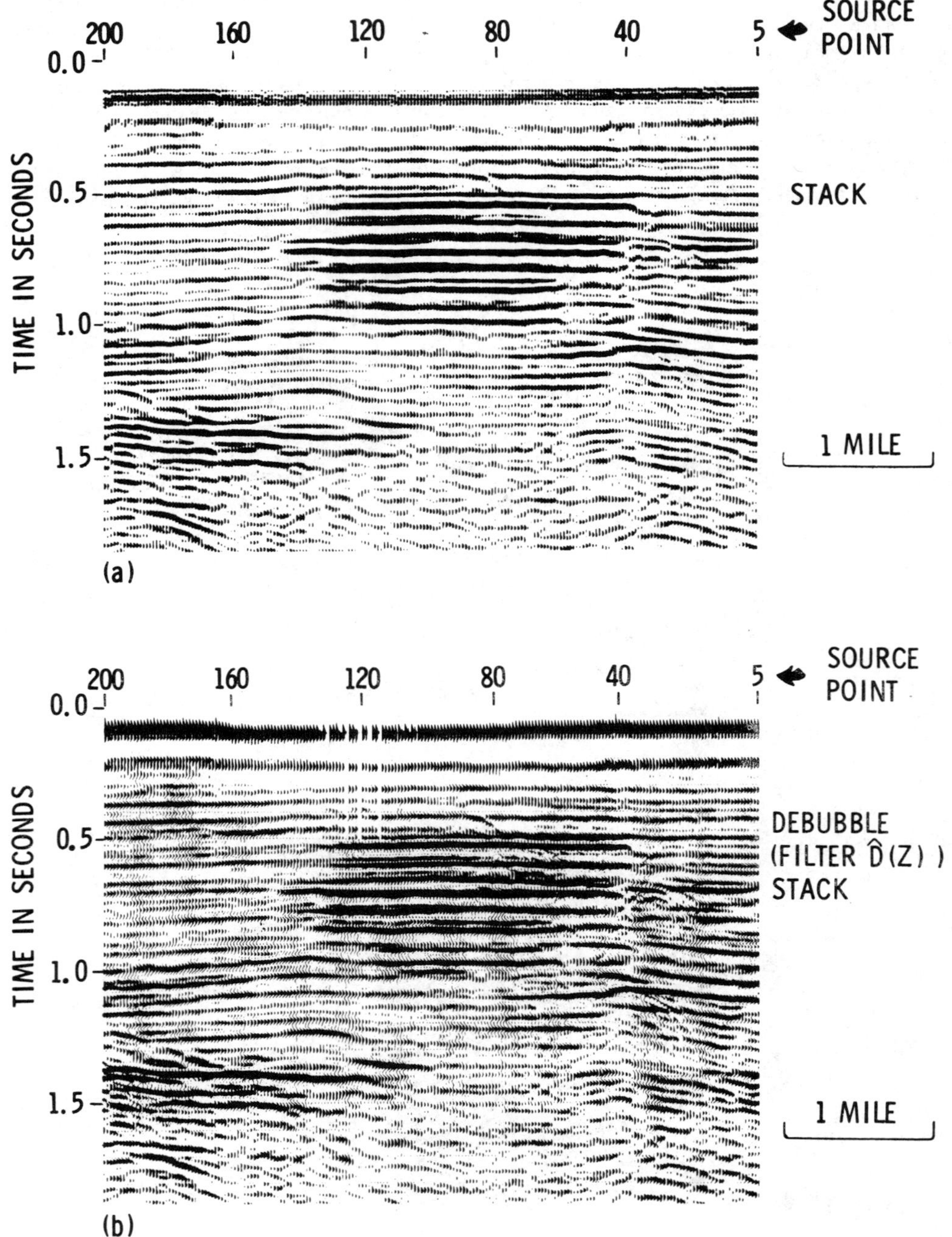

FIG. 8. A comparison of Aquapulse data processed (a) without debubbling and (b) with debubbling filter $\hat{D}(z)$.

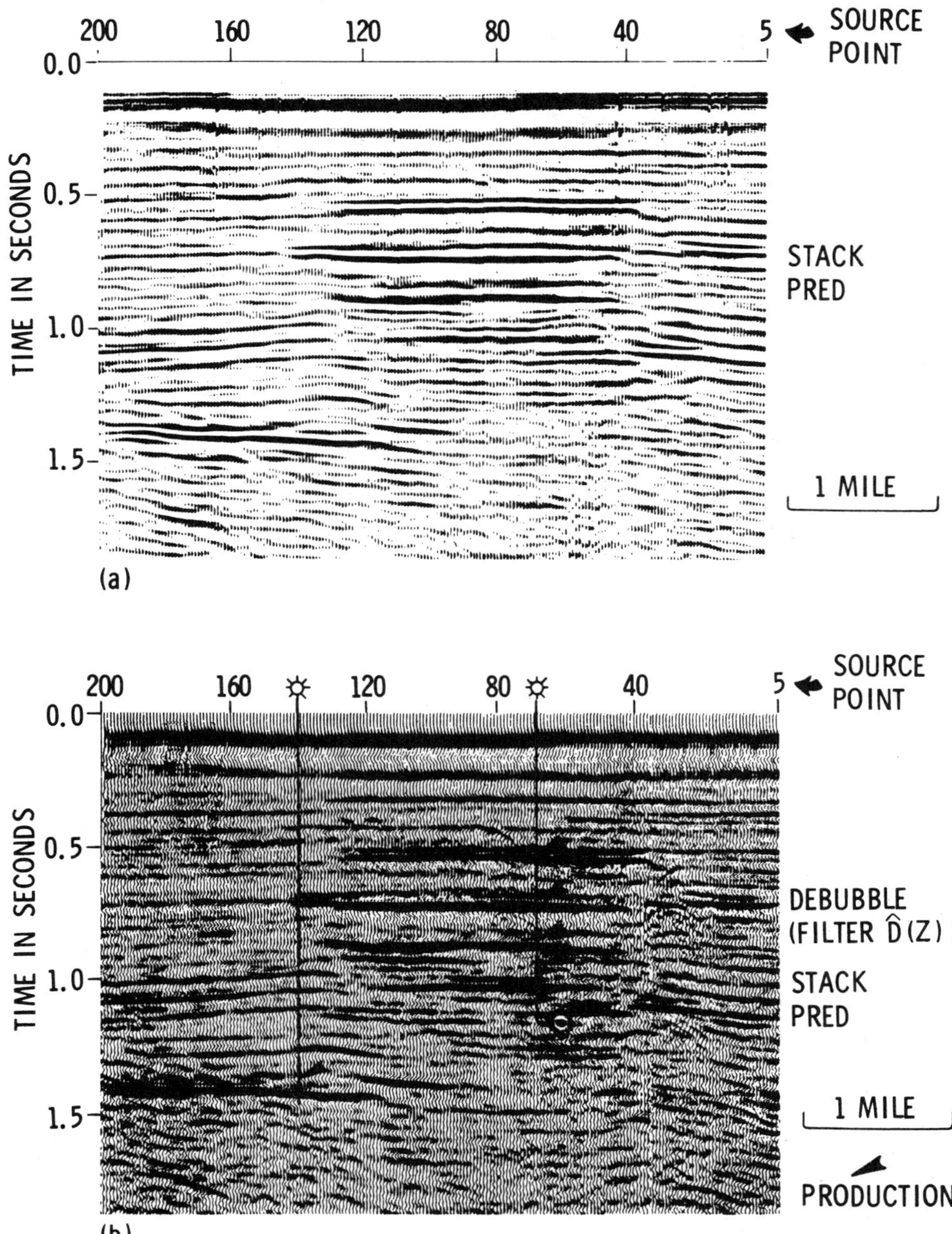

FIG. 9. Predictive deconvolution applied to the data shown in Figure 8.

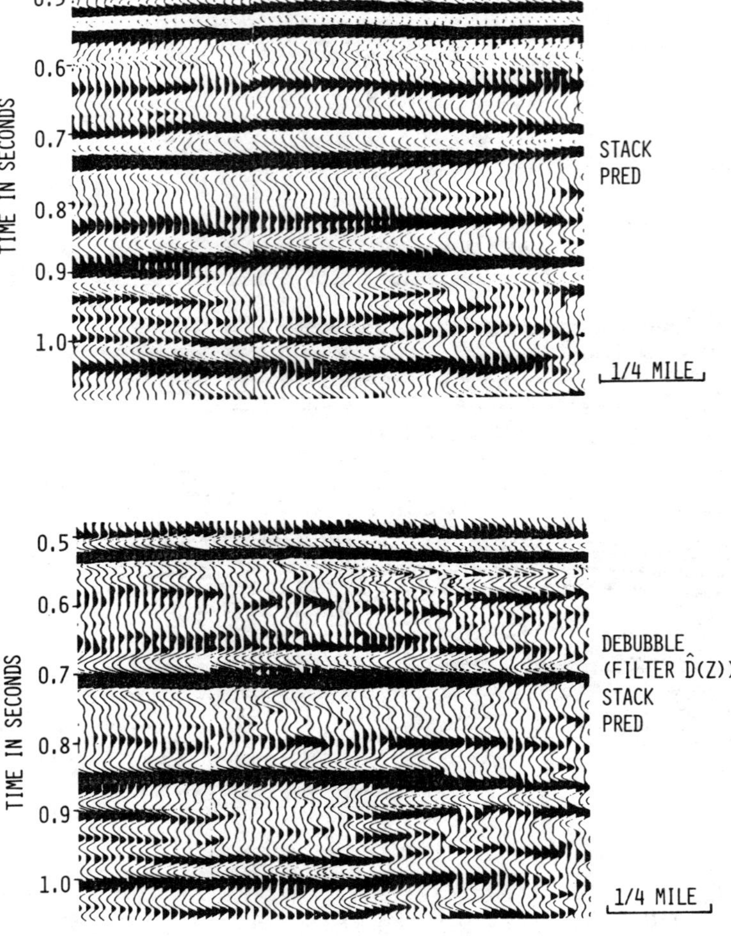

FIG. 10. Enlarged areas of the data shown in Figure 9.

two categories. There are those benefits accruing from the matched filtering (i.e., crosscorrelation) step and those relating to zero-phase filters.

The experiments of first crosscorrelating the source signature with the data and inverse filtering the autocorrelation of the source pulse to a zero-phase pulse of a small time duration using Maxipulse data led us to the following conclusions:

(1) Crosscorrelation helps to align all common-source records at a common reference time. Since each source pulse is recorded on a specified channel for reference, and this particular channel contains the source signature's autocorrelation function after the initial correlation step, it becomes a simple matter to scan these various reference channels for maximum amplitude values (i.e., zero-lag values).

These zero-lag determinations provide relative time differentials between common-source records with which to synchronize source initiations.

(2) Crosscorrelation may often compress signals by a significant factor (Geyer, 1969). In the case for Maxipulse, the signature's autocorrelation function has minor lobes that are of smaller amplitude and decay more rapidly than the corresponding bubble oscillations on the recorded signature itself.

(3) By crosscorrelating common-source records with specified signatures, the source wavelet present in the data after correlation should be the signature's autocorrelation and, therefore, have a known phase characteristic (i.e., zero-phase).

(4) Crosscorrelation automatically corrects for

phase distortions caused by recording instruments. The source signature, however, must itself be recorded with the same instrumentation as the reflection data. This statement follows from the zero-phase property of autocorrelation functions.

(5) Crosscorrelation achieves improved signal-to-noise ratios prior to designing Wiener inverse filters and compressing pulses. Matched filtering attenuates random noise and improves signal-to-noise ratios by a factor proportional to the ratio of signal power to noise power.

(The five points described above involve the crosscorrelation step, and therefore should be advantages unique to our debubbling procedure.)

(6) The zero-phase Wiener filters do not alter the phase spectrum of the correlated data. Our pulse compression method alters only amplitude spectra of the correlated data and does not affect phase spectra, thereby reducing signal distortion.

(7) A valid comparison of Maxipulse, Aquapulse, and air-gun sources cannot be made until more source-signature data have been recorded for each type of initiator, and the corresponding records subjected to similar debubbling procedures in areas with good geologic control.

We have described two new methods for debubbling seismic data when corresponding source signatures can be specified. We have also described some important advantages associated with match filtering and zero-phase filters. Our methods, however, are only as good as our knowledge of far-field source signatures, and this fact points out the need for reliable methods for measuring such signatures and/or for estimating wavelets in data processing so as to provide the geophysicist with the most reliable data for interpretation.

CONCLUSIONS

This paper describes simple and effective procedures for shaping and compressing source signatures to zero-phase pulses of small time duration. We have presented the mathematical relationships between both schemes. Source signatures, in the far-field, are neither minimum-phase nor zero-phase but, rather, mixed-phase, and we have observed in practice that many signatures are quite symmetrical, having phase response curves more closely approximating a zero-phase rather than a minimum-phase condition. It is a difficult problem to shape mixed-delay wavelets to minimum-delay wavelets whose

maximum amplitudes occur near the wavelet onset.

The examples just shown suggest that the performance of both of our new debubbling filters is about equal. However, the filter $\hat{D}(z)$ is somewhat easier to calculate in practice, and this method, therefore, constitutes our current preferred solution to the debubbling problem. In the limit of large filter lengths, state-of-the-art methods such as that of Mateker (1971) as well as the two proposed here produce similar estimates of the ideal bubble inverse, $1/X_n(z)$. In practice, however, the matched filter stages of both of our approaches enable us to solve systems of normal equations which are of much lower order than would otherwise be the case. Of course, additional experience needs to be gained with both appraoches in order to establish more accurate performance criteria. For example, a more detailed study of the effect of the filter length parameters k and p in the expression for $D(z)$ and $\hat{D}(z)$ [see equations (7) and (8)] is clearly desirable. In any event, it is gratifying to note that, in this instance at least, both roads have apparently led to Rome.

ACKNOWLEDGMENTS

The authors wish to thank Amoco Production Co. for permission to publish this study. Appreciation is given also to unknown referees for many constructive suggestions.

REFERENCES

Geyer, R. L., 1969, The Vibroseis system of seismic mapping: Seismograph Service Corp. monograph, 20 p.

Giles, B. F., 1968, Pneumatic acoustic energy source: Geophys. Prosp. v. 16, p. 21–53.

Kologinczak, J., 1974, Stagaray system improves primary pulse/bubble ratio in marine exploration: Offshore Technology Conference Paper No. OTC 2020, p. 801–808.

Kramer, F. S.; Peterson, R. A., and Walter, W. C., (editors), 1968, Seismic energy sources 1968 handbook, United Geophysical Corp. 57 p.

Mateker, E. J., 1971, Big benefits seen with Maxipulse system use: Oil and Gas Journal v. 69, no. 7, p. 116–120.

Peacock, K. L., and Treitel, S., 1969, Predictive deconvolution: theory and practice: Geophysics v. 34, p. 155–169.

Robinson, E. A. and Treitel, S., 1965, Dispersive digital filters: Rev. of Geophys. v. 3, p. 433–461.

Smith, S. G., 1975, Research note measurement of airgun waveforms: Geophys. J. Roy. Astr. Soc., v. 42, p. 273–280.

Treitel, S. and Robinson, E. A., 1966, The design of high-resolution digital filters: IEEE Trans. on Geosc. Electr., v. GE-4, p. 25–38.

Treitel, S. and Robinson, E. A., 1969, Optimum digital filters for signal to noise ratio enhancement: Geophys. Prosp. v. 17, p. 248–293.

Wielandt, E., 1975, Generation of seismic waves by underwater explosions: Geophys. J. Roy. Astr. Soc., v. 40, p. 421–439.

APPENDIX

THEORETICAL ANALYSIS OF DEBUBBLING PROCEDURES

Consider a known input signal x_t of length $(n + 1)$,

$$x_t = (x_0, x_1, \ldots, x_n),$$

with z-transform

$$X_n(z) = x_0 + x_1 z + \cdots + x_n z^n, \quad x_0 \neq 0,$$

where $t = 0, 1, 2, \ldots$, is the discrete time variable. Excluding the case of zeroes on the unit circle $|z| = 1$, $X_n(z)$ can be factored in the form

$$X_n(z) = P_m(z)Q_{n-m}(z), \quad m \leq n,$$

where

$$P_m(z) = p_0 + p_1 z + \cdots + p_m z^m, \quad P_0 \neq 0,$$

is the z-transform of the $(m + 1)$-length minimum-delay sequence $(p_0, p_1, \ldots, p_m)$, and where

$$Q_{n-m}(z) = q_0 + q_1 z + \cdots + q_{n-m} z^{n-m}, \quad q_0 \neq 0,$$

is the z-transform of the $(n - m + 1)$-length maximum-delay sequence $(q_0, q_1, \ldots, q_{n-m})$. We may take $m \leq n$ without loss of generality. Thus, if $m = n$, $Q_0(z) = q_0$ and $X_n(z) = q_0 P_n(z)$ is pure minimum delay. Alternatively, if $m = 0$, $P_0(z) = p_0$ and $X_n(z) = p_0 Q_n(z)$ is pure maximum delay. In general, therefore,

$$X_n(z) = P_m(z)Q_{n-m}(z)$$

is the z-transform of a known mixed-delay input signal.

Note that since $Q_{n-m}(z)$ is a maximum-delay polynomial, the reverse polynomial

$$z^{n-m}Q_{n-m}(z^{-1})$$

is a pure minimum-delay polynomial.

Let us now write $X_n(z)$ in the form,

$$X_n(z) = P_m(z)z^{n-m}Q_{n-m}(z^{-1}) \frac{Q_{n-m}(z)}{z^{n-m}Q_{n-m}(z^{-1})},$$

or

$$X_n(z) = S_n(z)A_{n-m}(z), \quad (A\text{--}1)$$

where

$$S_n(z) = P_m(z)z^{n-m}Q_{n-m}(z^{-1})$$

is pure minimum delay, and where

$$A_{n-m}(z) = \frac{Q_{n-m}(z)}{z^{n-m}Q_{n-m}(z^{-1})} \quad (A\text{--}2)$$

is a type 1 all-pass, or dispersive filter (Robinson and Treitel, 1965). In other words, a given $(n + 1)$-length input sequence x_t with z-transform $X_n(z)$ can be represented as the cascading of an $(n + 1)$-length minimum-delay filter $S_n(z)$ with an all-pass filter $A_{n-m}(z)$, whose numerator and denominator are both of degree $(n - m)$.

Let us compute the z-transform $R(z)$ of the autocorrelation of x_t,

$$R(z) = X_n(z)X_n(z^{-1})$$

$$= S_n(z)S_n(z^{-1})A_{n-m}(z)A_{n-m}(z^{-1}).$$

From (A–2), one easily verifies that

$$A_{n-m}(z)A_{n-m}(z^{-1}) = 1 \quad (A\text{--}3)$$

and, therefore,

$$R(z) = X_n(z)X_n(z^{-1}) = S_n(z)S_n(z^{-1}),$$

so that x_t and s_t have the same autocorrelation function. Consider now the calculation of the zero-delay Wiener spiking filter for the input wavelets x_t and s_t. The desired output d_t is, in both cases,

$$d_t = (1, 0, \ldots, 0).$$

But both x_t and s_t have the same autocorrelation function, so that the normal equations for either the input x_t or the input s_t can be written

$$\begin{bmatrix} r_0 \cdots r_k \\ \cdot \quad\quad \cdot \\ \cdot \quad\quad \cdot \\ \cdot \quad\quad \cdot \\ \cdot \quad\quad \cdot \\ r_k \cdots r_0 \end{bmatrix} \begin{bmatrix} b_0 \\ b_1 \\ \cdot \\ \cdot \\ \cdot \\ b_k \end{bmatrix} = \begin{bmatrix} v \\ 0 \\ \cdot \\ \cdot \\ \cdot \\ 0 \end{bmatrix},$$

where $b_t = (b_0, b_1, \ldots, b_k)$ is the $(k + 1)$-length zero-delay Wiener spiking filter, and where v is a constant for each value of k. The normal equations for the input s_t differ from those of input x_t by only a constant scale factor, which we shall neglect in what follows.

Let $B_k(z)$ be the z-transform of b_t, and let us apply it to $X_n(z)$:

$$B_k(z)X_n(z) = B_k(z)S_n(z)A_{n-m}(z).$$

Now for sufficiently large k,

$$B_k(z)S_n(z) \sim 1 \text{ and hence } X_n(z)B_k(z) \sim A_{n-m}(z).$$

This implies that

$$X_n(z)X_n(z^{-1})B_k(z)B_k(z^{-1})$$

$$\sim A_{n-m}(z)A_{n-m}(z^{-1}) = 1, \qquad (A\text{-}4)$$

where we have made use of the fact that the auto-correlation of an all-pass response is the unit spike [cp. equation (A–3)].

Let

$$D(z) = X_n(z^{-1})B_k(z)B_k(z^{-1}), \qquad (A\text{-}5)$$

so that

$$X_n(z)D(z) \sim 1.$$

We now see that the filter $D(z)$ spikes the mixed-delay input signal x_t, and that the performance of this spiking filter improves monotonically as k increases.

Next consider the debubbling filter

$$\hat{D}(z) = X_n(z^{-1})G_p(z),$$

where $G_p(z)$ is the z-transform of the approximate optimum-lag Wiener inverse filter for $X_n(z)X_n(z^{-1})$. Since by equation (A–4),

$$X_n(z)X_n(z^{-1})B_k(z)B_k(z^{-1}) \sim 1,$$

and because we also have

$$X_n(z)X_n(z^{-1})G_p(z) \sim 1,$$

we see that $B_k(z)B_k(z^{-1})$ and $G_p(z)$ will approach equality as k and p increase. In other words, the lagged Wiener inverse filter for the autocorrelation of a wavelet is equal to the autocorrelation of the zero-delay Wiener inverse filter for the same wavelet. It can thus be seen that the filters $D(z)$ and $\hat{D}(z)$ will become increasingly similar as the filter lengths k and p become larger.

The normal equations for the near optimum-lag Wiener inverse filter $G_p(z)$ will now be displayed. Let $r_t = (r_n, r_{n-1}, \ldots, r_1, r_0, r_1 \ldots, r_{n-1}, r_n)$ be the autocorrelation of x_t. Also, let $u_t = (u_{2n}, u_{2n-1}, \ldots, u_1, u_0, u_1, \ldots, u_{2n-1}, u_{2n})$ be the autocorrelation of r_t. Then the normal equations for $G_p(z)$ become

$$
\begin{bmatrix}
u_0\, u_1 & \cdots u_p \\
u_1\, u_0\, u_1 \cdots u_{p-1} \\
\cdot & \cdot \\
\cdot & \cdot \\
\cdot & \cdot \\
\cdot & \cdot \\
\cdot & \cdot \\
\cdot & \cdot \\
\cdot & \cdot \\
u_p & u_0
\end{bmatrix}
\begin{bmatrix}
g_0 \\
g_1 \\
\cdot \\
\cdot \\
g_{q-1} \\
g_q \\
g_{q+1} \\
\cdot \\
\cdot \\
g_{p-1} \\
g_p
\end{bmatrix}
=
\begin{bmatrix}
r_q \\
r_{q-1} \\
\cdot \\
\cdot \\
r_1 \\
r_0 \\
r_1 \\
\cdot \\
\cdot \\
r_{q-1} \\
r_q
\end{bmatrix}
$$

where $q = (p - 1)/2$ and p is usually taken to be an odd integer. If $q > n$, use zeros to fill locations where r_t and u_t are not defined.

6

A GEOPHYSICAL APPLICATION: DECONVOLUTION
OF SEISMIC DATA

V. K. Arya and H. D. Holden
Shell Development Corporation

1. INTRODUCTION

The impact of the digital computer on the processing of seismic data can
hardly be overemphasized. In a typical year domestic oil companies
acquire in excess of 300,000 line miles of data which constitutes
trillions of bits of information. The ability to handle a great volume
of data, along with the benefits gained by a dramatic increase in
dynamic range (greater than 70 db) make the transformation from analog
to digital processing the most significant technological advance in
geophysical prospecting. Subsequently many powerful digital signal
processing techniques became available to the seismic data analyst.
One such technique, deconvolution, is the topic of this paper. The
paper is necessarily limited in content due to the proprietory nature
of some of the material.

In seismic exploration of the earth's subsurface, an acoustic energy
waveform -- a seismic wavelet in geophysical terms -- is emitted and
then the reflections from changes in the velocity or density, with
depth of the rock packages in the subsurface are detected using
pressure or velocity sensitive detectors. If we assume, for now, that
the shape of the seismic wavelet remains constant along its propagation
path, then the reflected signal will be a superposition of delayed
wavelets. Each wavelet will be scaled according to the reflectivity
it encountered and also to the degree of divergence the wavefront has
undergone. Thus, we can represent the seismic recording (trace) $S(t)$
(Fig. 1) by

$$S(t) = \sum_i r(i)W(t-\tau_i) \tag{1.1}$$

where $W(t)$ is the wavelet, $r(i)$ are the scale factors and τ_i are the
delays. If we take reflectivity to be a continuous function of delay
time, then we have the widely used convolution model for a seismic
trace

$$S(t) = \int_0^\infty R(\tau)W(t-\tau)d\tau . \tag{1.2}$$

94

The goal of a deconvolution scheme is to remove the effect of the wavelet
from the seismic trace and then correct for wavefront divergence, thus
retrieving the earth's reflectivity function.

The modern geophysicist wants to know $r(i)$ or $R(\imath)$ with as much precision
and resolution as possible for a variety of reasons. For example, large
reflections may indicate sand layers which are potential hydrocarbon
reservoirs. The estimated thickness of a possibly hydrocarbon-bearing
sand layer strongly influences the amount of money one can afford to bid
for drilling rights on a given tract at a competitive offshore lease sale.

In this paper we wish to survey four different deconvolution schemes that
are being currently used in the industry for deconvolution of seismic
data. Section 2 describes the historical development of the predictive
deconvolution technique. Section 3 outlines homomorphic filtering, and
Section 4 describes Kalman filtering as applied to deconvolution. The
most direct and conceptually the simplest technique for the deconvolution
of seismic data is to measure the seismic wavelet and then remove its
effects. This technique is described in Section 5. Finally, Section 6
briefly discusses the above techniques. In representing the seismic
trace, reflectivity and wavelet, we will use upper-case letters for the
continuous domain and will use lower-case letters for the discrete
domain.

2. PREDICTIVE DECONVOLUTION

In 1952, the M.I.T. Geophysical Anlaysis Group [1],[2] began to apply
prediction techniques, developed by Norbert Wiener, to the deconvolution
of seismic data. Basically, the idea is that, if we consider the seismic
trace to be from a stationary random process, the predictable part of the
trace is due to the wavelet while the earth's reflectivity is completely
unpredictable (i.e., a white random process). Conceptually, a
statistically optimal prediction operator is derived, and at each point
the predicted value is subtracted from the actual data value. This is
shown schematically in Fig. 2. In Fig. 2a, a Wiener filter is used to
derive a prediction operator $a(n)$ which optimally predicts the value of
the seismic trace at the later time $m = n+\alpha$

$$\hat{s}(m) = \sum_n s(n) a(m-n). \tag{2.1}$$

Fig. 2b shows first the application of the prediction filter and then
the subtraction which yields the prediction error. The prediction
error, or unpredictable part of the seismic trace, is an estimate of the
reflectivity function

$$\hat{r}(n) = s(n) - \hat{s}(n). \tag{2.2}$$

In Fig. 2c it is seen that the prediction error filter can be formulated
as a single step operator. Indeed, one can show [3] that if the
prediction distance is one sample, then the prediction error filter is
equivalent to the optimal, zero-lag inverse filter. A zero-lag inverse
filter is most appropriate if the seismic wavelet is minimum phase which
is often a fair approximation. Predictive deconvolution has been a very
useful tool for several years. It becomes ineffective, however, when
its underlying assumptions -- the wavelet is minimum phase and the
reflectivity function is completely random -- are violated.

3. HOMOMORPHIC DECONVOLUTION

Homomorphic systems [4]-[8] are a class of nonlinear systems which
satisfy a generalized principle of superposition. Such systems are
particularly useful in separating signals which have been combined
through convolution. As discussed in Section 1, a seismic trace is
often represented as the convolution of a wavelet with the reflectivity
function (Eqn. 1.2). Homomorphic filtering can thus be used effectively
to recover the seismic wavelet $W(t)$ from the seismic trace $S(t)$. The
theoretical basis of homomorphic filtering is given by Oppenheim [4];
we will present only a summary of relevant points [7].

Consider the transformation defined by

$$y = T(x). \tag{3.1}$$

If T is a linear system, it satisfies the superposition relation defined
by

$$T(ax_1 + bx_2) = aT(x_1) + bT(x_2) \tag{3.2}$$

where a and b are constants. If we wish to generalize the notion
expressed by Equation (3.2) to a signal resulting from the convolution
of components $x = x_1 * x_2$, we look for a system with transformation H
such that

$$H(ax_1 * bx_2) = aH(x_1) * bH(x_2) \tag{3.3}$$

where a and b are constants.

Oppenheim et al.[5] have shown that a system of the class represented by
Equation (3.3) has the canonic representation shown in Figure 3a. The
system D, referred to as the characteristic system, is a homomorphic
system with transformation from a convolutional space to an additive
space. It is defined by the relation

$$D(ax_1 * bx_2) = aD(x_1) + bD(x_2). \tag{3.4}$$

The system L in Figure 3a is a linear system and the system D^{-1}, the
inverse of D, performs the transformation from an additive space back

to the convolutional space. Figures 3b and 3c show the canonic
representation of characteristic systems D and D^{-1} respectively.

It is clear from Figure 3 that the complex cepstrum, $\tilde{s}(n)$, contains
the additive contributions of the wavelet and the reflection
coefficients. These contributions may be separated by low pass and
high pass filters. Low pass filtering will lead to the complex cepstrum
of the wavelet, $\tilde{w}(n)$, whereas high pass filtering will result in the
complex cepstrum of the reflection coefficients, $\tilde{r}(n)$. The inverse
system D^{-1} performs the transformation from an additive space back to
the convolutional space and results in the output $w(n)$ or $r(n)$ as
desired.

Otis and Smith [9] discuss the improvement in recovering the wavelet by
averaging the log spectra of several reflection records. This is based
on the assumption that the source function $w(n)$ is considered stationary
and the earth's response, $r(n)$, spatially nonstationary. By averaging
the log spectra of several reflection records, the log spectrum of the
wavelet will be enhanced and the log spectrum of the reflection
coefficients will average out.

The important considerations in the computation of the complex cepstrum
are discussed in detail by Shafer [6].

<u>Example</u>

Stoffa et al.[8] describe the application of homomorphic deconvolution
to shallow-water marine seismology. This example is taken from their
paper. Figure 4a shows a mixed phase reflector series in which the
separation between the shot and the first reflector is greater than
18 milliseconds. Figure 4b shows a non-minimum phase wavelet,
Figure 4c shows the convolution of the reflector series with the wavelet
(weighted exponentially to make the reflector series, but not the trace,
minimum phase). The complex cepstrum is shown in Figure 4d. Figure 4e
shows the deconvolution which results from setting all complex cepstrum
contributions for -18 ms < T < 18 ms equal to zero (linear operator in
Figure 3a). The importance of this example is that it shows a good
deconvolution result when the wavelet was not minimum phase. Predictive
deconvolution would be ineffective in this example because the minimum
phase assumption is violated. Treitel and Robinson [10] have addressed
this question relating to differences in homomorphic and predictive
error filtering. They conclude that the method of homomorphic
deconvolution does not automatically yield the "correct" mixed-delay
source wavelet estimate. Two equivalent problems -- the determination
of the "correct" all-pass filter to convert the mixed-delay wavelet to
minimum phase wavelet for predictive deconvolution, and the determination
of the "correct" low-pass cutoff quefrencies (linear system L) in the

case of homomorphic deconvolution -- still await solution.

4. KALMAN FILTERING

Bayless and Brigham [11], Ott and Meder [12], and Crump [13] have
introduced a Kalman filtering approach for the deconvolution of seismic
signals. The motivation for this alternate approach has been that it
has the potential for permitting more flexible modeling assumptions
than the Wiener filtering approach. The Wiener filtering method is
conventionally applied under the assumption of a time invariant system
and stationary statistics; the Kalman filtering technique is suitable
for handling time varying models. Figure 5 shows a state-space
representation of a seismic trace. The deconvolution problem is
analogous to estimating r(n), the reflection sequence, from the seismic
trace.

Refering to Figure 5, we can describe the state space representation
for the convolution model of a seismic trace s(n) as follows:

$$\underline{x}(n+1) = \Phi \underline{x}(n) + \Gamma r(n) \tag{4.1}$$

$$y(n) = H\underline{x}(n) \tag{4.2}$$

$$s(n) = y(n) + v(n) \tag{4.3}$$

where r(n) is an input to the linear system with a state vector $\underline{x}$(n) and
v(n) is the measurement noise. r(n) and v(n) are uncorrelated zero mean
white sequences for which $E[r(k)r'(\ell)] = Q$ and $E[v(k)v'(\ell)] = R$, where
$E[r]$ represents the expected value of the random variable r. Φ, Γ and
H are matrices of proper dimensions and they can be either determined by
direct calculations from the known wavelet or can be identified from the
seismic trace s(n). The problem of estimating the white reflection
coefficient sequence r(n) is equivalent to the problem of estimating the
random disturbance in a state equation. Mendel [14] has pointed out
that this is a nonstandard problem in estimation theory which had led to
some difficulties in trying to do predictive deconvolution via Kalman
filtering. The difficulties are due to the facts that we are interested
in an optimal estimate of sequence r(n) but Kalman filter obtains
estimates of the state vector $\underline{x}$(n) and not r(n).

Let $\underline{s}(j) = [s(1),s(2),...,s(j)]'$ denote the measurement of seismic trace
and $\hat{\underline{x}}(n|j)$ denote the minimum variance estimate of $\underline{x}$(n) which uses all
the measurements in $\underline{s}(j)$. It is well known [15] that

$$\hat{\underline{x}}(n+1|n) = E[\underline{x}(n+1)|\underline{s}(n)] \tag{4.4}$$

$$\hat{\underline{x}}(n+1|n+1) = E[\underline{x}(n+1)|\underline{s}(n+1)] \tag{4.5}$$

and that $\hat{\underline{x}}(n+1|n)$ and $\hat{\underline{x}}(n+1|n+1)$ can be computed from the Kalman filter whose equations are:

$$\hat{\underline{x}}(n+1|n) = \Phi\hat{\underline{x}}(n|n) \tag{4.6}$$

$$P(n+1|n) = \Phi P(n|n)\Phi' + \Gamma Q \Gamma' \tag{4.7}$$

$$K(n+1) = P(n+1|n)H'[HP(n+1|n)H' + R]^{-1} \tag{4.8}$$

$$\tilde{s}(n+1|n) = s(n+1) - H\hat{\underline{x}}(n+1|n) \tag{4.9}$$

$$\hat{\underline{x}}(n+1|n+1) = \hat{\underline{x}}(n+1|n) + K(n+1)\tilde{s}(n+1|n) \tag{4.10}$$

$$P(n+1|n+1) = [I - K(n+1)H]\,P(n+1|n) \tag{4.11}$$

where $P(\cdot|\cdot)$ is an error covariance matrix for the estimate $\hat{x}(\cdot|\cdot)$ and $K(\cdot)$ is a time varying gain for the estimator of Equation (4.10). Mendel [14] points out that the first meaningful minimum variance linear estimate of $r(n)$ can be obtained as

$$\Gamma\hat{r}(n|n+1) = \hat{\underline{x}}(n+1|n+1) - \Phi\hat{\underline{x}}(n|n+1). \tag{4.12}$$

It should be noted that $\hat{r}(n|n+1)$ is a single-stage smoothed estimate of $r(n)$ which requires not only the optimal filtered estimate of $\underline{x}(n+1)$, but also the optimal single-stage smoothed estimate of $\underline{x}(n)$ for its implementation. Thus he concludes that the minimum variance estimates of the reflection coefficient $r(n)$ are optimal smoothed estimates and that an optimal filtered estimate of that sequence does not exist.

So far, there has not been any reported application of Kalman filtering for seismic deconvolution in which a time varying model of the seismic wavelet is considered or where more flexible modeling assumptions than those used in Wiener filtering are used to advantage.

5. DETERMINISTIC DECONVOLUTION

The most direct and conceptually the simplest technique for the deconvolution of seismic data is to first make a direct measurement of the outgoing wavelet and then remove it from the seismic data. Then, remaining effects such as earth absorption and receiver ghosting can be modeled and compensated for. These two steps comprise the technique of deterministic deconvolution. This technique has proved to be more effective on marine data than on land data for several reasons. The source wavelet is very repeatable for airgun arrays, and it can be accurately measured in the homogeneous water medium. The ghost from the water-air interface, while still a serious problem, is at least relatively constant along a seismic line.

One method of measuring the outgoing wavelet is to tow a deep hydrophone as data is being acquired (see Fig. 6). The water must be rather deep so that the reflection from the sea floor does not corrupt the wavelet measurements. Fortunately, the wavelet is quite repeatable; one can obtain an average outgoing wavelet in deep water and -- so long as the shooting parameters are not changed -- use that wavelet to derive a filter which will deconvolve data acquired in water of any depth.

If the wavelets are measured, as described earlier, by a deep towed hydrophone, then they lack one essential ingredient contained in the seismic wavelet. The measured wavelets do not contain the receiver ghosting operator; this operator must be either convolved with the wavelets prior to filter derivation or deconvolved from the filter after derivation. A simple model for the ghosting operator is

$$g(t) = \delta(t) - \beta\delta(t-\tau) \tag{5.1}$$

where β denotes the water-air reflection coefficient and τ denotes the ghost delay time. In the frequency domain the ghosting operator becomes

$$G(\omega) = (1 - \beta e^{-j\omega\tau}) \tag{5.2}$$

so that one must either simulate measured wavelets by

$$W_i'(\omega) = W_i(\omega) \cdot (1 - \beta e^{-j\omega\tau}) \tag{5.3}$$

or deconvolve the ghost from the filter

$$F'(\omega) = F(\omega)/(1 - \beta e^{-j\omega\tau}). \tag{5.4}$$

This compensation to the deconvolution filter requires that the depth of the streamer cable be known and relatively uniform.

A. Absorption Effects

After the effects of the source pulse, the source and receiver ghosts, and the recording filters have been deconvolved from the seismic data, the effect of the earth itself on the seismic wavelet must be accounted for. Inelastic absorption of seismic energy in the earth has been studied for many years and remains a viable research topic today. It has been determined experimentally [16] that the amplitude spectrum of a seismic wavelet as it travels through homogeneous earth material systematically loses high frequencies approximately according to $e^{-\gamma f r}$. That is, the amplitude spectrum of the wavelet at distance r_2 from the source is related to the spectrum at r_1 by

$$W(f,r_2) = W(f,r_1)\, e^{-\gamma f(r_2-r_1)} \tag{5.5}$$

A synthetic example illustrating the above effect of absorption in the time-space domain is shown in Figure 7. It is clear that the wavelet is losing its high frequency content and it becomes broader in time.

It is evident from Equation (5.5) that the proper deabsorption filter depends upon the distance that the wavelet has traveled in the earth. If the number of db which the higher frequencies are boosted is to remain relatively constant down the trace, then the high frequency cutoff of the deabsorption filters must steadily decrease. Amplitude spectra of a set of deabsorption filters are shown in Figure 8. Note that for greater distances, the filters have steeper slopes since γr is larger, but the high frequency cutoff decreases.

B. Bandwidth Considerations

In any seismic deconvolution scheme, the purpose is to obtain better resolution so that the reflectivity function of the subsurface can be faithfully generated. To achieve this, we need to increase the "useful" bandwidth of the seismic data. The bandwidth of the data is controlled by many factors -- the source spectrum, ghosting due to source and receiver, recording filter response, absorption, various types of noises, etc. A system's viewpoint must be taken in choosing the shooting geometry and recording filter so as to maximize the "usuable" bandwidth. Recording filters provide various functions such as locut-hicut and alias protection which are very important in the digital recording of seismic data. The sampling rate of seismic data recording and the amount of alias protection varies with particular geophysical objectives. Figure 9a shows the ghosting response for two shooting geometry configurations (30 ft/40 ft and 20 ft/20 ft combination for source and receiver depth). Figure 9b shows the responses of typical anti-alias filters for 2 ms and 3 ms sampling rates. It is clear from Figure 9a,b that the selection of a shooting geometry and anti-alias filters should not be made independently, but rather a systems approach should be taken so that maximum bandwidth is achieved.

6. CONCLUSION

We have described four techniques for the deconvolution of seismic data. The predictive deconvolution technique assumes that the wavelet is minimum phase and that the reflectivity spectrum is white. The success of homomorphic deconvolution depends upon the degree of separability of the wavelet and the reflectivity function in the complex cepstrum domain. The Kalman filtering approach to deconvolution is essentially an extension of Wiener filtering or predictive deconvolution to accomodate time varying processes. The deterministic deconvolution technique does not require assumptions and is based on the direct measurement of the seismic wavelet.

We have also discussed the physical effects governing the bandwidth of the seismic data, which include the shooting geometry, recording filters, and absorption. Application of the deconvolution technique described in this paper lead the geophysicist to a less ambiguous and more meaningful interpretation of the geological structure of the subsurface.

Acknowledgments

We wish to express our appreciation to the management of Shell Development Company for permission to publish this paper (Publication No. BRC 799, Release 401).

References

1. G. P. Wadsworth et al., "Detection of Reflections on Seismic Records by Linear Operators," Geophysics, v. 18, No. 3, pp. 539-586, July 1953.

2. E. A. Robinson, "Predictive Decomposition of Seismic Traces," Geophysics, v. 22, no. 4, pp. 767-778, Oct. 1957.

3. K. L. Peacock and S. Treitel, "Predictive Deconvolution: Theory and Practice," Geophysics, v. 34, no. 2, pp. 155-169, April 1969.

4. A. V. Oppenheim, "Superposition in a Class of Nonlinear Systems," Res. Lab. of Electronics, M.I.T., Technical Report 432, 1965.

5. ______, R. W. Shafer, and T. G. Stockham, "Nonlinear Filtering of Multiplied and Convolved Signals," Proc. IEEE, v. 56, pp. 1264-1291, 1968.

6. R. W. Shafer, "Echo Removal by Discrete Generalized Filtering," Res. Lab. of Electronics, M.I.T., Technical Report 466, 1969.

7. T. J. Ulrych, "Application of Homomorphic Deconvolution to Seismology," Geophysics, v. 36, no. 4, pp. 650-660, Aug. 1971.

8. P. L. Stoffa, P. Buhl, and G. M. Bryan, "The Application of Homomorphic Deconvolution to Shallow-Water Marine Seismology," Geophysics, v. 39, no. 4, pp. 401-416, Aug. 1974.

9. R. M. Otis and R. B. Smith, "Homomorphic Deconvolution by Log Spectral Averaging," Geophysics, v. 42, no. 6, pp. 1146-1157, Oct. 1977.

10. S. Treitel and E. A. Robinson, "Deconvolution - Homomorphic or Predictive?," IEEE Trans. on Geoscience Electronics, v. GE-15, no. 1, pp. 11-13, Jan. 1977.

11. J. W. Bayless and E. O. Brigham, "Application of the Kalman Filter to Continuous Signal Restoration," Geophysics, v. 35, no. 1, pp. 2-23, Feb. 1970.

12. N. Ott and H. G. Meder, "The Kalman Filter as a Prediction Error Filter," Geophysical Prospecting, v. 20, pp. 549-560, 1972.

13. N. Crump, "A Kalman Filter Approach to the Deconvolution of Seismic Signals," Geophysics, v. 39, no. 1, pp. 1-13, Feb. 1974.

14. J. M. Mendel, "White Noise Estimators for Seismic Data Processing in Oil Exploration," IEEE Trans. on Automatic Control, v. AC-22, no. 5, pp. 694-706, Oct. 1977.

15. J. S. Meditch, _Stochastic Optimal Linear Estimation and Control_, McGraw-Hill Co., 1969.

16. F. J. McDonal et al., "Attenuation of Shear and Compressional Waves in Pierre Shale," Geophysics, v. 23, no. 3, pp. 421-439, July 1958.

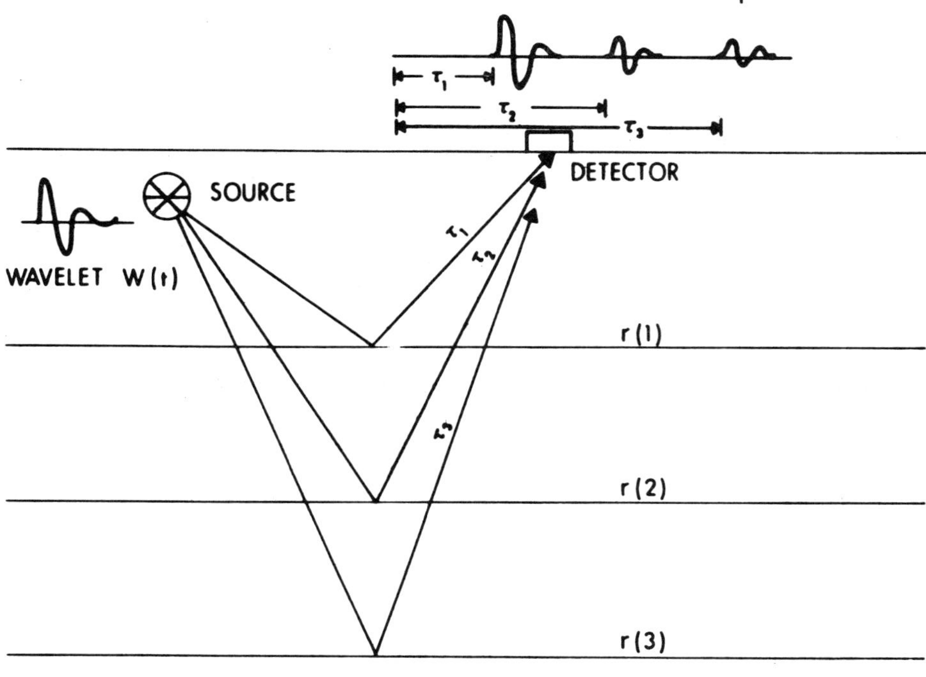

Fig. 1.   Representation of seismic recording.

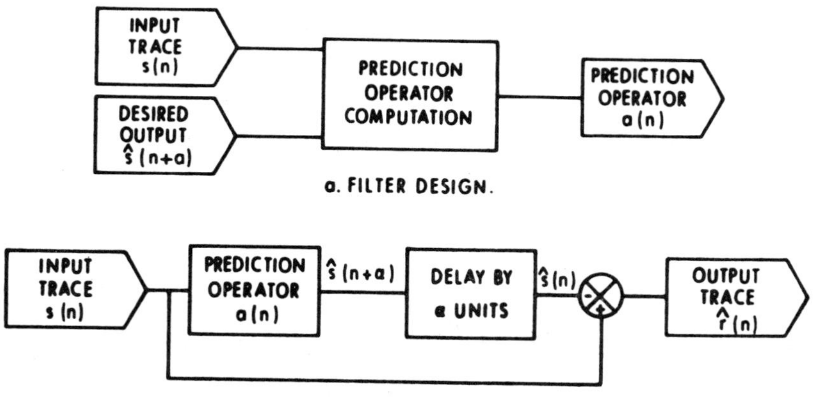

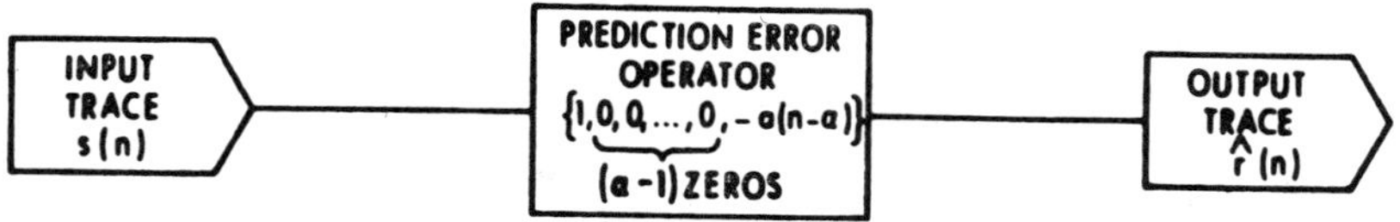

Fig. 2.   Predictive deconvolution.

                    V. K. Arya and H. D. Holden

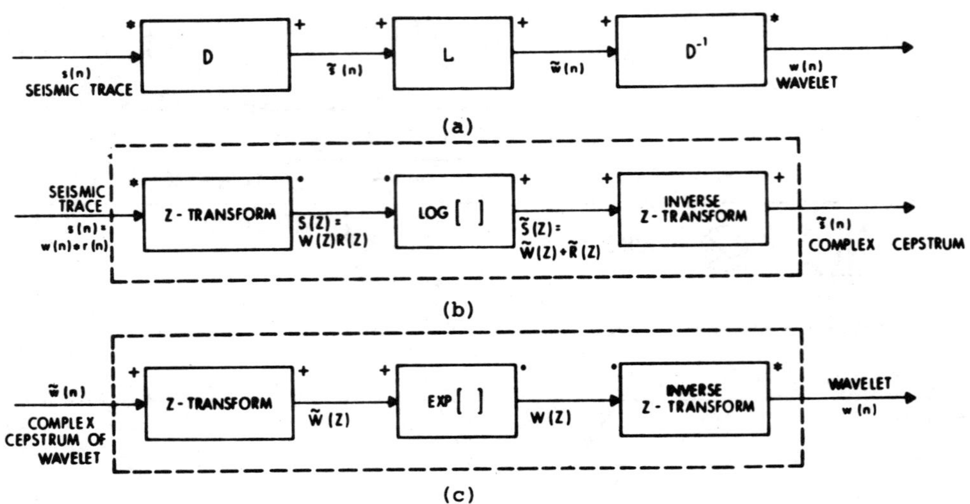

Fig. 3.   (a) Canonic representation of homomorphic system; (b) Canonic representation of D; (c) Canonic representation of D^{-1}.

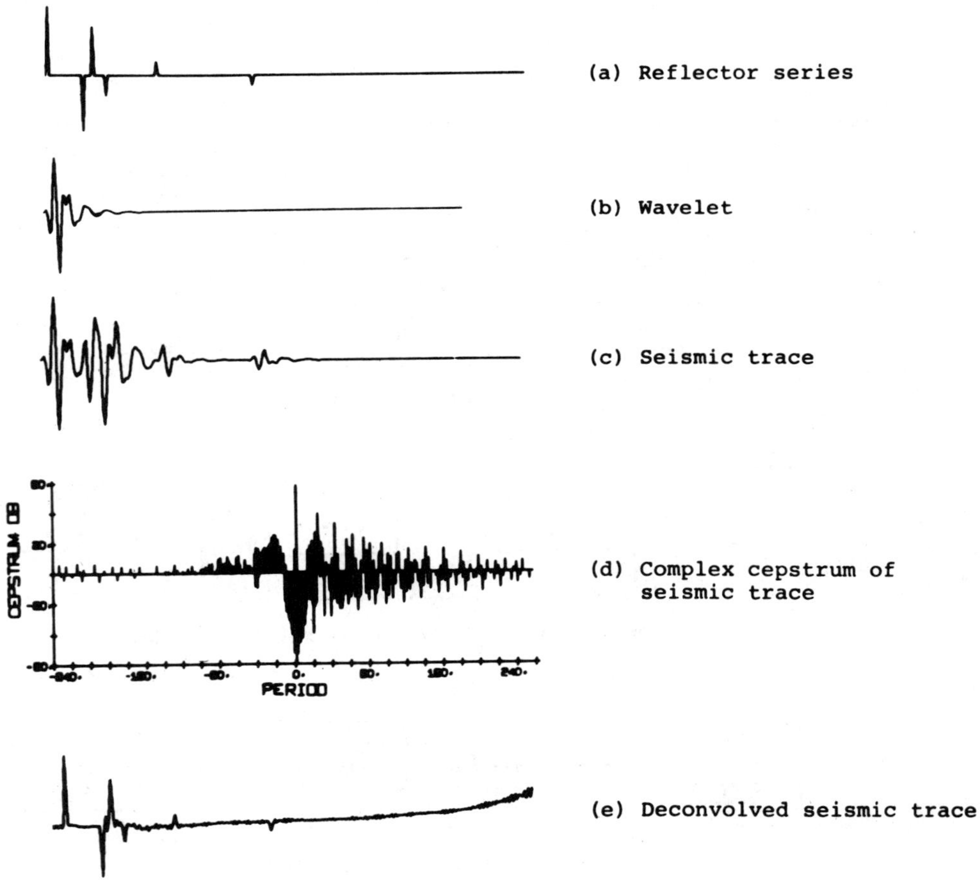

(a) Reflector series

(b) Wavelet

(c) Seismic trace

(d) Complex cepstrum of seismic trace

(e) Deconvolved seismic trace

Fig. 4.   Homomorphic deconvolution example (after Stoffa et al., August 1974, courtesy GEOPHYSICS).

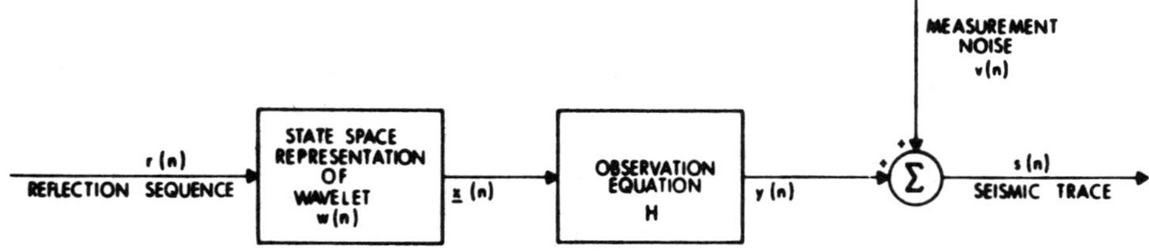

Fig. 5. State space representation of seismic trace.

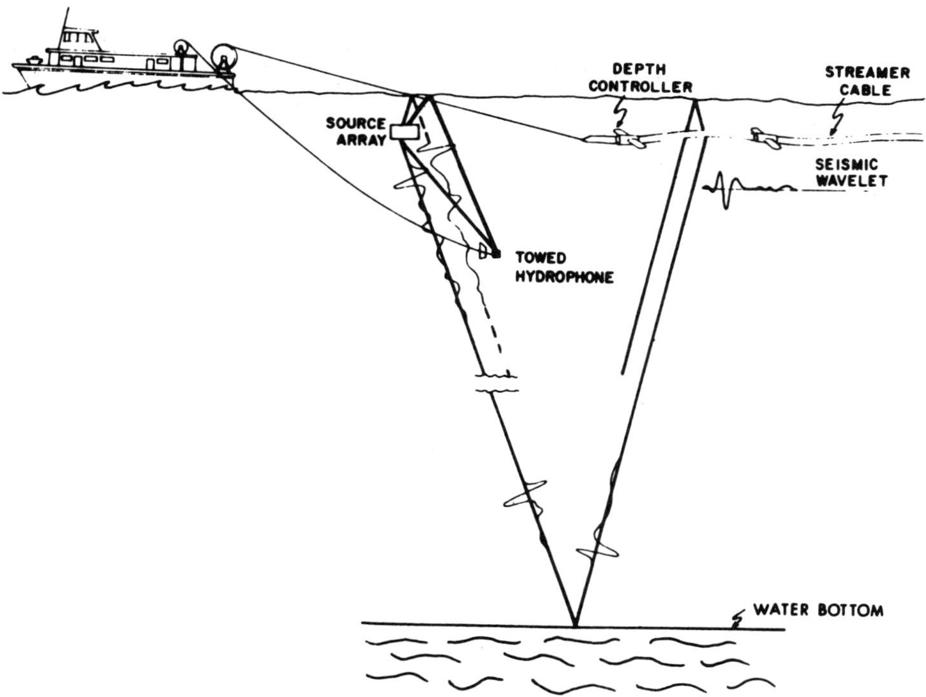

Fig. 6. Schematic of marine seismic data acquisition.

$$W(f, r_2) = W(f, r_1) e^{-\gamma f (r_2 - r_1)}$$

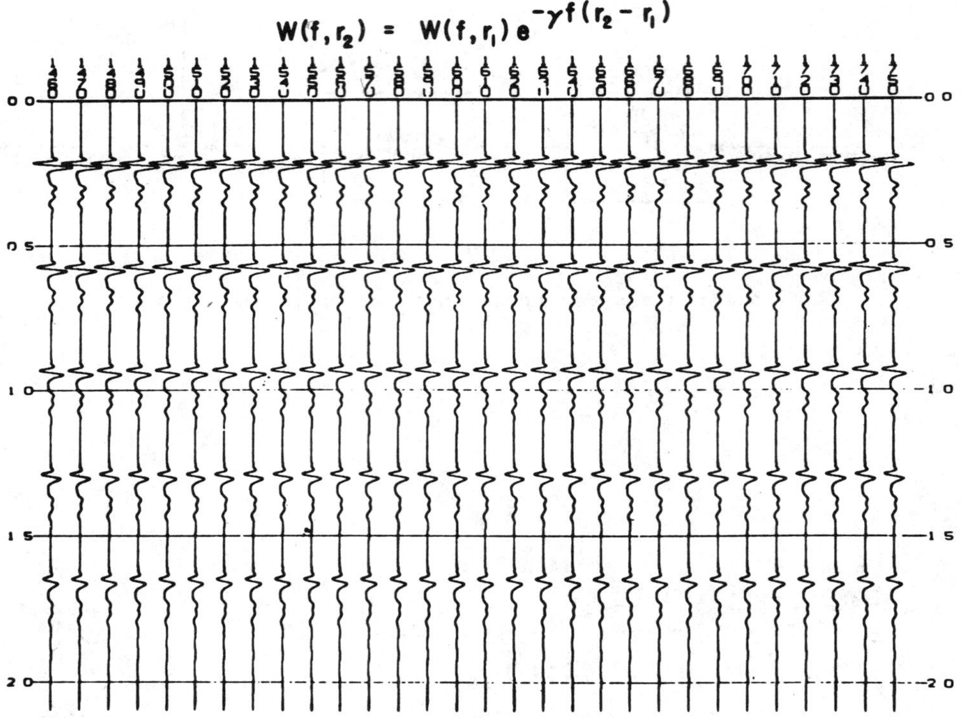

Fig. 7.   Measured wavelets absorbed.

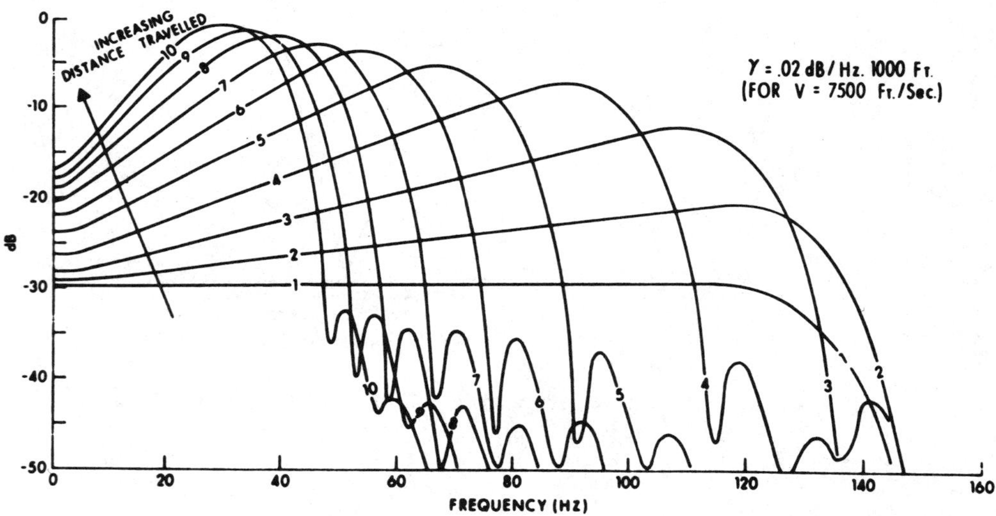

Fig. 8.   Amplitude spectra of deabsorption filters.

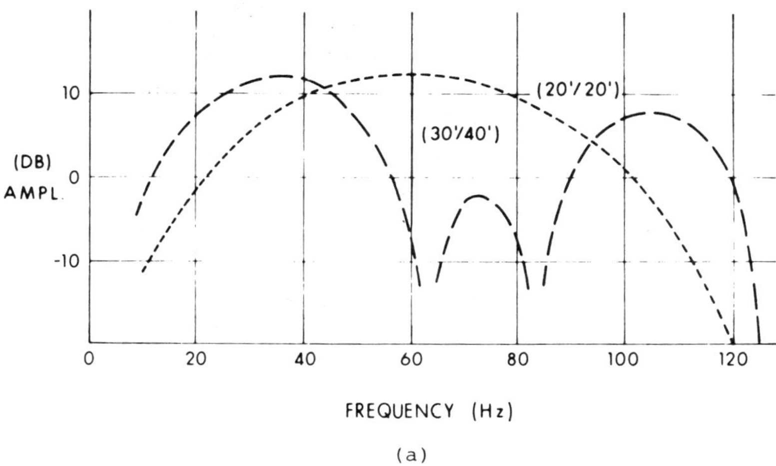

(a)

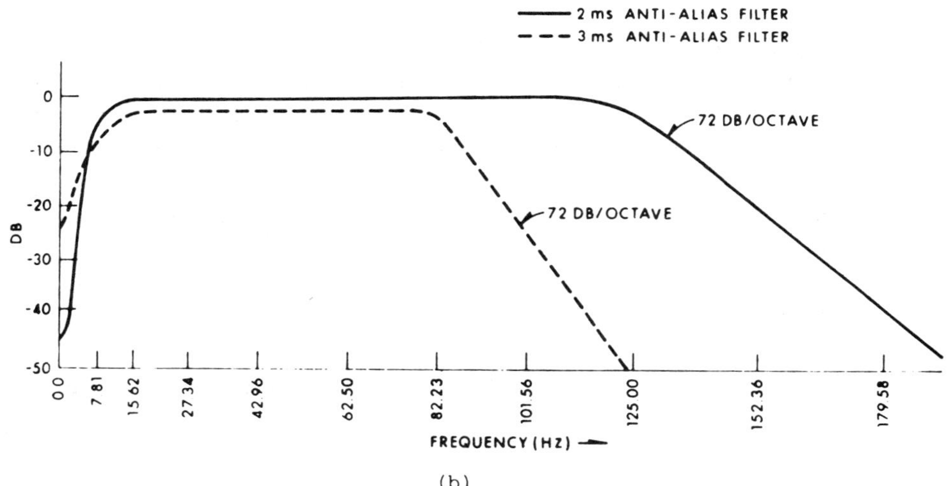

(b)

Fig. 9.　(a) Response of ghosting operators;
　　　　(b) Power spectrum of 2 ms and 3 ms anti-alias filters.

BIOGRAPHIES

<u>Vijay K. Arya</u> received the B.E. and M.E. degrees with honors from the University of Rajasthan in 1960 and 1961, respectively, and was a recipient of the gold medal awards for merit. He received the Ph.D. degree in electrical engineering from the Ohio State University, Columbus, in 1967. From 1961 to 1963 he was an Assistant Professor of Electrical Engineering at the Birla College of Engineering, Pilani, India. From 1963 to 1965 he was a Research Assistant and from 1965 to 1967 a Research Associate at the Communication and Control Systems Laboratory at Ohio State University, Columbus. During 1967-1972 he was a Research Engineer with Shell Oil Company, engaged in computer control of petro-chemical processes. Since 1972 he has been with the exploration research group of Shell Development Company, Houston, Texas, where presently he is a Staff Research Geophysicist. His current research interests are in the areas of digital signal processing, estimation theory, and mathematical modeling. Dr. Arya is a member of Sigma Xi and Society of Exploration Geophysicists.

<u>H. D. Holden</u> received the B.S. degree in electrical engineering from The University of Texas, Austin, in 1959. After greduation, he worked at Collins Radio Company in the field of microwave communications and then at Ryan Aeronautical Company specializing in airborne electronics. In 1962 he returned to The University of Texas where he received the M.S. and Ph.D. degrees in electrical engineering in 1964 and 1968, respectively. Since that time he has been employed by Shell Development Company, a subsidiary of Shell Oil Company, devoted to research and development. Current interests include the acquisition of high resolution seismic data and the application of digital signal processing to well log data. Dr. Holden is a member of Eta Kappa Nu and the Society of Exploration Geophysicists.

Part II

WIENER FILTERING
(TIME-VARYING CASE)

Editors' Comments
on Papers 7, 8, and 9

Seismic data is inherently time-varying in nature due to inelastic absorption of seismic energy in the earth as well as other time-varying phenomena like ghosting, near-surface effects, and receiver-array filtering. The seismic wavelet loses high frequencies as it travels through the earth sediments; therefore, proper deconvolution of seismic data requires time-varying filtering to account for the time-varying nature of the seismic wavelet. Clarke (1968) considers time-varying deconvolution in order to account for the earth's absorption and suggests estimating the time-varying autocorrelation function from the data to account for other time-varying effects. Thompson and Cooper (1972) present an iterative scheme to estimate nonstationary autocorrelation functions. In the time-invariant case, their algorithm reduces to the usual time-average procedure applied to several different windows along the trace.

Most existing time-varying filter techniques involve arbitrary division of a seismic trace into a number of gates of given length and assume time stationarity over these gates. A time-invariant filter is determined for each such gate and is applied to the proper gate. Usually an interpolation of two filters is used in the transition zone of two adjacent gates. Wang (Paper 7) describes a technique to establish optimum gate lengths directly from the input trace. The technique is based on the approach of Berndt and Cooper (1965), which minimizes an upper bound for the mean-square error between the true, and a given, approximated time-varying autocorrelation function.

Griffiths, Smolka, and Trembly (Paper 8) introduce an adaptive deconvolution procedure based on the use of a continuously adaptive linear-prediction operator in which the operator coefficients are updated using a steepest-descent method to minimize the mean square error between a given desired response signal and the adaptively filtered signal. Critical parameters of the algorithm are described, and the results of an adaptive deconvolution procedure are illustrated using synthetic and real seismic data to remove multiples with varying periods. Wang (1977) describes another adaptive-predictive deconvolution method that combines the time-varying difference equation model with the adaptive method of Nagumo and Noda (1967) for removing both the long- and short-period reverberations.

Ristow and Kosbahn (Paper 9) also consider time-varying prediction filtering by adaptive updating of the predictive operator. Their updating technique is essentially similar to that of Griffiths et al. (Paper 8). However, they provide more complete details of the updating procedure. The scheme is based on the adaptive algorithm of Widrow et al. (1976). They also develop a sequential updating formula for the prediction operator and for the error matrix of the estimated prediction operator, with the help of Kalman filtering.

REFERENCES

Berndt, H., and G. R. Cooper, 1965, An Optimum Observation Time for Estimates of Time-Varying Correlation Functions, *IEEE Trans. Inf. Theory* **IT-11**:307–310.

Clarke, G. K. C., 1968, Time-Varying Deconvolution Filters, *Geophysics* **33**:936–944.

Nagumo, J., and A. Noda, 1967, A Learning Method for System Identification *IEEE Trans. Autom. Control* **AC-12**:282–287.

Thompson, D. D., and G. R. Cooper, 1972, Estimation of a Time-Varying Seismic Autocorrelation Function, *Geophysics* **37**:947–952.

Wang, R. J., 1977, Adaptive Predictive Deconvolution of Seismic Data, *Geophys. Prospect.* **25**:342–381.

Widrow, B., J. M. McCool, M. G. Larimore, and C. R. Johnson, Jr., 1976, Stationary and Nonstationary Learning Characteristics of the LMS Adaptive Filter, *IEEE Proc.* **64**:1151–1162.

7

THE DETERMINATION OF OPTIMUM GATE LENGTHS FOR TIME-VARYING WIENER FILTERING†

R . J . W A N G*

The response function of a time-varying filter changes with the output signal, or observation time. Most existing time-varying filter techniques involve the empirical division of a seismic trace into a number of gates (or time windows) of given length, and a time-invariant filter is determined for each such gate. Few treatments have dealt with analytical methods to establish the gate lengths according to some optimum criterion.

This paper describes a technique for the determination of optimum gate lengths. It is based on the work of Berndt and Cooper, which is here applied to the calculation of time-varying Wiener filters. The Berndt and Cooper technique produces an upper bound for the mean-square error between the true and a given approximated time-varying correlation function. The minimization of this upper bound leads to a relation which enables one to establish gate lengths directly from the input trace. Thereafter, ordinary time-invariant Wiener filters can be computed for each gate. The overall filtered trace is obtained in the form of a suitably combined version of the individually filtered gates.

Experimentally it is shown that, with the Berndt and Cooper technique to determine optimum gate lengths, time-varying Wiener filters can be better than a time-invariant filter.

INTRODUCTION

In a nonstationary process, the usual entities that serve to characterize the input, such as mean, variance, correlation functions, etc., vary with time. If the process is nonstationary and if one wishes to design a filter which yields minimum mean-square error between some desired output and an actual output, Wiener's (1949) solution based on a stationary process is not applicable. Instead, one must consider an integral equation such as Booton's (1952) for a nonstationary process. Unfortunately, the general solution to Booton's integral equation is not yet known. A method of solving this for some special cases has been proposed by Shinbrot (1957, 1958). While Shinbrot's solution has the advantage that it is in closed form, the assumptions needed are rather restrictive.

For a stationary input, Wiener (1949) has shown that a necessary and sufficient condition for the mean-square error between some desired output $z(t)$ and an actual output $y(t)$ to be a minimum, the integral equation (called the Wiener-Hopf integral equation of the first kind)

$$\phi_{zx}(\tau) = \int_0^\infty g(\sigma)\phi_{xx}(\tau - \sigma)d\sigma \qquad (1)$$

must be satisfied, where $g(\sigma)$ is the impulse response function of the filter to be determined,

† Presented at the 38th Annual International SEG Meeting in Denver, Colorado, October 1, 1968. Manuscript received by the Editor October 16, 1968; revised manuscript received February 11, 1969.

* Research Center, Pan American Petroleum Corporation, Tulsa, Oklahoma 74102.

112

$\phi_{xx}(\tau)$ is the autocorrelation function of the input $x(t)$, and $\phi_{zx}(\tau)$ is the crosscorrelation function between the desired output $z(t)$ and the input $x(t)$. If the response time of the filter σ and the lag time of the correlation function τ take on only discrete values, equation (1) may be written as a set of linear algebraic equations called the normal equations (Robinson and Treitel, 1967). An efficient method of solving these equations was first given by Levinson (1947). The digital solution for $g(\sigma)$ with the aid of z-transform theory was illustrated by Robinson and Treitel (1964), and the recursive scheme to speed up digital computations was presented by Shanks (1967).

Once $g(\sigma)$ is determined, the output $y(t)$ is found from the well-known convolution equation

$$y(t) = \int_0^t g(t - \sigma)x(\sigma)d\sigma. \qquad (2)$$

A simpler form of equation (2) is

$$y(t) = g(t) * x(t), \qquad (3)$$

where * denotes convolution. In equations (2) and (3) it has been assumed that $x(t)$, $y(t)$, and $g(t)$ are all zero for $t<0$, i.e., they are causal.

Let us now turn our attention to a nonstationary input. In this case, equation (1) is no longer valid since both the autocorrelation and the crosscorrelation functions vary with output or observation time. We introduce the new notations $\phi_{xx}(t, \gamma)$ and $\phi_{zx}(t, \gamma)$ instead of $\phi_{xx}(\tau)$ and $\phi_{zx}(\tau)$, respectively. The time γ equals $\tau+t$, where τ, as before, represents the lag time, and t denotes the output, or observation time. In other words, for a nonstationary process, the correlation functions depend on both t and γ, whereas for a stationary process, the correlation functions depend only on the difference $\gamma-t=\tau$. For a nonstationary input, Booton (1952) has shown that a necessary and sufficient condition for the mean-square error between some desired output and an actual output to be a minimum is

$$\phi_{zx}(t, \gamma) = \int_0^\infty g(t, \sigma)\phi_{xx}(\sigma, \gamma)d\sigma, \qquad (4)$$

where $g(t,\sigma)$ is the time-varying impulse response function of the filter to be determined, $\phi_{xx}(t,\gamma)$ is the time-varying autocorrelation function of the input $x(t)$, and $\phi_{zx}(t,\gamma)$ is the time-varying

crosscorrelation function of the desired output $z(t)$ and the input $x(t)$. Given $g(t,\sigma)$, the actual output $y(t)$ is

$$y(t) = \int_0^\infty g(t, \sigma)x(\sigma)d\sigma. \qquad (5)$$

Equation (4) is called Booton's integral equation, or the modified Wiener-Hopf integral equation of the second kind. (See Appendix I for its derivation.)

THE ESTIMATION OF TIME-VARYING CORRELATION FUNCTIONS AND THE IMPLEMENTATION OF TIME-VARYING WIENER FILTERING

Since the general solution of equation (4) is not known, one can only approximate $g(t,\sigma)$ in some way. The first obstacle in an attempt to approximate $g(t, \sigma)$ according to equation (4) is encountered in the determination of time-varying correlation functions $\phi_{xx}(t, \gamma)$ and $\phi_{zx}(t, \gamma)$. This problem does not arise in the stationary, ergodic[1] case, where the corresponding correlation functions $\phi_{xx}(\tau)$ and $\phi_{zx}(\tau)$ are time-invariant.

The time-varying correlation functions $\phi_{xx}(t,\gamma)$ and $\phi_{zx}(t, \gamma)$ are defined, respectively, as

$$\phi_{xx}(t, \gamma) = E[x(t)x(\gamma)]$$

and

$$\phi_{zx}(t, \gamma) = E[z(t)x(\gamma)],$$

where $E[\ ]$ denotes the expected value, or ensemble average, of the quantity within brackets. Before proceeding further, we note that we are here dealing only with a single channel, i.e., there is only one input and one output. Under these conditions it is impossible to determine $\phi_{xx}(t, \gamma)$ and $\phi_{zz}(t, \gamma)$ rigorously by computing the ensemble average of the quantities $[x(t) x(\gamma)]$ and $[z(t) x(\gamma)]$, respectively. Rather, it is necessary to assume that the process is "piecewise stationary." In other words, we propose to divide both the input $x(t)$ and the desired output $z(t)$ into sections, where each section may be considered to be one realization of some stationary and ergodic pro-

---

[1] A stationary random process is said to possess the ergodic property if one can equate ensemble averages and averages with respect to time performed on a single "representative" function of the ensemble. For a definition of ensemble averages, see Lee (1960), Chapter 7. Ergodicity will be assumed for the stationary process in the sequel.

cess. Thus, $x(t)$ may be expressed as a sum of stationary inputs $x_1(t)$, $x_2(t)$, $\cdots$, $x_N(t)$, namely,

$$x(t) = \sum_{k=1}^{N} x_k(t) \tag{6}$$

where

$$x_k(t) = x(t)\{u[t - (k-1)T] - u(t - kT)\}.$$

In the above expression, $u(t)$ denotes a unit step function, defined by

$$u(t) = \begin{cases} 1 & \text{if } t \geq 0 \\ 0 & \text{if } t < 0, \end{cases}$$

and T represents an interval over which $x(t)$ is assumed stationary. This operation is illustrated in Figure 1. The desired output $z(t)$ is similarly divided into N such sections of duration T. Once $x_k(t)$ and $z_k(t)$, $k = 1, 2, \cdots, N$, are determined, the time-invariant autocorrelation and crosscorrelation functions are found from the relations

$$\phi_{xx}^{[k]}(\tau) = \frac{1}{T} \int_{(k-1)T}^{kT} x_k(t)x_k(t + \tau)dt$$

and

$$\phi_{zx}^{[k]}(\tau) = \frac{1}{T} \int_{(k-1)T}^{kT} z_k(t)x_k(t + \tau)dt, \tag{7}$$

$$k = 1, 2, \cdots, N.$$

This is illustrated in Figure 2.

We next propose to split equation (4) into the components

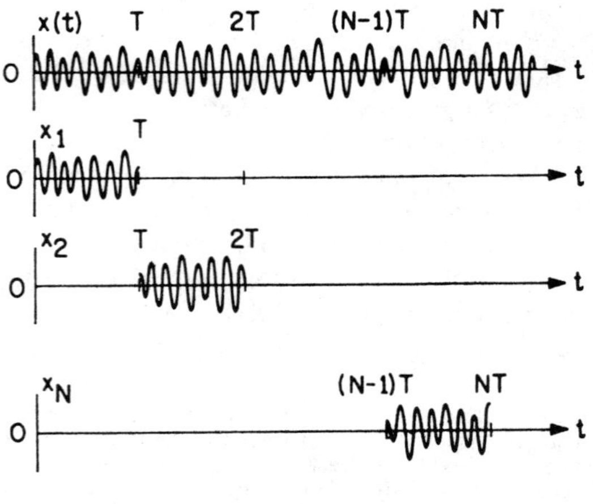

FIG. 1. Subdivision of a seismic trace.

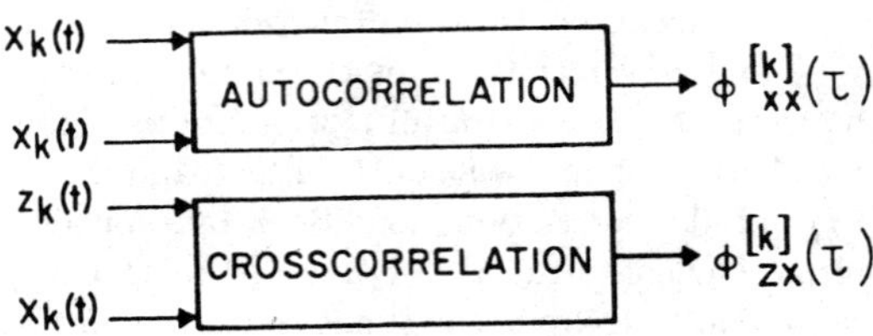

FIG. 2. Time-invariant correlation.

$$\phi_{zx}^{[k]}(\tau) = \int_0^{\infty} g_k(\sigma)\phi_{xx}^{[k]}(\tau - \sigma)d\sigma,$$

$$k = 1, 2, \cdots, N, \tag{8}$$

where the $g_k(\sigma)$ are the time-invariant components of $g(t,\sigma)$, i.e., we express $g(t,\sigma)$ in the form

$$g(t, \sigma) = \sum_{k=1}^{N} g_k(\sigma)\{u[t - (k-1)T] - u(t - kT)\}. \tag{9}$$

As can be seen, the form of equation (8) is exactly the same as that of equation (1).

Given $g_k(\sigma)$, $k = 1, 2, \cdots, N$, the actual output $y(t)$ is computed by means of the formula

$$y(t) = \sum_{k=1}^{N} y_k(t) \tag{10}$$

where

$$y_k(t) = \int_0^t g_k(t - \sigma)x_k(\sigma)d\sigma. \tag{11}$$

Equation (10) implies that $y_k(t)$ may also be expressed in terms of $y(t)$ in the form

$$y_k(t) = y(t)\{u[t - (k-1)T] - u(t - kT)\}.$$

Let us show that equation (4) reduces to equation (8) for the stationary case. In this event we have

$$\begin{aligned} \phi_{xx}(t, \gamma) &= \phi_{xx}(\tau) \\ g(t, \sigma) &= g(\sigma) \end{aligned}$$

and

$$\phi_{xx}(\sigma, \gamma) = \phi_{xx}(\tau - \sigma) \tag{12}$$

Because of the assumption that $x(t)$ is stationary over the interval $(k-1)T \leq t < kT$, $k = 1, 2, \cdots$, N, we substitute equation (12) into equation (4) and obtain

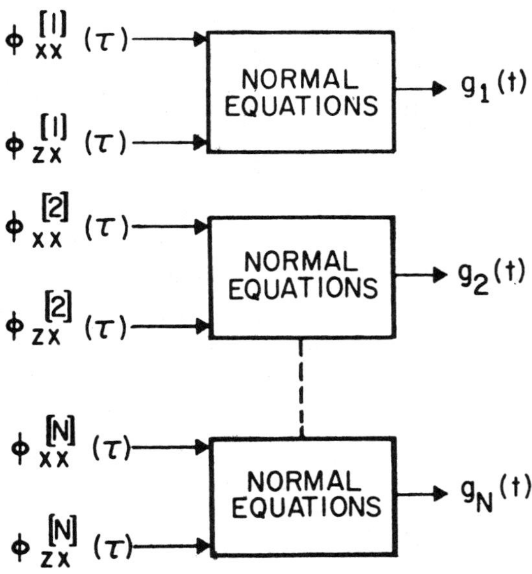

FIG. 3. Time-varying impulse response function.

$$\phi_{zx}^{[k]}(\tau) = \int_0^\infty g_k(\sigma)\phi_{xx}^{[k]}(\tau - \sigma)d\sigma,$$

$$k = 1, 2, \cdots, N. \qquad (8)$$

For the discrete case, the functions $g_k(\sigma)$, $k=1$, $2, \cdots, N$, are determined through solution of the N sets of normal equations as illustrated in Figure 3. The combined operation of equations (10) and (11) is illustrated in Figure 4.

Thus far, the concept involved is quite simple and straightforward. However, one question remains outstanding, namely: how closely do the time-invariant correlation functions $\phi_{xx}^{[k]}(\tau)$ and $\phi_{zx}^{[k]}(\tau)$, $k=1, 2, \cdots, N$, approximate the time-varying correlation functions $\phi_{xx}(t,\gamma)$ and $\phi_{zx}(t,\gamma)$? Evidently, there is some optimum value of T for which these time-invariant correlation functions best approximate the time-varying correlation functions. Now, for an ergodic process the variance of such an estimate vanishes as the observation interval T approaches infinity. Thus, it is desirable to make T as large as possible. On the other hand, the expected error between an estimated correlation function and a true correlation function decreases as T decreases [see equation (A-18), Appendix II]. What then is this optimum value of T? An equation derived by Berndt and Cooper (1965) seems to provide a promising criterion for the optimum selection of T. The following section describes certain aspects of Berndt and Cooper's theory and its application to seismic trace analysis.

THE DETERMINATION OF OPTIMUM GATE LENGTHS

Let us assume that a time-varying autocorrelation function may be expressed in the form

$$\phi_{xx}(t, \tau) = \sum_{i=1}^{n} a_i(t)b_i(\tau),^2 \qquad (13)$$

and that the function $a_i(t)$ may be expanded in a Taylor series about a point $t=t_0$, up to the mth terms; that is,

$$a_i(t) = \sum_{l=0}^{m} c_{il}(t - t_0)^l,$$

$$i = 1, 2, \cdots, n, \qquad (14)$$

where the Taylor coefficients are given by

$$c_{il} = \frac{1}{l!} a_i^{(l)}(t_0). \qquad (15)$$

The superscript (l) denotes the lth derivative with respect to t. Furthermore, let $^0\phi_{xx}(t_0,\tau,T)$ denote the estimated autocorrelation function at $t=t_0$. For a Gaussian process, Berndt and Cooper (1965) have shown that the upper bound of the mean-square error defined by

$$S^2 = E\{[^0\phi_{xx}(t_0, \tau, T) - \phi_{xx}(t_0, \tau)]^2\} \qquad (16)$$

is minimized when the relation

---

^{2} For clarity and in conformity with the notations used in Appendix II, the time-varying autocorrelation function will be written as $\phi_{xx}(t, \tau)$ instead of $\phi_{xx}(t, \gamma)$, where τ and t denote lag and observation times, respectively.

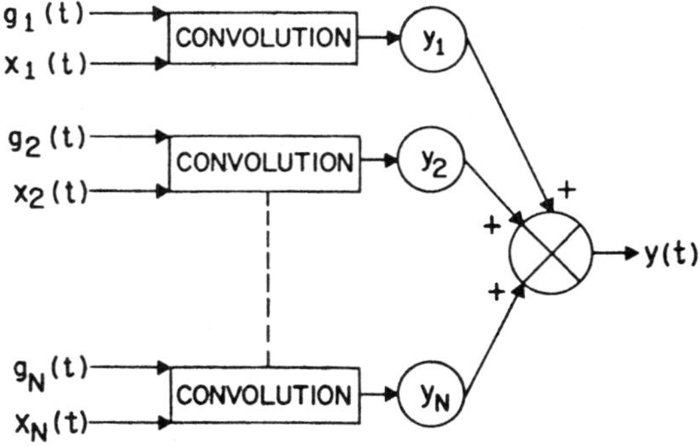

FIG. 4. Time-varying convolution.

$$\sum_{i=1}^{n} \sum_{j=1}^{n} \sum_{p=1}^{\mu} \sum_{q=1}^{\mu} c_{i2p}c_{j2q}$$

$$\cdot \frac{(p+q)b_i(0)b_j(0)\,T^{2(p+q)+1}}{(2p+1)(2q+1)2^{2(p+q)}}$$

$$= \int_{-\infty}^{\infty} [\phi_{xx}(t_0, \tau)]^2 d\tau \qquad (17)$$

is satisfied, where the estimated autocorrelation function at t_0 is given by

$$^0\phi_{xx}(t_0, \tau, T) = \frac{1}{T} \int_{-T/2}^{T/2} x\left(t + t_0 + \frac{\tau}{2}\right)$$

$$\cdot x\left(t + t_0 - \frac{\tau}{2}\right) dt. \qquad (18)$$

Relation (17) is derived in Appendix II. For the discrete case, equation (18) becomes

$$^0\phi_{xx}(t_0, \tau, T) = \frac{1}{T} \sum_{t=-T/2}^{T/2} x(t + t_0 + \tau/2)$$

$$\cdot x(t + t_0 - \tau/2). \qquad (19)$$

In equation (17), n is the number of terms used in the expansion of $\phi_{xx}(t,\tau)$ as given in equation (13), and,

$$\mu = \begin{cases} \dfrac{m}{2} & \text{if } m = \text{even} \\[2ex] \dfrac{m-1}{2} & \text{if } m = \text{odd.} \end{cases}$$

In particular, if the Taylor expansion for $a_i(t)$ can be terminated after the second term, m is 2 and $\mu = 1$. Then equation (17) reduces to

$$\sum_{i=1}^{n} \sum_{j=1}^{n} c_{i2}c_{j2} \frac{b_i(0)b_j(0)\,T^5}{72}$$

$$= \int_{-\infty}^{\infty} [\phi_{xx}(t_0, \tau)]^2 d\tau. \qquad (20)$$

Thus the optimum gate length for this case is

$$T = \left\{ \frac{72 \displaystyle\int_{-\infty}^{\infty} [\phi_{xx}(t_0, \tau)]^2 d\tau}{\displaystyle\sum_{i=1}^{n} \sum_{j=1}^{n} c_{i2}c_{j2}b_i(0)b_j(0)} \right\}^{1/5}. \qquad (21)$$

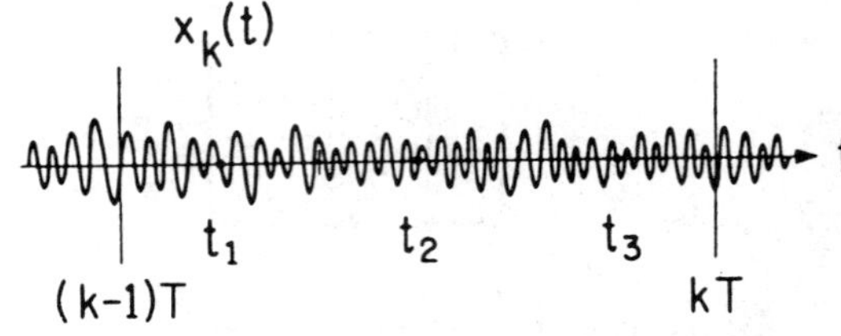

Fig. 5. Portion of a nonstationary time series.

Consider a portion of a nonstationary time series of length T, as shown in Figure 5. Divide T into three equal sections and label the center points of these three sections as t_1, t_2, and t_3. Furthermore, let[3]

$$a_i(t) = d_i + e_i(kT - t) + f_i(kT - t)^2$$

and

$$b_i(\tau) = \cos(i-1)\frac{2\pi\tau}{T} \qquad (22)$$

Then the time-varying autocorrelation function (13) may be approximated by three time-invariant autocorrelation functions

$$\phi_{xx}^{[j]}(\tau) = \sum_{i=1}^{n} a_i(t_j)b_i(\tau), \quad j = 1, 2, 3, \qquad (23)$$

where

$$a_i(t_j) = d_i + e_i(kT - t_j) + f_i(kT - t_j)^2. \qquad (24)$$

The letter k in expressions (22) and (24) is an index which identifies the kth section of the trace, where each section is of equal length T. The coefficients d_i, e_i, and f_i can be determined by solving the three simultaneous equations (24), which result when one successively sets $j = 1$, 2, and 3. (See Appendix III for a numerical example, and the reason why the gate is further divided into three short sections.)

Now from equation (15) by differentiating

[3] The form

$$a_i(t) = d_i + e_i t + f_i t^2 \qquad (24)$$

could also have been chosen for the function $a_i(t)$. For a given τ, the autocorrelation functions of seismic traces tend in general to decrease as t increases. The form of $a_i(t)$ in equations (22) will satisfy this condition if d_i, e_i, and f_i are all positive. Similarly, other forms for the function $b_i(\tau)$ could also have been chosen. With $b_i(\tau) = \cos(i-1)2\pi\tau/T$, the functions $a_i(t)$, $i = 1, 2, \cdots$, n, simply become the coefficients of the cosine transform of the autocorrelation function (23).

$a_i(t)$ with respect to t and setting $t = t_j$ we obtain

$$c_{i0} = a_i(t_j) = d_i + c_i(kT - t_j) + f_i(kT - t_j)^2,$$

$$c_{i1} = a_i^{(1)}(t_j) = -c_i - 2f_i(kT - t_j),$$

$$c_{i2} = \tfrac{1}{2}a_i^{(2)}(t_j) = f_i,$$

$$c_{i3} = c_{i4} = \cdots = c_{im} = 0.$$

Since the Taylor expansion of $a_i(t)$ terminates after the second term ($m = 2$), equation (21) is applicable for the determination of optimum observation time T_0. Also from equation (22), $b_i(0) = 1$. Hence, equation (21) may be rewritten in the form,

$$T = \left\{ \frac{144 \int_0^\infty [\phi_{xx}(t_2, \tau)]^2 d\tau}{\sum_{i=1}^{n} \sum_{j=1}^{n} f_i f_j} \right\}^{1/5}, \qquad (25)$$

where the symmetry of $\phi_{xx}(t,\tau)$ about $\tau = 0$ has been utilized to obtain the above expression. As previously stated, it is impossible to determine $\phi_{xx}(t,\tau)$ rigorously by computing the ensemble average of the quantity $[x(t)x(\gamma)]$ in the event that only one member of the ensemble is available. We must therefore find $\phi_{xx}(t_2,\tau)$ by taking a time average of the quantity $[x(t_2)x(\gamma)]$, i.e.,

$$\phi_{xx}(t_2, \tau) = \frac{1}{T} \int_{-T/2}^{T/2} x\left(t + t_2 + \frac{\tau}{2}\right)$$

$$\cdot x\left(t + t_2 - \frac{\tau}{2}\right) dt. \qquad (26)$$

It is now clear that the right-hand side of expression (25) is a function of the selected gate length T. Thus we must resort to a trial-and-error method in determining the optimum gate length T_0, which satisfies the condition (25). An arbitrary but reasonable value of T is first selected and the right-hand side of equation (25) is then evaluated. If the value of the right-hand side of equation (25) thus established deviates by more than a predetermined amount from the value that was first assumed, another value of T is selected and the right-hand side is again computed. This procedure is repeated until the condition (25) is almost satisfied, i.e., until $T \cong T_0$.

A NUMERICAL STUDY OF BERNDT AND COOPER'S EQUATION APPLIED TO TIME-VARYING WIENER FILTERING

A synthetic trace was formed with the aid of the equation

$$x(t) = s(t) + n(t), \qquad (27)$$

where $x(t)$ is a nonstationary time series, $s(t)$ represents a signal trace consisting of time-varying wavelets, and $n(t)$ is white noise. The functions $x(t)$, $s(t)$, and $n(t)$ (see Figure 6) are each described by 420 equally spaced sample values. $s(t)$ is composed of six signal wavelets whose onsets are separated by 70 sampling increments.

In order to evaluate the performance of Berndt and Cooper's theory, we chose the mean-square error as a measure of error. For example, assume that the 420-length trace is divided into six segments of 70 sample values each. Then we will generally have six different errors from equation (25), each corresponding to one of the six segments. Let $\phi_{xx}(t_k,\tau)$ and f_i^k denote the autocorrelation function and the Taylor coefficients corresponding to the kth segment respectively. Then the normalized error between the right- and left-hand sides of equation (25) is defined to be

$$E_k = 1 - \frac{T_k}{T},$$

where T_k is computed from

$$T_k = \left\{ \frac{144 \sum_{\tau=0}^{T-1} [\phi_{xx}(t_k, \tau)]^2}{\sum_{i=1}^{n} \sum_{j=1}^{n} f_i^k f_j^k} \right\}^{1/5} \qquad (28)$$

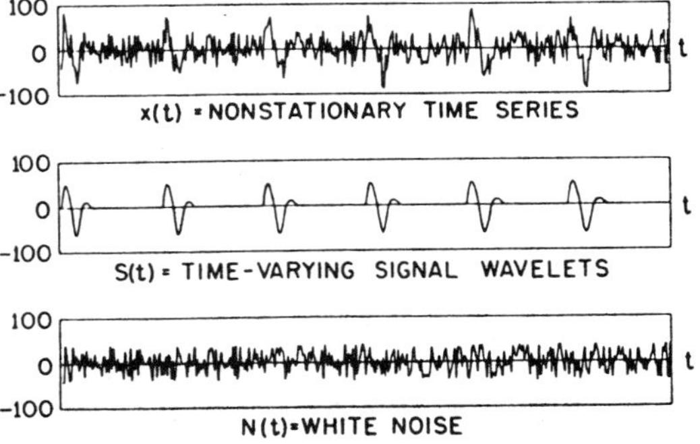

FIG. 6. Composition of a nonstationary time series. $x(t) = s(t) + n(t)$.

and where T is the selected gate length (see previous section). Since there are six segments ($T = 70$), the mean-square error is defined to be

$$Q_T = \frac{1}{6} \sum_{k=1}^{6} E_k^2. \qquad (29)$$

A plot of Q_T versus T for the synthetic trace is shown in Figure 7. The values of T ranged from 30 to 100 sample points. The minima for Q_T occur at $T = 35$ and in the neighborhood of $T = 70$. Time-varying Wiener filtering[4] has also been applied to the trace for various values of T. The mean-square errors between the desired and the actual outputs versus T for Wiener filter lengths of 40 and 100 sample points are shown in Figure 8. The desired output is the signal trace, or $s(t)$, in this case. T ranges from 30 to 100 sample points for Figure 8, and one unit of the abscissa corresponds to five points in both Figures 7 and 8. It should be pointed out that the maximum number of lags used in the autocorrelation and cross-correlation functions has been made identical to the corresponding gate lengths. It has been observed empirically that the mean-square error tends to decrease as T decreases even though the filter length is unchanged.

[4] By time-varying Wiener filtering we mean that different Wiener filters are applied to different segments and the resultant segmented outputs are combined to form the actual output.

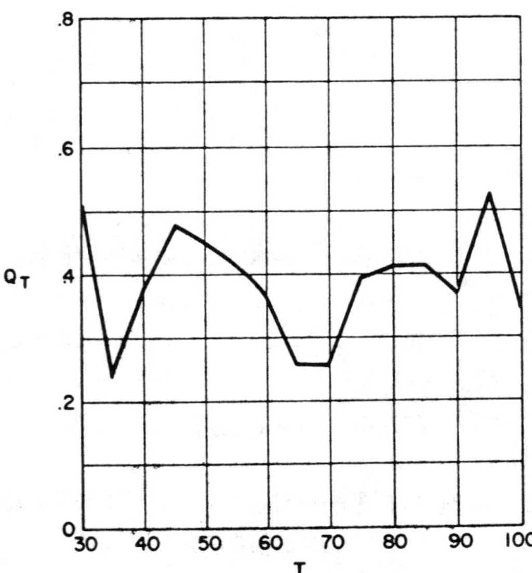

FIG. 7. Q_T (mean-square error from Berndt and Cooper's equation) versus T (gate length).

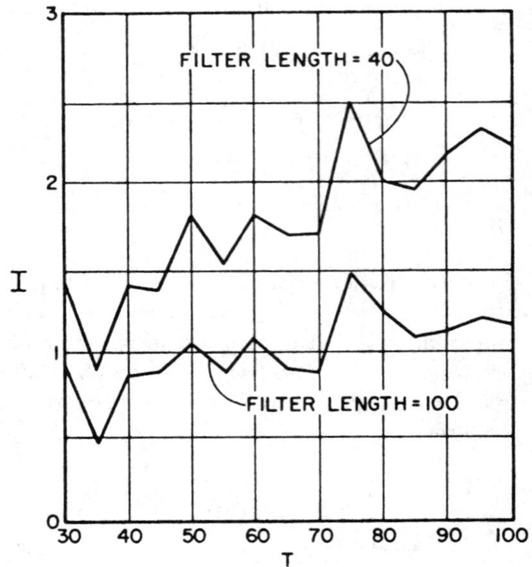

FIG. 8. I (mean-square error between the actual and desired outputs) versus T (gate length).

In Figure 8, even though the mean-square errors at $T = 70$ are considerably greater than those at $T = 35$, they both fall in troughs of the error curves. Since $T = 35$ is one-half of $T = 70$ and also turns out to be the average length of the signal wavelets, small mean-square errors should be expected at $T = 35$. A comparison of mean-square errors at $T = 70$ on Figure 8 showed that the mean-square error decreased by 49.3 percent when the filter length was changed from 40 to 100 sample points, while those at $T = 65$ and 75 decreased by 47.3 percent and 43.3 percent, respectively.

Figure 9 shows the results of time-invariant

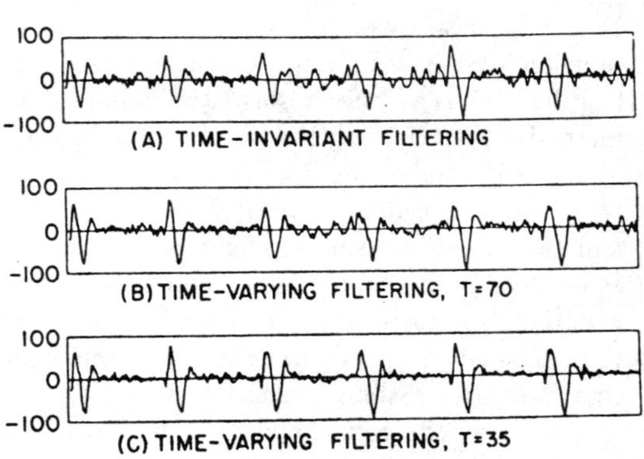

FIG. 9. Comparison of time-invariant and time-varying Wiener filtering. Filter length = 40 in all cases.

Wiener filtering [see Robinson and Treitel (1964)] and time-varying Wiener filtering for $T = 35$ and $T = 70$. The Wiener filter length used was 40 for all cases, and the mean-square errors for $T = 35$, $T = 70$, and time-invariant filtering were found to be 0.888, 1.713, and 2.808, respectively. This result indicates that time-varying Wiener filtering can be better than time-invariant Wiener filtering, when the gates are properly chosen.

CONCLUSIONS

The results of the previous section indicate that the Berndt and Cooper criterion is suitable for the selection of optimum gate lengths for which the mean-square error in Wiener filtering is small. Ideally, we would like the Q_T versus T plot of Figure 7 and the mean-square error versus T plot of Figure 8 to show even better agreement than that which we actually have been able to obtain. Among the reasons that we can give to explain the observed lack of agreement are the following:

a) The equation of Berndt and Cooper is derived for a Gaussian process (see Appendix II). The synthetic trace to which this equation was applied is of course not strictly Gaussian.

b) The performance criterion for the optimum observation time used in the derivation of equation (17) is that the upper bound of the mean-square error for the estimated autocorrelation function be minimum (see Appendix II). There is no assurance that this mean-square error itself is minimum at any time.

c) The function $a_i(t)$ was assumed to be of the form given by equation (22), so that equation (21) rather than equation (17) could be used. The approximation for $\phi_{xx}(t,\tau)$ can be improved if we retain higher order terms in $a_i(t)$ and then apply equation (17).

d) In the example of the previous section the lengths of the Wiener filter were kept unchanged while the gate lengths varied. A study of Figure 8 suggests that better agreement between the minimum mean-square error and the minimum Q_T might be obtained by a suitable choice of the ratio of filter length to gate length. A further investigation to determine the optimum number of filter coefficients for each value of T is therefore in order.

ACKNOWLEDGEMENTS

The author wishes to thank Dr. Sven Trietel of Pan American Petroleum Corporation for many helpful suggestions and discussions on the subject and the text of this paper. The author also wishes to express his thanks to Pan American Petroleum Corporation for permission to publish this paper.

REFERENCES

Berndt, H. and Cooper, G. R., 1965, An optimum observation time for estimates of time-varying correlation functions: IEEE Trans, on Inform. Theory, v. IT-11, p. 307–310.

Booton, R. C., 1952, An optimization theory for time-varying linear systems with nonstationary inputs: Proc. IRE, v. 40, p. 977–981.

Laning, J. H., and Battin, R. H., 1956, Random processes in automatic control: New York, McGraw-Hill.

Lee, Y. W., 1960, Statistical theory of communication: New York, John Wiley and Sons, Inc.

Levinson, N., 1947, The Wiener RMS (Root Mean Square) error criterion in filter design and prediction: J. Math. and Phys., v. 25, p. 261–278.

Robinson, E. A., and Treitel, S., 1964, Principles of digital filtering: Geophysics, v. 29, p. 395–404.

——— 1967, Principles of digital Wiener filtering: Geophys. Prosp., v. 15, p. 311–333.

Shanks, J. L., 1967, Recursion filters for digital processing: Geophysics, v. 32, p. 33–51.

Shinbrot, M., 1957, On the integral equation occurring in optimization theory with nonstationary inputs: J. Math. and Phys., v. 26, p. 121–129.

——— 1958, Optimization of time-varying linear systems with nonstationary inputs: Trans. ASME, v. 80, p. 457–462.

Wiener, N., 1949, Extrapolation, interpolation, and smoothing of stationary time series: New York, John Wiley and Sons, Inc.

APPENDIX I

DERIVATION OF BOOTON'S EQUATION

Let

$x(t) =$ input,
$y(t) =$ actual output,
$z(t) =$ desired output,
$g(t, \sigma) =$ time-varying impulse response of the filter to be determined,
$I(t) =$ mean-square error between the actual output and the desired output.

Then the actual output $y(t)$ and the mean-square error $I(t)$ are respectively expressed as

$$y(t) = \int_0^\infty g(t, \sigma)x(\sigma)d\sigma \qquad \text{A-1}$$

and

$$I(t) = E\{[z(t) - y(t)]^2\}, \qquad \text{A-2}$$

where the notation $E\{\ \}$ denotes the expected value of the function within the braces.

Substituting A-1 into A-2, we have

$$I(t) = E\left\{\left[z(t) - \int_0^\infty g(t, \sigma)x(\sigma)d\sigma\right]^2\right\}$$

$$= \phi_{zz}(t, t) - 2\int_0^\infty g(t, \sigma)\phi_{zx}(t, \sigma)d\sigma \quad \text{A-3}$$

$$+ \int_0^\infty g(t, \sigma)\int_0^\infty g(t, \gamma)\phi_{xx}(\gamma, \sigma)d\gamma d\sigma,$$

where

$$\phi_{zz}(t, t) = E\{[z(t)]^2\}$$
$$\phi_{xx}(\gamma, \sigma) = E\{x(\gamma)x(\sigma)\}$$
$$\phi_{zx}(t, \sigma) = E\{z(t)x(\sigma)\}.$$

Assume now that there exists an impulse response function of the system $h(t, \sigma)$ different from that of the optimum system $g(t, \sigma)$. Then $h(t, \sigma)$ may be written in the form

$$h(t, \sigma) = g(t, \sigma) + \epsilon f(t, \sigma), \quad \text{A-4}$$

where $f(t, \sigma)$ is any impulse response function, and ϵ any arbitrary number. If we denote the mean-square error due to the impulse response $h(t, \sigma)$ by $N(t)$, and replace $g(t, \sigma)$ by $h(t, \sigma)$ in A-3, we have

$$N(t) = \phi_{zz}(t, t) - 2\int_0^\infty h(t, \sigma)\phi_{zx}(t, \sigma)d\gamma$$

$$+ \int_0^\infty h(t, \sigma)$$

$$\cdot \int_0^\infty h(t, \gamma)\phi_{xx}(\gamma, \sigma)d\gamma d\sigma. \quad \text{A-5}$$

Substituting A-4 into A-5 and simplifying yield

$$N(t) = M(t) - 2\epsilon\int_0^\infty f(t, \sigma)\phi_{zx}(t, \sigma)d\sigma$$

$$+ \epsilon\int_0^\infty f(t, \sigma)\int_0^\infty g(t, \gamma)\phi_{xx}(\gamma, \sigma)d\gamma d\sigma$$

$$+ \epsilon\int_0^\infty g(t, \sigma)\int_0^\infty f(t, \gamma)\phi_{xx}(\gamma, \sigma)d\gamma d\sigma$$

$$+ \epsilon^2\int_0^\infty f(t, \sigma)$$

$$\cdot \int_0^\infty f(t, \gamma)\phi_{xx}(\gamma, \sigma)d\gamma d\sigma, \quad \text{A-6}$$

where $M(t)$ is the minimum mean-square error, given by A-3.

If we recognize that

$$\int_0^\infty f(t, \sigma)\int_0^\infty g(t, \gamma)\phi_{xx}(\gamma, \sigma)d\gamma d\sigma$$

$$= \int_0^\infty g(t, \sigma)\int_0^\infty f(t, \gamma)\phi_{xx}(\gamma, \sigma)d\gamma d\sigma,$$

A-6 can be reduced to

$$N(t) = M(t) - 2\epsilon I_1(t) + \epsilon^2 I_2(t), \quad \text{A-7}$$

where

$$I_1(t) = \int_0^\infty f(t, \sigma)$$

$$\cdot \left\{\phi_{zx}(t, \sigma) - \int_0^\infty g(t, \gamma)\phi_{xx}(\gamma, \sigma)d\gamma\right\}d\sigma$$

$$I_2(t) = E\left\{\left[\int_0^\infty f(t, \sigma)x(\sigma)d\sigma\right]^2\right\}.$$

Let us suppose that

$$I_1(t) \neq 0.$$

Then for some value of ϵ, and some $f(t, \sigma)$, it is possible to have

$$\left[\epsilon I_1(t) - \frac{\epsilon^2}{2}I_2(t)\right] > 0.$$

But this implies

$$M(t) > N(t).$$

By hypothesis, however,

$$N(t) > M(t) = \text{minimum mean-square error};$$

that is, for any nonoptimum impulse response function $h(t, \sigma)$, the actual mean-square error $N(t)$ must be greater than the minimum mean-square error $M(t)$. Therefore, the inequality

$$\left[\epsilon I_1(t) - \frac{\epsilon^2}{2}I_2(t)\right] > 0$$

must not hold. This can only mean that

$$I_1(t) = 0,$$

if $g(t, \sigma)$ is to be the optimum impulse response function. Thus, for an optimum system,

$$I_1(t) = \int_0^\infty f(t, \sigma) \left\{ \phi_{zx}(t, \sigma) \right.$$

$$\left. - \int_0^\infty g(t, \gamma)\phi_{xx}(\gamma, \sigma)d\gamma \right\} d\sigma = 0, \quad \text{A-8}$$

for any impulse response function $f(t, \sigma)$. But equation A-8 is satisfied if and only if

$$\phi_{zx}(t, \sigma) = \int_0^\infty g(t, \gamma)\phi_{xx}(\gamma, \sigma)d\gamma. \quad \text{A-9}$$

Equation A-9 is called Booton's integral equation or the modified Wiener-Hopf linear integral equation of the second kind; this is also equation (4) of the main text. Equation A-9 is the necessary and sufficient condition for the system to be optimum in the sense that the mean-square error

$$I(t) = E\left\{ [z(t) - y(t)]^2 \right\}$$

be minimum. By use of equations A-3 and A-9, the minimum mean-square error finally reduces to

$$M(t) = \phi_{zz}(t, t)$$

$$- \int_0^\infty g(t, \sigma)\phi_{zx}(t, \sigma)d\sigma. \quad \text{A-10}$$

APPENDIX II

AN OPTIMUM OBSERVATION TIME FOR ESTIMATES OF TIME-VARYING CORRELATION FUNCTIONS

Let us assume that a time-varying autocorrelation function can be expressed as

$$\phi(t, \tau) = \sum_{i=1}^n a_i(t)b_i(\tau). \quad \text{A-11}$$

Let us further assume that $a_i(t)$, $i = 1, 2, \cdots, n$ can be expanded in a Taylor series about an estimation point t_0, up to an mth term, or

$$a_i(t) = \sum_{l=0}^m c_{il}(t - t_0)^l,$$

$$i = 1, 2, \cdots, m, \quad \text{A-12}$$

where m is sufficiently large so that the remainder can be neglected. The Taylor coefficients are given by

$$c_{il} = \frac{1}{l!} a_i^{(l)}(t_0), \quad \text{A-13}$$

where the superscript (l) denotes the lth derivative of $a_i(t)$ with respect to t.

Substituting A-12 into A-11, we obtain

$$\phi(t, \tau) = \sum_{i=1}^n \sum_{l=0}^m c_{il}(t - t_0)^l b_i(\tau). \quad \text{A-14}$$

As an initial approximation, let us choose

$$^0\phi(t_0, \tau, T) = \frac{1}{T} \int_{-T/2}^{T/2} x\left(t + t_0 + \frac{\tau}{2}\right)$$

$$\cdot x\left(t + t_0 - \frac{\tau}{2}\right) dt. \quad \text{A-15}$$

The function $^0\phi(t_0, \tau)$ yields the desired estimate of $\phi(t_0, \tau)$ when T equals the optimum observation time T_0. Now the expected value of this function is

$$E\left\{ ^0\phi(t_0, \tau, T) \right\}$$

$$= \frac{1}{T} \int_{-T/2}^{T/2} \phi(t + t_0, \tau)dt, \quad \text{A-16}$$

where

$$\phi(t + t_0, \tau)$$

$$= E\left\{ x\left(t + t_0 + \frac{\tau}{2}\right) x\left(t + t_0 - \frac{\tau}{2}\right) \right\}.$$

Substitution of A-14 into A-16 gives

$$E\left\{ ^0\phi(t_0, \tau, T) \right\}$$

$$= \frac{1}{T} \int_{-T/2}^{T/2} \sum_{i=1}^n \sum_{l=0}^m c_{il}t^l b_i(\tau)dt$$

which can be written as

$$E\left\{ ^0\phi(t_0, \tau, T) \right\}$$

$$= \sum_{i=1}^n \sum_{l=0}^m c_{il}b_i(\tau)\left[\frac{1}{T} \int_{-T/2}^{T/2} t^l dt\right]. \quad \text{A-17}$$

But,

$$\frac{1}{T} \int_{-T/2}^{T/2} t^l dt = \begin{cases} \dfrac{1}{l + 1}\left(\dfrac{T}{2}\right)^{l+1} & \text{if } l \text{ is even} \\ 0 & \text{if } l \text{ is odd.} \end{cases}$$

Hence A-17 becomes

$$E\{{}^{0}\phi(t_0, \tau, T)\}$$

$$= \sum_{i=1}^{n} \sum_{l=0}^{\mu} c_{i2l} \frac{T^{2l}}{2^{2l}(2l+1)} b_i(\tau)$$

$$= \phi(t_0, \tau) + \sum_{i=1}^{n} \sum_{l=1}^{\mu} c_{i2l} \frac{T^{2l} b_i(\tau)}{2^{2l}(2l+1)}, \quad \text{A-18}$$

where

$$\mu = \begin{cases} m/2 & \text{if } m \text{ is even} \\ (m-1)/2 & \text{if } m \text{ is odd.} \end{cases}$$

The double summation in the last term of A-18 is the expected error in the approximation. As can be seen, it is a function of T^2. Thus, the observation interval must be small for the expected error to be small.

The variance σ^2 of ${}^{0}\phi(t_0, \tau, T)$ is given by

$$\sigma^2 = E\{[{}^{0}\phi(t_0, \tau, T) - {}^{0}\bar{\phi}(t_0, \tau, T)]^2\}$$

or

$$\sigma^2 = E\{{}^{0}\phi^2(t_0, \tau, T)\} - \{E[{}^{0}\phi(t_0, \tau, T)]\}^2, \quad \text{A-19}$$

where

$${}^{0}\bar{\phi}(t_0, \tau, T) = E\{{}^{0}\phi(t_0, \tau, T)\},$$

and

$${}^{0}\phi(t_0, \tau, T) = \frac{1}{T} \int_{-T/2}^{T/2} x\left(t + t_0 + \frac{\tau}{2}\right)$$

$$\cdot x\left(t + t_0 - \frac{\tau}{2}\right) dt.$$

Substituting the above expressions for ${}^{0}\phi(t_0, \tau, T)$ and ${}^{0}\bar{\phi}(t_0, \tau, T)$ in equation A-19 yields

$$\sigma^2 = \frac{1}{T^2} \int_{-T/2}^{T/2} \int_{-T/2}^{T/2} \psi^2(t_1, t_2, t_0, \tau) dt_1 dt_2$$

$$- \frac{1}{T^2} \left[\int_{-T/2}^{T/2} \phi(t + t_0, \tau) dt\right]^2, \quad \text{A-20}$$

where $\psi^2(t_1, t_2, t_0, \tau)$ is the fourth product moment of $x(t)$, which is given by

$$\psi^2(t_1, t_2, t_0, \tau)$$

$$= E\left\{x\left(t_1 + t_0 + \frac{\tau}{2}\right) x\left(t_1 + t_0 - \frac{\tau}{2}\right)\right.$$

$$\left. \cdot x\left(t_2 + t_0 + \frac{\tau}{2}\right) x\left(t_2 + t_0 - \frac{\tau}{2}\right)\right\} \quad \text{A-21}$$

In equations A-20 and A-21, t_1 and t_2 are the dummy variables for the integration. t_0 and τ are, respectively, the observed time and the lag time of the autocorrelation function. If $x(t)$ is a Gaussian process, $\psi^2(t_1, t_2, t_0, \tau)$ can be written as [see Laning and Battin (1956), p. 160–163.]

$$\psi^2(t_1, t_2, t_0, \tau)$$

$$= \phi(t_1 + t_0, \tau)\phi(t_2 + t_0, \tau)$$

$$+ \phi(t_1 + t_0, t_1 - t_2)$$

$$\cdot \phi(t_1 + t_0 - \tau, t_1 - t_2)$$

$$+ \phi(t_1 + t_0, t_1 - t_2 + \tau)$$

$$\cdot \phi(t_1 + t_0 - \tau, t_1 - t_2 - \tau). \quad \text{A-22}$$

Equation A-20 may now be written in the form

$$\sigma^2 = \frac{1}{T^2} \int_{-T/2}^{T/2} \int_{-T/2}^{T/2} \{\phi(t_1 + t_0, t_1 - t_2)$$

$$\cdot \phi(t_1 + t_0 - \tau, t_1 - t_2)$$

$$+ \phi(t_1 + t_0, t_1 - t_2 + \tau)$$

$$\cdot \phi(t_1 + t_0 - \tau, t_1 - t_2 - \tau)\} dt_1 dt_2. \quad \text{A-23}$$

Let us assume that the observation time can be made small enough so that the time average of the autocorrelation function over T does not differ appreciably from the true value at t_0. Then A-23 can be approximated by

$$\sigma^2 \cong \frac{1}{T^2} \int_{-T/2}^{T/2} \int_{-T/2}^{T/2}$$

$$\cdot \{\phi^2(t_0, t_1 - t_2) + \phi(t_0, t_1 - t_2 + \tau)$$

$$\cdot \phi(t_0, t_1 - t_2 - \tau)\} dt_1 dt_2. \quad \text{A-24}$$

Equation A-24 is derived from equation A-23 by setting:

$$\phi(t_0, t_1 - t_2) \cong \phi(t_1 + t_0, t_1 - t_2),$$

$$\phi(t_0, t_1 - t_2) \cong \phi(t_1 + t_0 - \tau, t_1 - t_2),$$

$$\phi(t_0, t_1 - t_2 + \tau) \cong \phi(t_1 + t_0, t_1 - t_2 + \tau),$$

$$\phi(t_0, t_1 - t_2 - \tau) \cong \phi(t_1 + t_0 - \tau, t_1 - t_2 - \tau).$$

An upper bound of σ^2 may be established if τ is small compared to T, but T is large compared to the significant duration of the autocorrelation function. Replacing $(t_1 - t_2)$ by λ in equation A-24, we have [see Berndt and Cooper (1965)],

$$\sigma^2 \cong \frac{1}{T} \int_{-T}^{T} \left(1 - \frac{|\lambda|}{T}\right) \qquad \text{A-25}$$
$$\cdot \left\{ \phi^2(t_0, \lambda) + \phi(t_0, \lambda + \tau)\phi(t_0, \lambda - \tau) \right\} d\lambda.$$

By using the Schwarz inequality and assuming that the "energy" of the autocorrelation function is bounded, the upper bound of σ^2 can be obtained from equation A-25 in the form

$$\sigma^2 \leq \frac{2}{T} \int_{-\infty}^{\infty} \phi^2(t_0, \lambda) d\lambda. \qquad \text{A-26}$$

Let S^2 be the mean-square error of the approximation, which is

$$S^2 = E\left\{ [^0\phi(t_0, \tau, T) - \phi(t_0, \tau)]^2 \right\},$$

or

$$S^2 = \sigma^2 + \left\{ ^0\bar{\phi}(t_0, \tau, T) - \phi(t_0, \tau) \right\}^2. \qquad \text{A-27}$$

However, from equation A-18 we have

$$^0\bar{\phi}(t_0, \tau, T) - \phi(t_0, \tau)$$
$$= \sum_{i=1}^{n} \sum_{l=1}^{\mu} c_{i2l} \frac{T^{2l}}{2^{2l}(2l + 1)} b_i(\tau),$$

and thus the use of inequality A-26 yields

$$S^2 \leq \frac{2}{T} \int_{-\infty}^{\infty} \phi^2(t_0, \lambda) d\lambda$$
$$+ \sum_{i=1}^{n} \sum_{j=1}^{n} \sum_{p=1}^{\mu} \sum_{q=1}^{\mu} c_{i2p}c_{j2q}$$
$$\cdot \frac{T^{2(p+q)}b_i(0)b_j(0)}{(2p + 1)(2q + 1)2^{2(p+q)}} \cdot \qquad \text{A-28}$$

Let R denote the upper bound of S^2. Then the optimum observation time is attained when the upper bound R is a minimum. This requires that the conditions

$$\left.\begin{array}{l} \dfrac{\partial R}{\partial T}\bigg|_{T=T_0} = 0 \\[1.5em] \dfrac{\partial^2 R}{\partial T^2}\bigg|_{T=T_0} > 0 \end{array}\right\} \qquad \text{A-29}$$

be satisfied.

Differentiating the right-hand side of inequality A-28, R with respect to T yields

$$\sum_{i=1}^{n} \sum_{j=1}^{n} \sum_{p=1}^{\mu} \sum_{q=1}^{\mu} c_{i2p}c_{j2q}$$
$$\frac{(p + q)b_i(0)b_j(0)T^{2(p+q)+1}}{(2p + 1)(2q + 1)2^{2(p+q)}}$$
$$= \int_{-\infty}^{\infty} \phi^2(t_0, \lambda) d\lambda. \qquad \text{A-30}$$

In particular, if the Taylor expansion of $a_i(t)$ can be terminated after the second term, equation A-30 reduces to

$$\sum_{i=1}^{n} \sum_{j=1}^{n} c_{i2}c_{j2} \frac{b_i(0)b_j(0)T^5}{72}$$
$$= \int_{-\infty}^{\infty} \phi^2(t_0, \lambda) d\lambda \qquad \text{A-31}$$

or

$$T = \left\{ \frac{72 \displaystyle\int_{-\infty}^{\infty} \phi^2(t_0, \lambda) d\lambda}{\displaystyle\sum_{i=1}^{n} \sum_{j=1}^{n} c_{i2}c_{j2}b_i(0)b_j(0)} \right\}^{1/5} \qquad \text{A-32}$$

which is equation (21) of the main text.

APPENDIX III

A NUMERICAL EXAMPLE FOR DETERMINING THE COEFFICIENTS OF $a_i(t)$.

Let us assume that we have determined the three time-invariant autocorrelation functions $\phi_{xx}^{[1]}(\tau)$, $\phi_{xx}^{[2]}(\tau)$, and $\phi_{xx}^{[3]}(\tau)$, corresponding to t_1, t_2, and t_3, respectively (see Figure 5). We can then expand these autocorrelation functions in finite cosine series, that is,

$$\phi_{xx}^{[j]}(\tau) = \sum_{i=1}^{n} a_i(t_j) \cos (i - 1) \frac{2\pi\tau}{T} \cdot$$
$$j = 1, 2, 3, \qquad \text{A-33}$$

where n is an appropriate number of terms to be used in the expansion and the $a_i(t_j)$ are the ith coefficients of the cosine transforms of $\phi_{xx}^{[j]}(\tau)$. Furthermore, if we assume that $a_i(t)$ in equation (7) of the main text varies as a second-order equation in t, and let

123

$$a_i(t) = d_i + e_i(kT - t) + f_i(kT - t)^2,$$

$$i = 1, 2, \cdots, n, \quad \text{A-34}$$

d_i, e_i, and f_i can easily be found by solving three simultaneous algebraic equations, namely,

$$a_i(t_1) = d_i + e_i(kT - t_1)$$
$$+ f_i(kT - t_1)^2,$$
$$a_i(t_2) = d_i + e_i(kT - t_2)$$
$$+ f_i(kT - t_2)^2, \quad \text{A-35}$$
$$a_i(t_3) = d_i + e_i(kT - t_3)$$
$$+ f_i(kT - t_3)^2,$$

where k is an index identifying the kth section of the trace and T is the gate length.

For example, let $k=1$, $T=30$, $t_1=5$, $t_2=15$, $t_3=25$, and $i=1$. Furthermore, assume that $a_1(t_1)=5$, $a_1(t_2)=2$, and $a_1(t_3)=1$. Then, from equations A-35, we have

$$5 = d_1 + e_1(25) + f_1(25)^2,$$
$$2 = d_1 + e_1(15) + f_1(15)^2,$$
$$1 = d_1 + e_1(5) + f_1(5)^2.$$

Solving the above equations for d_1, e_1, and f_1 yields $d_1 = 1.25$, $e_1 = -0.10$, $f_1 = 0.01$, and $a_1(t)$ can now be expressed as

$$a_1(t) = 1.25 - 0.10(30 - t) + 0.01(30 - t)^2.$$

$a_1(t)$ tells us how the first coefficient in the cosine expansion of the time-varying autocorrelation function varies with observation time within the interval $(k-1)T \leq t < kT$.

From equation (25) of the main text, however, one sees that only the coefficients f_i are needed for determination of the optimum gate lengths Cramer's rule, f_i can be found from A-35 as

$$f_i = \frac{1}{\Delta} \begin{vmatrix} 1 & (kT - t_1) & a_i(t_1) \\ 1 & (kT - t_2) & a_i(t_2) \\ 1 & (kT - t_3) & a_i(t_3) \end{vmatrix},$$

where

$$\Delta = \begin{vmatrix} 1 & (kT - t_1) & (kT - t_1)^2 \\ 1 & (kT - t_2) & (kT - t_2)^2 \\ 1 & (kT - t_3) & (kT - t_3)^2 \end{vmatrix}.$$

In the above expressions the vertical bars denote the determinant. By now it should be evident that given the form of $a_i(t)$ as A-34, three equations are needed to compute f_i. This, of course, is the reason why the gate is further divided into three short sections and the autocorrelation functions are then found at t_1, t_2, and t_3.

8

ADAPTIVE DECONVOLUTION: A NEW TECHNIQUE FOR PROCESSING TIME-VARYING SEISMIC DATA

L. J. GRIFFITHS*, F. R. SMOLKA‡ AND L. D. TREMBLY‡

A time-varying deconvolution method has been developed which is based upon adaptive linear filtering techniques. This adaptive deconvolution is applicable for use in processing reflection seismic data which contain multiples with periods that vary with traveltime.

Filter coefficients are designed for each sample of the input trace using an adaptive algorithm. Convergence properties of the adaptive processor are discussed and compared to conventional deconvolution techniques. The adaptive deconvolution method is illustrated using both synthetic and field reflection seismic data. By proper selection of parameters and by procesing the data in both time-reverse and time-forward directions, adaptive deconvolution removes multiples with varying periods while leaving primary reflections relatively undistorted.

INTRODUCTION

This paper presents a new time-varying deconvolution method for use in processing reflection seismograms. The technique is based on the use of a continuously adaptive linear prediction operator in which the operator coefficients are updated using a simple adaptive algorithm. New coefficient values are computed for each data sample in the seismic record so as to minimize a mean-square error criterion. This procedure differs significantly from time-varying deconvolution methods described by Clarke (1968), Wang (1969), and others. Previous methods have employed the well-known three-stage processes of first, computing autocorrelation estimates from the data; second, solving a set of appropriate normal equations to determine the operator coefficient values; and third, applying the operator to the data to obtain the deconvolved output trace. In the adaptive deconvolution procedure proposed in this paper, new coefficient values are computed directly from the seismic data values as the operator is applied to the data. In effect, the operator is designed as the deconvolved output is produced.

Deconvolution operators which employ a minimum mean-square error criterion—i.e., Wiener filters—have been widely used in reflection seismogram processing. The method generally assumes that the statistical structure of the data is stationary over the design gate used to generate the deconvolution operator. This is true even for the piecewise stationary time-varying methods discussed by Clarke (1968) and Wang (1969). In some seismic processing applications, the piecewise stationary assumption may be overly restrictive. An example of such data can be found in the marine, shallow, hard water-bottom environment. It is readily shown (Middleton and Whittlesey, 1968) that the time between successive water bottom reflections, for source-receiver spacings other than zero, changes markedly with reflection number. Because these multiples are relatively close together, the resulting seismogram exhibits nonstationary statistical properties over a short time interval, and conventional Wiener filtering methods do not yield satisfactory results.

Theoretical extensions of Wiener filtering theory to include time-varying systems have been proposed by Boonton (1952) but the resulting integral equations are extremely difficult to establish without a detailed a priori knowledge of the

Presented at the 45th Annual International SEG Meeting, October 15, 1975 in Denver. Manuscript received by the Editor June 10, 1976.
* University of Colorado, Boulder, CO 80302.
‡ Marathon Oil Co., Littleton, CO 80120.

125

nature of the underlying time variations. Kalman filtering methods have also been applied to the time-varying deconvolution problem (Crump, 1974), but the state-variable model used in this formulation also requires accurate estimates of the underlying time-varying parameters. Reliable estimates of the parameters can only be obtained under conditions of extremely high signal-to-noise ratio (SNR). Another approach, involving the use of homomorphic techniques, was recently proposed by Buhl et al (1974). Although this method is shown to handle successfully the time-varying multiple problem, the authors have not considered the effects of additive noise on the process and the procedure is also restricted to high SNR environments.

The adaptive deconvolution method we describe in this paper uses a time-continuous adaptive procedure similar to that used extensively in the communications and antenna array processing field (Widrow et al, 1967; Lacoss, 1968; Griffiths, 1969; Frost, 1972; Riegler and Compton, 1973; Widrow et al, 1975; Gabriel, 1976; Griffiths, 1976). These applications have demonstrated a substantial improvement in processed SNR in the presence of time-varying noise and interference. The algorithms used to obtain this performance can be grouped into two broad classes: those which adapt to minimize the average squared error between a given desired response signal and the adaptively filtered signal (Widrow et al, 1967; Wang and Treitel, 1971;

trace γ time samples later in a manner directly analogous to the predictive deconvolution procedure described by Peacock and Treitel (1969). The algorithm is then used to update the prediction coefficients as the filter moves along the seismogram. Details of the algorithm and processing examples illustrating its use on both synthetic and field-recorded seismic data are presented in the sections following.

ADAPTIVE DECONVOLUTION

Our adaptive deconvolution procedure utilizes a continuously time-varying operator to process the seismic trace. Thus, a different operator is used at each data point in the trace as the processor moves along the reflection seismogram. A simple adaptive algorithm, based on the method of steepest descent, is used to update the operator coefficients as each data point is deconvolved. In this section, we first present some background material on predictive deconvolution of stationary traces and on the development of adaptive techniques in general. We then describe the specific application of adaptive processing to the deconvolution problem.

We denote the set of N data points in the seismogram to be processed by $x(t)$, $t = 0,1,\cdots N - 1$. As shown by Peacock and Treitel (1969), predictive deconvolution of this trace for the special case of statistically stationary $x(t)$ is achieved by first solving the normal equations for an L-point prediction operator denoted by the vector F_o. The appropriate equations are given by

$$\begin{bmatrix} r_x(0) & r_x(1) \cdots & r_x(L-1) \\ r_x(1) & r_x(0) & \cdot \\ \cdot & & \cdot \\ \cdot & & \cdot \\ r_x(L-1) & \cdots & r_x(0) \end{bmatrix} \begin{bmatrix} f_0(0) \\ f_0(1) \\ \cdot \\ \cdot \\ f_0(L-1) \end{bmatrix} = \begin{bmatrix} r_x(\gamma) \\ r_x(\gamma+1) \\ \cdot \\ \cdot \\ r_x(\gamma+L-1) \end{bmatrix} \qquad (1)$$

Griffiths, 1975) and those which adapt to minimize the adapted output power, subject to a linear constraint on the filter coefficients (Booker and Ong, 1971; Frost, 1972).

The algorithm proposed for time-varying deconvolution in this paper is a member of the first class described above. An adaptive filter L samples in length is used to linearly combine L successive samples of the recorded reflection seismogram. The coefficients used in this filter are then adapted to minimize the mean-square difference between the filter output and the data value in the seismic trace which occurs γ samples later in time. In effect, L data values are used to predict the

where $r_x(l)$ is the trace autocorrelation at lag l, $f_0(l)$ is the lth prediction coefficient and $\gamma \geq 1$ is the prediction distance. This equation may be expressed more compactly in matrix notation as

$$\mathcal{R}_{xx}F_0 = P_x(\gamma), \qquad (2)$$

where $\mathcal{R}_{xx}$ is the expected value of $\mathbf{X}(t)\mathbf{X}^{\mathrm{T}}(t)$ and $P_x(\gamma)$ is the expected value of $x(t + \gamma)\mathbf{X}(t)$.

In the usual application of predictive deconvolution methods to field recorded data, the autocorrelation values required for $\mathcal{R}_{xx}$ and $P_x(\gamma)$ are computed using time averages and the resulting prediction coefficients are then computed from equation (2) using the Levinson method—see

Robinson (1967). A deconvolved trace $y(t)$ is computed from the input $x(t)$ using the resulting Value of F_0. One method for visualizing this computation which is particularly convenient for describing adaptive deconvolution proceeds as follows:

(a) The operator F_0 is first used to generate a predicted or estimated trace $\hat{x}(t)$ from the data

$$\hat{x}(t + \gamma) = \sum_{l=0}^{L-1} f_0(l)x(t - l) \cdot \quad (3)$$

In matrix notation,

$$\hat{x}(t + \gamma) = F_0^T X(t) = X^T(t)F_0, \quad (4)$$

where

$$X^T(t) = [x(t), x(t - 1), \cdots x(t - L + 1)]. \quad (5)$$

(b) The deconvolved trace $y(t)$ is then the difference between the input and predicted traces

$$y(t) = x(t) - \hat{x}(t). \quad (6)$$

A heuristic development of an adaptive method for obtaining $y(t)$ may be given directly from equation (2). We begin by assuming that a gradient descent procedure is used to find F_0 which solves equation (2), rather than the computationally much more efficient Levinson algorithm. Using gradient descent, we start with an arbitrary initial guess $F(0)$ for the prediction coefficients and then update $F(t)$ using the following algorithm:

$$F(t + 1) = F(t) + \mu[P_x(\gamma) - \mathcal{R}_{xx}F(t)]. \quad (7)$$

In this case, we denote the lth coefficient in $F(t)$ by $F(l; t)$.

It is readily shown [see Wilde (1964)] that the term in square brackets is the negative of the gradient of the mean-square error between $x(t + \gamma)$ and $\hat{x}(t + \gamma)$ when $F(t)$ is used to produce the estimate $\hat{x}(t + \gamma)$. Further, if $0 < \mu < 2/\lambda_{max}$, where λ_{max} is the largest eigenvalue of $\mathcal{R}_{xx}$, it can be shown that $F(t)$ assymptotically approaches F_0, i.e.,

$$\lim_{t \to \infty} F(t) = F_0 \cdot \quad (8)$$

In 1960, Widrow and Hoff suggested that a simple, noisy gradient descent algorithm could be derived from equation (7) by replacing the autocorrelation terms $P_x(\gamma)$ and $\mathcal{R}_{xx}$ by their appropriate instantaneous values. Thus,

$$P_x(\gamma) \to x(t + \gamma) X(t),$$
$$\mathcal{R}_{xx} \to X(t)X^T(t),$$

and the algorithm in equation (7) becomes,

$$F(t + 1) = F(t) + \mu[x(t + \gamma) \\ - \hat{x}(t + \gamma)] X(t), \quad (9)$$

where $\hat{x}(t + \gamma) = X^T(t)F(t)$ is the current estimate of the trace value $x(t + \gamma)$ based on the coefficients $F(t)$.

Equation (9) has been termed the LMS algorithm by Widrow (1966) and others. Its simplicity is immediately apparent. Whereas the original gradient method in (7) required the order of L^2 operations per adaptation, the LMS algorithm in (9) requires only L operations because $x(t + \gamma)$ and $\hat{x}(t + \gamma)$ are scalar. An additional L operations are also required to compute the prediction $\hat{x}(t + \gamma)$, but these are also required in the normal equation method of predictive filtering described above.

Although the LMS algorithm requires less computation than the direct gradient approach, inspection of (9) reveals that the sequence of prediction coefficients $F(t)$ are stochastic and, in general, correlated in time due to the fact that $x(t)$ is a stochastic process. In fact, the LMS algorithm is closely related to the stochastic approximation methods described by Robbins and Monroe (1951), Blum (1954), and Dupac (1965). The primary difference is that in stochastic approximation methods, the adaptive step size μ is made inversely proportional to t or to a power of t. It turns out that this difference is of significant practical importance in terms of the convergence properties of the algorithm. When μ decreases appropriately with increasing t, it can be shown (see the stochastic approximation references above) that $F(t)$ approaches F_0 with probability one when $x(t)$ is a stationary time series. Equivalently, the deconvolved output $y(t)$ will converge exactly to that obtained via the normal equations approach.

For the case of fixed μ, convergence analysis is considerably more difficult. Since the introduction of the LMS algorithm, a variety of authors, including Senne (1968), Daniell (1970), Gersho (1968), Kim and Davisson (1975), and Widrow et al (1976) have studied this problem and the interested reader is referred to these papers for a detailed discussion. Briefly, at this point it has been shown that for stationary input data, if μ is constant and satisfies $\mu < 2/\lambda_{max}$, then the mean coefficient vector converges to the optimal, i.e.,

$$\lim_{t \to \infty} E[F(t)] = F_0. \tag{10}$$

In addition, the variances of the individual coefficients in F(t) converge to bounded values which decrease with decreasing μ.

At this point, one may question the use of the LMS adaptive algorithm in equation (7) for use as a predictive deconvolution tool, in that the normal equation approach avoids all of the previously mentioned convergence difficulties. Indeed, for those cases in which the input trace $x(t)$ can be modeled as a stationary time series, the best overall approach is undoubtedly the use of the normal equations. However, for those applications which occur all too frequently in practice (i.e., for the case of continuously time-varying input statistics), the LMS algorithm offers a potential advantage. Consider a simple reverberation model in which the period between successive reverberations is changing slowly in some unknown manner throughout the trace. The correlation measurements required for the normal equation approach necessitate time averages over window lengths and may involve averaging over significant time changes in the data. The deconvolution operator so obtained will then not provide effective multiple removal. (A numerical example illustrating this effect is presented below.)

If an adaptive deconvolution operator can converge in a time which is short compared with the time scale of the reverberation changes, it can then track these changes and provide multiple removal in the time-varying environment. It is shown in this paper that adaptive deconvolution can realize this potential advantage in practical reflection seismogram processing.

The specific adaptive deconvolution procedure suggested here is based directly on the LMS algorithm defined in equation (9). Two methods are employed: forward-time adaptation in which the iteration proceeds from earlier to later times along the input trace; and reverse-time adaptation corresponding to iteration in the opposite direction. In both cases, the predicted data point $\hat{x}(t + \gamma)$ is at a later time in the trace than the data which are presently stored in the $F(t)$ operator. The corresponding algorithms are given by

$$F(t + 1) = F(t) + \mu[x(t + \gamma) - \hat{x}(t + \gamma)]X(t), \tag{11}$$

for forward-time adaptation, and

$$F(t - 1) = F(t) + \mu[x(t + \gamma) - \hat{x}(t + \gamma)]X(t). \tag{12}$$

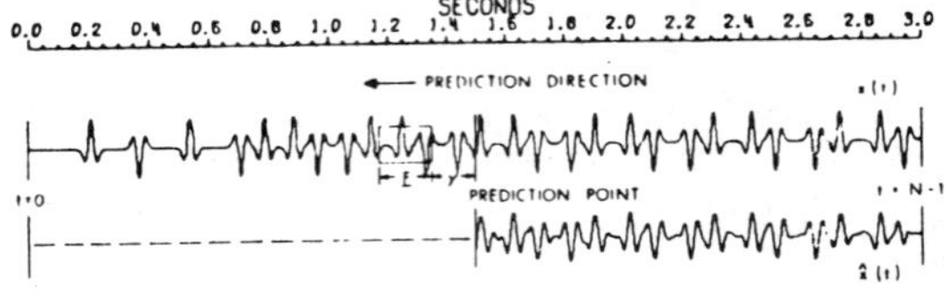

FIG. 1. Schematic representation of time-reverse adaptive deconvolution.

for reverse-time processing. In these expressions, the predicted trace value, $\hat{x}(t + \gamma) = F^T(t)X(t)$ is computed using the previous operator $F(t)$ prior to adaptation. Note that the term in square brackets is the deconvolved trace which is generated as adaptation proceeds.

After updating the operator coefficients in this manner, the operator is moved either one sample further down the trace (time-forward processing) or one sample closer to the trace origin (time-reverse processing) and a new prediction value is generated using these coefficients. The prediction results in a new error which is then used to update the coefficients. Adaptation proceeds in this manner until the entire trace has been processed. In order to avoid having the operator $F(t)$ overrun the ends of the trace, backward-time processing is carried out for $t = (N - 1 - \gamma), \cdots, L, L - 1$. Selection of the proportionality constant μ for practical data is discussed below.

Evidence obtained using both experimental and synthetic data has shown that the best results are achieved when the trace is first processed in time-reverse order using an initial operator $F^T(N - 1 - \gamma) = [0, 0, \cdots 0]$. Figure 1 illustrates time-reverse processing schematically. A deconvolved reverse-time output $y_r(t)$ is formed as the difference between the input trace $x(t)$ and the predicted trace $\hat{x}(t)$. This processing is then followed by a forward-time pass in which the initial operator $F(L - 1)$ is set equal to the final set of coefficients obtained from time-reverse adaptation. Note that the forward-time processing uses the original input trace $x(t)$ and not the deconvolved output produced by the reverse-time procedure. Denoting the deconvolved reverse and forward-time traces by $y_r(t)$ and $y_f(t)$, respectively, the final deconvolved trace $Y(t)$ is obtained as a simple stack.

$$y(t) = 1/2[y_f(t) + y_r(t)]. \tag{13}$$

It should be noted that the double-deconvolution and stacking method just described has been found to provide the best overall processing. In many cases, however, the additional improvement offered by including the forward-time pass is minimal and satisfactory results can often be achieved using only reverse-time adaptive deconvolution. Examples illustrating this effect are presented in sections following.

One additional property of adaptive deconvolution merits discussion. The prediction coefficients are continually updated using the noisy gradient algorithm in (9) as the input trace is processed. As a result, each coefficient acts as a noisy modulator of the input trace and this coefficient noise appears directly at the filter output as an additional term in the prediction $\hat{x}(t + \gamma)$. Widrow (1966, 1976), Senne (1968), Daniell (1970), and others have studied this effect, which is termed misadjustment noise. It can be shown that the noise level is directly related to the value of the adaptive proportionality constant μ. A simple relationship has been given by Widrow et al (1976) for the case of adaptation on stationary data. If $\sigma_y^2(\mu)$ is used to denote the power in the deconvolved trace when adaptive processing is used, and $\sigma_y^2(\min)$ represents the output power achieved using a fixed, optimal, predictive operator, then

$$\frac{\sigma_y^2(\mu) - \sigma_y^2(\min)}{\sigma_y^2(\min)} = \mu L \sigma_x^2, \tag{14}$$

where $\sigma_x^2 = r_x(0)$ is the average power level of the input trace.

Clearly, the additional output noise power due to coefficient fluctuation can be made arbitrarily small by sufficiently small choice of μ. However, as shown in the following section, the rate of convergence of the adaptive processor is proportional to μ, and small values may lead to excessively long convergence times. The selection of appropriate values for μ for practical applications is the subject of the following section.

SELECTION OF ALGORITHM PARAMETERS

Inspection of the adaptive algorithm defined by equations (11) and (12) above shows that in addition to the initial coefficient vector, a total of three parameters are required to completely specify the procedure. These are: operator length L, prediction distance γ, and proportionality constant μ. The first two parameters are identical with those required for conventional predictive methods and the criteria used to select them are not changed in the adaptive procedure. Briefly, the filter length should be sufficiently long to encompass the basic

wavelet, a value which is typically between 20 and 100 msec. If wavelet whitening is desired, a prediction distance of unity may be selected. For predictive deconvolution, γ should be chosen such that the distance from the midpoint of the operator to the prediction point is approximately equal to the multiple spacing.

The determination of an appropriate value for the proportionality constant μ depends upon the desired adaptive convergence properties. Previous theoretical work by Daniell (1970) and Kim and Davisson (1975) has described the convergence of this algorithm for the case of stationary input statistics. Griffiths (1975) has applied the algorithm with $\gamma = 1$ for purposes of short-term spectral estimation and has shown that the mean value of the operator converges to the solution given by the normal equations, provided that μ satisfies the following inequalities:

$$\mu = \frac{\alpha}{L \cdot \sigma_x^2}, \tag{15}$$

and

$$0 < \alpha < 2, \tag{16}$$

where σ_x^2 is the average power level of the input trace. Substituting equation (15) into the formula given in the previous section for misadjustment, i.e., equation (14), yields

$$\frac{\sigma_y^2(\mu) - \sigma_y^2(\min)}{\sigma_y^2(\min)} = \alpha. \tag{17}$$

Thus, $\alpha = 1$ will produce a 100 percent increase in excess output noise power, and $\alpha = 0.1$ produces a 10 percent increase (for the case of stationary input data). In practical data containing amplitude fluctuations, however, the value of α must be set to ensure stability on the amplitude peaks, and the actual misadjustment levels observed in lower amplitude sections of the data will be lower by an amount determined by the peak-to-trough power ratios.

Values of α near 2 will provide rapid, noisy

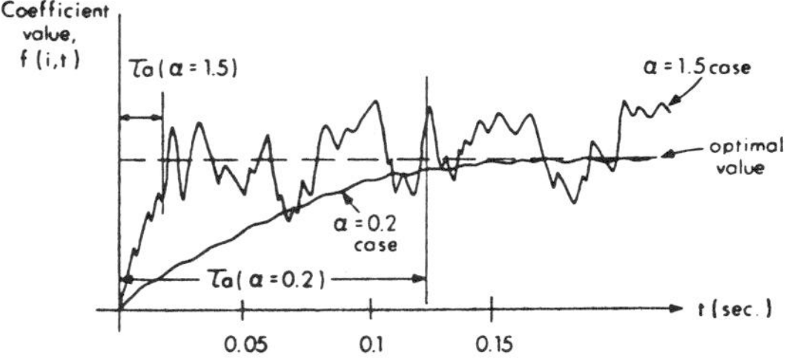

FIG. 2. Time behavior of adaptive operator coefficient for stationary trace statistics.

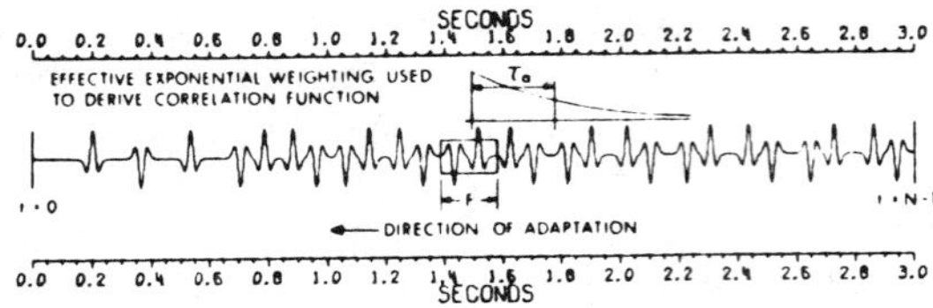

FIG. 3. Effective exponential weighting provided by adaptive deconvolution.

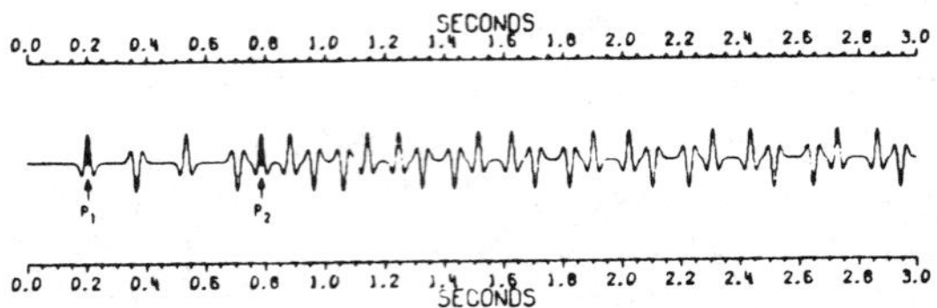

FIG. 4. Synthetic time-varying multiple trace with two primary events.

adaptation while those closer to zero produce slower, smoother results. This behavior is illustrated schematically in Figure 2 which shows the value of a hypothetical adaptive operator coefficient $f(i; t)$ as a function of adaptation time t (equivalently, time along the trace) for a stationary input example. Also shown for reference, is the optimal coefficient value, as would be computed from the normal equations for this case. Experimental results have indicated that α values in the range 0.05 to 0.2 are appropriate for stationary, constant power data and that little change in deconvolution output is observed for α in this range.

The time constant τ_a indicated in Figure 2 can be computed exactly if the input statistics are known exactly. The value obtained is related to both α and the number of coefficients L. An approximate value which has been verified experimentally by Griffiths (1975) is given by

$$\tau_a \simeq \frac{-\Delta T}{\ln(1 - \mu\sigma_x^2)} \qquad (18)$$

$$= \frac{-\Delta T}{\ln(1 - \alpha/L)}, \qquad (19)$$

where ΔT is the sampling interval.

In the case of stationary input statistics, adaptive deconvolution provides no advantages over conventional methods in which autocorrelation properties of the input waveform are computed and used to solve for the optimal operator using the normal equations. For inputs containing time-varying properties, however, adaptive methods offer a computationally simple procedure for tracking these variations. Although no theoretical results are available to date regarding the nature of convergence of the adaptive algorithm for the case of time-varying statistics, a great deal of experimental results have been accumulated in nonseismic applications which demonstrate the convergence properties in these cases. The results show that if the period of the time variations is the order of the convergence time shown in Figure 2, or longer, the algorithm can successfully track the variations. In effect, since the operator coefficients are continually updated, their value represents an average, with exponential weighting, over the statistics of previous trace samples. The time constant of this exponential is identical with the adaptation time constant τ_a defined by equation (19). Figure 3 shows this weighting diagrammatically.

Because the adaptive proportionality constant μ in equation (15) depends upon the average power level σ_x^2 of the input trace, care must be taken for those input waveforms which contain power level changes along the trace. One procedure for handling such waveforms is to first compute a power average using an averaging window which is equal to the length of the deconvolution operator. The averaging procedure is conducted for all such windows in the input trace. The value of σ_x^2 in equation (15) is then set equal to the maximum power average obtained in this manner. As a result, the algorithm will be guaranteed to provide stable processing, regardless of the amplitude variations present in the input trace. When α is set on data peaks in this manner, values in the range 0.3 to 0.8 have been shown to provide appropriate processing results for seismic data. This procedure was used for all processing results presented in the sections following.

SYNTHETIC DATA PROCESSING RESULTS

The adaptive deconvolution procedure described herein is designed for use in those seismic processing applications containing nonstationary statistical properties over relatively short-time intervals. A synthetic digital trace (4 msec sample period) representing an example of nonstationary behavior is presented in Figure 4. The trace contains two primary events, denoted by P_1 and P_2. The constant-amplitude multiples for each event have a spacing which increases by four msec at each order of the multiple. Thus, while the first

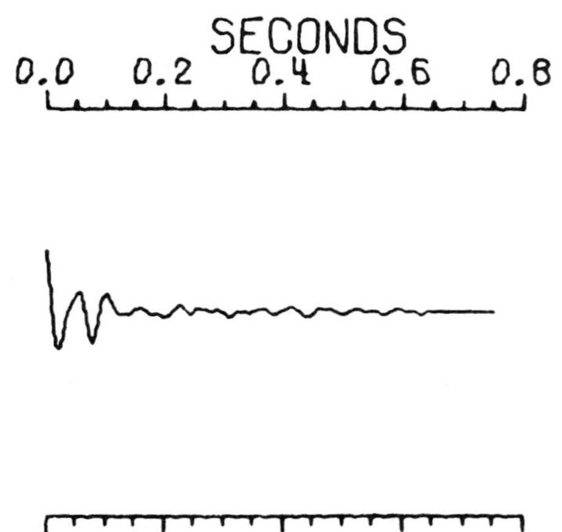

FIG. 5. Autocorrelation function calculated using 0 to 3.0 sec design window on synthetic time-varying trace.

multiple arrives about 160 msec after the primary, the multiple spacing near the end of the trace is 212 msec.

When conventional, fixed design gate deconvolution methods are applied to data of this type, the results are generally less than completely satisfactory. Because the correlation statistics of the trace vary with time along the trace, some form of time varying processor must be employed. For example, Figure 5 illustrates the autocorrelation function calculated using a 0 to 3.0 sec window on

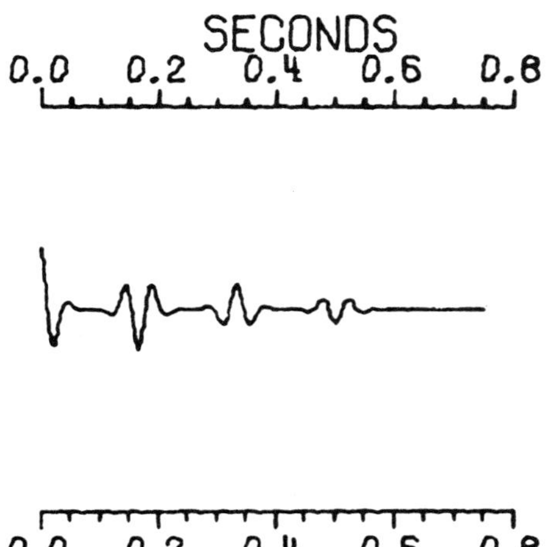

FIG. 6. Autocorrelation function calculated using 0 to 0.75 sec design window on synthetic time-varying trace.

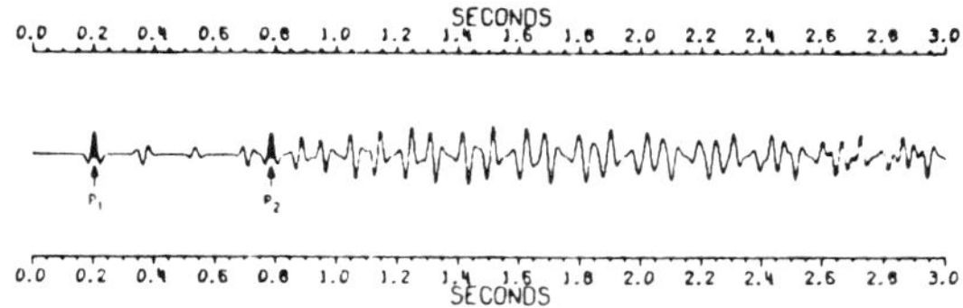

FIG. 7. Conventionally deconvolved trace obtained using 0 to 0.75 sec design window (80 msec operator with 120 msec prediction distance).

the trace in Figure 4. Little evidence of the multiples is observed due to their time-varying nature. Shorter design gates provide better evidence, however, as illustrated in Figure 6 using a 0 to 0.75 sec design gate. A predictive deconvolution operator of length 80 msec and prediction distance 120 msec was designed from this autocorrelation function using the normal equation solution method given by Peacock and Treitel (1969). The resulting operator was then applied to the entire trace and the deconvolved output is presented in Figure 7. As expected, the result indicates superior deconvolution results over the design gate relative to other regions of the trace. Better overall deconvolution may be achieved using the piecewise stationary deconvolution method described by Wang (1969). Figure 8 illustrates results obtained using a total of five overlapping design windows, each one sec in length, and an appropriately chosen prediction distance for each window.

Although the primary events are more readily distinguishable in Figure 8 than in Figure 7, a considerable amount of residual multiple energy remains in the deconvolved trace. Presumably, shorter design gates would provide improved performance. The computational load of this procedure, however, increases geometrically as shorter gates are selected and may well prevent a completely satisfactory deconvolution procedure.

The results obtained for this synthetic seismogram using adaptive deconvolution with reverse-time only, forward-time only, and reverse/

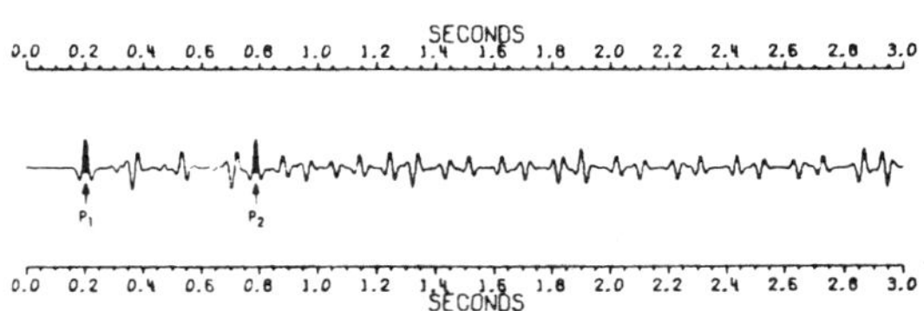

FIG. 8. Conventionally deconvolved trace using five design windows (80 msec operator with appropriate prediction distance for each window).

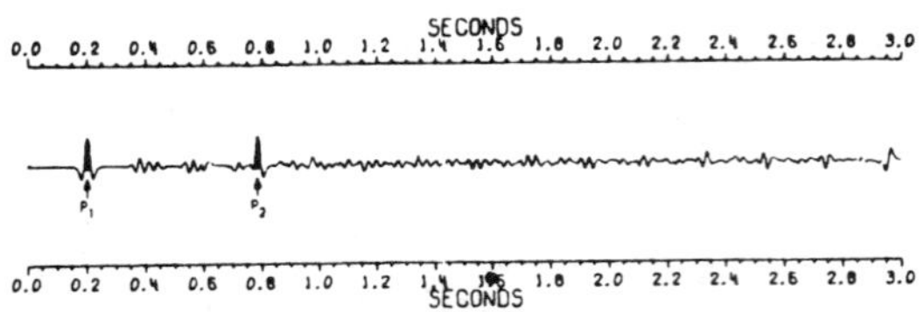

FIG. 9. Adaptively deconvolved trace obtained using only time-reverse processing.

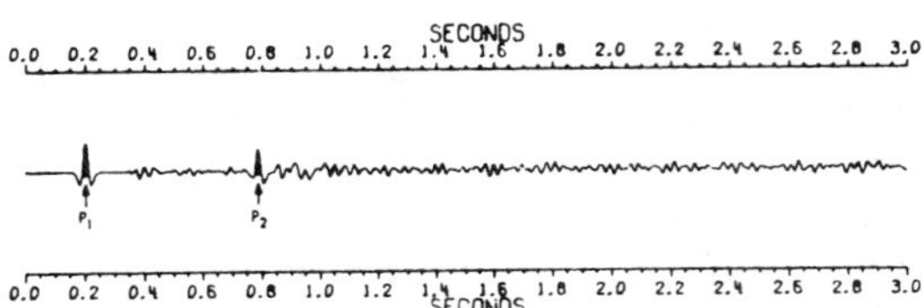

FIG. 10. Adaptively deconvolved trace obtained using time-forward processing.

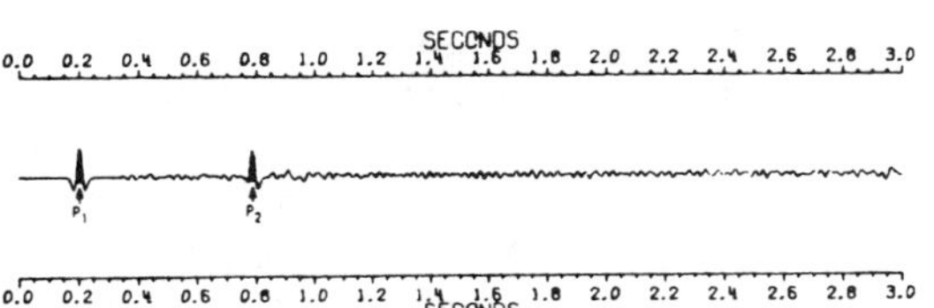

FIG. 11. Adaptively deconvolved trace obtained using time-reverse and time-forward, then sum processing.

forward stack are shown in Figures 9, 10, and 11, respectively. These results were achieved using an 80 msec length operator and constant 150 msec prediction distance throughout the trace. The value of α used to generate the deconvolved traces was 0.5. A comparison of Figures 9 and 11 indicates clearly the degree to which multiple rejection can be improved using the stacking procedure.

When adaptive deconvolution is applied to a trace containing noise only (i.e., unpredictable data), little modification of the input trace occurs—provided that the proportionality constant remains within the range previously given. This effect is illustrated here using the synthetic bandpass noise example shown as the original trace in Figure 12. This trace was generated by filtering white noise with a 5–60 Hz bandpass filter. The output traces observed after applying adaptive deconvolution to this example with different values of α are also shown in Figure 12. The operator parameters used for these calculations were similar to those used in the previous synthetic experiments, which were an operator length of 80 msec and a prediction distance of 120 msec.

Inspection of Figure 12 shows that little change in output is observed as α varies over a 50 to 1 range (0.01 to 0.50). However, at $\alpha = 0.50$ the effects of misadjustment noise are barely discernible which is to be expected since equation

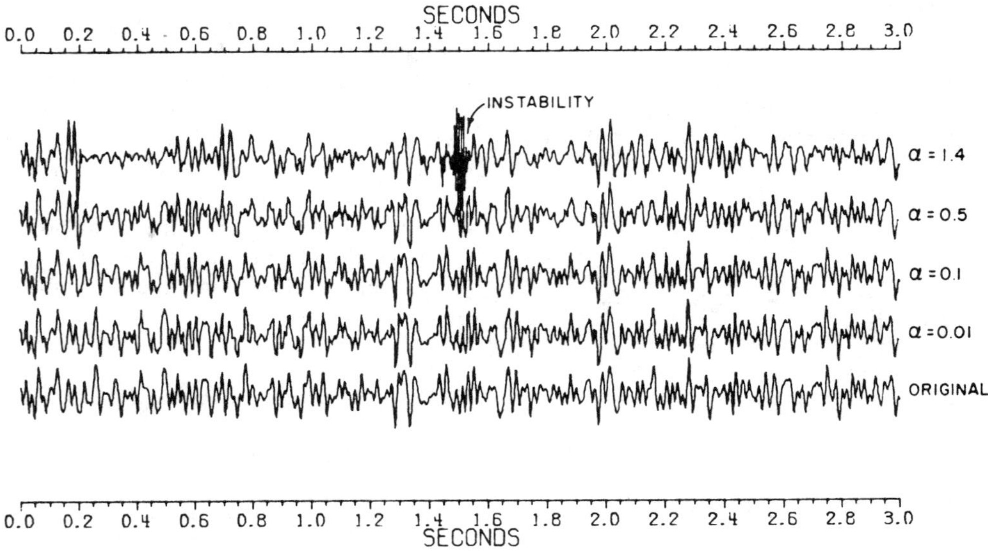

FIG. 12. The effects of adaptive deconvolution on filtered (6–60 Hz) white noise (i.e., unpredictable data) using varying values of α. The adaptive filter length was 80 msec and the prediction distance was 120 msec.

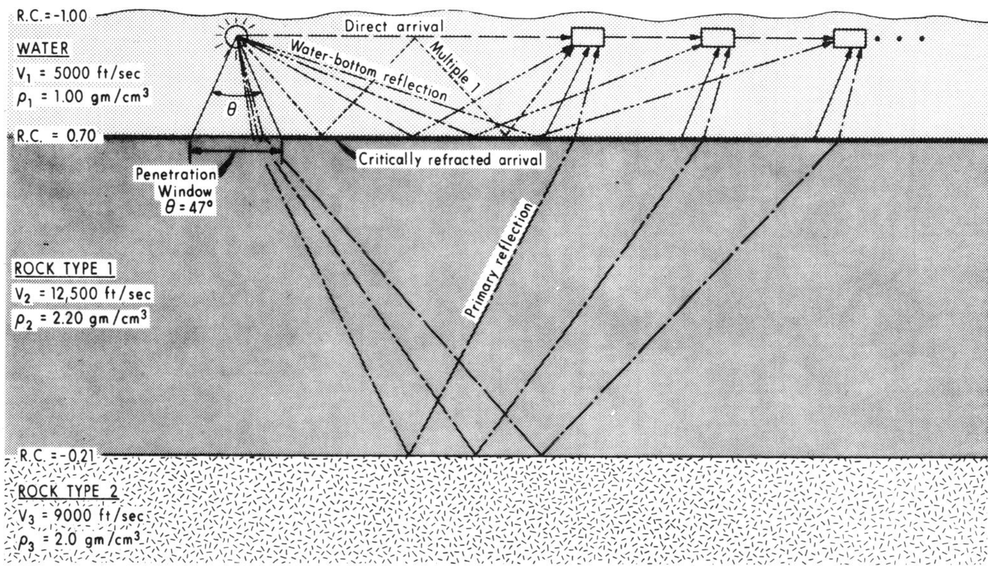

FIG. 13. A simplified diagram of a hard water-bottom model showing the raypaths for some of the arrivals.

(17) predicts a 50 percent increase in noise power at α = 0.5. The output trace that results using a value of α = 1.4 shows the filter response when driven to instability at 1.50 sec. Note that this value of α is well above the values recommended for practical application as discussed earlier. It is important to realize that the character of the unstable filter response is typically a rapid oscillating of the polarity of the output trace. We still have much to learn about the filter behavior outside the range on α which guarantees stability of the adaptive deconvolution filter.

In summary, synthetic experiments have demonstrated that adaptive deconvolution can provide effective multiple removal from a time-varying input trace, provided that these variations are slower than the time constant of adaptation. In addition, for recommended filter parameters it

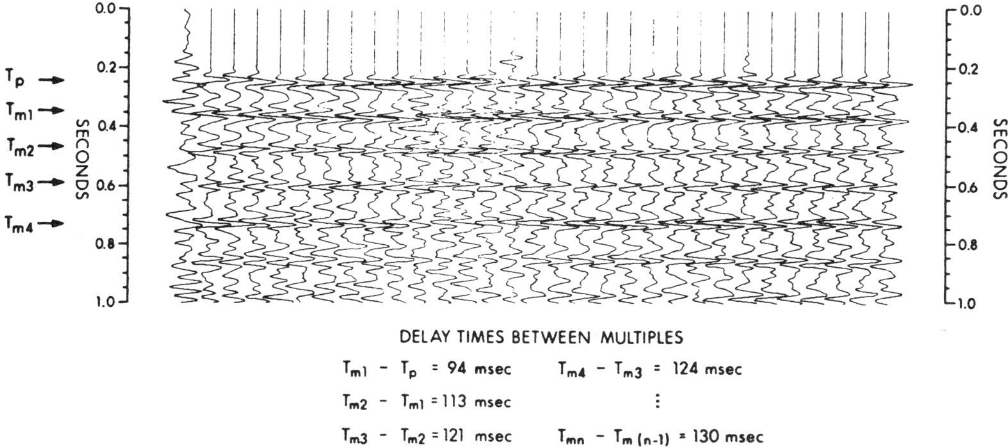

FIG. 14. A plot showing the near-offset traces for several adjacent shotpoints from line A with offset equal to 914 ft and water depth equal to 300 ft. T_p = traveltime of primary reflection and T_{M1} = traveltime of the first water-bottom multiple, etc.

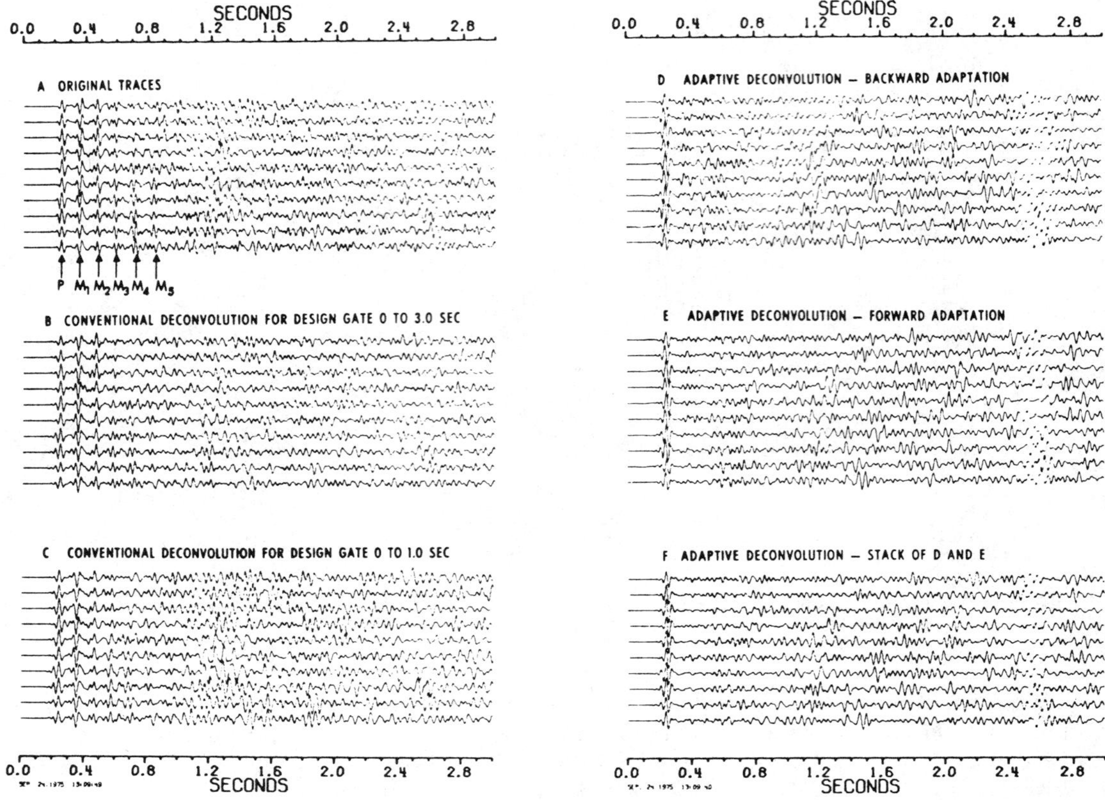

FIG. 15. Comparison of conventional and adaptive deconvolution results for near-offset traces from line B for ten adjacent shotpoints. (Offset = 914 ft, water depth = 300 ft, operator length = 80 msec, prediction distance = 120 msec, and α = 0.5.)

does not appreciably change noise-like (or primaries only) input traces which contain no multiple energy. We have also illustrated that adaptive deconvolution is a relatively robust procedure. That is, the output trace is not a critical function of the adaptive step size α, and increasing or decreasing its value by a factor of two will not significantly affect performance. Of course, many unanswered questions remain regarding the characteristics of adaptive processors in general. These include an appropriate performance model for those cases in which the input trace time variations are of the order of an adaptive time constant or less. Another relatively unknown area relates to the effects of gain variations on adaptive performance. While research on these problems is continuing, we feel that we have demonstrated that adaptive filtering is an effective deconvolution process. The next section completes the demonstration using field-recorded data.

FIELD-RECORDED DATA PROCESSING RESULTS

The purpose of this section is to illustrate the use of adaptive deconvolution methods on field-recorded marine seismic data which contain time-varying water-bottom reverberations. The seismic data shown are from the Celtic Sea offshore Ireland.

Reflection seismic data acquired in areas where the sediments at the water bottom possess high velocity and/or density are notoriously poor quality. Two of the main reasons for the poor quality of such reflection data are: (1) the small amount of energy transmitted through the water-sediment interface; and (2) the large amount of energy reflected between the water-sediment interface and the surface of the water which is nearly a perfect reflector. Under these conditions, the water layer is a very efficient wave guide where many types of

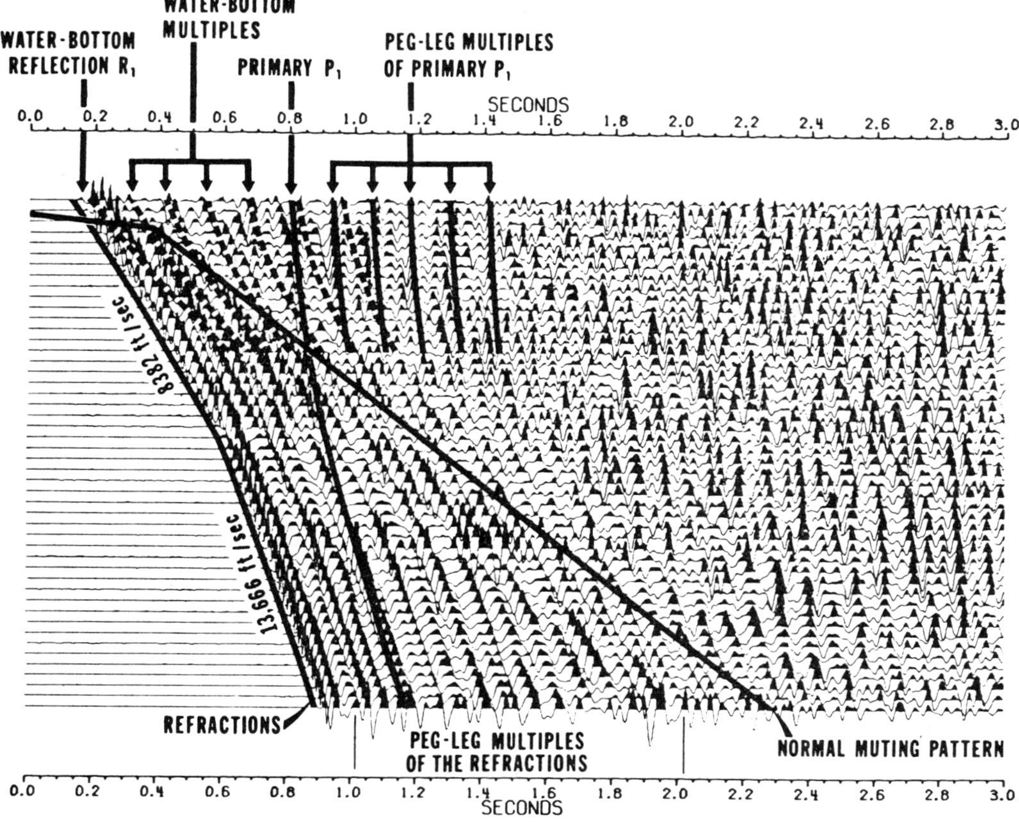

FIG. 16. An unprocessed 48-trace record from line C showing several of the primary reflection arrivals and refraction arrivals and their multiples. Near-offset distance = 669 ft, group interval = 164 ft, and water depth = 310 ft.

source-generated noises can propagate. The result of these conditions is usually low-amplitude primary reflections present in a background of very high amplitude and very coherent noise. Two of the most serious sources of noise are the refraction arrivals (and their multiples) and the water-bottom reverberations. These arrivals are shown on the simplified diagram in Figure 13. The reflection from the water bottom, the critical refraction along the water-sediment interface, and each of the primary reflections from interfaces in the sub-surface materials are followed by multiples within the water layer. The simple reflection coefficients for compressional waves at each of the interfaces are shown in Figure 13. One can compute, using Snell's law, the angular extent of the penetration window, which for the simplified model is 47 degrees.

The water-bottom multiples referred to in this paper are those arrivals which follow the primary reflection from the water-sediment interface making additional two-way trips through the water layer (see for example, the multiple 1 in Figure 13). The character of the water-bottom multiples is illustrated in Figure 14 which shows the near-offset traces only from several adjacent records in line A. The offset of these traces was 914 ft, and the water depth was approximately 325 ft. Clearly, the water-bottom reverberations dominate these near-offset traces. The most troublesome characteristic of these multiples is that the period of multiples is not constant. That is, the delay time between each multiple is a function of the order of the multiple resulting in a time series of reverberations which is time variant. One can easily calculate what the variations in the periods

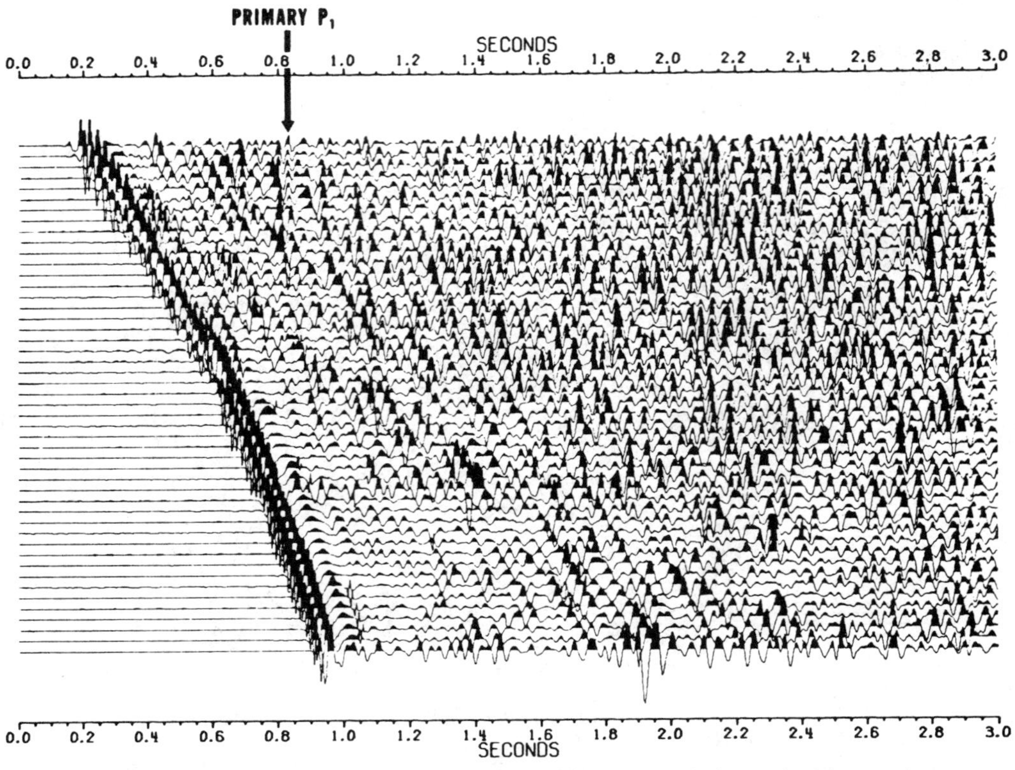

FIG. 17. An unmarked version of the unprocessed record shown in Figure 16.

of the multiples will be by assuming a cable geometry and a water depth and calculating the distances traveled by the various orders of multiples. The results of such calculations are shown in the lower portion of the figure. Note that the delay times increase with the order of the multiple and eventually become equal to the two-way vertical traveltime in the water layer. While the differences in the delay times of the multiples are not large, they are great enough to violate the assumption of stationary statistics used in conventional deconvolution and thus cause conventional deconvolution to be ineffective.

A comparison of conventional and adaptive deconvolution results for field seismic data is shown in Figure 15 for the nearest offset traces from several adjacent shotpoints of field data on line B. As shown in Figure 15a, the delay time between the primary P and the first order multiple M_1 is shorter than the period between M_1 and M_2, etc. The results of conventional deconvolution with autocorrelation design windows 0.0 to 3.0 sec and 0.0 to 1.0 sec are shown in Figures 15b and 15c, respectively. The reverberations have not been satisfactorily removed by either of the conven-

tional deconvolution filters. Experiments have also been conducted applying conventional deconvolution filters after normal-moveout corrections, but the results were also unsatisfactory.

The same near-offset traces are shown after the backward, forward, and summing implementations of adaptive deconvolution in Figures 15d, 15e, and 15f, respectively. Clearly, the water-bottom multiples have been satisfactorily removed. The parameter used in the processing shown in Figures 15d, 15e, and 15f were operator length = 40 msec, prediction distance = 84 msec, and α = 0.5. The data at traveltimes greater than 1.2 sec have not been altered by the filter because there were no periodic events present. That is, the adaptive deconvolution technique does not alter the data in regions not containing significant multiple structure.

Naturally, adaptive deconvolution must be applied to all traces of a common-depth-point (CDP) gather. An unprocessed 48-trace CDP gather is shown in Figure 16 with several of the events of interest identified. In this case, the sediments beneath the water layer are made up of two layers, one rock type having compressional veloci-

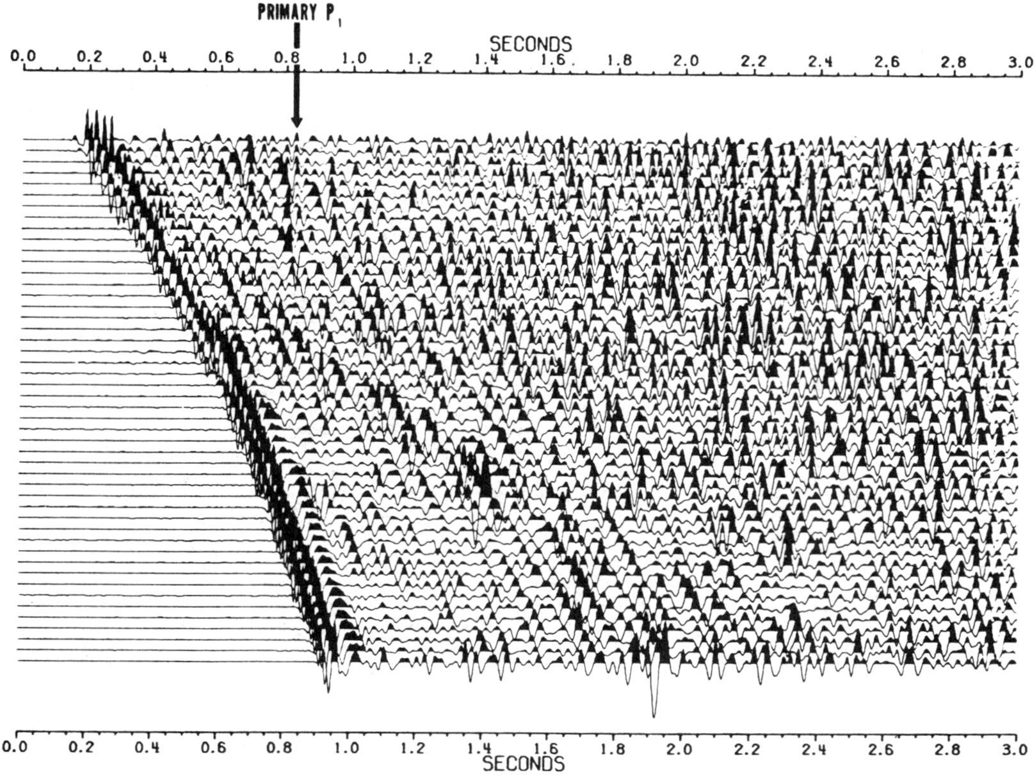

Fig. 18. The same 48-trace record from line C after adaptive deconvolution was applied. (40 msec operator length, 84 msec prediction distance, and $\alpha = 0.5$).

ties of 8382 ft/sec and the other having compressional wave velocity equal to 13,666 ft. The most obvious events on the record are the peg-leg multiples of the refraction arrivals (i.e., refraction arrivals which have made additional trips through the water layer). Note that the critical distance of the peg-leg multiples of the refraction appear to increase with the order of the multiples. This signature implies that each of the water-bottom multiples generate new refractions which are, in effect, closer to the detector array with each order of the multiple. The result of these peg-leg refraction multiples is that the data at offsets larger than the normal muting pattern shown on the record must be omitted from the stack. The result of this muting is to reduce the fold of stack at small traveltimes, which reduces the ability of the stacking process to remove the water-bottom multiples shown in Figure 16. Thus, it become imperative to remove the water-bottom multiples within the usable "data window" before stack. If these multiples are not removed by deconvolution, any

primary reflection within the reduced-stack data window will be masked by the multiples.

If one calculates the normal-moveout hyperbola of a primary event near the base of the high-velocity layer (e.g., the event at 800 msec in Figure 16), the reflection approaches asymptotically the linear moveout pattern of the refraction multiples. Clearly it would be desirable to remove the refraction multiples while leaving the primary event, if possible, so as to increase the fold of effective CDP stack on the primary. Finally, the peg-leg multiples following the primary event at 800 msec are also shown on the 48-trace record. Peg-leg multiples are difficult to remove by conventional predictive deconvolution if there are only a few primary events in the data. This characteristic is due to the lack of evidence of these multiples in the autocorrelation function calculated from the data. Adaptive deconvolution, on the other hand, appears to recognize the presence of these peg-leg multiples and effectively removes them as shown in Figure 18 when compared to the unmarked

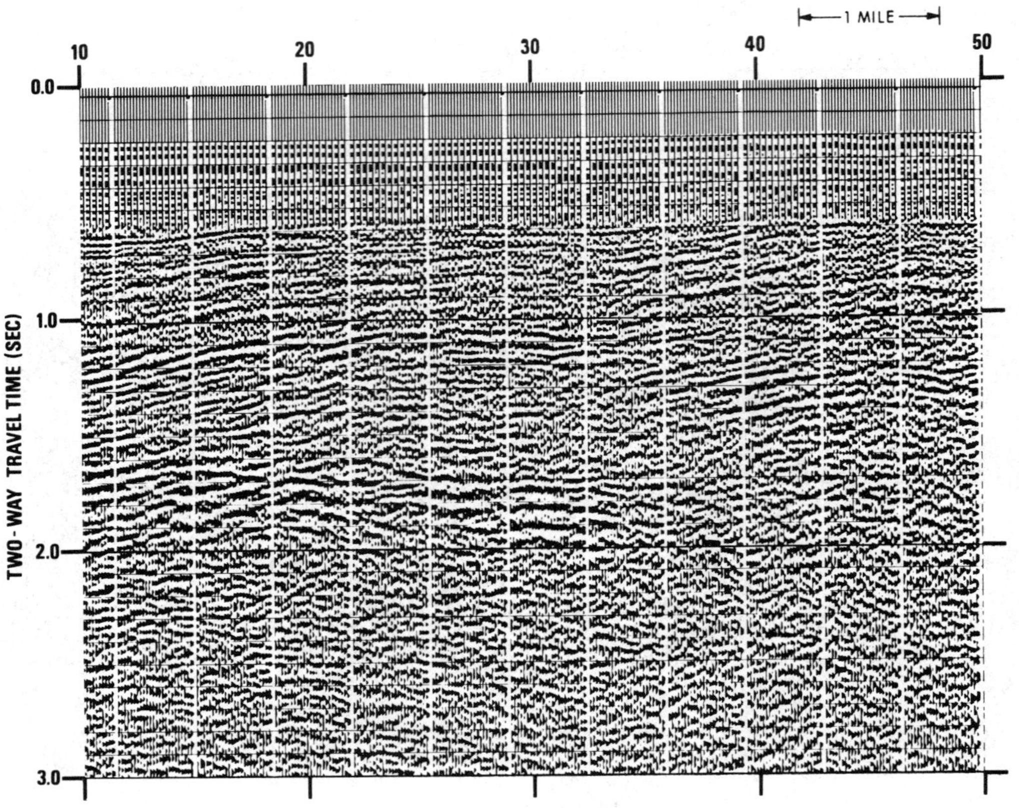

FIG. 19. 48-fold stacked section of line C with conventional deconvolution applied before stack.

original record shown in Figure 17. The reverse-then forward-time and stack implementation of adaptive deconvolution was used in processing the record shown in Figure 18. The values used for operator length, prediction distance, and α were 40 msec, 84 msec, and 0.5, respectively. It should be noted that, in some cases, the prediction distance may need to be made shorter for traces at increased offset distance (e.g., when arrivals with converging normal-moveout patterns, such as wa-ter-bottom multiples, are the dominant noise to be removed at large offsets). For the record shown in Figure 18, however, the refraction multiples are the most dominant arrivals to be removed and their linear moveout pattern requires a prediction distance which is constant with offset distance.

Clearly the adaptive deconvolution has re-moved most of the multiples of the refraction arrivals. However, it is questionable that any wide-angle primary reflections are left in the data. The conclusion drawn from these studies is that it is best to mute these far-offset data and not take

the chance of stacking a refraction multiple and mistaking it for a primary reflection.

Comparison of the traces within the usable data window in Figures 16, 17, and 18 shows that adaptive deconvolution has effectively removed the water-bottom multiples and the peg-leg multi-ples from the primary event at 800 msec. It is also clear that the primary event P has been left undis-torted in both amplitude and waveform.

The final stacked sections obtained using con-ventional deconvolution before stack are shown in Figures 19 and 21, and those obtained using adaptive deconvolution before stack are shown in Figures 20 and 22. While the plotting scales on the stacked sections may not be exactly the same, comparison of the amplitudes of primary events and first arrivals show they are fairly close. Thus, the appearance of lower overall amplitude on the sections is mostly due to the removal of predict-able energy by the adaptive deconvolution proc-ess. The adaptive deconvolution parameters were the same as those used in Figure 18. We feel that

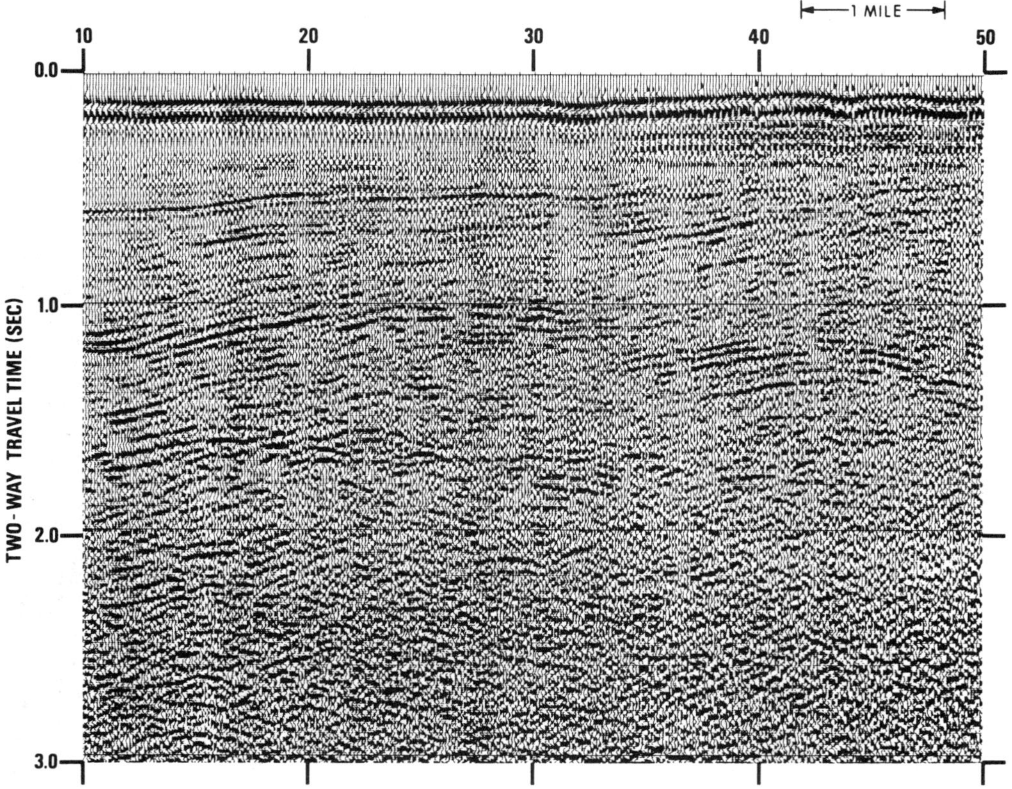

FIG. 20. 48-fold stacked section of line C with time-reverse, time-forward, then sum adaptive deconvolution applied before stack.

the use of adaptive deconvolution has made the sections more interpretable, especially at less than 1.0 sec traveltime where the base of the high-velocity surface layer is found. For example, the event at 650 msec at shotpoint 10 can be confidently mapped in Figure 20, at least to shotpoint 35, while it was not clearly mappable in Figure 19. The improvement in the shallow reflections on line D (compare Figures 21 and 22) is less apparent, but the water-bottom multiples have definitely been removed by the adaptive deconvolution. Comparisons on both lines C and D show that the peg-leg multiples associated with the deeper events have been more effectively removed by the adaptive deconvolution than by conventional deconvolution, thus making interpretation less ambiguous.

DISCUSSION AND CONCLUSIONS

This paper has presented a new time-varying deconvolution method for processing seismic re-flection data. The method is based on the use of a simple adaptive algorithm which allows continuous updating of the deconvolution operator as the seismic trace is processed. Previously published work relating to the use of this algorithm, particularly in the communications field, has established that the procedure offers significant improvements in processed signal-to-noise ratio. In the present paper, the algorithm was applied to both synthetic and field-recorded reflection seismic data which contained a significant number of time-varying multiple reflections. The results obtained demonstrate that the multiple energy was effectively removed by the adaptive deconvolution procedure.

These experimental results have also demonstrated several advantages of the method. First, it was found that the effectiveness of the technique in removing undesired multiple energy is not a sensitive function of the parameters required to implement the procedure. Specifically, the scalar

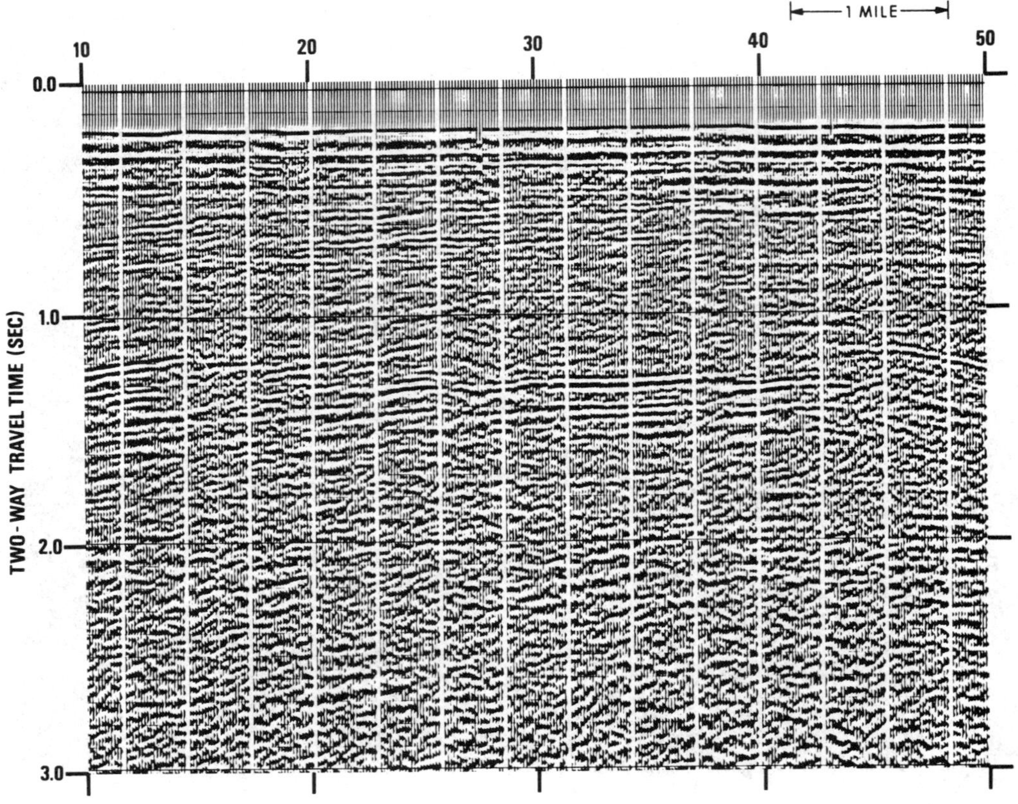

FIG. 21. 48-fold stacked section of line D with conventional deconvolution applied before stack.

constant α which controls the adaptive time constant can be varied between at least 0.3 and 0.8 without significantly affecting the quality of the deconvolved trace. A second advantage of adaptive deconvolution is that its use is not restricted to high signal-to-noise ratio input data. Of course, the ability of the procedure to remove time-varying multiples will be degraded as the quality of the input data is reduced. A third advantage noted is that the number of arithmetic operations required to implement adaptive deconvolution is actually smaller than the number required to implement conventional deconvolution methods. To illustrate, an input trace containing N data points can be deconvolved with an L point adaptive operator using $2NL$ multiplies and adds. Conventional deconvolution with five design gates would require the order of $2NL + 5L^2$ multiplies and adds.

Possibilities for further research on adaptive deconvolution include:

1) Implementation of the algorithm parameters (filter length, prediction distance, and adaptive time constant) in a time-varying manner.
2) A theoretical study of the convergence properties of the algorithm in nonstationary statistical environments.
3) The extension of adaptive deconvolution methods to multi-channel processing.

In summary, this paper has given a method for removing water-bottom multiples and refraction multiples. Clearly, the removal of these multiples is not the complete solution to obtaining good quality reflection seismic data in hard water-bottom areas. Much work remains to be done to develop the acquisition and processing techniques necessary to overcome all of the problems in such areas.

ACKNOWLEDGMENTS

We wish to thank Marathon Oil Co. for permission to publish this paper. In addition, we thank R. R. Burke, B. E. Merchant, and C. R.

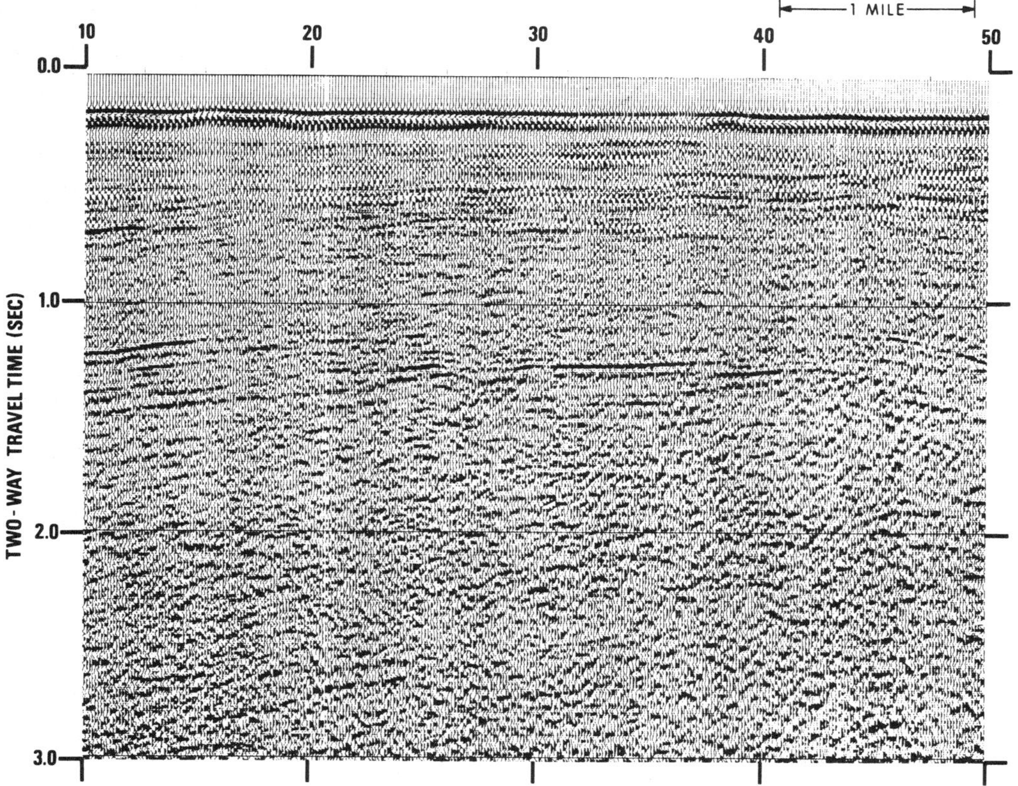

FIG. 22. 48-fold stacked section of line D with time-reverse, time-forward, then sum adaptive deconvolution applied before stack.

Harwood of Marathon Oil Co. Production International for supplying us with the seismic data.

REFERENCES

Blum, J. R., 1954, Multidimensional stochastic approximation methods: Annals Math. Stat., v. 25, p. 737–744.

Booker, A., and Ong, C., 1971, Multiple-constraint adaptive filtering: Geophysics, v. 36, p. 498–509.

Boonton, R. C., 1952, An optimization theory for time-varying linear systems with non-stationary statistical inputs: Proc. IRE, v. 40, p. 977–981.

Buhl, P., Stoffa, P. L., and Bryan, G. M., 1974, Application of homomorphic deconvolution to shallow-water marine seismology: Geophysics, v. 39, p. 401–426.

Clarke, G. K. C., 1968, Time-varying deconvolution filters: Geophysics, v. 33, p. 936–944.

Crump, N. D., 1974, A Kalman filter approach to the deconvolution of seismic signals: Geophysics, v. 39, p. 1–13.

Daniell, T. P., 1970, Adaptive estimation with mutually correlated training sequences: IEEE Trans. Systems Sci. and Cybern., p. 12–19.

Dupac, V., 1965, A dynamic stochastic approximation method: Annals Math. Stat., v. 36, p. 1695–1702.

Frost, O. L., III, 1972, An algorithm for linearly-constrained adaptive array processing: Proc. IEEE, v. 60, p. 926–935.

Gabriel, W. F., 1976, Adaptive arrays: An introduction: Proc. IEEE, v. 64, p. 239–271.

Gersho, A., 1968, Convergence properties of an adaptive filtering algorithm: Proc. 2nd Asilomar Conf. Circuits and Systems, p. 302–304.

Griffiths, L. J., 1969, A simple adaptive algorithm for real-time processing in antenna arrays: Proc. IEEE, v. 57, p. 1696–1704.

——— 1975, Rapid measurement of digital instantaneous frequency: IEEE Trans., v. ASSP-23, p. 207–222.

——— 1976, Time-domain adaptive beamforming of H F back-scatter radar signals: IEEE Trans., v. AP-24, no. 5.

Kim, J. K., and Davisson, L. D., 1975, Adaptive linear estimation for stationary M-dependent processes: IEEE Trans., v. IT-21, p. 23–31.

Lacoss, R. T., 1968, Adaptive combining of wideband array data for optimum reception: IEEE Trans. Geosci. Electron., v. 6, p. 78–86.

Middleton, D., and Whittlesey, J. R., 1968, Seismic models and deterministic operators for marine reverberation: Geophysics, v. 33, p. 557–583.

Peacock, K. L., and Treitel, S., 1969, Predictive deconvolution: Theory and practice: Geophysics, v. 34, p. 155-169.

Riegler, R. L., and Compton, R. T., Jr., 1973, An adaptive array for interference rejection: Proc. IEEE, v. 61, p. 748-758.

Robbins, H., and Monroe, S., 1951, A stochastic approximation method: Annals Math. Stat., v. 22, p. 400-407.

Robinson, E. A., 1967, Multichannel time series analysis with digital computer programs: San Francisco, Holden-Day, Inc.

Senne, K., 1968, Adaptive linear discrete-time estimation: Ph.D. dissertation, Stanford Univ.

Wang, R. J., 1969, The determination of optimum gate length for time-varying Wiener filtering: Geophysics, v. 34, p. 683-695.

Wang, R. J., and Treitel, S., 1971, Adaptive signal processing through stochastic approximation: Geophys. Prosp., v. 19, p. 718-727.

Widrow, B., 1966, Adaptive filters I: Fundamentals: Stanford Electronics Lab., rept. SEL-66-126.

Widrow, B., and Hoff, M. E., 1960, Adaptive switching circuits: IRE WESCON Conven. Rec., part 4, p. 96-104.

Widrow, B., Mantey, P. E., Griffiths, L. J., and Goode, B. B., 1967, Adaptive antenna systems: Proc. IEEE, v. 55, p. 2143-2159.

Widrow, B., et al, 1975, Adaptive noise cancelling: Principles and applications: Proc. IEEE, v. 63, p. 1692-1716.

Widrow, B., et al, 1976, Stationary and non-stationary learning characteristics of the LMS adaptive filter: Proc. IEEE, v. 64, p. 1521-1529.

9

TIME-VARYING PREDICTION FILTERING BY MEANS OF UPDATING *

BY

D. RISTOW ** and B. KOSBAHN **

ABSTRACT

RISTOW, D., and KOSBAHN, B., 1979, Time-varying Prediction Filtering by Means of Updating, Geophysical Prospecting 27, 40-61.

In contrast to the conventional deconvolution technique (Wiener-Levinson), the spike-, predictive-, and gap-deconvolution is realized with the help of an adaptive up-dating technique of the prediction operator. As the prediction operator will be updated from sample to sample, this procedure can be used for time variant deconvolution. Updating formulae discussed are the adaptive updating formula and the sequential algorithm for the sequential estimation technique. This updating technique is illus-trated using both synthetic and real seismic data.

INTRODUCTION

This paper describes a practical time-variant updating procedure for the prediction filter operator or for the prediction error operator in applied seismics. Updating procedures are discussed in literature on signal theory and control theory under the terms "adaptive filter" and "sequential filter". Several authors, for instance Zypkin (1972), prefer the term "learning process". In the field of automatic control theory the procedure falls under the general term "identification techniques" (see for instance Alam and Sage (1977)).

A detailed description on adaptive filters was recently published by Widrow, McCool, Lavimore, and Johnson (1976) in a special issue on adaptive systems. Gibson, Melsa, and Jones (1975), Mehra (1971), Alam (1974), and Kashyap (1974) discussed this procedure under the term sequential filter. Further articles published on system identification can be found in a special issue of IEEE-Transactions on automatic control (1974).

Papers dealing with applied geophysics were published by Wang and Treitel (1971) on adaptive signal processing, by Alam (1974) on sequential adaptive deconvolution, by Griffiths, Smolka and Trembly (1977) on adaptive deconvolution and by Wang (1977). Berkhout and Zaanen (1976) published

---

* Presented at the thirty-eighth meeting of the European Association of Exploration Geophysicists, The Hague, June 1976.
** Prakla-Seismos GmbH, Haarstr. 5, 3000 Hanover 1, Federal Republic of Germany.

143

a detailed theoretical work on the Wiener Filter, the Kalman Filter, on auto-regressive estimation, and on recursive least-mean-squares estimation. The last part of their paper is closely related to updating methods and gives valuable theoretical references. Furthermore, many theoretical articles are found in journals on automatic control and many practice-orientated articles in journals on acoustics and speech analysis (e.g. Morgan and Craig 1976).

This paper describes the updating of the prediction operator for every sample of a seismic trace. In contrast to the conventional Wiener-Levinson method, the autocorrelation is not directly required and calculated. The updating procedure is outlined stepwise, starting from simple considerations. This method leads to the adaptive algorithm of Widrow et al (1976) and to the algorithm of the sequential estimation technique. By application of the adaptive sequential procedure to synthetic and real data the efficiency of the spike-, predictive-, and gap-deconvolution is outlined and demonstrated.

FORMULATION OF THE PROBLEM

The deconvolution process has become a generally accepted and theoretically well-founded standard process and has been applied to seismic data for more than fifteen years. In an abstract form it can be described as follows:

A seismic trace is the result of a convolution of a reflectivity function with an operator whereby this operator can be a wavelet, a reverberation operator, or a ghost operator. Only the reflectivity function is of primary importance for the interpretation of seismic measurements. The operator is a disturbing factor, and the aim of the deconvolution process is to eliminate the operator by means of an anti-operator. A linear anti-operator can be calculated by means of a least-mean-squares method from the estimated auto- and cross-correlation functions of the seismic trace. This "Wiener-Levinson"-filter operator is applied to the given seismic trace and regenerates, in a statistical sense, the reflectivity function. The deconvolution process described above assumes that the seismic trace has stationary statistical properties. As this assumption is practically never satisfied, one arrives at the time-variant deconvolution.

Problems arise in the realization of a time-variant deconvolution (see fig. 1). A non-stationary seismic trace is subdivided into time-segments in which stationarity is assumed. A specific operator is applicable to each time-segment. Usually, an interpolation of the two operators takes place in the transition zone of two neighbouring segments.

The problem of segmenting a trace has been theoretically studied but for practical problems it is not yet sufficiently clarified. A theoretically feasible, but practically useless, solution would be the following one: operator $\mathbf{f}_n$ is calculated for a given time-gate n; the time-gate is then shifted by one sample,

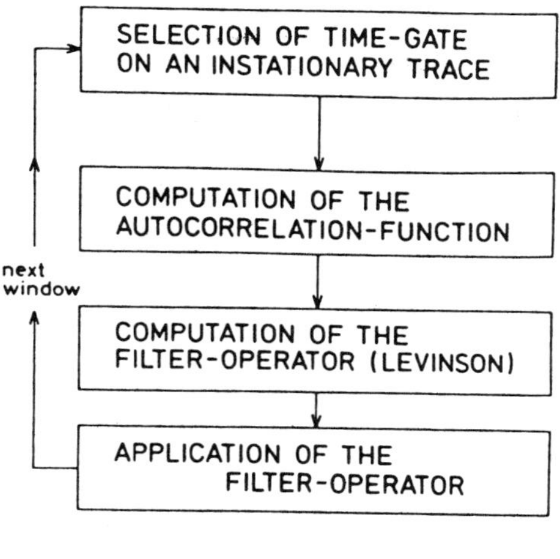

Fig. 1. Flow chart of conventional time-varying-deconvolution.

and a new operator $\mathbf{f}_{n+1}$ is calculated for time-gate $n + 1$ etc. The operators calculated with great mathematical expenditure will differ only by a small amount from one time-gate to the next. One question arises automatically: Will it be possible to calculate an operator according to a least-mean-squares fit for every sample of a seismic trace by digressing from the above mentioned Wiener-Levinson method and by making use of the small variation of the operator from one sample to the next? This problem will be solved with an updating procedure that will be explained in the next section.

UPDATING PROCEDURE

Basic Principles

Fig. 2 outlines the set-up for the solution of the problem. The given seismic trace x_t is filtered with the prediction operator $\mathbf{f}_n$ which is valid at the time $t = n \cdot \Delta t$ in order to calculate the prediction value x'_{n+p}.

We have

$$x'_{n+p} = \sum_{\nu=0}^{M} f_\nu x_{n-\nu}, \tag{1}$$

where

x_n — samples of a given seismic trace,

$\mathbf{f}_n$ — prediction operator, valid at the time $t = n \cdot \Delta t$, $(M+1)$-dimensional vector,

Δt — sampling interval,

$(M+1)$ — length of the prediction operator,

p — prediction distance.

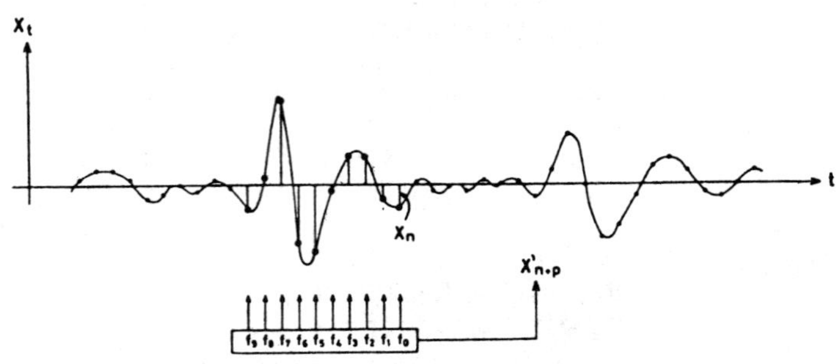

Fig. 2. Illustration of the prediction-operator. For updating the predicted value x'_{n+p} will be compared with the actual value x_{n+p}.

In vector notation (1) reads:

$$x'_{n+p} = \mathbf{f}_n^T \cdot \mathbf{x}_n = \mathbf{x}_n^T \cdot \mathbf{f}_n, \tag{2}$$

where

$$\mathbf{x}_n = (x_n, x_{n-1}, x_{n-2}, \ldots, x_{n-M})^T \tag{3}$$

$$\mathbf{f}_n = (f_0, f_1, f_2, \ldots, f_M)^T.$$

In order to check and to update the operator it is necessary to compare the predicted value x'_{n+p} with the actually measured value x_{n+p}. If the difference $x_{n+p} - x'_{n+p}$, the prediction error, is small, it is of no use to modify the operator essentially. If the difference is large, the performance of the operator has to be improved.

The set-up for the modification of the filter operator from time t to time $t + \Delta t$ reads as follows (with p short for $p \cdot \Delta t$):

$$\mathbf{f}_{t+\Delta t} = \mathbf{f}_t + \mathbf{C} \, (x_{t+p} - x'_{t+p}), \tag{4}$$

where

$$x'_{t+p} = \mathbf{f}_t^T \cdot \mathbf{x}_t = \mathbf{x}_t^T \cdot \mathbf{f}_t. \tag{5}$$

The filter operator at the time $(t+\Delta t)$ is equal to the filter operator at the time t plus a correction which is controlled by the prediction error. $\mathbf{C}$ is an unknown gain-vector that still needs to be specified and the structure of which becomes evident from the following considerations (after Widrow et al 1976).

Equation (4) is solved by the method of the steepest descent. The change of the operator is proportional to the negative of a gradient vector still to be defined with a proportionality factor μ:

$$\mathbf{f}_{t+\Delta t} = \mathbf{f}_t + \mu \cdot (-\mathbf{grad}\ \Phi\ (f_0, f_1, \ldots, f_M)). \tag{6}$$

The gradient vector is:

$$\mathbf{grad}\ \Phi = \left(\frac{\partial \Phi}{\partial f_0}, \frac{\partial \Phi}{\partial f_1}, \frac{\partial \Phi}{\partial f_2}, \ldots, \frac{\partial \Phi}{\partial f_M} \right). \tag{7}$$

Φ represents any differentiable function of the operator coefficients f_i.

The least-mean-squares technique uses for Φ the square of a single error sample:

$$\Phi = \varepsilon_t^2 = (x_{t+p} - \sum_{\nu=0}^{M} f_\nu x_{t-\nu})^2; \tag{8}$$

the components of the gradient vector are:

$$\frac{\partial \Phi}{\partial f_j} = 2 \left(x_{t+p} - \sum_{\nu=0}^{M} f_\nu x_{t-\nu} \right) \cdot (- x_{t-j}) \tag{9}$$

with $j = 0, 1, 2, \ldots, M$.

Inserting (9) into (6) we obtain:

$$\mathbf{f}_{t+\Delta t} = \mathbf{f}_t + 2\mu \cdot \mathbf{x}_t \ (x_{t+p} - \mathbf{f}_t^T \cdot \mathbf{x}_t). \tag{10}$$

Comparison of (10) with (4) yields

$$\mathbf{C} = 2\mu \cdot \mathbf{x}_t. \tag{11}$$

Obviously, the actual samples $x_t, \ldots x_{t-M}$ contribute to the updating of the operator.

The scalar parameter 2μ is a convergence factor. It affects the stability of the procedure. If it is too small, updating is done too slowly; if it is too large, there is no stable solution. Two questions concerning the convergence factor remain open: for which values of μ does the procedure converge, and towards which value does it converge?

Widrow et al (1976) have outlined the following details for the stationary case: from theoretical considerations the gradient should not be calculated from one single error sample, but rather from its expected value. The necessary condition for convergence yields (Widrow et al 1976):

$$0 < \mu < \frac{1}{\lambda_{max}} \tag{12}$$

where λ_{max} is the largest eigenvalue of the input correlation matrix:

$$\mathbf{R} = \mathrm{E}\,[\mathbf{x}_t \mathbf{x}_t^T] \tag{13}$$

As long as (12) is valid, the solution converges towards the Wiener solution.

In the derivation of formula (10), however, the gradient of a single square error was used. It can be shown that the gradient of a single square error sample is equal to the unbiased estimate of the gradient of the mean-square-error.

If the sequence of the values x_t, x_{t-1}, ... x_{t-M} is uncorrelated, i.e. if

$$\mathrm{E}\,[\mathbf{x}_t \mathbf{x}_{t+p}^T] = \mathbf{0} \quad \text{for} \quad p \neq 0 \tag{14}$$

is valid, the expected value of the gradient estimate is equal to the true gradient, and the solution vector converges, in a statistical sense, towards the Wiener solution (Widrow et al 1976). The condition (12), however, must be fulfilled.

In order to get a feeling for the largest eigenvalue of correlation-matrices, we have calculated the autocorrelation functions of the signals with different spectral band-width. The autocorrelation function consisted of forty-one samples at a sampling rate of $\Delta t = 2$ ms. The largest eigenvalues ($\lambda_i > 1$) are presented in fig. 3. The center frequency was 50 Hz. Δf is the bandwidth. The signal energy is concentrated in the frequency band from $\left(50 - \dfrac{\Delta f}{2}\right)$ Hz to $\left(50 + \dfrac{\Delta f}{2}\right)$ Hz.

Fig. 3 illustrates the fact—already well known in filter theory—that the distribution of the eigenvalues clusters with growing Δf, and all eigenvalues converge to unity in the limiting case of a white spectrum.

For practical purposes it is rather cumbersome to determine the largest eigenvalue of $\mathbf{R}$. The following estimate is valid for a positive definite matrix $\mathbf{R}$

$$\text{trace } \mathbf{R} > \lambda_{max} \tag{15}$$

trace $\mathbf{R}$ can be estimated from the actual samples:

$$\text{trace } \mathbf{R} \approx \sum_{i-1}^{t-M} x_i^2 = \mathbf{x}_t^T \cdot \mathbf{x}_t. \tag{16}$$

From (15) and (16) follows for the updating formula (10)

$$\mathbf{f}_{t+\Delta t} = \mathbf{f}_t + \frac{\alpha \cdot \mathbf{x}_t}{\mathbf{x}_t^T \cdot \mathbf{x}_t} \cdot (x_{t+p} - x_{t+p}'), \tag{17}$$

where

$$x'_{t+p} = \mathbf{f}_t^T \cdot \mathbf{x}_t. \tag{18}$$

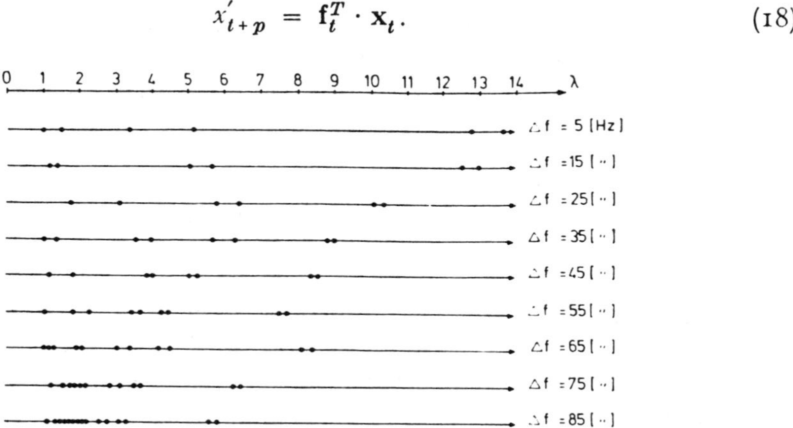

Fig. 3. Plot of eigenvalues $(\lambda_i \geq 1)$ depending on the bandwidth Δf.

The remaining factor α is of the order 1 due to the inequality (15). Griffiths et al (1975) used this formula for the adaptive deconvolution. This formula is an updating formula that can be well used for seismic measurements and easily programmed on a computer.

It has been outlined by many authors, e.g. Griffiths et al (1977), that the updating-formula (17) can be used for stationary input data as well as for instationary input data.

To discuss formula (17) we apply once again the updated filter operator $\mathbf{f}_{t+\Delta t}$ to the data vector $\mathbf{x}_t$.

We multiply (17) by $\mathbf{x}^T$ to obtain

$$x''_{t+p} = \mathbf{x}_t^T \cdot \mathbf{f}_{t+\Delta t} = x'_{t+p} + \alpha\,(x_{t+p} - x'_{t+p}) \tag{19}$$

x_{t+p} is the value predicted with the updated operator. It can clearly be seen that for $\alpha = 1$ the operator changes so strongly that after repeated prediction of the value x_{t+p} the following value is exactly obtained:

$$x''_{t+p} = x_{t+p} \text{ for } \alpha = 1 \tag{20}$$

We thus obtain for $\alpha = 1$ an extremely strong updating procedure. For smaller values of α (< 1) x''_{t+p} is, according to (19), a linear combination of x_{t+p} and x'_{t+p}.

Improvement of the Updating-Formulae

Up to now, the updating formulae, starting from (4) via (10) to (17) have been steadily improved by taking into account certain mathematical criteria. By means of intuitive, heuristic procedures further modifications can be applied to the gain vector in formula (17).

We replace for instance in (17)

$$\frac{\alpha \cdot \mathbf{x}_t}{\mathbf{x}_t^T \cdot \mathbf{x}_t} \quad \text{by} \quad \frac{\alpha\,(t) \cdot \mathbf{x}_t}{\mathbf{x}_t^T \cdot \mathbf{x}_t + R_t}. \tag{21}$$

The effect of both parameters is as follows: function $\alpha\,(t)$ should be a monotonously decreasing function, for instance $1/t$. This factor $\alpha\,(t)$ causes the updating to decrease with increasing time and the prediction operator to change only slightly.

The summand R_t in the denominator of formula (21) can be interpreted as a noise term. If there is strong noise on the seismic traces, R_t is large, and vice versa.

Introducing R_t and $\alpha\,(t)$ into formula (21) seems to be somewhat arbitrary; it results, however, in a further improvement of the updating formulae. The formulae (17), taking (21) into account, reads:

$$\mathbf{f}_{t+\Delta t} = \mathbf{f}_t + \frac{\alpha\,(t) \cdot \mathbf{x}_t}{\mathbf{x}_t^T \cdot \mathbf{x}_t + R_t}\,(x_{t+p} - \mathbf{f}_t^T \cdot \mathbf{x}_t) \tag{22}$$

Gibson et al (1975) used this updating algorithm for speech analysis.

The updating formulae (4), (10), and (17) up to now contain no information on the improved performance of the updated filter operators. An improvement of the updating formulae is achieved if, in addition to the updating formula of the filter operator, the error of the filter operator is updated; thus, two values are updated from sample to sample, the coefficients of the prediction filter and the covariance-matrix of the prediction filter. This is realized by means of the Kalman filter the formal properties of which are described in the appendix. The first updating formula for the filter operator reads:

$$\mathbf{f}_t = \mathbf{f}_{t-\Delta t} + \mathbf{k}_t\,[x_{t+p} - \mathbf{f}_{t-\Delta t}^T \cdot \mathbf{x}_t], \tag{A 14a}$$

where the gain vector is given by:

$$\mathbf{k}_t = \frac{\mathbf{P}_t'}{\mathbf{x}_t^T \cdot \mathbf{P}_t' \cdot \mathbf{x}_t + R_t} \cdot \mathbf{x}_t. \tag{A 16a}$$

$\mathbf{P}_t'$ is an $(m \times m)$-matrix, the predicted covariance-matrix. The second updating formula for the error-matrix (covariance-matrix) reads:

$$\mathbf{P}_t = \mathbf{P}_t'\,(\mathbf{I} - \mathbf{k}_t \cdot \mathbf{x}_t^T), \tag{A 15a}$$

where

$$P'_t = P_{t-\Delta t} + Q_{t-\Delta t},\qquad\qquad\text{(A 12a)}$$

e.g.

$$Q_t = q_t \cdot I \quad\text{or}\quad Q_{t-\Delta t} = q_{t-\Delta t} \cdot I \qquad\qquad\text{(A 19)}$$

where q_t is the plant-noise term and I is the unit matrix. Q_t is called plant-noise matrix. In a similar way the formulae (A 14a) to (A 19) were discussed by Gibson et al. (1975).

The two updating formulae (A 14a) and (A 15a) are coupled, and at a first glance it seems to be difficult to recognize their mode of operation. We therefore consider several extreme cases and start with the assumption that the updating process already progressed so far that the operator nearly reached a stable state.

This means that the operator—in a statistical sense—predicts well the actual values. In this case the difference in the right bracket of formula (A 14a) is small on the average, and the operator at time t is practically equal to the operator at time $t - \Delta t$ plus a small amount.

If we assume that from a certain t on the measured actual values x_{t+p} are very noisy, parameter R_t must be increased drastically; then the gain vector, according to formula (A 16a) is nearly equal to the zero vector, and from (A 14a) follows that the operator $f_{t-\Delta t}$ is equal to the operator f_t. In this case the operator is only extrapolated.

The opposite case is valid if parameter R_t is decreased so that both the operator $f_{t-\Delta t}$ and the actual samples contribute to update the new operator f_t.

As we know that the signal-to-noise ratio on seismic data will decrease with increasing time, the parameter R_t should increase with time.

The influence of another parameter, given by the scalar q_t (in A 19) can be clarified in the following way: we increase q_t (e.g. $q_0 = 0.05$) to a large value (e.g. $q = 100$). Consequently, the covariance-matrix P'_t, according to formula (A 12a) practically corresponds to the matric $q \cdot I$, thus from formula (A 16a) follows for the gain vector:

$$k_t \cong \frac{x_t}{x_t^T \cdot x_t + \dfrac{R_t}{q}} \qquad\qquad\text{(23)}$$

Neglecting $\dfrac{R_t}{q}$ in (23) we obtain the updating formula (17), in which α is set equal to one. But from (19) and (20) we know that this updating formula (19) with $\alpha = 1$ includes strong adaptability. With parameter q_t we are able to control the updating formula (A 14a, etc.) for use with non-stationary data.

Initial Values

Some parameters, vectors, and matrices, must be set or initialized for the updating process. The prediction operator, consisting of $(M + 1)$-samples, can be initialized in different ways. The first and most simple method is to set all samples equal to zero. However, a learning period ensues so that the first predicted results become unreliable. The second method, theoretically the best (one), consists in deriving the prediction operator for the initial segment of a seismic trace from the conventional Wiener-Levinson algorithm. The third method is the most convenient: it consists in using the prediction operator from a seismic trace $x_n(t)$ for the time $t = T$ as an initial operator for the seismic trace $x_{n+1}(t)$ for the time $t = 0$.

The second initialization has to be done for the covariance matrix of the prediction operator. The samples of the prediction operator itself are of magnitude 1 or smaller. Therefore, we may set the covariance matrix equal to the unit matrix $\mathbf{I}$ if the zero operator was chosen for the initial operator $\mathbf{f}_0$.

$$\mathbf{P}_0 = \mathbf{I}. \tag{24}$$

If, however, an initial operator is chosen which was computed from the preceding seismic trace, this unit matrix is multiplied with a factor smaller than 1 (for instance $p = 0.05$), thus decreasing the variance of the initial filter operator:

$$\mathbf{P}_0 = p \cdot \mathbf{I} \quad \text{with} \quad p < 1. \tag{25}$$

Two noise parameters follow which are closely related to the noise statistics of the Kalman filter. First of all, the accuracies of the measured samples x_t are estimated. A seismic trace is suitably normalized (e.g. maximum = 1000). The assumed additive noise probably lies—after experimental studies and tests—in a range from $R_t = 50$ to $R_t = 500$. The choice of this parameter depends on the quality of the data and can be chosen in a time-varying manner. To update the prediction operator, parameter R_t plays the following part: if it is too small, for instance $R_t = 0$, an extraordinary accuracy is assigned to the seismic data, the operator's learning process is incorrect and unstable. If parameter R_t is too large, the initial operator changes slightly and approaches slowly its stationary solution in the stationary case.

The second noise parameter is given by the plant noise (see A 19). This parameter $q_{t-\Delta t}$ is added to the diagonal members of the covariance matrix $\mathbf{P}_{t-\Delta t}$ (see A 12a) at every extrapolationstep of the filter operator from $t - \Delta t$ towards t. That means that the accuracy of the filter operator is decreased in every interpolation according to (A 12a). Thus, the actual samples, at constant R_t, get a higher weight, and the filter operator is updated the faster the larger the values that are assigned to q. For $q = 0$ we have the stationary case, the operator learns and at the same time the error of the operator is steadily diminished. From a certain point of time on, the variance of the filter operator has become so small that the actual samples of the trace with their specified accuracy are no longer important for the updating of the filter operator. Parameter q can also be interpreted as a learning readiness; large q gives the operator the possibility of being able to better adapt to the data in the non-stationary case.

Examples

In the following we present several examples for synthetic and real seismic data. As the predicted sample x'_{t+p} has to be calculated at every updating step, all predicted samples can be displayed for illustration.

Basically, three time functions can thus be represented:

1. Input trace
2. Predicted trace
3. Input trace minus predicted trace = output trace.

The third representation represents the application of the prediction error operator. The representation of the predicted portion is quite plausible and reveals the degree of reliability with which the predictable portion can be determined by the prediction operator.

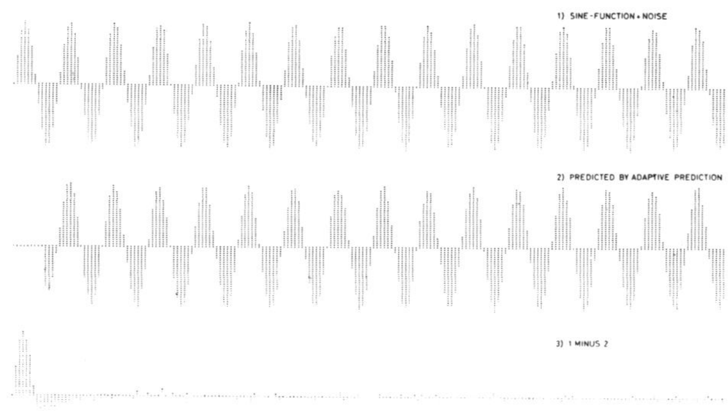

Fig. 4. Prediction of a deterministic function; 1) Sine-function and noise as input function; 2) Predicted function; 3) Input function minus predicted function as output function.

Fig. 4 presents a simple example for a deterministic predictable function. We have a sine function with additive white noise; the predicted portion is presented in the middle of fig. 4. It can clearly be seen how the operator learns and how the operator is able, after several samples, to predict with sufficient accuracy. The subtraction result is shown in the lower portion of fig. 4. It can theoretically be demonstrated that already a two-point operator exactly predicts a sine-function with the prediction distance $p = 1$.

Fig. 5 (upper partion) also presents a sine function as an input function; however, there is a hidden spike. The difference between input and predicted trace is shown in the lower portion of fig. 5. A large prediction error occurs where the non-predictable spike is located. The predicted time-function is not presented in this example.

Fig. 6 presents an example for spike deconvolution. Fig. 6-1 represents a part of a dynamically corrected seismogram. The result of spike deconvolution according to Wiener-Levinson is shown in the middle portion of fig. 6. The updating procedure yields a result that is comparable to conventional procedures. Its better adaptibility to the steadily varying frequency contents is obvious at the beginning of the traces. Parameters used for the updating procedure were: filter length $M = 20$, prediction distance $p = 1$, noise parameters $R_t = 100$, $q_t = 0.001$.

A further example for a predictive deconvolution is presented in fig. 7. Parameters used were: $M = 40$, $p = 24$ ms, $q_t = 0.001$. The prediction deconvolution was applied to a stacked section. Fig. 8 only shows a portion of fig. 7.

The length of the prediction operator is an important parameter and depends on the multiples to be removed. With increasing filter length, however, computing time will be enhanced drastically.

With certain multiple reflections, in particular long-period multiples, it is necessary to choose a large length for the operator, see for instance fig. 9 upper part.

It can also be shown that many samples of the operator are unnecessary. This is especially true for the elimination of sea bottom multiples. This consideration leads us to the well-known gapped operator, presented in fig. 9 (lower part). This becomes still more evident if one starts from the well-known three-point-anti-ringing operator after Backus (Peacock and Treitel 1969; see fig. 10 (upper part)). These three-point-operators can be applied if the time-interval between the multiples is known and remains constant. This, however, is not always the case in practice, so that after Kunetz and Fourmann (1968) a modification is needed (see fig. 10, below). The number of samples that differ from zero is increased. The gap, however, is still clearly recognizable.

The computation of such a gapped operator (see fig. 10, lower portion) according to the Wiener-Levinson method after Kunetz and Fourmann involves some mathematical complications but can be simply realized with the updating procedure. Fig. 11 (above) presents a reflectivity function $r(t)$ which was convolved with a known ringing operator $w(t)$:

$$x(t) = w(t) * r(t) \tag{26}$$

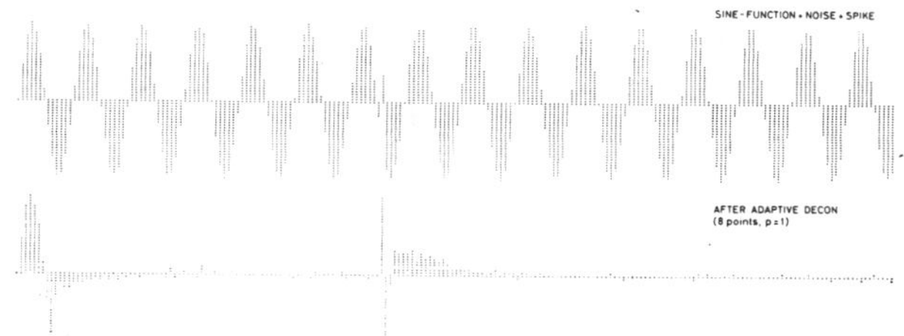

Fig. 5. Prediction of a deterministic function, built up by a sine function, a single spike and additive noise (top). The difference between the input-function (top) and the predicted function is shown below.

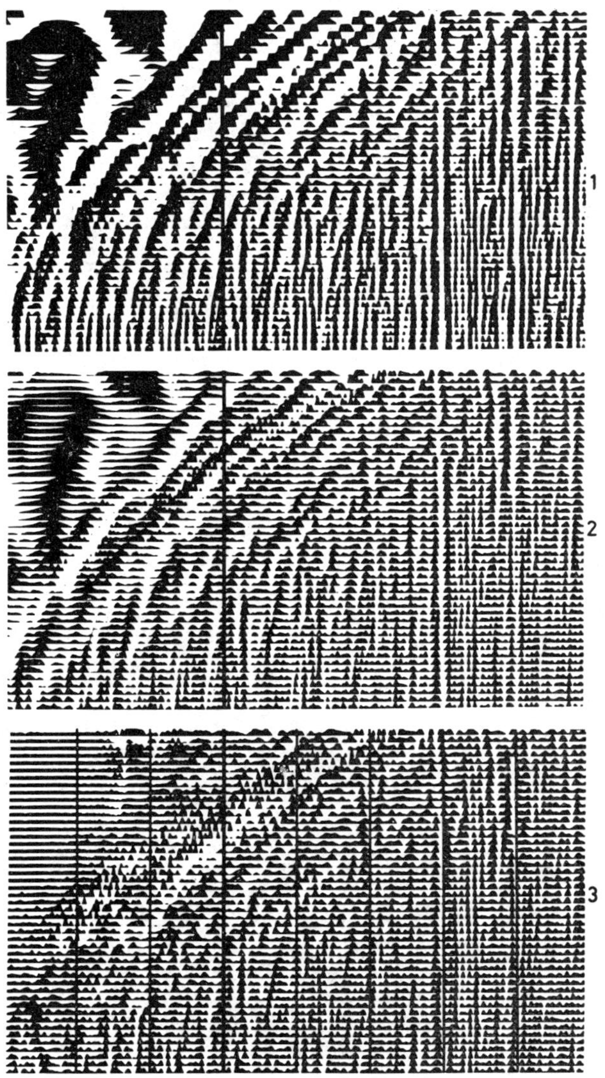

Fig. 6. 1) Single seismogram after dynamic correction. 2) Spike-deconvolution after Wiener-Levinson. 3) Spike-deconvolution after adaptive deconvolution.

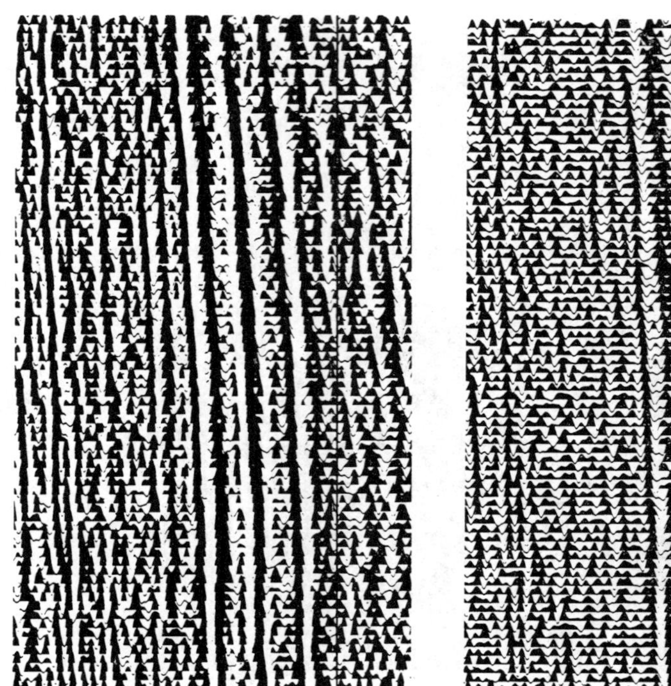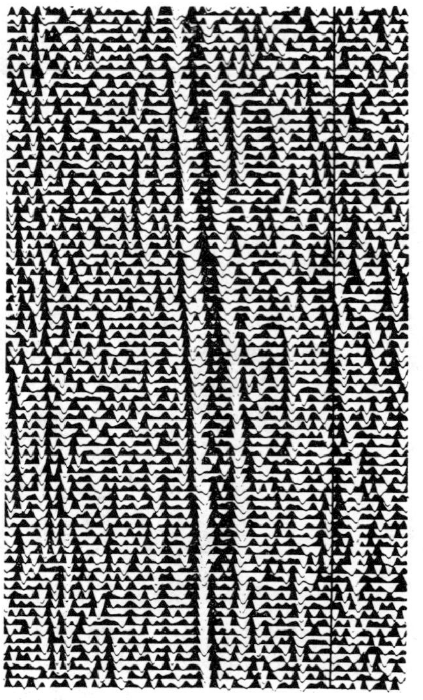

Fig. 8. Above: Portion of fig. 7 above, without any deconvolution. Below: Portion of fig. 7 below, adaptive predictive deconvolution.

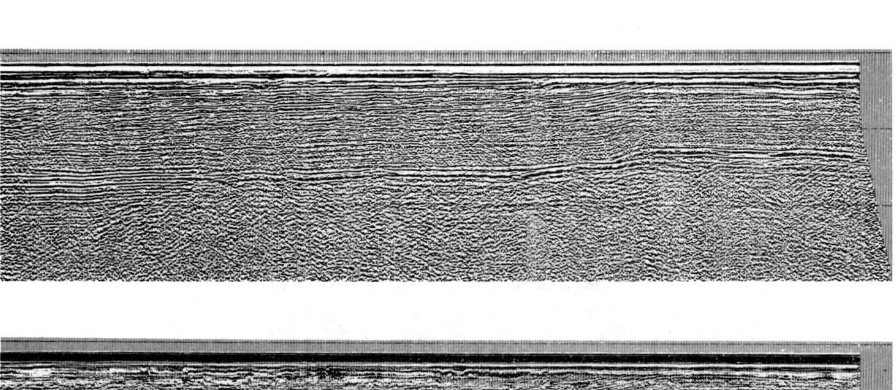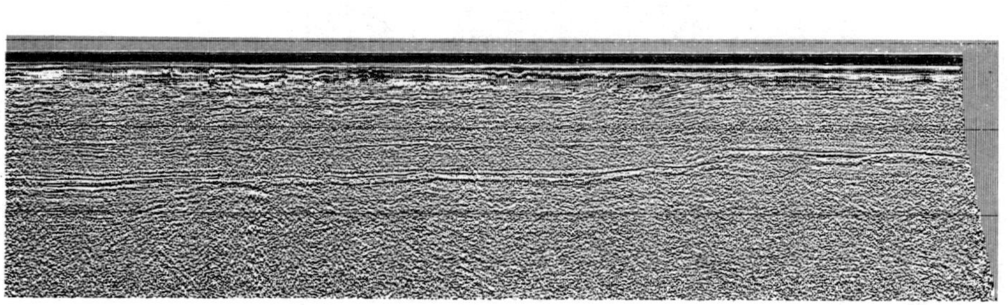

Fig. 7. Above: Stacked section without any deconvolution. Below: stacked section after adaptive predictive deconvolution, prediction distance $p = 24$ms.

            D. RISTOW AND B. KOSBAHN

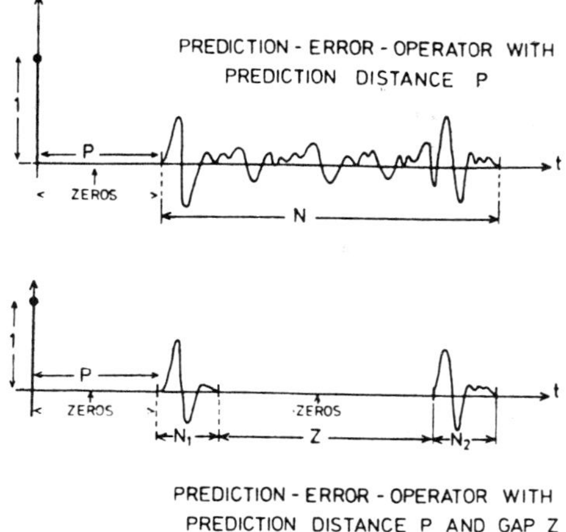

Fig. 9. Above: Conventional prediction-error-operator. Below: Gapped-prediction-error-operator.

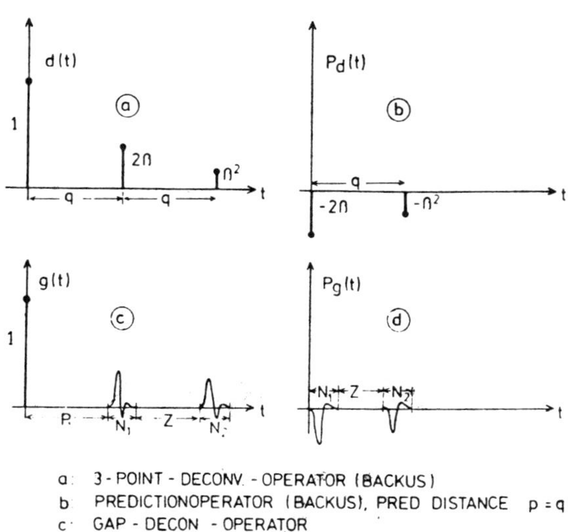

Fig. 10. The parameters of the gapped-operator.

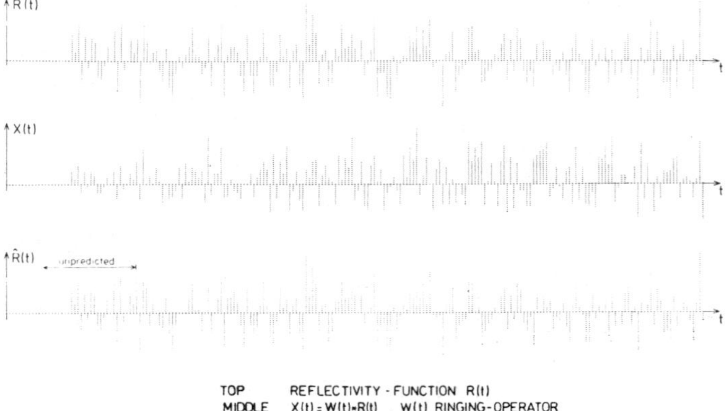

Fig. 11. The estimation of the reflectivity-function with help
of the adaptive gap deconvolution.

The z-transform of $w(t)$ is given by

$$W(z) \; = \; \frac{\alpha}{(1 + \beta \cdot z^q)^2} \tag{27}$$

where:

α is a scaling constant

β is a constant factor with $|\beta| < 1$

$q \cdot \Delta t$ is the two-way traveltime from the surface of the sea bottom.

The application of the gapped operator using the updating method yields the result presented in fig. 11 (below). The conformity with the reflectivity function is at first rather striking, but can simply be explained. The prediction operator was applied as shown in fig. 10 (above); β was determined adaptively.

The autocorrelation functions corresponding to fig. 11 are presented in fig. 12. Calculation of the autocorrelation function was only carried out for control purposes and was not at all used for the updating procedure.

Fig. 13 shows the effectiveness of the adaptive gap deconvolution. On the left hand side is a stacked marine section; direct water multiples are recognizable. The application of an adaptive-prediction gapped operator yields a predicted section (middle of fig. 13). This predicted section is an intermediate result of the procedure. By subtracting the predicted section from the input section we obtain a section which is free of some multiples. This is especially the case with the lower part of the section. It is rather striking to see how the direct water multiples have only slightly been affected by the procedure.

In fig. 14 two different procedures for multiple-attack were used. A coherency procedure after Buttkus (1978) for the suppression of the direct water-

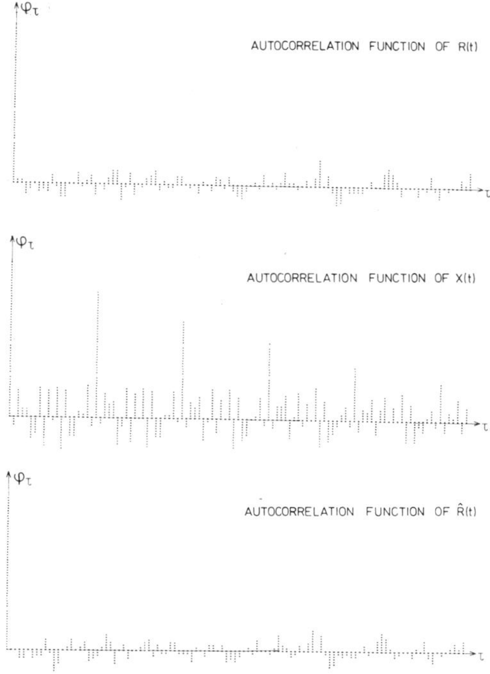

Fig. 12. The autocorrelation functions of the corresponding time series in fig. 11.

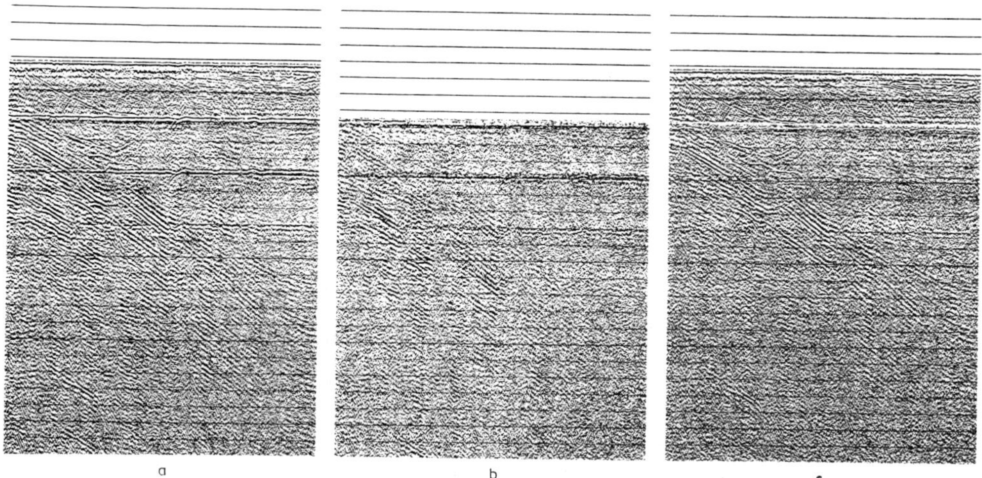

Fig. 13. a) Stacked section (marine) as input section; b) Predicted section with help of the adaptive gapped-operator; c) The output-section is equal to the input-section minus the predicted section.

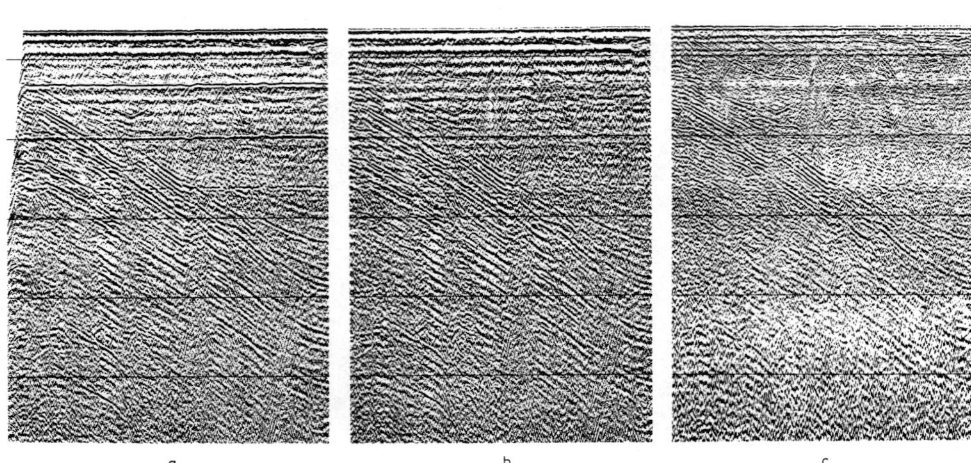

Fig. 14. a) Stacked section (marine) with strong direct water multiples; b) Stacked section (marine) after the suppression of the direct water multiples. The suppression was performed on the single traces by a special coherency procedure after Buttkus (1976); c) Stacked section (marine) after the adaptive gap deconvolution on the stacked section b).

multiples was applied on the single traces before stack, the result is to be seen in the middle of fig. 14. On this section the adaptive gap deconvolution was performed (on the right of fig. 14), the combination of both procedures leads to a result, in which direct multiples and other reverberations have been suppressed.

Conclusion

Spike-, predictive, and gap deconvolution can be realized by means of the adaptive sequential method. Since the operator is updated from sample to sample, time-variant deconvolution can be carried out by choosing corresponding parameters. Because of the samplewise updating, the computer time is larger than that required by Wiener-Levinson deconvolution. However, programmable array processors are most convenient to carry out adaptive processes rapidly.

Using the adaptive sequential method, new frequency analyses can be derived after Griffiths and Prieto-Diaz (1977), multichannel filters can be realized, and after Kashyap (1974) parameters of the process ARMA can be estimated.

Acknowledgement and Disclaimer

The development of the method described in this paper was supported with public funds by the Ministry of Research and Technology, Federal Republic of Germany, under the code ET-3115 A.

The ministry is not responsible either for the correctness, the accuracy, and completeness of the data, or for any infringement of private rights of third parties.

APPENDIX

In this appendix the sequential updating formula for the prediction operator and for the error matrix of the estimated prediction operator are derived with the help of the Kalman filter. For the basic description of the Kalman filter we refer to Sorenson (1966).

A dynamic system is described by two equations: first by a linear vector difference equation, describing the dynamic system

$$\mathbf{x}_k = \mathbf{\Phi}_{k,\,k-1} \cdot \mathbf{x}_{k-1} + \mathbf{w}_{k-1}, \tag{A 1}$$

where:

$\mathbf{x}_k$ is the state vector at the time k (n-dimensional column vector)
$\mathbf{\Phi}_{k,\,k-1}$ is the transition matrix ($n \times n$)
$\mathbf{w}_{k-1}$ is the white noise vector (plant noise, n-dimensional).

The linear dependence of the measurement data on the state vector is described by the second equation:

$$\mathbf{z}_k = \mathbf{H}_k \cdot \mathbf{x}_k + \mathbf{v}_k, \tag{A 2}$$

where:

$\mathbf{v}_k$ is a white noise vector (measurement noise, n-dimensional)
$\mathbf{H}_k$ is the mapping matrix ($m \times n$),
$\mathbf{z}_k$ is the measurement vector (m-dimensional).

The following should be valid for the noise vectors:

$$E\,[\mathbf{w}_k] = 0, \tag{A 3}$$

$$E\,[\mathbf{v}_k] = 0, \tag{A 4}$$

$$E\,[\mathbf{v}_k \mathbf{v}_j^T] = \mathbf{R}_k \cdot \delta_{k,j}, \tag{A 5}$$

$$E\,[\mathbf{w}_k \mathbf{w}_j^T] = \mathbf{Q}_k \cdot \delta_{k,j}, \tag{A 6}$$

and

$$E\,[\mathbf{v}_k \mathbf{w}_j^T] = 0 \quad \text{for all } k,j. \tag{A 7}$$

An estimated value $\hat{\mathbf{x}}_k$ of the state vector $\mathbf{x}_k$ is to be calculated from the measured vectors $\mathbf{z}_0, \mathbf{z}_1, \ldots, \mathbf{z}_k$ according to the least-mean-squares principle. The criterion for the determination of the estimated value is:

$$E\,[(\hat{\mathbf{x}}_k - \mathbf{x}_k)^T\,(\hat{\mathbf{x}}_k - \mathbf{x}_k)] = min. \tag{A 8}$$

The covariance matrix plays an important part, since it describes the accuracy of the estimated value $\hat{\mathbf{x}}_k$:

$$\mathbf{P}_k = \mathrm{E}\left[(\hat{\hat{\mathbf{x}}}_k - \mathbf{x}_k)(\hat{\hat{\mathbf{x}}}_k - \mathbf{x}_k)^T\right]. \tag{A 9}$$

In order to review Kalman filter computation we will decompose the processing sequence into logical steps (after Sorenson 1966).

1. The following estimates are given:

   $\hat{\mathbf{x}}_{k-1}$ is the optimal estimate at $k-1$,

   $\mathbf{P}_{k-1}$ is the covariance matrix at $k-1$. $\hspace{2cm}$ (A 10)

2. Prediction of the estimates at time k without using the measurement vector $\mathbf{z}_k$:

$$\mathbf{x}'_k = \mathbf{\Phi}_{k,k-1} \cdot \hat{\mathbf{x}}_{k-1}, \tag{A 11}$$

$$\mathbf{P}'_k = \mathbf{\Phi}_{k,k-1} \cdot \mathbf{P}_{k-1} \cdot \mathbf{\Phi}^T_{k,k-1} + \mathbf{Q}_{k-1}. \tag{A 12}$$

3. Measurement of the vector $\mathbf{z}_k$:

$$\mathbf{z}_k = \mathbf{H}_k \cdot \mathbf{x}_k + \mathbf{v}_k. \tag{A 13}$$

4. Optimal estimates, considering the measurement vector $\mathbf{z}_k$:

$$\hat{\mathbf{x}}_k = \mathbf{x}'_k + \mathbf{k}_k \cdot [\mathbf{z}_k - \mathbf{H}_k \cdot \mathbf{x}'_k], \tag{A 14}$$

$$\mathbf{P}_k = \mathbf{P}'_k - \mathbf{k}_k \mathbf{H}_k \mathbf{P}'_k, \tag{A 15}$$

where

$$\mathbf{k}_k = \mathbf{P}'_k \mathbf{H}^T_k \cdot [\mathbf{H}_k \mathbf{P}'_k \mathbf{H}^T_k + \mathbf{R}_k]^{-1} \tag{A 16}$$

(Kalman-gain-matrix).

After index k is incremented to $k+1$ the loop starts again at (A 10).

For the sequential adaptive procedure the following definitions are valid:

a) The state vector $\mathbf{x}_k$ is given by the $(M+1)$-dimensional prediction operator

$$\mathbf{f}_t = \{\hat{f}_0, f_1, \ldots, f_M\}^T$$

b) The transition matrix $\mathbf{\Phi}_{k,k-1}$ is equal to the unity matrix $\mathbf{I}$.

c) The mapping matrix $\mathbf{H}_k = \mathbf{H}_t$ is a row vector and contains the actual samples $x_t, x_{t-1}, \ldots, x_{t-M}$:

$$\mathbf{H}_t = (x_t, x_{t-1}, \ldots, x_{t-M}) = \mathbf{x}^T_t$$

d) The measurement vector $\mathbf{z}_k$ is a scalar and is given by the actual sample x_{t+p}.

At the beginning of the process, the unknown state vector $\mathbf{x}_0$ and the covariance matrix $\mathbf{P}_0$ must be initialized:

The state vector $\mathbf{f}_0$ is set equal to the zero vector:

$$\mathbf{f}_0 = \mathbf{0}. \tag{A 17}$$

The initial covariance matrix $\mathbf{P}_0$ is set equal to a matrix in which only the members on the main diagonal are different from zero:

$$\mathbf{P}_0 = p_0 \cdot \mathbf{I} \quad \text{e.g.} \quad p_0 = 1, \tag{A 18}$$

where $\mathbf{I}$ is the unit matrix.

The matrix $\mathbf{Q}$ of the plant noise is simply defined by

$$\mathbf{Q}_t = q_t \cdot \mathbf{I}. \tag{A 19}$$

q_t can also depend on t.

The matrix $\mathbf{R}$ of the measurement noise is a scalar:

$$\mathbf{R}_t = \mathrm{E}\,[\mathbf{v}_t^2] \tag{A 20}$$

Using assumptions a-d and (A 19) and (A 20) we write the corresponding formulae from (A 11) to (A 16):

$$\mathbf{f}_t' = \mathbf{f}_{t-1}. \tag{A 11a}$$

The equation corresponding to (A 12) reads

$$\mathbf{P}_t' = \mathbf{P}_{t-1} + \mathbf{Q}_{t-1} = \mathbf{P}_{t-1} + q_{t-1} \cdot \mathbf{I}. \tag{A 12a}$$

The variances on the main diagonal are enhanced by summing up the term q_{t-1}; that means the accuracy of the operator is reduced by the extrapolation from $(t-1)$ towards (t).

The measurement equation (A 13) reads:

$$z_k = x_{t+p}. \tag{A 13a}$$

To apply equations (A 14) and (A 15) the gain vector according to (A 16) must first be calculated.

In our case the general matrix inversion in (A 16) is a simple division. We obtain:

$$\mathbf{k}_t = \frac{\mathbf{P}_t' \cdot \mathbf{x}_t}{\mathbf{x}_t^T \cdot \mathbf{P}_t' \cdot \mathbf{x}_t + \mathbf{R}_t}. \tag{A 16a}$$

From (A 14) and (A 15) we obtain the two updating formulae:

$$\mathbf{f}_t = \mathbf{f}_{t-1} + \mathbf{k}_t \cdot [x_{t+p} - \mathbf{x}_t^T \cdot \mathbf{f}_{t-1}] \tag{A 14a}$$

and

$$\mathbf{P}_t = \mathbf{P}_t' [\mathbf{I} - \mathbf{k}_t \cdot \mathbf{x}_t^T]. \tag{A 15a}$$

Further reduction of computing time can be achieved by the simplifying assumption that $\mathbf{P}_t$ has non-zero elements only on the main diagonal, e.g.:

$$\mathbf{P}_t = \alpha_i\,\delta_{i,j}. \qquad (\text{A } 21)$$

Such simplifications reduce computing time considerably and lead to a suboptimal system.

If we set $\mathbf{P}'_t = \mathbf{I}$ and $\mathbf{R}_t = \mathbf{0}$ in (A 16a), we obtain from (A 14a) the updating formula according to formula (17) in the chapter on the updating procedure of this paper.

REFERENCES

ALAM, M. A., and SAGE, A. P., 1974, Error and sensitivity analysis of stochastic approximation algorithms for linear system identification, IEEE-Transactions Automatic Control AC-19, 810-815.

ALAM, M. A., 1974, A sequential adaptive deconvolution algorithm based on Kalman-filter approach, SEG meeting in Dallas.

BERKHOUT, A. J., and ZAANEN, P. R., 1976, A comparison between Wiener filtering, Kalman filtering, and deterministic least squares estimation, Geophysical Prospecting 24, 141-197.

BUTTKUS, B., 1979, Coherency weighting—an effective approach to the suppression of long leg multiples, Geophysical Prospecting 27, 29-39.

GIBSON, J. D., MELSA, J. L., and JONES, S. K., 1975, Digital speech analysis using sequential estimation techniques, IEEE-Transact. Acoustics, Speech, and Signal Processing ASSP 23, 362-368.

GRIFFITHS, L. G., SMOLKA, F. R., and TREMBLY, L. D., 1977, Adaptive deconvolution: A new technique for processing time-varying seismic data, Geophysics 42, 742-759.

GRIFFITHS, L. J., and PRIETO-DIAZ, R., 1977, Spectral analysis of natural seismic events using autoregressive techniques, IEEE-Transact. Geoscience Electronics GE-15, 13-25.

IEEE-Transactions on Automatic Control, 1974, Special issue on system identification and time-series analysis, AC-19, Nr. 6.

KASHYAP, T. L., 1974, Estimation of parameters in a partially whitened representation of a stochastic process, IEEE-Transact. Automatic Control AC-19, 13-21.

KUNETZ, G., and FOURMANN, J. M., 1968, Efficient deconvolution of marine seismic records, Geophysics 33, 412-423.

MEHRA, R. K., 1971, Approaches to adaptive filtering, IEEE-Trans. Automatic Control, AC-10, 693.

MORGAN, D. R., and CRAIG, S. E., 1976, Real time adaptive linear prediction using the least mean square gradient algorithm, IEEE-Transactions on Acoustics, Speech, and Signal Processing ASSP-24, 494-506.

PEACOCK, K. L., and TREITEL, S., 1969, Predictive deconvolution: theory and practice, Geophysics 34, 155-169.

SORENSON, H. W., 1966, Advances in control systems, theory and applications, Leondes, (ed), New York, Academic Press, v. 3, 219-292.

WANG, R. J., and TREITEL, S., 1971, Adaptive signal processing through stochastic approximation, Geophys. Prosp. 19, 718-728.

WANG, R. J., 1977, Adaptive predictive deconvolution of seismic data, Geophys. Prosp. 25, 342-381.

WIDROW, B., McCOOL, J. M., LAVIMORE, M. G., and JOHNSON, C. R., 1976, Stationary and nonstationary learning characteristics of the LMS-adaptive filter, Proceedings of the IEEE 64, 1151-1161.

ZYPKIN, J. S., 1972, Grundlagen der Theorie lernender Systeme, VEB Verlag Technik Berlin.

Part III

HOMOMORPHIC FILTERING

Editors' Comments
on Papers 10 Through 14

Homomorphic systems (Oppenheim, 1965; Oppenheim, Schafer, and Stockham, 1968; Schafer, 1969; Oppenheim and Schafer, 1975; Tribolet, 1979) are a class of nonlinear systems that satisfy a generalized principle of superposition. Such systems are particularly useful in separating signals that have been combined through convolution. As described earlier, a seismic trace, $[s(n)]$, is often represented as the convolution of a wavelet, $[w(n)]$, with the reflectivity function $[r(n)]$. Thus homomorphic filtering can be used effectively to recover the seismic wavelet from the seismic trace. This method does not require the seismic wavelet to be minimum phase or the reflector series to be random in nature, which are essential requirements for Wiener filtering and predictive deconvolution. The complex cepstrum $[S(n)]$ of the seismic trace contains the additive contributions of the wavelet and the reflection function. Seismic wavelets, in general, have a smooth power spectrum that causes the complex cepstrum of the wavelet $[W(n)]$ to concentrate around the time origin. However, the complex cepstrum of reflectivity function $[R(n)]$ is usually quite complicated. This complicated structure can be simplified by exponentially weighting the input sequence. This technique is suggested by Schafer (1969) and is commonly used in seismic applications to make the reflector series minimum phase. Shensa (1976) has described a modified complex exponential-weighting scheme that can be used to determine delay times in complex cepstrum for homomorphic deconvolution. For a minimum-phase reflector series, the reflectivity function will contribute to the complex cepstrum only for positive times greater than or equal to the interval between the first two arrivals. Thus the contributions of the wavelet cepstrum $[W(n)]$ and reflectivity cepsturm $[R(n)]$ may be separated, in principle, by low-pass and high-pass filters. Low-pass filtering will lead to the complex cepstrum of the wavelet $[W(n)]$ whereas high-pass filtering will result in the complex cepstrum of the reflectivity function $[R(n)]$. This can be accomplished equivalently by the proper choice of cepstral windowing. The inverse canonic system performs the tranformation from an additive space back to the convolutional space and results in the output $[w(n)]$ or $[r(n)]$, as desired.

The use of homomophic filtering for seismic deconvolution was first reproted by Ulrych (Paper 10) in the context of earthquake seismology. Knowledge of the shape of the seismic wavelet allows the determination of the attenuation and dispersion properties of the transmission path. This wavelet is derived by short-time cepstral gating (near the origin) of the complex cepstrum. Examples of minimum-phase, mixed-phase, and exponentially weighted sequences are given to illustrate the concepts in homomorphic filtering. The effects of additive and convolutional noise on the deconvolution results are described. In particular, it is observed that the additive noise complicates the phase spectrum and influences the long-time portion of the complex cepstrum to a much larger degree than the short-time portion. Thus the seismic wavelet can be recovered with good accuracy even in the presence of additive noise. It is recommended that the estimation of the impulse train, for low signal to noise ratio, be obtained preferably by spectral division as opposed to estimation from the high-time portion of complex cepstrum of teleseismic events. Ulrych et al. (1972) and Clayton and Wiggins (1977) have reported further successful applications of homomorphic filtering to earthquake seismology.

Stoffa et al. (Paper 11) describe the application of homomorphic deconvolution to shallow-water marine seismology for wavelet deconvolution and dereverberation. They describe the relation of the complex cepstrum to the original time function in terms of the perodicities of the time function and explain the effect of exponentially weighting the input data. An example with a mixed-phase reflector series and non-minimum-phase wavelet is used to show the potential effectiveness of homomorphic deconvolution where predictive deconvolution will be ineffective. Long-time cepstral gating is used to eliminate the reverberations.

Buttkus (Paper 12) investigates a similar procedure for deepwater reverberations. He suggests simple notch filters in the complex-cepstrum domain for the suppression of multiples because of the simplicity and predictability of these components. He also discusses the influence of random noise on the computation of the complex cepstrum and suggests prefiltering the phase to minimize the adverse effect in computation. The same technique is suggested by Ulrych (Paper 10) whereby the limitations of seismic deconvolution by cepstral gating, due to its dependence on overlapping of the cepstrum components of the wavelet and reflectivity functions, are described in detail.

Otis and Smith (Paper 13) discuss the improvement in recovering the wavelet by averaging the log spectra of several reflection records. This is based on the assumption that the wavelet [$w(n)$] is stationary while the reflectivity function [$r(n)$] is spatially nonstationary. By averaging several reflection records, the log spectrum of the wavelet will be enhanced, and the log spectrum of the reflection coefficients will be averaged out. This technique is an extension of blind deconvolution (Stockham, Cannon, and Ingebretsen, 1975) and statistical deconvolution (Eisenstein and Cerrato, 1976). Angeleri (1980) also discusses a statistical approach to the extraction of the wavelet by averaging the unwrapped phases of the traces.

Tribolet (Paper 14) reports several novel results in homomorphic filtering. He discusses a class of homomorphic systems matched to the bandpass nature of seismic signals and introduces homomorphic bandpass systems as opposed to the fullband homomorphic systems used in other papers. Also the concept of short-time wavelet estimation is introduced that specifically accounts for the time-varying nature of seismic wavelet. An improved and more reliable algorithm for phase unwrapping is also discussed. A detailed treatment of seismic applications of homomorphic signal processing is given in Tribolet (1979).

In all these papers, it is emphasized that homomorphic filtering of seismic data does not require the wavelet to be minimum phase or the reflector series to be random as in prediction deconvolution and Wiener filtering. Treitel and Robinson (1977) have addressed the question relating to the differences in homomorphic and prediction error filtering. They conclude that the method of homomorphic deconvolution does not automatically yield the correct mixed-delay source wavelet estimate. Still awaiting solution are two equivalent problems: the determination of the correct all-pass filter to convert the mixed-delay wavelet to minimum-phase wavelet for predictive deconvolution and the determination of the correct low-pass cutoff quefrencies in the case of homomorphic deconvolution.

REFERENCES

Angeleri, G. P., 1980, A Statistical Approach to the Extraction of the Seismic Propagating Wavelet, paper presented at the 42nd Annual Meeting of EAEG, Istanbul.

Clayton, R. W., and R. A. Wiggins, 1977, Source Shape Estimation and Deconvolution of Teleseismic Bodywaves, *R. Astron. Soc. Geophys. J.* **47**:151–177.

Eisenstein, B. A., and L. R. Cerrato, 1976, Statistical Deconvolution, *Franklin Inst. J.* **302**:147–157.

Oppenheim, A. V., 1965, Superposition in a Class of Non-linear Systems, M.I.T. Research Laboratory of Electronics, Technical Report 432.

Oppenheim, A. V., and R. W. Schafer, 1975, *Digital Signal Processing,* Prentice-Hall, Englewood Cliffs, N. J.

Oppenheim, A. V., R. W. Schafer, and T. G. Stockhan, 1968, Nonlinear Filtering of Multiplied and Convolved Signals, *IEEE Proc.* **56**:1264–1291.

Schafer, R. W., 1969, Echo Removal by Discrete Generalized Filtering, M.I.T. Research Laboratory of Electronics, Technical Report 466.

Shensa, M. J., 1976, Complex Exponential Weighting Applied to Homomorphic Deconvolution, *R. Astron. Soc. Geophys. J.* **44**:379–387.

Stockham, T. G., T. M. Cannon, and R. B. Ingebretsen, 1975, Blind Deconvolution through Digital Signal Processing, *IEEE Proc.* **63**:678–692.

Tribolet, J. M., 1979, *Seismic Applications of Homomorphic Signal Processing,* Prentice-Hall, Englewood Cliffs, N. J.

Treitel, S., and E. A. Robinson, 1977, Deconvolution—Homomorphic or Predictive?, *IEEE Trans. Geosci. Electron.* **GE-15**:11–13.

Ulrych, T., O. G. Jensen, R. M. Ellis, and P. G. Sommerville, 1972, Homomorphic Deconvolution of Some Teleseismic Events, *Seismol. Soc. Am. Bull.* **62**:1269–1281.

10

Reprinted from *Geophysics* **36**:650–660 (1971)

APPLICATION OF HOMOMORPHIC DECONVOLUTION TO SEISMOLOGY†

T. J. ULRYCH*

Homomorphic systems (Oppenheim, 1965a and 1965b) are a class of nonlinear systems which satisfy a generalized principle of superposition. Such systems are particularly useful in separating signals which have been combined through convolution. This paper deals with the application of homomorphic deconvolution to the recovery of the seismic wavelet from a time series formed by the convolution of this wavelet with an impulse train. The unique point about this approach is that it does not require the usual assumptions of a minimum-phase wavelet and a random distribution of impulses.

INTRODUCTION

A seismic record is often represented as the convolution of a wavelet with the impulse response of the transmission path. The process of separating these two components of the convolution is termed deconvolution and finds considerable application in seismology. Deconvolution as commonly performed by means of inverse filtering or optimum zero-lag Wiener filtering (Rice, 1962; Robinson and Treitel, 1967) suffers from the limitation that either the shape of the seismic wavelet to be removed must be known or the assumption that the wavelet is minimum-phase (Robinson, 1966; Ulrych and Lasserre, 1966) must be made. Although predictive deconvolution with a prediction distance greater than unity (Peacock and Treitel, 1969) does not require this assumption, that process is effective in removing repetitive events and would not be applied to the deconvolution problem considered in this paper. Peacock and Treitel (1969) have pointed out that the unit prediction deconvolution filter is equivalent within a scale factor to the least squares, zero-lag inverse filter and does require the minimum-phase assumption.

It is the purpose of this paper to investigate the application to seismic signals of a nonlinear deconvolution technique originally proposed by Oppenheim (1965a, 1965b) as an application of the theory of generalized superposition. This method, known as homomorphic deconvolution and recently applied by Schafer (1969) and by Oppenheim et al (1968) to the problem of echo removal, offers the considerable advantage that no prior assumption about the nature of the seismic wavelet or the impulse response of the transmission path need be made. The seismic wavelet which is recovered by homomorphic deconvolution is of importance in studies of elastic wave attenuation and dispersion in the earth (Futterman, 1962; Strick, 1970).

THEORY

The theoretical basis of homomorphic deconvolution has been dealt with fully in the publications cited above; hence, only a summary of some relevant points will be presented here.

The analytical convenience of linear filtering is a result primarily of the principle of superposition which linear systems satisfy. Oppenheim (1965a) has suggested the generalization of this principle to a certain class of nonlinear systems for which linear filtering in the usual sense is not meaningful. We will restrict ourselves in this paper to the consideration of signals which have been combined by means of convolution and specifically to

† Manuscript received by the Editor September 23, 1970; revised manuscript received February 22, 1971.
* University of British Columbia, Vancouver, Canada.

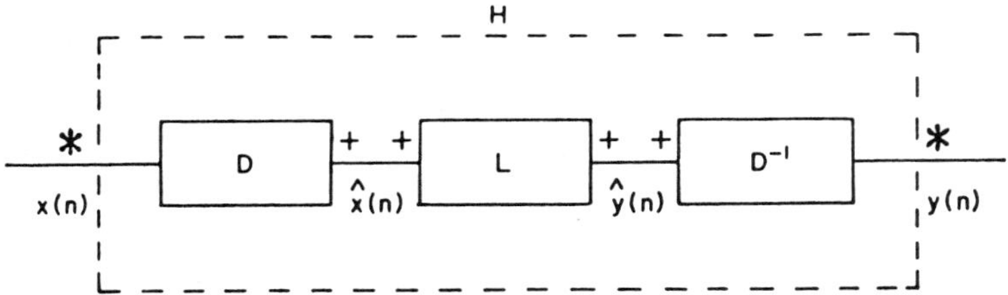

FIG. 1. Canonic representation for homomorphic deconvolution.

the case where one of the signals is an impulse train.

Consider the transformation defined by

$$y = T[x].$$

If T is a linear system, it satisfies the superposition relationship defined by

$$T[ax_1 + bx_2] = aT[x_1] + bT[x_2], \quad (1)$$

where a and b are constants. We can see from equation (1) why linear systems are particularly convenient for separating signals which have been additively combined.

If we wish to generalize the notion expressed by equation (1) to a signal resulting from the convolution of components $x = x_1 * x_2$, we look for a system with transformation H such that

$$H[^{(a)}x_1 * {}^{(b)}x_2] = {}^{(a)}H[x_1] * {}^{(b)}H[x_2], \quad (2)$$

where $^{(a)}$ denotes scalar multiplication. The formalism for this representation, which has been studied in detail by Oppenheim (1965a), lies in interpreting the system inputs and outputs as vector spaces where vector addition is defined as convolution. $^{(a)}$ denotes a rule for combining inputs with scalars in the convolutional space. For example, if a is an integer, $^{(a)}x$ signifies the convolution of x with itself a times. That a system H, known as a homomorphic system and defined by equation (2), does satisfy the postulates of vector addition has been formally shown by Oppenheim (1965a).

The advantage in the representation of a system by equation (2) is that systems of this class have been shown by Oppenheim (1965a) to have the canonic representation shown in Figure 1.

The system D, referred to as the characteristic system, is a homomorphic system with transformation from a convolutional space to an additive space. It is defined by the relationship

$$D[^{(a)}x_1 * {}^{(b)}x_2] = aD[x_1] + bD[x_2]. \quad (3)$$

The system L is a familiar linear system, and the system D^{-1}, the inverse of D, is a homomorphic system and performs the transformation from an additive space back to the output convolutional space.

The great flexibility of the arrangement shown in Figure 1 is that, once the characteristic system D has been determined, it remains fixed for all deconvolution problems; and the process reduces to one of linear filtering.

The characteristic system D

Since the realization of homomorphic deconvolution is performed on sampled data, it will be convenient in this discussion to deal with z transforms of input sequences $x(n)$ as defined in equation (4).

$$X(z) = \sum_{n=-\infty}^{\infty} x(n)z^{-n}. \quad (4)$$

The function of the characteristic system D is to transform signals from a convolutional space to an additive space. Since the z transform of two convolved signals is equal to the product of their z transforms, the required transformation may be accomplished as illustrated in Figure 2. The relevant steps are as follows:

The input sequence $x(n)$ is z transformed to obtain $X(z)$.

The logarithm of $X(z)$, $\hat{X}(z)$, separates the product into a sum.

The inverse z transform of $\hat{X}(z)$ gives $\hat{x}(n)$, the input to the linear system L in Figure 1.

$\hat{x}(n)$ has been termed the complex cepstrum. The word cepstrum originates from the work of Bogert et al (1962), who termed the power spectrum of the logarithm of the power spectrum the cepstrum by direct paraphrase of the word spec-

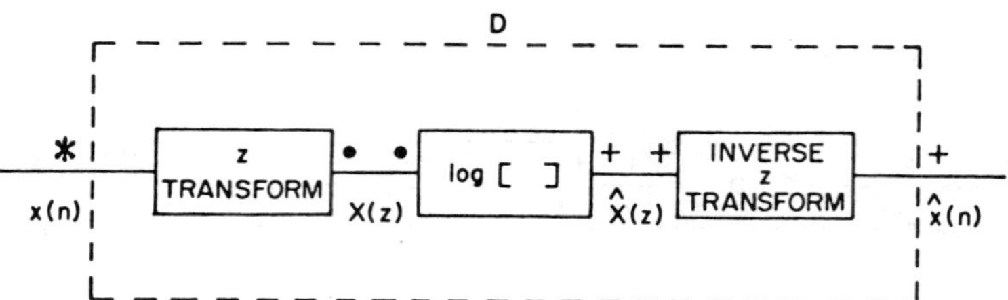

FIG. 2. Canonic representation of the characteristic system D.

trum. Complex has been added by Oppenheim et al (1968) to emphasize the point that the cepstrum has been computed utilizing both the amplitude and phase information contained in $x(n)$.

The complex cepstrum is defined by

$$\hat{x}(n) = \frac{1}{2\pi j} \int_C \log[X(z)]z^{n-1}dz, \quad (5)$$

where C is a circular contour specified by

$$z = e^{\sigma+j\omega}, \quad -\pi < \omega < \pi.$$

The complex cepstrum $\hat{x}(n)$

There are some important considerations in regard to the actual computation of the complex cepstrum that result from the requirement the transformation D be unique. However, before discussing these, let us consider a simple example which will illustrate the concept of the complex cepstrum as applied to homomorphic deconvolution.

Consider a sampled signal $x(n)$, which is composed of a wavelet $s(n)$ and an echo n_o samples later.

$$x(n) = s(n) + as(n - n_o),$$

where a is a constant.

If $\delta(n)$ is the Dirac delta function,

$$x(n) = s(n) * [\delta(n) + a\delta(n - n_o)]. \quad (6)$$

Evaluating the z transform of $x(n)$ on the unit circle gives

$$X(e^{j\omega}) = S(e^{j\omega})[1 + ae^{-j\omega n_o}],$$

and taking the logarithm of $X(e^{j\omega})$, we find

$$\hat{X}(e^{j\omega}) = \log S(e^{j\omega}) + \log[1 + ae^{-j\omega n_o}]. \quad (7)$$

Therefore, since as seen from equation (7) the echo is represented in the log spectrum as an additive periodic component, the complex cepstrum $\hat{x}(n)$ will exhibit a peak at the echo delay.

Figure 3 shows the seismic wavelet which has been used in the examples which follow. It is approximately 35 samples long; the only point to note about this wavelet is that it is not a minimum-phase function. Figure 4(a) represents $x(n)$ of equation (6). The peak-to-peak separation of the wavelet and its echo is 12 samples; $a=0.9$. The complex cepstrum corresponding to this input sequence is shown in Figure 4(b).

This example illustrates several important points. In the first place, we can see that the contribution of the wavelet $s(n)$ in equation (6) to the complex cepstrum is concentrated near $n=0$, whereas the contribution of $[\delta(n)+a\delta(n-n_o)]$

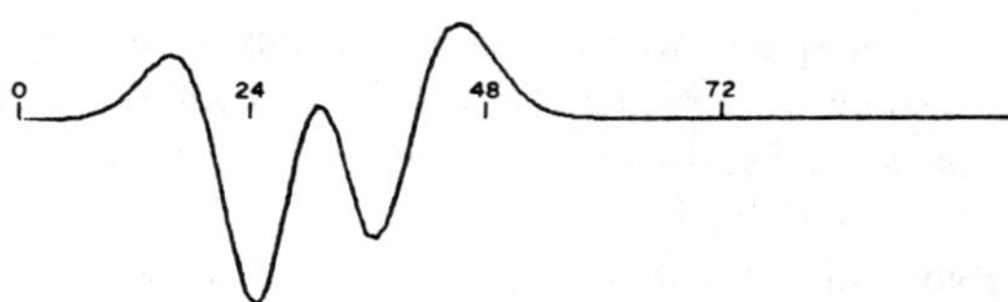

FIG. 4(a). A simple echo.

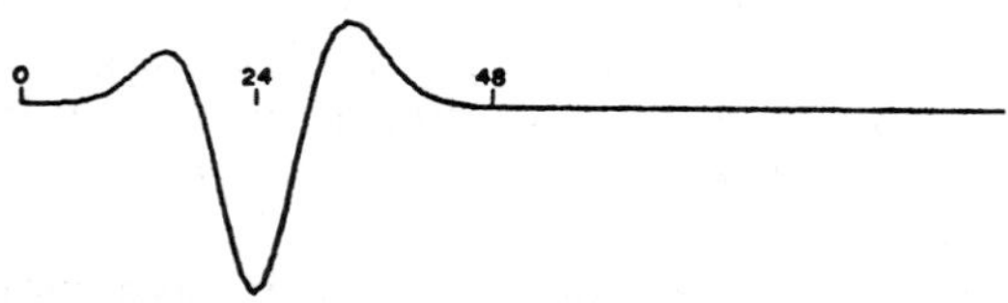

FIG. 3. A mixed-phase seismic wavelet.

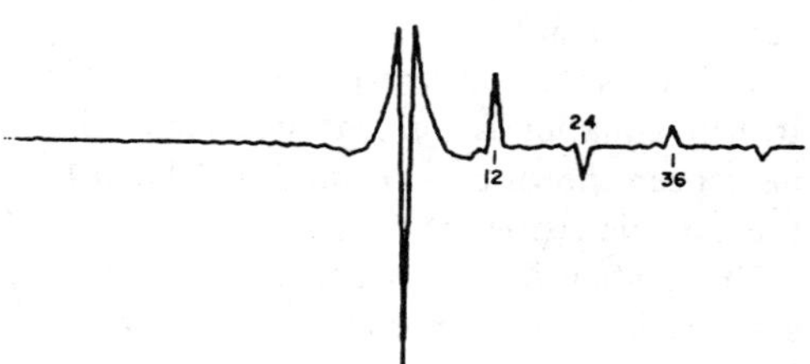

FIG. 4(b). Complex cepstrum of Fig. 4(a).

occurs for higher n values and is well separated from the wavelet component. We will see that, for seismic signals which are considered to be the result of the convolution of a wavelet with an impulse train, this observation is generally true. Secondly, it is clear from Figure 4(b) that, depending on the length of the delay, the echo may be simply removed by means of linear filtering.

Computational considerations.—There are four important considerations in the computation of the complex cepstrum:

I. The complex cepstrum involves the computation of the inverse z transform of a logarithmic function. We may write this function in terms of its magnitude and argument as

$$\log [X(z)] = \log | X(z) | + j\, arg\, [X(z)]. \quad (8)$$

In turn,

$$arg[X(z)] = ARG[X(z) \pm j2\pi k], \quad (9)$$

where

$$k = 0, 1, 2 \cdots \quad \text{and}$$
$$-\pi < ARG[X(z)] < \pi.$$

It can be seen from equations (8) and (9) that the complex logarithm is multivalued. Further, since $ARG[X(z)]$ is a discontinuous function, $\log[X(z)]$ will not, in general, be an analytic function. The homomorphic system D can, however, be unique only if in equation (5) $\log[X(z)]$ is analytic in an annular region containing the contour C. We can achieve this condition by computing $ARG[X(z)]$ and then unwrapping it to produce $arg[X(z)]$, which is continuous, providing that the phase curve has been sampled at sufficiently small intervals.

II. The requirement that the complex cepstrum $\hat{x}(n)$ be real for real input sequences $x(n)$ implies that (a) $arg[X(z)]$ is an odd function of ω and periodic in ω with a period of 2π and (b) $\log| X(z)|$ is an even function of ω and periodic in ω with a period of 2π.

III. The input sequences with which we will be concerned are always restricted by the condition

$$x(n) = 0, \quad M > n > 0.$$

Such sequences are characterized by z transforms which have no singularities in the z plane and

which are polynomials in z^{-1}. The z transform may be represented by equation (10) (Schafer, 1969).

$$X(z) = A_z{}^{-M_0} \prod_{k=1}^{m_i} (1 - a_k z^{-1}) \prod_{k=1}^{m_o} (1 - b_k z),$$
$$| a_k | < 1, \; | b_k | < 1, \quad (10)$$

where the a_k's are the m_i zeroes inside the unit circle, the b_k's are the m_o zeroes outside the unit circle, and z^{-m_o} represents a shift of the input sequence.

Let us consider the effect of the term z^{-m_o} on the computation of the complex cepstrum. Since the contour C in equation (5) is specified by $z = e^{\sigma+j\omega}$, equation (5) may be written as

$$\hat{x}(n) = \frac{1}{2\pi} \int_{-\pi}^{\pi} \log [X(e^{\sigma+j\omega})]e^{\sigma n}e^{j\omega n}d\omega. \quad (11)$$

Let the contribution of z^{-m_o} to $\hat{x}(n)$ be $\hat{\phi}(n)$. Then if for convenience we choose the contour C in equation (5) to be the unit circle,

$$\hat{\phi}(n) = -\frac{1}{2\pi} \int_{-\pi}^{\pi} m_0 \log [e^{j\omega}]e^{j\omega n}d\omega,$$

which integrates to

$$\hat{\phi}(n) = -\frac{m_o \cos \pi n}{n}. \quad (12)$$

In a practical case, $X(z)$ may have many zeroes outside the unit circle. m_o is thus large and the effect of $\hat{\phi}(n)$ is to swamp the interesting information contained in the complex cepstrum. It is, therefore, of importance to remove the linear phase component prior to the computation of $\hat{x}(n)$. This is easily achieved and, in fact, the computation of the unwrapped phase curve is combined with this operation. Since the removal of the linear phase component is equivalent to a shift of the output sequence, the final step in the deconvolution process is to reposition the output sequence.

IV. The input sequences of interest are always of finite length. $X(z)$ thus has a region of convergence which includes the unit circle. This allows $X(z)$ and the inverse transform to be evaluated for $z = e^{j\omega}$; therefore, implementation of the z transform and its inverse is by means of the discrete Fourier transform pair computed using

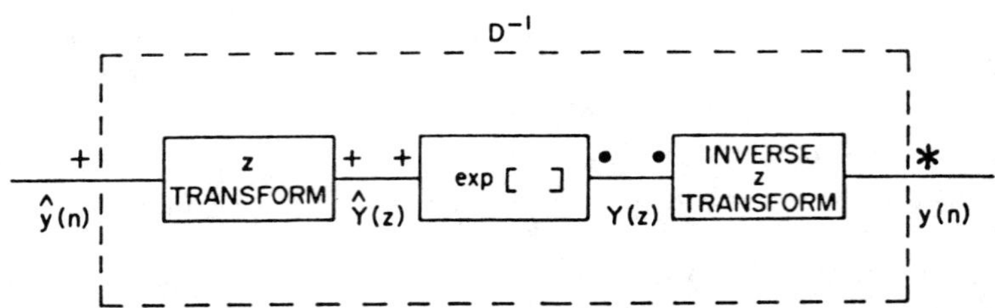

FIG. 5. Canonic representation of the inverse system D^{-1}.

the Cooley-Tukey fast Fourier transform algorithm (Cooley and Tukey, 1965).

The System D^{-1}

The transformation performed by the homomorphic system D^{-1} is from an additive to a convolutional space. It is the inverse system to D and its canonic representation is shown in Figure 5.

Linear filtering of the complex cepstrum

As we have seen in the example illustrated in Figure 4, the complex cepstrum contains the additive contributions of the wavelet and of the impulse response of the transmission channel. In this particular case, these contributions may be very easily separated by means of ideal low-pass and high-pass filters. The results of low-pass and high-pass filtering of the cepstrum of Figure 4(b) with a cutoff length equal to 12 samples, followed by processing with the system D^{-1}, are shown in Figure 6(a) and Figure 6(b). A comparison of the input and deconvolved wavelets shows that even in the case of this rather crude filtering, the shape of the wavelet is essentially preserved. A "comb" filter designed to have zero response at the echo peaks would, of course, provide a more exact component separation. For convenience, however, simple filtering has been used throughout the paper.

Examples of complex cepstra

The example of Figure 4 illustrates a very simple case and one to which linear filtering may be very easily applied. In seismology, we are generally concerned with an input sequence $x(n) = s(n)*i(n)$ where $i(n)$ may be a very complex impulse series. It turns out that minimum-phase impulse sequences play a very important part in homomorphic deconvolution. Let us therefore briefly consider the properties of the complex cepstra of such sequences. [Some analytical expressions for complex cepstra of impulse trains are given by Schafer (1969).]

Minimum-phase sequences.—Let $i(n)$ be a finite-length sequence such that $I(z)$, the z-transform of $i(n)$, has all its zeroes inside the unit circle. Then,

$$I(z) = A \prod_{k=1}^{m_i} (1 - a_k z^{-1}), \text{ where } |a_k| < 1,$$

and taking the contour C to be the unit circle, we have

$$\hat{i}(n) = \frac{1}{2\pi}$$
$$\cdot \int_{-\pi}^{\pi} \log \left\{ A \prod_{k=1}^{m_i} (1 - a_k z^{-1}) \right\} e^{j\omega n} d\omega, \tag{13}$$

where $z = e^{j\omega}$.

Each of the terms inside the integral sign may be expanded in a Laurent series about $z = 0$ to give

$$\log (1 - a_k z^{-1}) = - \sum_{n=1}^{\infty} \frac{a_k^n}{n} z^{-n} \tag{14}$$
$$\text{for } |z| > |a_k|.$$

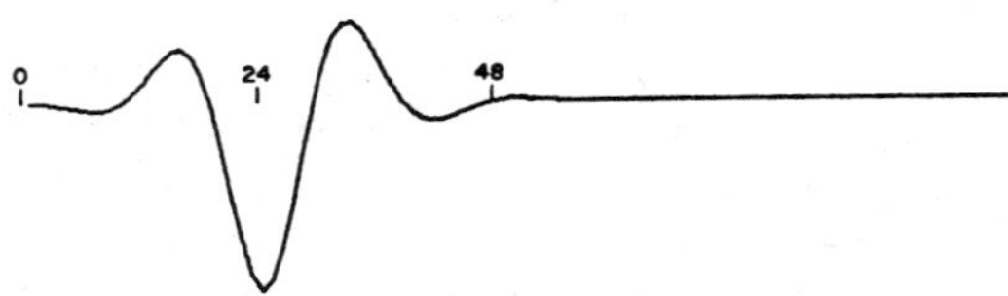

FIG. 6(a). Result of the low-pass filtering of the complex cepstrum of Fig. 4(b).

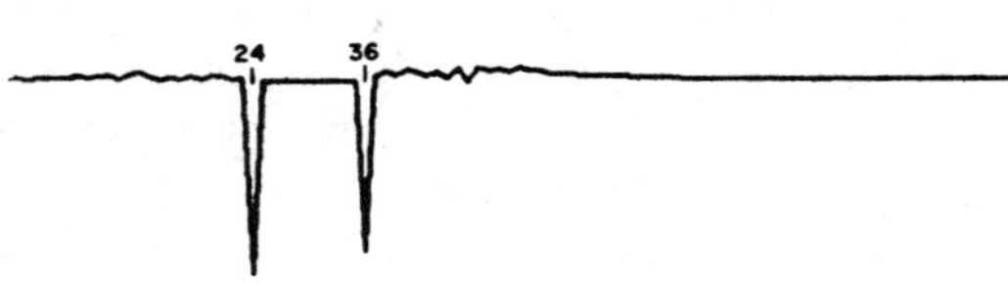

FIG. 6(b). Result of the high-pass filtering of the complex cepstrum of Fig. 4(b).

We see from equations (13) and (14) that $\hat{i}(n) = 0$ for $n < 0$. This property of minimum-phase sequences is illustrated in Figure 7. Figure 7(a) shows the input trace, which is the convolution of the wavelet of Figure 3 with a minimum-phase impulse train composed of four unequally spaced impulses. The complex cepstrum, Figure 7(b), shows clearly the contributions of the two components which may be easily separated by linear filtering as shown in Figures 7(c) and 7(d). As a general comment, we can say that when the impulse train is minimum-phase, the two convolved components of the seismic trace are separated in the complex cepstrum by an amount equal to the separation of the first two impulses.

Mixed-phase sequences.—For finite-length mixed-phase sequences, $I(z)$ may be expressed by equa-

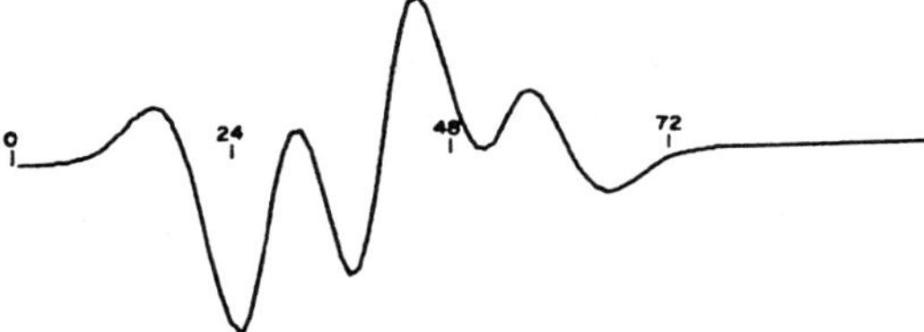

FIG. 7(a). A minimum-phase input sequence.

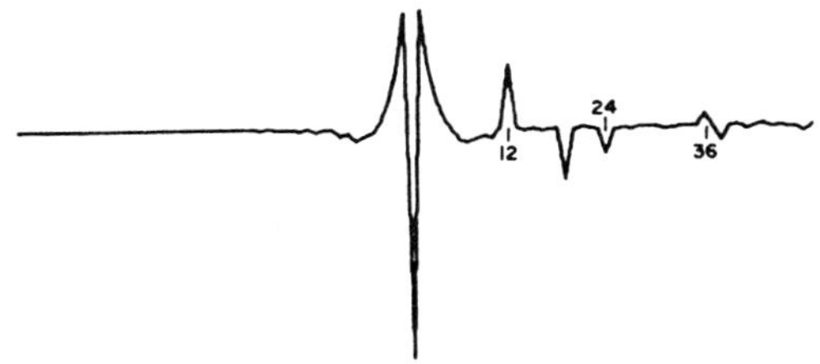

FIG. 7(b). Complex cepstrum of Figure 7(a).

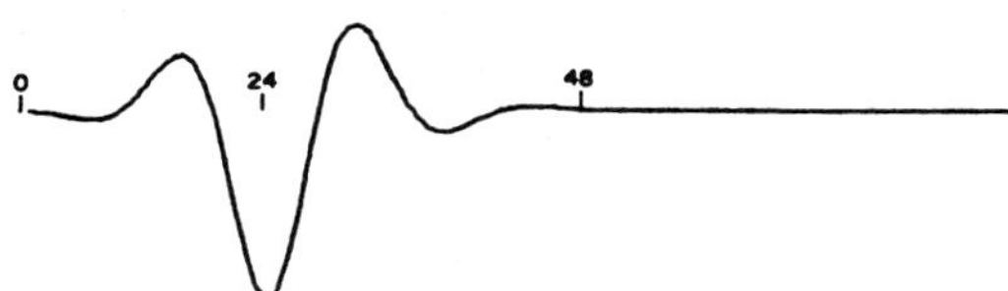

FIG. 7(c). Low-pass output.

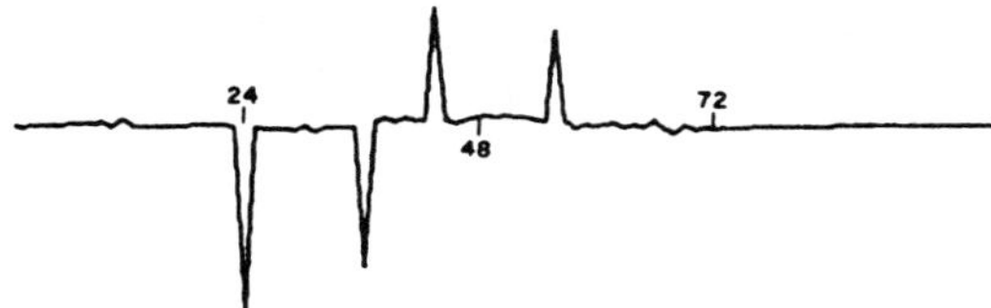

FIG. 7(d). High-pass output.

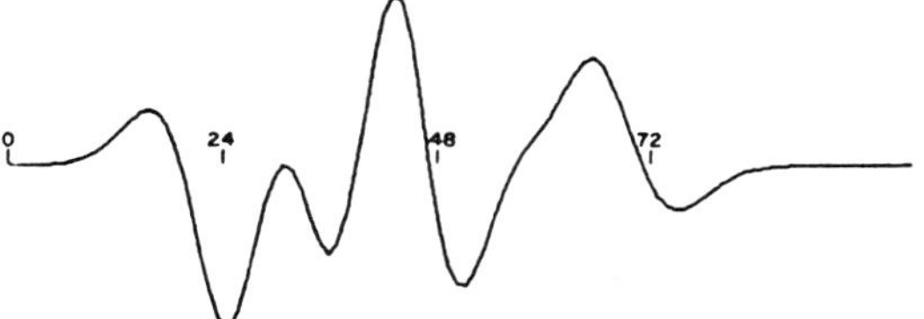

FIG. 8(a). A mixed-phase sequence.

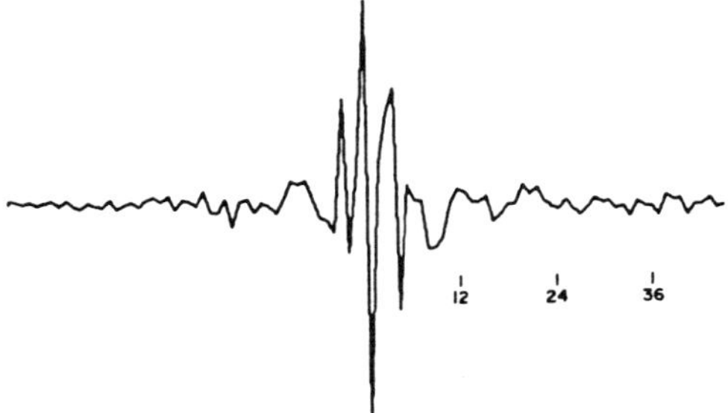

FIG. 8(b). Complex cepstrum of Figure 8(a).

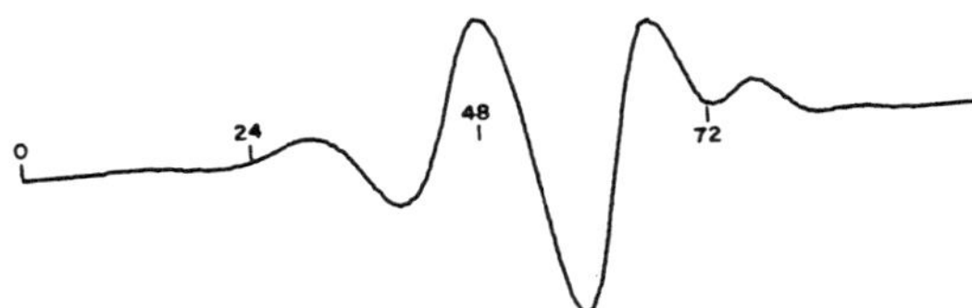

FIG. 8(c). Low-pass output.

tion (10) and, since the Laurent series about $z=0$ for terms log $(1 - b_k z)$ is

$$\sum_{n=-\infty}^{-1} \frac{b_k^{-n}}{n} z^{-n} \quad \text{for } |z| < |b_k^{-1}|,$$

we can see that $\hat{i}(n)$ will have values in the range $-\infty < n < \infty$. The complex cepstra of sequences of unequally spaced impulses which are mixed phase are generally very complicated and the two components of the convolution are no longer separated. Figure 8(a) shows a mixed-phase input trace obtained by convolving the wavelet of Figure 3 with a mixed-phase series of five unequally spaced impulses. The separation of the first and second impulse is the same as in Figure 7. The complex cepstrum of this trace, shown in Figure 8(b), illustrates the complexity which may arise. Low-pass filtering to recover the seismic wavelet, with the same filter used to obtain the wavelet in Figure 7(c), produces a wavelet, Figure 8(c), which bears little resemblance to the

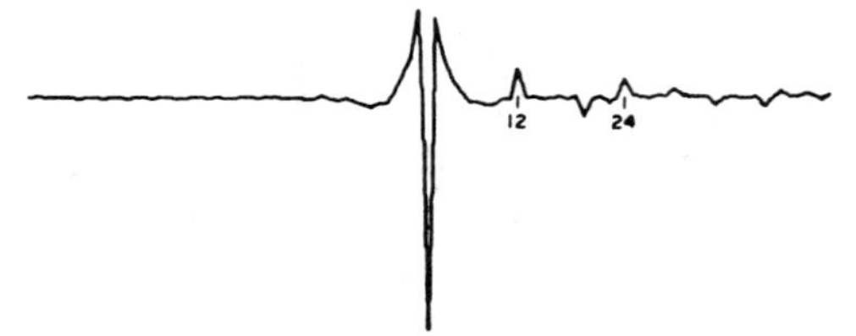

FIG. 9(a). Complex cepstrum of the sequence of Figure 8(a) exponentially weighted with $\alpha=0.965$.

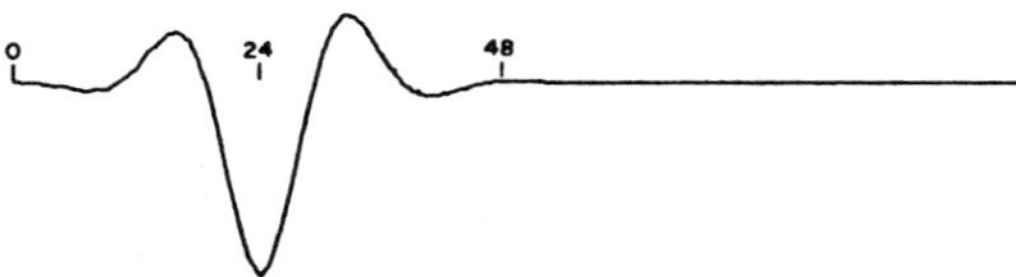

FIG. 9(b). Low-pass output.

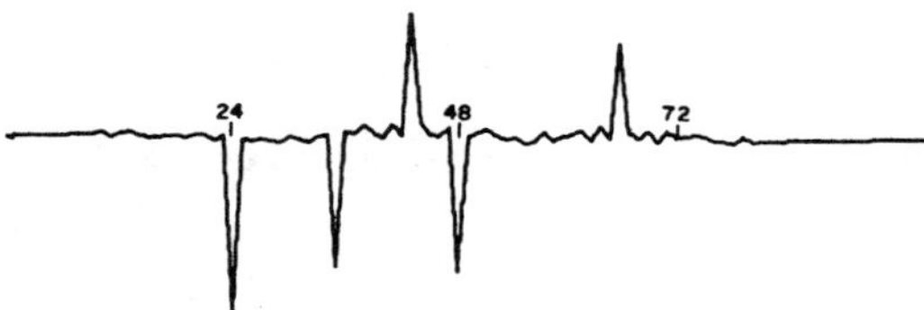

FIG. 9(c). High-pass output.

original. This is due to the fact that the region of the cepstrum near $n=0$ now contains the combined contributions of the wavelet and the impulse train.

Exponential weighting

Schafer (1969) has suggested an ingenious method by which, in order to exploit the special properties of minimum-phase sequences, a mixed-phase sequence may be converted to a minimum-phase sequence.

Suppose that the furthest zero of $I(z)$, the z transform of the mixed-phase impulse train $i(n)$, is at z_o, where $|z_o|>1$. We wish to transform $i(n)$ into a minimum-phase impulse train $j(n)$. Thus, the furthest zero of $J(z)$ must be at αz_o, where $|\alpha z_o|<1$, i.e.,

$$J(z) = I(\alpha^{-1}z);$$

and, therefore,

$$j(n) = \alpha^n i(n), \quad \alpha < 1.$$

In other words, a mixed-phase sequence may be made into a minimum-phase sequence by means of exponential weighting. To illustrate this point, the input sequence of Figure 8(a) was ex-

ponentially weighted with $\alpha=0.965$. The resulting complex cepstrum is shown in Figure 9(a) and the deconvolved wavelet and impulse train are shown in Figures 9(b) and 9(c).

Effect of noise on homomorphic deconvolution

The examples considered thus far are ideal in the sense that the input sequences are noise free. An actual seismic trace, $x(n)$, may be represented as

$$x(n) = s(n) * i(n) + m(n),$$

where the noise $m(n)$ may be decomposed into $m(n)=\eta(n)*s(n)+\mu(n)$. $\eta(n)$ is the part of the noise convolved by the wavelet $s(n)$ and $\mu(n)$ is a noise superimposed on the seismic trace. We will consider each component in turn.

Additive noise component $\mu(n)$.—The addition of $\mu(n)$ complicates the computation of a smooth-phase curve, and since the complex cepstrum depends on the contribution of the phase component of the input sequence, the simplicity of the complex cepstrum of a noise-free sequence, such as illustrated in Figure 7, is destroyed. However, providing that the impulse train has been made minimum-phase by exponential weighting, the portion of the complex cepstrum near $n=0$ (or the "short time" portion) may still be used to recover the seismic wavelet. Figure 10(a) shows the input sequence of Figure 7(a), together with white noise with a signal-to-noise amplitude ratio of 10 to 1. After exponential weighting with $\alpha=0.965$, the recovered seismic wavelet is shown in Figure 10(c). A considerably better result is achieved if the input sequence is filtered first of all. An important consideration in homomorphic deconvolution of filtered signals is the rate at which the continuous time signal is sampled (Schafer, 1969). This point is discussed below.

The high-frequency content of a filtered seismic signal is low and the Fourier transform of such signals may be considered to be zero for frequencies greater than f_c say. If the signal is sampled at the Nyquist rate $1/(2f_c)$, aliasing is avoided and $|X(e^{j\omega})|$ is finite at all frequencies. If the sampling interval is less than $1/(2f_c)$, aliasing is, of course, also avoided but $|X(e^{j\omega})|$ is zero and, hence, $\log|X(e^{j\omega})|$ is undefined over a finite interval. In general, if the sampling rate is greater than the Nyquist rate, there exists an interval in

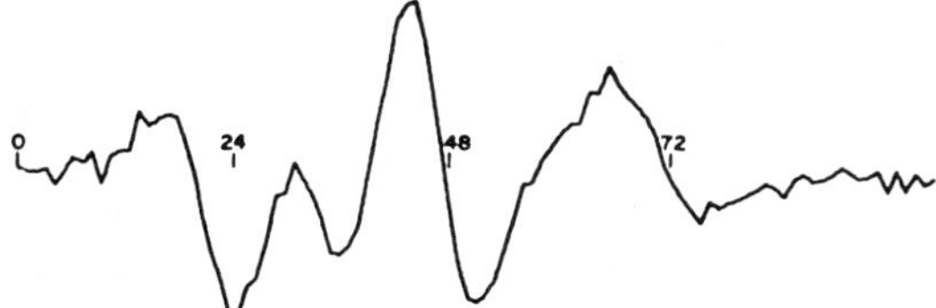

FIG. 10(a). Input sequence of Figure 8(a) with an added noise component.

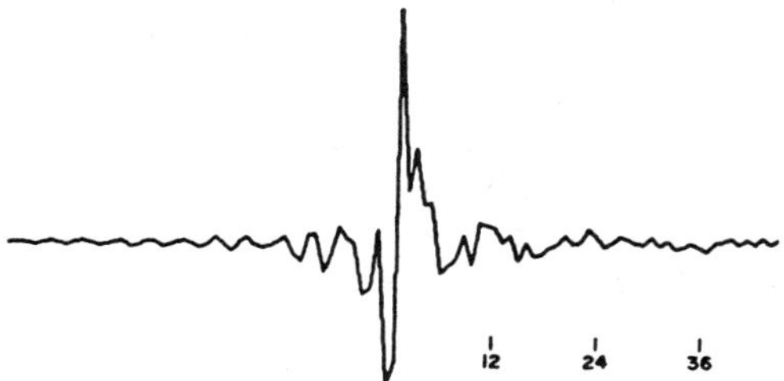

FIG. 10(b). Complex cepstrum of the sequence of Figure 10(a) exponentially weighted with $\alpha = 0.965$.

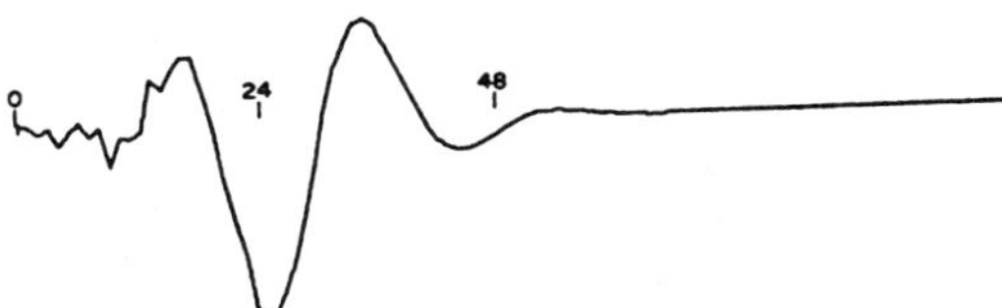

FIG. 10(c). Low-pass output.

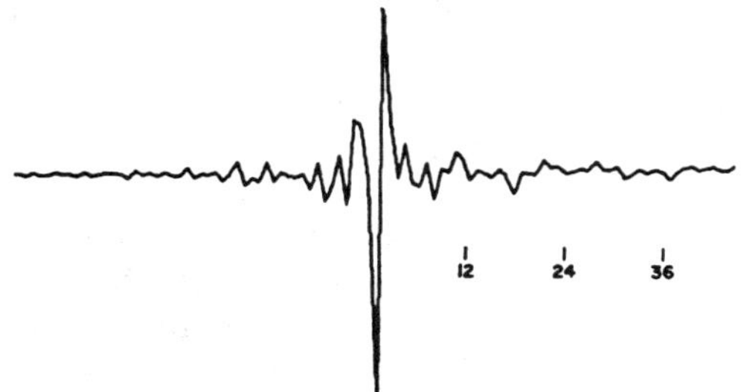

FIG. 10(d). Complex cepstrum of sequence of Figure 10(c) after optimum filtering and exponential weighting.

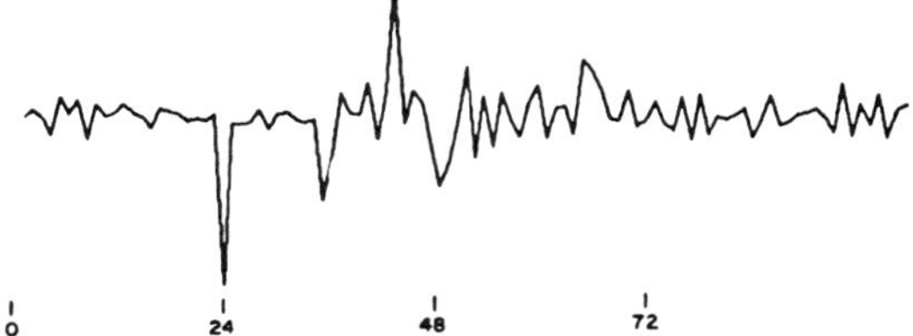

FIG. 10(e). High-pass output.

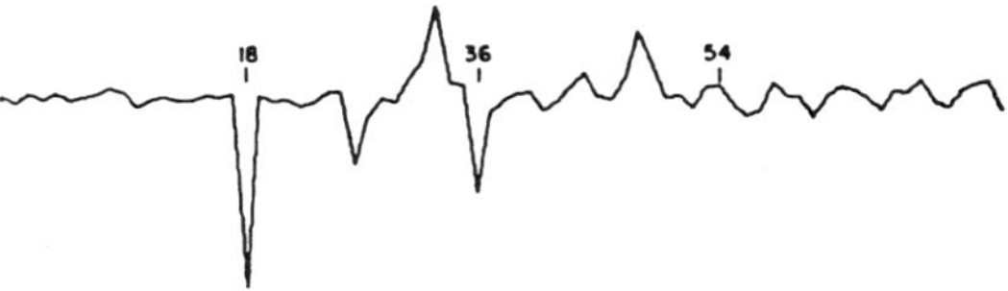

FIG. 10(f). Deconvolved impulse train showing the effect of increasing sample interval by 50 percent.

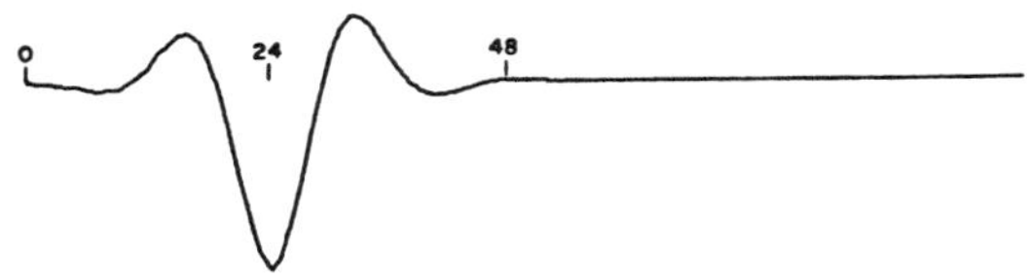

FIG. 10(g). Low-pass output of Figure 10(d).

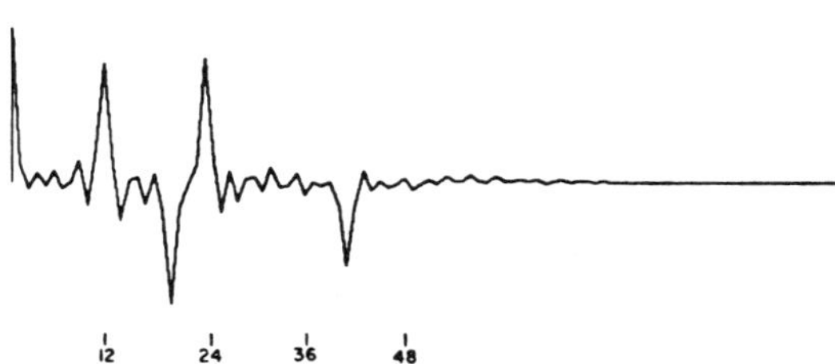

FIG. 10(h). Impulse train obtained by means of division in frequency domain.

which the real and imaginary parts of $X(e^{j\omega})$ are small and considerable errors may arise in the computation of $\log|X(e^{j\omega})|$ and $ARG|X(e^{j\omega})|$. It is important, therefore, that, following prefiltering of the signal, the sampling rate be made equal to or slightly higher than the Nyquist rate.

The above discussion is illustrated in Figure 10. The input sequence of Figure 10(a) was filtered using an optimum Wiener filter.[1] The complex cepstrum corresponding to this filtered sequence appears in Figure 10(d). Due to irregularities in the unwrapped phase curve, the deconvolved impulse train, shown in Figure 10(e), is very noisy. These irregularities are decreased by increasing the sampling interval, as discussed above;

[1] The Wiener filter, $H_{opt}(\omega)$, was designed on the basis of the power spectrum of the input signal.

$$H_{opt}(\omega) = \frac{P_s(\omega)}{P_s(\omega) + P_n(\omega)}$$

The noise power spectrum $P_n(\omega)$ was estimated by inspection of the input power spectrum to be 0.4 percent of the maximum input power. Since the noise is white, the noise power is a constant.

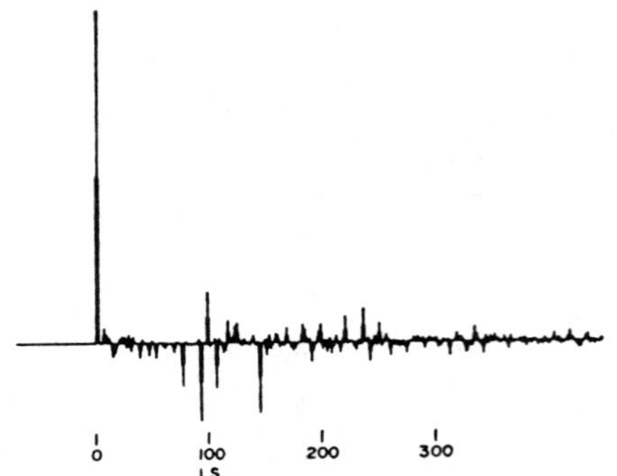

FIG. 11(a). Theoretical impulse response of crust near Leduc, Alberta (after O. Jensen).

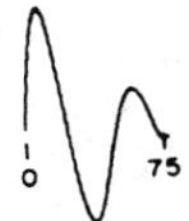

FIG. 11(b). An assumed seismic wavelet.

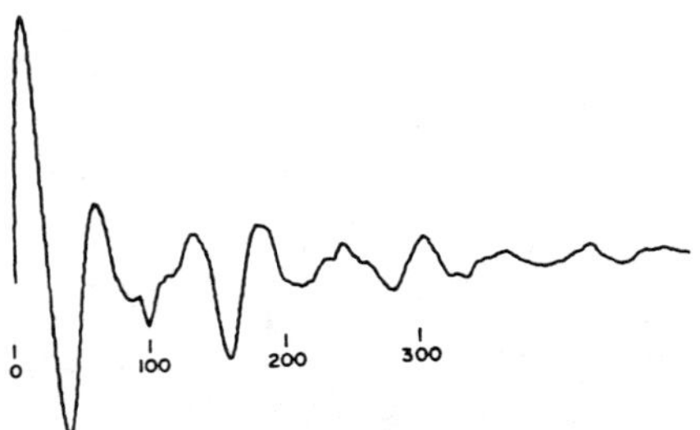

FIG. 11(c). Synthetic seismogram.

and the deconvolution is improved. Figure 10(f) shows the deconvolved impulse train after filtering but with a sampling interval 50 percent greater than that of the previous example. The improvement is marked.

The seismic wavelet which is recovered from the short-time portion of Figure 10(d) [Figure 10(g)] illustrates an important aspect of homomorphic filtering. It appears that noise in the phase curve influences the long-time portion of the complex cepstrum to a much larger degree than it does the short-time portion. Consequently, it is possible and may be preferable, once the wavelet has been recovered, to obtain the impulse train by means of division of Fourier transforms. The impulse train recovered in this manner using the wavelet of Figure 10(g) is shown in Figure 10(h). We remark, in summary, that the homo-

morphically deconvolved seismic wavelet is much less sensitive to additive noise than is the deconvolved impulse train. For this reason, for low signal-to-noise ratios, the impulse train is better recovered by means of division of Fourier transforms. A comparison of Figures 10(e) and 10(h) illustrates this point.

The complex cepstra of sequences which have been filtered to remove an additive noise component may not enjoy the simplicity of the complex cepstra of noise-free sequences. In such cases, it may be difficult to estimate the required length of the linear filter. We have found, however, that good results are obtained if the length of the seismic wavelet is approximately estimated from the input sequence and the cutoff length of the ideal filter is taken to be a third of this length.

Convolutional noise $\eta(n)$.—To illustrate the effect of convolutional noise, homomorphic deconvolution has been applied to a synthetic seismogram prepared by O. Jensen and thought to represent the response of the earth's crust in the vicinity of Leduc, Alberta to low-frequency teleseismic events. Figures 11(a), 11(b), and 11(c) show the impulse response calculated by Jensen, an assumed wavelet, and the resulting synthetic seismogram. The impulse response contains all minor impulses which would result from multiple reflections. The length of the wavelet was estimated to be 75 samples from Figure 11(a) and, consequently, the cutoff length of the simple low-pass filter was chosen to be 25 samples. Exponential weighting with $\alpha = 0.985$ was used. Figure 12(a) shows the complex cepstrum of the weighted input sequence and Figures 12(b) and 12(c) show the recovered impulse response and wavelet, respectively. It is apparent that the effect of convolutional noise is confined to the first 75 samples, i.e., the length of the wavelet, of the deconvolved impulse response. The striking similarity between the actual impulse response and the deconvolved response for $n > 75$ is extremely encouraging, although it must be remembered that this example is free of additive noise.

DISCUSSION

A major problem in exploration and earthquake seismology is the identification of the seismic wavelet. A knowledge of the shape of this wavelet allows the determination of the attenuation and dispersion properties of the transmission path, a problem of considerable interest in seismology (Strick, 1970). Although it is probable that in exploration seismology the wavelet may be often assumed to be minimum-phase, this assumption may not be made in the case of earthquake seismology. Homomorphic deconvolution appears to be a very powerful method of recovery of the seismic wavelet and, hence, also of the impulse response. Most important, homomorphic deconvolution obviates the necessity of making the usual assumptions of a minimum-phase wavelet and a random impulse train. Failure of these assumptions, which are required by methods commonly used in the deconvolution of sequences of the type considered in this paper, may lead to gross errors.

We are at present (Ulrych et al, 1971) applying homomorphic deconvolution to a series of teleseismic events recorded at Leduc, Alberta with encouraging results. A preliminary example of this work is presented here.

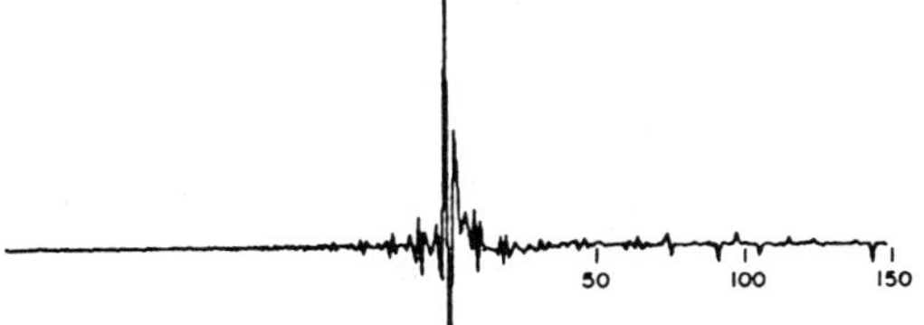

Fig. 12(a). Complex cepstrum of the trace of Figure 11(c) exponentially weighted with $\alpha=0.985$.

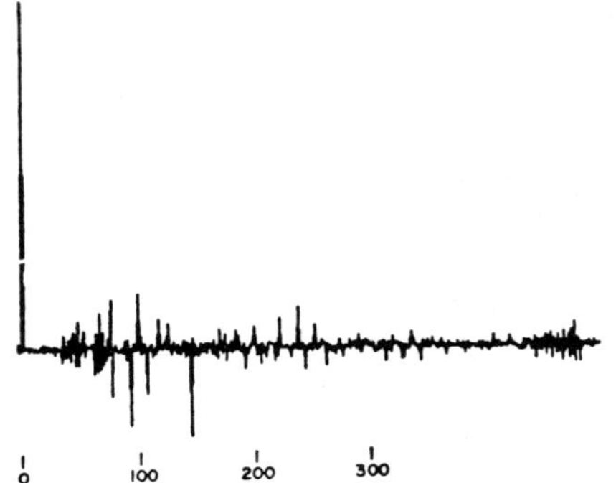

Fig. 12(b). High-pass output.

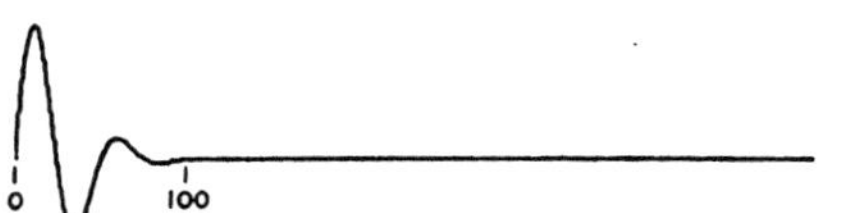

Fig. 12(c). Low-pass output.

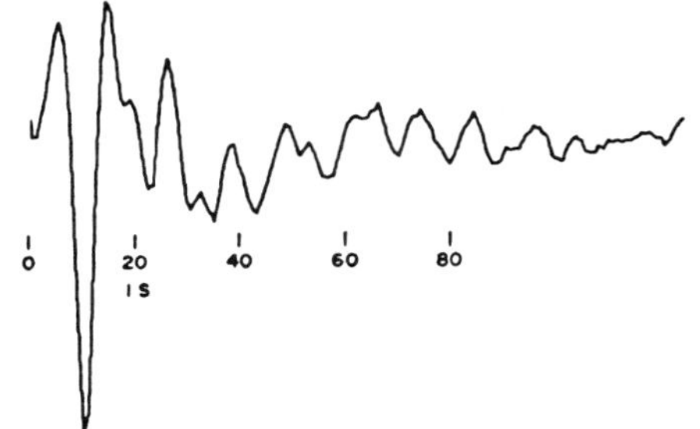

Fig. 13(a). Teleseismic event recorded in 1968 at Leduc, Alberta and originating in Venezuela.

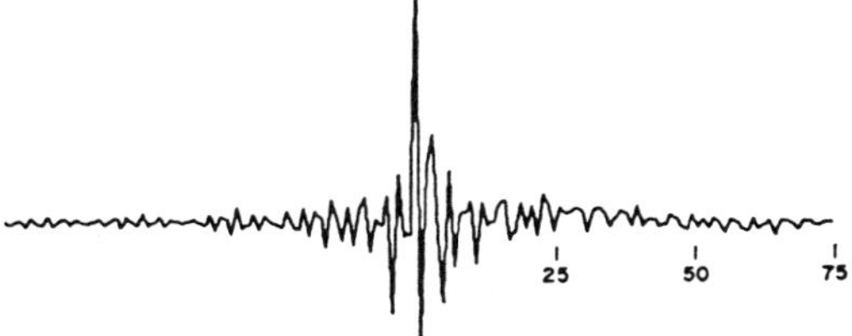

Fig. 13(b). Complex cepstrum of Figure 13(a) after exponential weighting with $\alpha=0.985$.

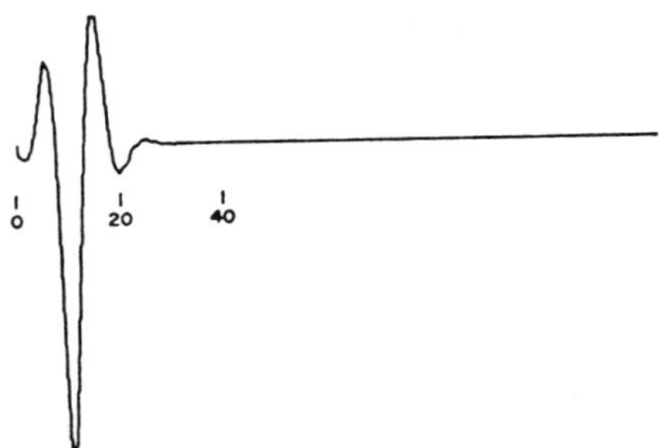

Fig. 13(c). Deconvolved seismic wavelet

Figure 13(a) shows an event which originated in Venezuela in 1968. The estimated length of the seismic wavelet is approximately 20 samples and the cutoff length for the simple low-pass filter which was used in the linear filtering was 8 samples. Following exponential weighting with $\alpha=0.985$, we recovered the seismic wavelet shown in Figure 13(c). An approximate check on the deconvolution is provided by convolving the wavelet in Figure 13(c) with the synthetic impulse response of Figure 11(a). The resultant "earthquake" shown in Figure 13(d) compares favorably with the observed event of Figure 13(a).

Many interesting questions remain to be explored. For example, perhaps some type of phase filtering may be employed to overcome the effect

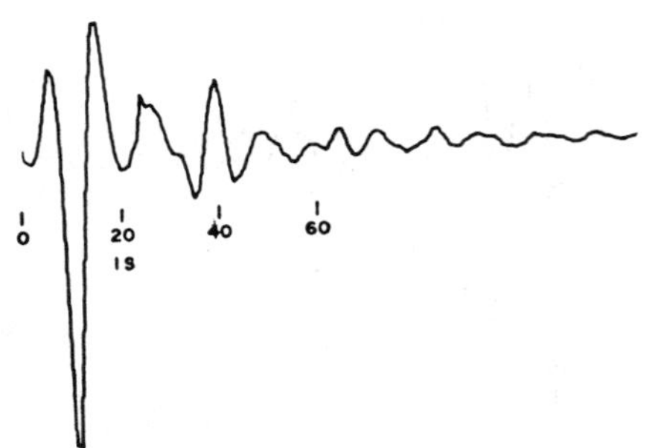

Fig. 13(d). Trace resulting from the convolution of Figure 13(c) with Figure 11(a).

of additive noise on the unwrapped phase curve. The choice of α depends at present on experience only, whereas a quantitative method of determining it is obviously preferable.

Schafer (1969) has suggested certain problems to which homomorphic deconvolution may be usefully applied. Our research has indicated the importance of this technique in processing seismic signals; it is hoped that this work will encourage further research into the application of homomorphic filtering to problems of importance in geophysics.

ACKNOWLEDGMENTS

I am extremely grateful to Oliver Jensen of the University of British Columbia for providing me with the synthetic seismogram used in this work, for stimulating thoughts, and for critically reading this manuscript. I am also obliged to Sven Treitel for an enlightening discussion of predictive deconvolution. This research was financed by the National Research Council grant-in-aid.

REFERENCES

Bogert, B. P., Healey, M. J., and Tukey, J. W., 1963, The quefrequency analysis of time series for echoes; cepstrum, pseudo-autocovariance, cross-cepstrum and saphe cracking: Proc. Symp. on Time Series Analysis, M. Rosenblatt, Ed., New York, Wiley, p. 209–243.

Cooley, J. W., and Tukey, J. W., 1965, An algorithm for the machine calculation of complex Fourier series: Math. of Comput., v. 19, p. 297–301.

Futterman, W. I., 1962, Dispersive body waves: J. Geophys. Res., v. 67, p. 5279–5291.

Oppenheim, A. V., 1965a, Superposition in a class of non-linear systems: Research Lab. of Electronics MIT, Tech. Rep. 432.

———— 1965b, Optimum homomorphic filters: Research Lab. of Electronics MIT, Quart. Progr. Rep. 77, p. 248–260.

Oppenheim, A. V., Schafer, R. W., and Stockham, T. G., 1968, Nonlinear filtering of multiplied and convolved signals: Proc. IEEE, v. 65, p. 1264–1291.

Peacock, K. L., and Treitel, S., 1969, Predictive deconvolution: Theory and practice: Geophysics, v. 34, p. 155–169.

Rice, R. B., 1962, Inverse convolution filters: Geophysics, v. 27, p. 4–18.

Robinson, E. A., 1966, Multichannel z transforms and minimum delay: Geophysics, v. 31, p. 473–500.

Robinson, E. A., and Treitel, S., 1967, Principles of digital Wiener filtering: Geophys. Prosp., v. 15, p. 311–333.

Schafer, R. W., 1969, Echo removal by discrete generalized linear filtering: Research Lab. of Electronics MIT, Tech. Rep. 466.

Strick, E., 1970, A predicted pedestal effect for pulse propagation in constant-Q solids: Geophysics, v. 35, p. 387–403.

Ulrych, T. J., and Lasserre, M., 1966, Minimum-phase: J. Can. Soc. Expl. Geophysicists, v. 2, p. 22–32.

Ulrych, T. J., Jensen, O., Ellis, R. M., and Summerfield, P. S., 1971, to be submitted to Bull. Seism. Soc. Am.

11

THE APPLICATION OF HOMOMORPHIC DECONVOLUTION TO SHALLOW-WATER MARINE SEISMOLOGY—PART I: MODELS

PAUL L. STOFFA,*§ PETER BUHL,*§ AND GEORGE M. BRYAN*

The complex cepstrum is investigated mathematically and through models for functions of interest in shallow-water marine seismology. Association of the slowly varying components of the phase spectrum with the source replaces the usual minimum-phase assumption. This is analogous to the usual treatment of the amplitude spectrum. Complex cepstrum expressions are developed for an arbitrary (but minimum-phase) reflector series, water-column multiple generator, and simplified bubble-pulse oscillation. While the complex cepstrum of all functions is of infinite extent, removing only the first n nonzero complex-cepstrum contributions of a decaying, impulsive, periodic time function (such as the water-column multiple generator) serves to eliminate the first n multiples entirely in the time domain and reduces the remaining multiples to at most $1/(n+1)$ of their original value.

A new method of computing the continuous, ramp-free phase spectrum required for complex-cepstrum analysis is developed on the basis of the derivative of the phase curve.

INTRODUCTION

In least-squares, time-domain inverse filtering (Robinson, 1957; Rice, 1962), removal of the seismic source begins with the autocorrelation function (ACF) of the seismic trace. The trace ACF is assumed equal to the ACF of the source. By taking only a small number of lags, relative to the trace length, we are effectively restricting our source to a short time duration. Since the ACF and the power spectrum are Fourier transform pairs, this is equivalent to saying that compared to the reflector series the power spectrum of the source varies slowly with respect to frequency. Thus time-domain deconvolution removes the slowly varying components from the power spectrum leaving the rapidly varying components, which presumably represent the reflector series. Thus we see that least-squares inverse filtering exploits a difference between the source and the trace, namely, the way their power spectra vary with frequency, and its success is in part determined by the size of this difference. It should be pointed out that a "random" reflector series has a white or flat amplitude spectrum only in a statistical sense. The spectrum of a particular random series, however, will not be white; it has both slowly and rapidly varying amplitude spectrum components.

In least-squares inverse filtering we must make an assumption about the phase spectrum of the source, since by taking the ACF we have destroyed all phase information. The usual realizations assume the source to be minimum phase. It might be better to treat the phase spectrum as we treat the power spectrum, i.e., associate the slowly varying components with the source and

Lamont-Doherty Geological Observatory contribution No. 2088.

Presented at the 43rd Annual International SEG Meeting, October 24, 1973, Mexico City. Manuscript received by the Editor July 23, 1973; revised manuscript received September 11, 1973.

* Lamont-Doherty Geological Observatory, Palisades, NY 10964.

§ Columbia University, New York, N.Y. 10027.

181

the rapidly varying components with the reflector series. It will be shown that this is indeed the case only if the reflector series is minimum-phase.

The above discussion leads us to the cepstrum (Bogert et al, 1963) and then the complex cepstrum which is one realization of a homomorphic system (Oppenheim, 1965; Schafer, 1969). This approach treats the complex natural logarithm of the amplitude and phase spectra, $\log \left[A(\omega)e^{i\phi(\omega)} \right]$, as a complex time series and takes the inverse Fourier transform of this series to produce the complex cepstrum. Then by removing the values of the complex cepstrum near the origin, we can eliminate the slowly varying components of both the phase and log-amplitude spectra.

In this paper, model studies are discussed which indicate that homomorphic deconvolution, as developed by Oppenheim (1965) and Schafer (1969), has application to the marine seismic deconvolution problem. [Ulrych (1971, 1972) has already had success with this method in extracting the source function from teleseismic events.] The marine seismic source and short-period, water-column multiples may be removed from marine seismic records by fairly simple operations in the complex cepstrum without the minimum-phase source assumption. Aliasing in the complex cepstrum, produced by the nonlinear complex logarithm operation, can be suppressed by judicious exponential weighting of the seismic trace.

Two methods of computing the continuous, ramp-free phase spectrum required for complex-cepstrum analysis are discussed. The first method is an iterative application of Schafer's (1969) algorithm, which "unwraps" the principal value of the phase spectrum. The second method is based on the derivative of the phase spectrum.

In Part II we will apply the methods of complex-cepstrum analysis outlined in this paper to real data.

THE COMPLEX CEPSTRUM—WEIGHTING,
ALIASING, AND MINIMUM PHASE

Using z-transforms, Schafer (1969, p. 12–13) gives the following three-step definition of the complex cepstrum for discrete functions of unit sample interval:

$$X(z) = \sum_{t=-\infty}^{\infty} x(t)z^{-t}, \quad \text{where } z = e^{\sigma+i\omega}, \quad (1a)$$

$$\hat{X}(z) = \log X(z)$$
$$= \log |X(z)| + i \arg [X(z)], \quad (1b)$$

$$\hat{x}(T) = \frac{1}{2\pi i} \oint_C \hat{X}(z)z^{T-1}dz, \quad (1c)$$

$$\text{where } T = 0, \pm 1, \pm 2, \cdots.$$

The three-step inverse definition for return to the time domain is:

$$\hat{X}(z) = \sum_{T=-\infty}^{\infty} \hat{x}(T)z^{-T}, \quad (2a)$$

$$X(z) = \exp [\hat{X}(z)], \quad (2b)$$

$$x(t) = \frac{1}{2\pi i} \oint_{C'} X(z)z^{T-1}dz. \quad (2c)$$

The complex natural logarithm defined in equation (1b) is a multivalued function since $\arg [X(z)]$ has a multiplicity of $2n\pi$ where $n = 0, 1, 2 \cdots$. Since $\hat{X}(z)$ must be continuous, $\arg [X(z)]$ cannot be restricted to its principal value ($n = 0$). In equation (1c) the values T define the quefrencies of Bogert et al (1963). Because of analogies developed later between the time domain and the complex cepstrum for impulsive time series, we prefer to associate the word period with the complex-cepstrum variable T. To emphasize this, we use the variable T in equation (1c). When performing the inverse transform [equations (2)], we choose the same contour of integration for equation (2c) as was used in equation (1c). Then the forward transform [equations (1)], followed by the reverse transform [equations (2)], will yield the original time series $x(t)$.

Weighting

In equation (1a) we have transformed a real-time function into its z-transform $X(z)$, which exists throughout the complex z-plane as does $\hat{X}(z)$ from equation (1b). Equation (1c) involves only those values of $\hat{X}(z)$ which lie on the closed-path integration contour C. Thus $\hat{x}(T)$ is a function of the particular contour. In computing the complex cepstrum, we use the discrete Fourier transform (DFT) for real frequencies in place of the z-transform, thus restricting the contour to the unit circle ($\sigma = 0$). However, if we multiply our original time function by the weighting function a^t, where $0 < a < 1$, we have effectively moved our integration contour to a circle of radius e^σ,

where

$$\sigma = -\log a. \qquad (3)$$

Since $a < 1$, σ is positive, and $e^{-\sigma t}$ is an exponentially decaying function. In equation (1a) we have defined $z = e^{\sigma + i\omega}$ to include the weighting function as part of the definition so that the contour in equation (1c) will be determined by our selection of σ. This is equivalent to off-axis integration in the complex ω-plane. In returning to the time domain via equations (2), we should unweight the result by a^{-t} which guarantees that $C = C'$ in equations (1c) and (2c). In the DFT, since we are confined to evaluating $\hat{X}(z)$ around the unit circle, we can consider that the effect of weighting is to move the poles and zeroes of $\hat{X}(z)$ radially inward by the factor e^{σ}; and unweighting with the inverse function restores them to their original location.

The logarithm introduces an additional number of zeros and poles into our function. If z_0 is a zero of the original function $X(z)$, it becomes a pole of $\hat{X}(z)$. In addition, if for z_1, $X(z_1) = 1$, then z_1 becomes a zero of $\hat{X}(z)$. Thus poles and zeros become poles, and "ones" become zeros.

If a function is minimum-phase, its Fourier transform will have all its poles and zeros in the left-half complex ω-plane (Ulrych and Lasserre, 1966). Under the definition of the z-transform used here, this corresponds to having all poles and zeros within the unit circle. Thus, if the zero of $X(z)$ farthest from the origin is at $z_0 = e^{\tau_0 + i\omega_0}$, then for $\sigma > \tau_0$, we will have moved all the zeros of $X(z)$ inside the unit circle, and the weighted function will be minimum-phase. Conversely, a maximum-phase function will have all its poles and zeros outside the unit circle. Schafer (1969) has shown that the complex cepstrum of a minimum-phase function is zero for $T < 0$, and the complex cepstrum of a maximum-phase function is zero for $T > 0$.

Aliasing

Aliasing is introduced into the complex cepstrum when the nonlinear logarithm operation of equation (1b) is followed by the discrete Fourier transform. Although $X(z)$ is adequately sampled, the nonlinear operations: logarithm, absolute value, and arctangent introduce harmonics into $\hat{X}(z)$. Thus $\hat{X}(z)$ is undersampled when $z = e^{in 2\pi/N}$, $n = 0, \pm 1, \pm 2, \cdots, \pm N/2$, as in the DFT. Since all harmonics out to infinite quefrencies or periods

are present, the complex cepstrum will in general be nonzero out to infinity. The effect of using the DFT is to alias these periods into the principal period range: $-1/2\Delta f < T \leq 1/2\Delta f$. Schafer (1969) has shown that an a^T weighting is impressed on the complex cepstrum by the a^t time-domain weighting. Thus weighting can help suppress aliasing. This is because $e^{-\sigma t}$ smooths $\hat{X}(z)$, when it is evaluated on the unit circle, by moving its poles (if the function is already minimum-phase) further inward away from the unit circle. This reduces the amplitude of the rapid fluctuations in $\hat{X}(z)$. Thus the high quefrencies or periods of $\hat{x}(T)$ and their harmonics are reduced. The shorter periods are less suppressed, but their harmonics have fallen off considerably by the folding period, $1/2\Delta f$. Figures 1a and b show how weighting reduces aliasing for a simple three-impulse time series. Figure 1a[1] is the complex cepstrum of the weighted trace, $a = 0.975$, which is minimum-phase. The aliasing is severe. Increasing the weighting to $a = 0.960$ (Figure 1b), results in a complex cepstrum which is not aliased. It has been our experience in complex-cepstrum analysis that it is convenient to weight heavily initially (e.g., $a = 0.94$), to guarantee an unaliased complex cepstrum, and then try less weighting until aliasing becomes a problem.

DEVELOPMENT OF THE COMPLEX CEPSTRUM

Before exploiting the properties of the complex cepstrum for deconvolution, we shall investigate the way in which the complex cepstrum develops for some signals of interest.

The reflector series

We model the reflector structure by a causal series of n impulses distributed arbitrarily in time at the unit sampling interval.

Let

$$r_o(t) = \sum_{i=1}^{n} \alpha_i \delta(t - t_i) \qquad t_i \geq 0. \qquad (4)$$

---

[1] The complex cepstrum is a real function. All complex-cepstrum plots are in terms of power, where we maintain the sign of the original real function. Thus a point at -60 db is a negative complex-cepstrum contribution whose power is 60 db. All contributions, positive and negative, whose power is less than zero db plot to zero db. The db axis is arbitrary and scaled with respect to the maximum absolute value power point of each complex cepstrum.

The first step in finding a complex cepstrum of this function is to weight it exponentially.

$$r(t) = r_0(t)a^t \qquad (5)$$

$$= \sum_{i=1}^{n} \alpha_i a^{t_i}\delta(t - t_i), \qquad (6)$$

where $0 < a \le 1$.

The z-transform is

$$R(z) = \sum_{i=1}^{n} \alpha_i a^{t_i}z^{-t_i}. \qquad (7)$$

We leave the weighting a^t as an integral part of the function and do not absorb it into z, so that $\sigma = 0$, and $z = e^{i\omega}$. Factoring $R(z)$ gives:

$$R(z) = \alpha_1 a^{t_1}z^{-t_1}\left[1 + \sum_{i=2}^{n} \beta_i a^{T_i}z^{-T_i}\right], \qquad (8)$$

where

$$\beta_i = \alpha_i/\alpha_1$$

and

$$T_i = t_i - t_1; \qquad i = 2, 3, \cdots, n.$$

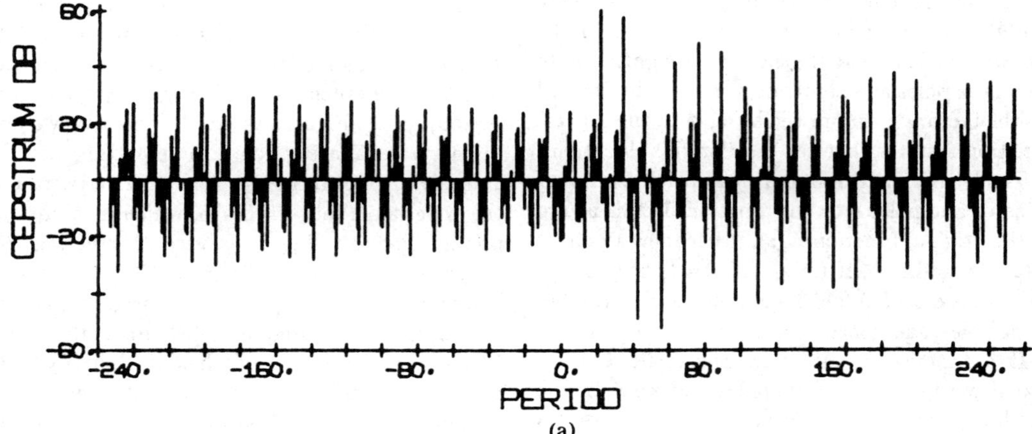

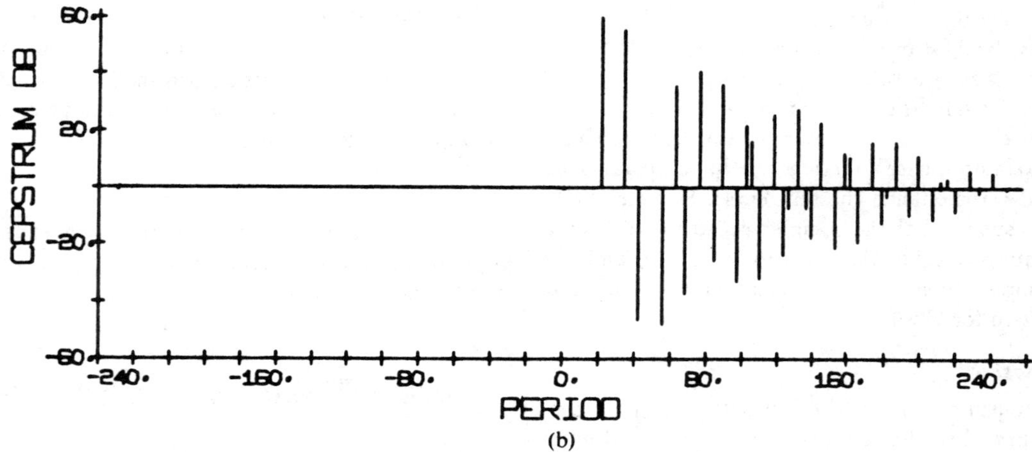

FIG. 1. Complex cepstra of the three impulse time series: $r_o(t) = \delta(t) + \delta(t-21) + \delta(t-34)$. (a) Weighting with a = 0.975 is sufficient to make the time series purely minimum phase (complex cepstrum fall off is toward increasing periods), but, the complex cepstrum is severely aliased. (b) Weighting with a = 0.96 reduces the complex cepstrum aliasing so that for all negative periods it is at least 60 db down from the largest cepstrum value. In this complex cepstrum it is quite clear that we get contributions at $T_2 = 21$, $T_3 = 34$, their multiples, and all combinations of their multiples. (All complex-cepstrum plots are scaled to the maximum complex-cepstrum contribution and plotted in db. However, we retain the sign of each complex-cepstrum contribution and plot it accordingly as a positive or negative contribution. Any complex-cepstrum contribution, positive or negative, which is 60 db or more down from the maximum plots to 0 db.)

Next, take the complex natural logarithm

$$\hat{R}(z) = \log\left(\alpha_1 a^{t_1} z^{-t_1}\right)$$
$$+ \log\left[1 + \sum_{i=2}^{n} \beta_i a^{T_i} z^{-T_i}\right]. \qquad (9)$$

We remove a linear phase term (ramp), if present, since its transform has contributions for all complex-cepstrum periods and may swamp the desired information. The z^{-t_1} term is the linear phase shift. We can interpret its removal as shifting our weighted function t_1 locations toward the time origin, so that the first impulse is now at the time origin (i.e., $t_1 = 0$). With respect to the z-transform $R(z)$ [equation (8)], this is equivalent to removing the t_1 zeros at infinity in the z-plane. Thus, we have

$$\hat{R}(z) = \log\left(\alpha_1\right)$$
$$+ \log\left[1 + \sum_{i=2}^{n} \beta_i a^{T_i} z^{-T_i}\right]. \qquad (10)$$

The final step of the transform is the contour integration of $\hat{R}(z)$ as indicated by the defining equation (1c). However, rather than perform the indicated integration, we follow the method outlined by Schafer (1969). If we can manipulate $\hat{R}(z)$ into a form such that it is recognizable as the z-transform of a known function, we then know that function. In this case, finding the Laurent-series expansion about $z = 0$ for the second logarithm of equation (10) is the proper step. The Laurent-series expansion of $\log(1+x)$ about $x = 0$ is:

$$\log(1 + x) = \sum_{m=1}^{\infty} \frac{(-1)^{m+1}}{m} x^m, \qquad (11)$$
$$\text{for } |x| < 1.$$

We let

$$x = \sum_{i=2}^{n} \beta_i a^{T_i} z^{-T_i}, \qquad (12)$$

and require that

$$\left| \sum_{i=2}^{n} \beta_i a^{T_i} z^{-T_i} \right| < 1. \qquad (13)$$

This requirement insures that the reflector series will be minimum-phase. While this appears overly restrictive, we are always able to satisfy equation (13) as we are free to choose an appropriate weighting. For real data, due to the geometric spreading loss and attenuation, little weighting should be necessary to make the reflector series (but not necessarily the trace) minimum-phase. We have then,

$$\hat{R}(z) = \log\left(\alpha_1\right) + \sum_{m=1}^{\infty} \frac{(-1)^{m+1}}{m}$$
$$\cdot \left(\sum_{i=2}^{n} \beta_i a^{T_i} z^{-T_i} \right)^m. \qquad (14)$$

The expression within parentheses is a polynomial of $n-1$ terms raised to the power m. We can expand this by use of the multinomial expansion (Morse and Feshbach, 1953, p. 412). The resulting $\hat{R}(z)$ is now recognizable as the z-transform of the following function:

$$\hat{r}(T) = \log\left(\alpha_1\right)\delta(T) + \sum_{m=1}^{\infty} \frac{(-1)^{m+1}}{m} \sum_{l_2,l_3,\cdots,l_n} \left[\frac{m!}{l_2!l_3! \cdots l_n!} \beta_2^{l_2}\beta_3^{l_3} \cdots \beta_n^{l_n} a^{\sum_{j=2}^{m} T_j l_j} \right]$$
$$\cdot \delta\left(T - \sum_{j=2}^{m} T_j l_j \right), \qquad (15a)$$

where

$$\sum_{j=2}^{n} l_j = m. \qquad (15b)$$

The second sum in equation (15a) is over all possible combinations of the l_j which satisfy (15b). Thus for $m = 1$, we will have a sum over $n-1$ terms in which each of the l_j is equal in turn to one, and the other l_j are zero. Little insight into the complex cepstrum is gained from equation (15a), except for some picture of the way the original impulse locations, individually and in combination, generate impulses in the complex cepstrum. We see from the δ-function in equation (15a) that we have complex-cepstrum contributions for all the original periods and all their multiples and also for all combinations of these multiples. The complex cepstrum is zero for negative T since the weighted reflector series is minimum-phase. Note that the complex cepstrum is zero between $T = 0$ and $T = T_2$, a property we shall use later when deconvolving the source. The complex cepstrum is of infinite extent because of the logarithm operation. Therefore, calculation of the complex cepstrum using the DFT results in an aliased version of equation (15a). However, the weighting will reduce the contribution of periods greater than the folding period, $|T| > T_n$. Thus the weighting operation has served a dual purpose: it guarantees that our reflector series will be minimum-phase and it reduces the aliasing.

Consider a specific example, the three-impulse case. Let

$$r_0(t) = \sum_{i=1}^{3} \alpha_i \delta(t - t_i) \qquad t_i > 0. \qquad (16)$$

Proceeding as above and using the binomial expansion, we obtain the complex cepstrum of a ramp-free, weighted version of equation (16):

$$\begin{aligned}
\hat{r}(T) = \; & \log(\alpha_1)\delta(T) \\
& + \sum_{m=1}^{\infty} \frac{(-1)^{m+1}}{m} \sum_{l=0}^{m} \frac{m!}{l!(m-l)!} \\
& \cdot \left(\frac{\alpha_2}{\alpha_1}\right)^{m-l} \left(\frac{\alpha_3}{\alpha_1}\right)^{l} a^{T_2(m-l)+T_3 l} \\
& \cdot \delta(T - (m-l)T_2 - lT_3),
\end{aligned} \qquad (17)$$

where

$$T_2 = t_2 - t_1$$
$$T_3 = t_3 - t_1.$$

Figure 1a illustrates this case for $T_2 = 21$, $T_3 = 34$, $\alpha_1 = \alpha_2 = \alpha_3$, and $a = 0.975$. Notice that we have chosen "a" such that the weighted function is minimum-phase, but aliasing is still present. In Figure 1b the same impulse sequence is weighted with $a = 0.96$, and there is no visible aliasing present. There are nonzero complex-cepstrum contributions at all multiples of the two original time-domain periods, T_2 and T_3, as well as all combinations of the *sums* of these two periods, but not the differences. It is important to notice that in all cases where we have weighted the original function sufficiently to make it minimum-phase there are no unaliased complex-cepstrum con-

tributions for $0 < T < T_2$. When performing deconvolution, which is subtraction in this domain, we can remove all complex-cepstrum contributions for $0 < T < T_2$, without affecting the impulse train, provided we have weighted the trace sufficiently to make the impulse train minimum-phase.

Water column reverberation

Following the discussion by Pfleuger (1972), we represent the multiple generator in the water column by:

$$m(t) = \sum_{n=0}^{\infty} (-1)^n R^n \delta(t - nT_w), \qquad (18)$$

where R is the reflection coefficient, $0 < R < 1$, and T_w is the two-way traveltime through the water column. This is only the source multiple generator and would correspond to a 2-point filter (Backus, 1959). The function is minimum-phase since R is less than unity.

The z-transform of $m(t)$ is:

$$M(z) = \sum_{m=0}^{\infty} (-1)^m R^m z^{-mT_w}, \qquad (19)$$

which can be written as

$$M(z) = (1 + Rz^{-T_w})^{-1}. \qquad (20)$$

Taking the complex logarithm and using the Laurent-series expansion about $z = 0$ gives

$$\hat{M}(z) = \sum_{m=1}^{\infty} (-1)^m \frac{R^m}{m} z^{-mT_w}, \qquad (21)$$

so that the complex cepstrum is

$$\hat{m}(T) = \sum_{m=1}^{\infty} (-1)^m \frac{R^m}{m} \delta(T - mT_w). \qquad (22)$$

Figure 2 illustrates this case for $R = 0.8$ and $T_w = 13\Delta t$. If Δt is 4 msec, this corresponds to a water depth of 39 m.

The bubble pulse

We shall represent the periodic part of the bubble-pulse oscillation by the following function:

$$b(t) = \sum_{n=0}^{\infty} R^n \delta(t - nT_b), \qquad (23)$$

where $0 < R < 1$; T_b is the period of the oscillations. Since R is less than unity, the oscillations are damped. The convolution of $b(t)$ [equation (23)], with a short-duration seismic source function would produce a reasonable approximation to the theoretical and experimental results of Ziolkowski (1970) and Schulze-Gattermann (1972). While equation (23) is a simplification even for the periodic part of the bubble-pulse function, it is sufficient for the purposes of our discussion.

Following exactly the method used above for the source water-column multiple generator, we obtain the complex cepstrum of $b(t)$:

$$\hat{b}(T) = \sum_{m=1}^{\infty} \frac{R^m}{m} \delta(T - mT_b). \qquad (24)$$

Table 1 summarizes some important transform pairs. The weighting has been chosen to insure that all functions are minimum-phase and un-

aliased. The last transform pair illustrates the case of the positive, impulsive, periodic, decaying time function described above.

DECONVOLUTION

Deconvolution in the time domain becomes subtraction in the complex cepstrum. An exact deconvolution of a convolutional component results in that component's complex-cepstrum contributions being set equal to zero. We have emphasized the correspondence between the time domain and the complex cepstrum for appropriately weighted traces. In the discussion which follows we assume that the trace has been weighted sufficiently so that the *reflector series* (but not necessarily the source) is minimum-phase and contains the shot at $t = 0$. Since the minimum-phase reflector series has contributions in the

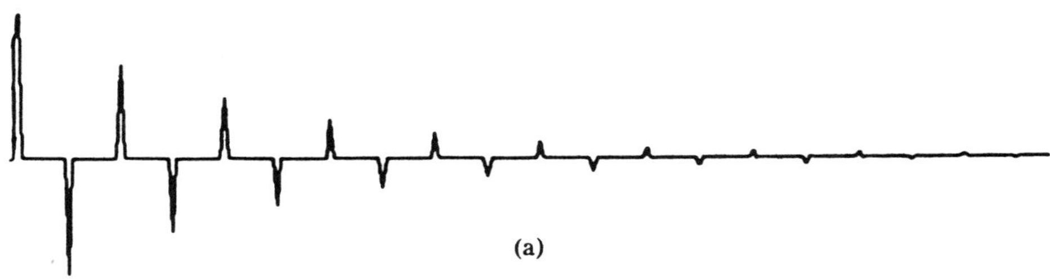

(a)

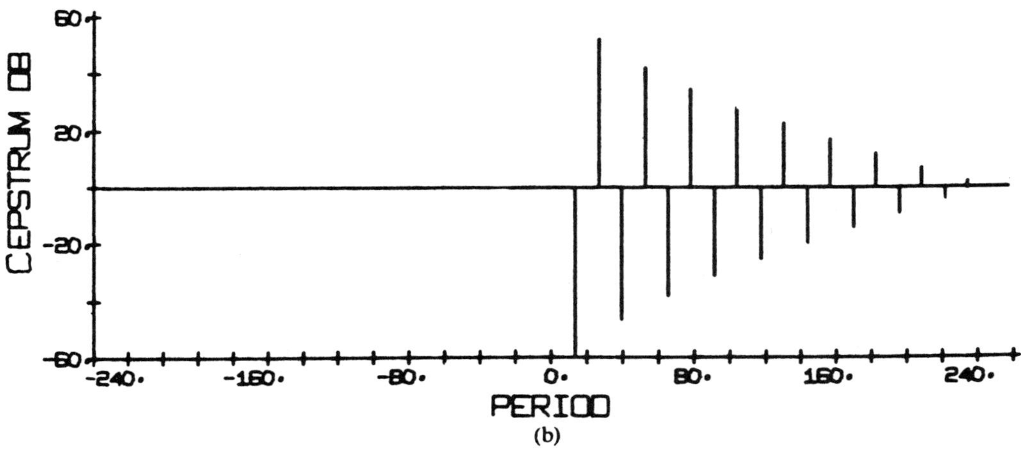

(b)

FIG. 2. (a) Multiple generator, where $T_w = 13$ and $R = 0.8$. (b) Corresponding complex cepstrum.

Stoffa et al

Table 1. Time and complex cepstrum transform pairs.[1]

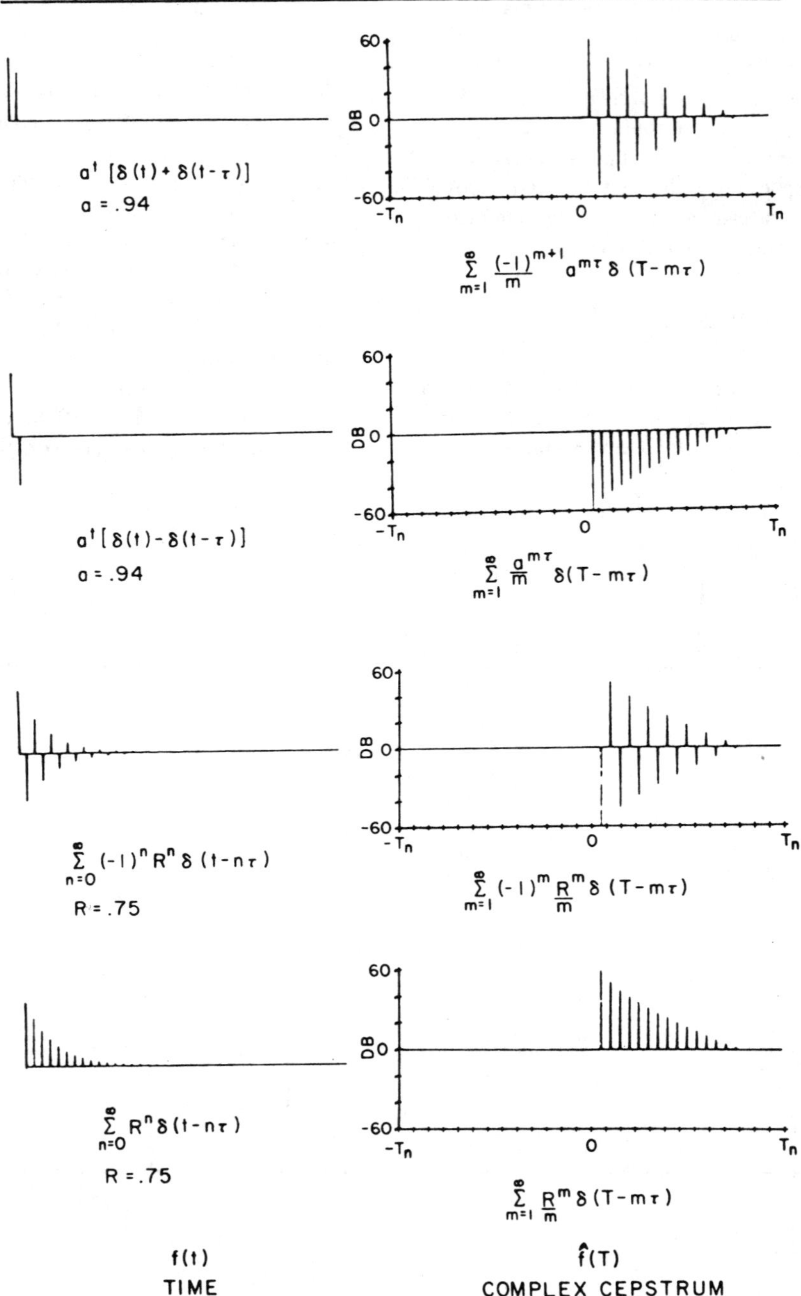

[1]It is apparent that a Backus two-point filter (first time function) is the convolutional inverse to an alternating sign multiple generator (the third time function) since the addition of the first and third complex cepstra would be zero for $R = a^\tau$. Similarly, the second and fourth transform pairs are inverses of each other.

complex cepstrum which are confined to periods greater than the two-way traveltime between the shot and the first reflector, one deconvolution we can perform is to set all these periods equal to zero. While this is a useful first attempt, we have shown that the bubble pulse and multiple generator of the water column are of infinite extent in the complex cepstrum. The higher-order complex-cepstrum terms of the source and multiple generator will be in the region of the reflectors.

Deconvolving a band-pass wavelet

First, let us consider a time function, Figure 3a, whose phase spectrum, power spectrum, and com-

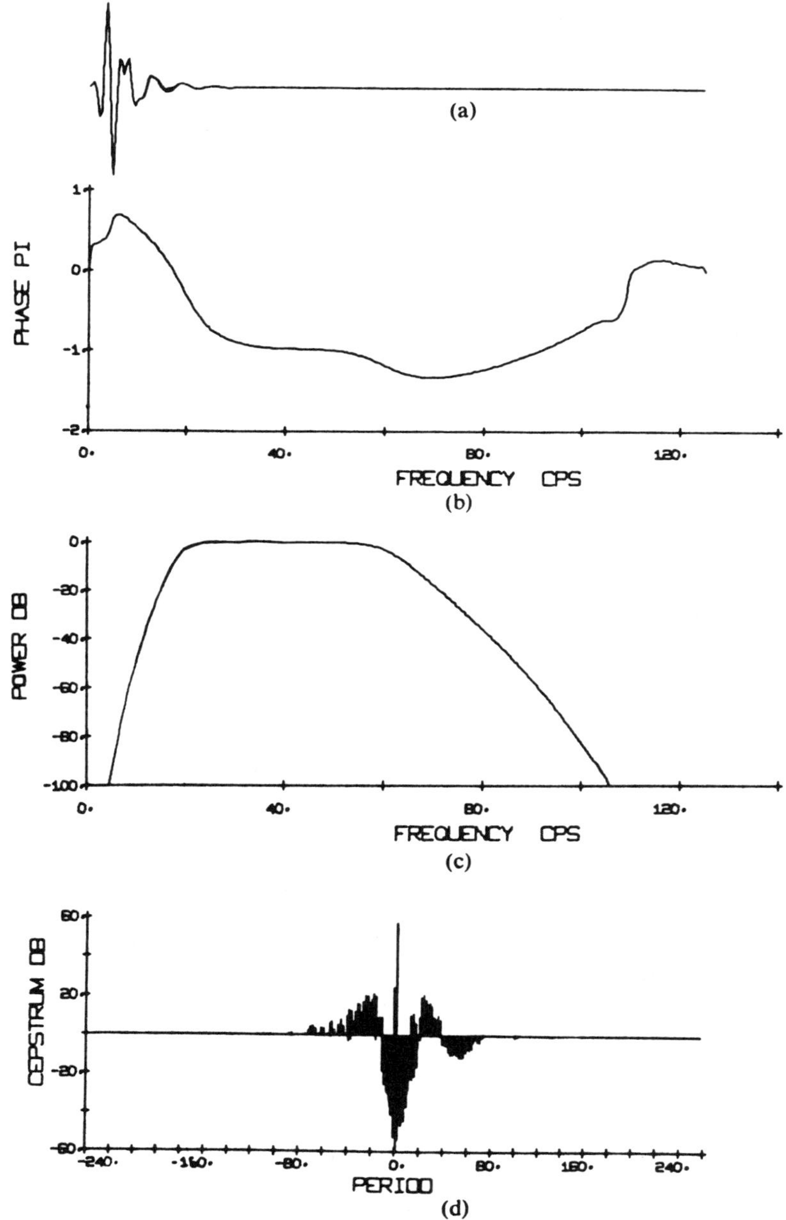

FIG. 3. (a) 20-60 hz, 8-pole band-pass wavelet. (b) Phase spectrum. (c) Power spectrum. (d) Corresponding complex cepstrum. (Note the nonimpulsive nature of the complex cepstrum in contrast to Figures 1 and 2.)

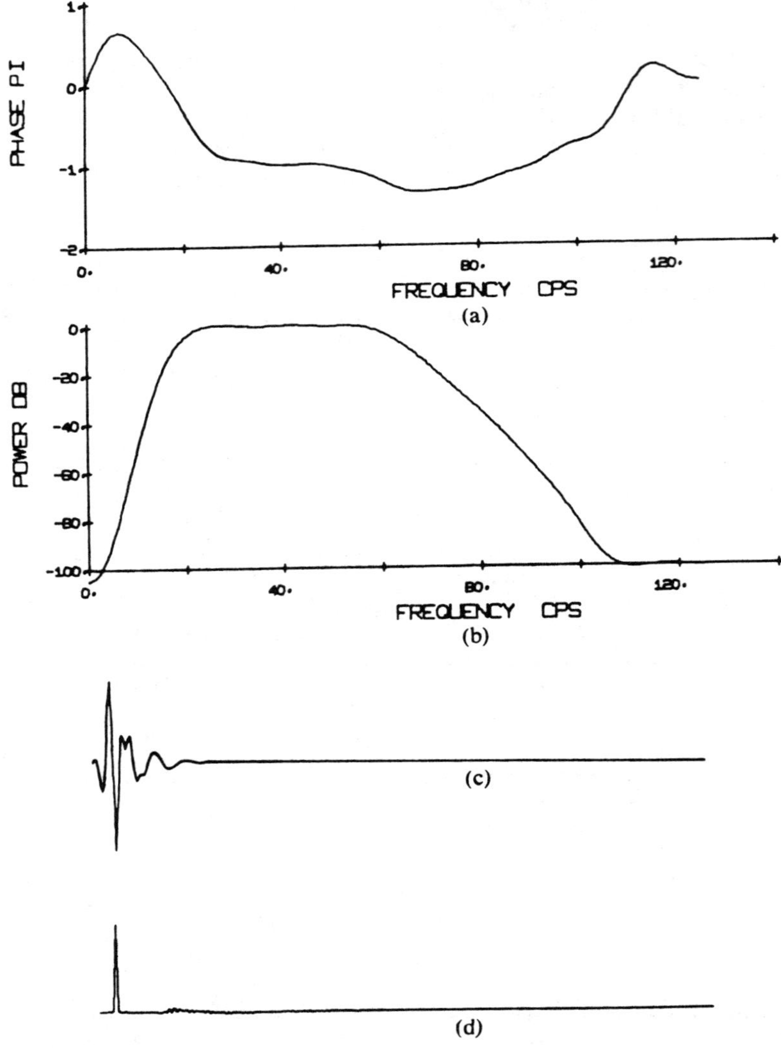

FIG. 4. (a) Phase spectrum obtained by defining the complex cepstrum in Figure 3d equal to zero for $|T| > 18$. (b) Power spectrum for same. (Note the similarity to Figures 3b and c indicating the dominance of the central complex-cepstrum contributions for this wavelet.) (c) Corresponding time domain wavelet. (d) Wave train given by zeroing the complex cepstrum of Figure 3d in the range $-18 \leq T \leq 18$. This wave train is the convolutional noise introduced by neglecting the complex cepstrum contribution of the wavelet for $|T| > 18$.

plex cepstrum are shown in Figures 3b, c and d. The function is the impulse response of a 20–60 hz, 8-pole band-pass filter representing a possible seismic source function. Note that the complex cepstrum has a significant contribution at all negative periods indicating a substantial maximum-phase component. Any attempt to recover a reflector series from a trace formed by convolving this function with the reflector series will fail if one uses a minimum-phase source assumption.

We remove all complex-cepstrum contributions at periods greater than 18. That is, all contributions at periods greater than $+18$ and less than -18 are set equal to zero. In Figure 4a and b we see the resulting phase and power spectra and in Figure 4c the time-domain function, which is al-

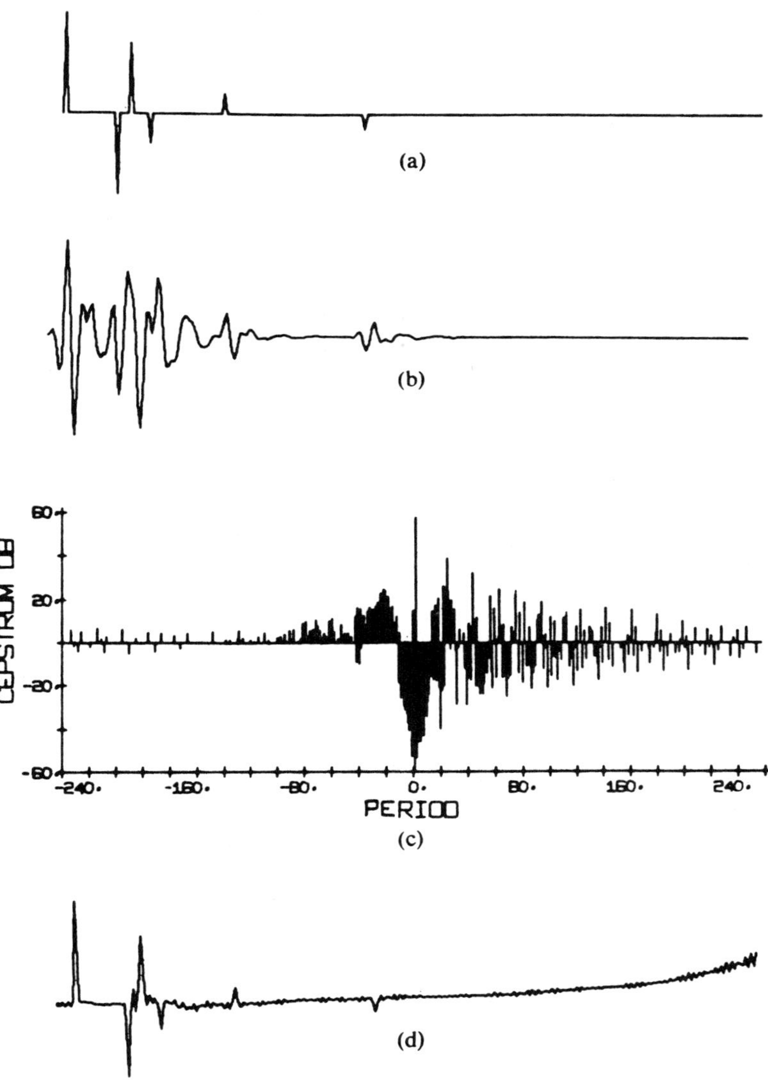

FIG. 5. (a) Mixed phase impulse train which is arbitrary except that separation of the first two impulses is 19. (b) Convolution of this impulse train with the wavelet of Figure 3a. (c) Complex cepstrum of the time series of Figure 5b, weighted with $a = 0.98$. This weighting is sufficient to make the reflector series, but not the trace, minimum phase. (Note that a small amount of aliasing is present.) (d) Deconvolved time series obtained by defining all complex cepstrum contributions for $-18 \leq T \leq 18$ as zero. (Note the convolutional noise, introduced as in Figure 4d, gets amplified at the end of the time series due to the unweighting operation.)

most identical to the original in Figure 3a. Therefore, removing complex-cepstrum contributions at locations $-18 \leq T \leq 18$ should remove most of the effects of a convolution with this function as shown in Figure 4d, which is our deconvolved source. Of course, any reflector whose first complex-cepstrum contribution lies between 0 and 18 will be removed along with the source (and, incidentally, will introduce a small amount of convolutional noise). Hence, it may be necessary to sacrifice a small amount of reflector information in order to adequately deconvolve the source.

For purposes of illustration, we choose a mixed-phase reflector series, Figure 5a, where the separation between the shot and first reflector is greater than 18. (Weighting with $a=0.98$ is sufficient to make this reflector series minimum-phase, hence, its first complex-cepstrum contribution is at $T>18$.) In Figure 5b it is shown convolved with the source function of Figure 3 and then weighted with $a=0.98$. The complex cepstrum is shown in Figure 5c. We deconvolve by setting all complex-cepstrum contributions for $-18 \leq T \leq 18$ equal to zero. The trace resulting from this deconvolution is then unweighted and shown in Figure 5d. The importance of this example is that it shows a convolution in which a wavelet, which was not minimum-phase, has been easily deconvolved. It is apparent from Figure 5d that we have introduced noise. This "filter" noise arises from the contributions of the band-pass wavelet which are still present in the periods we have retained. While the amplitude of the noise in this example is less than the amplitude of the impulses recovered, it is apparent in all cases that the success of the deconvolution is related directly to the degree of separation of all components in the complex cepstrum. In addition, the unweighting has amplified the noise at the end of the trace.

Deconvolving an impulsive, periodic function of time

Consideration of a multiple generator will serve as an example of deconvolving a damped, periodic, impulsive, time-domain function. The deconvolution of the bubble-pulse oscillations as developed above follows an analogous argument.

Consider removing the first n contributions from the complex cepstrum of the multiple generator, equation (22).

$$\hat{p}_n(T) = \hat{m}(T) - \sum_{l=1}^{n} (-1)^l \frac{R^l}{l} \delta(T - lT_w). \quad (25)$$

Performing the subtraction and taking the z-transform gives

$$\hat{P}_n(z) = \sum_{l=n+1}^{\infty} \frac{\zeta^l}{l}, \quad (26)$$

where $\zeta = -Rz^{-T_w}$.

Exponentiating and expanding the exponential in a power series, we obtain

$$P_n(z) = \prod_{l=n+1}^{\infty} \sum_{j=0}^{\infty} \frac{1}{j! l^j} \zeta^{lj} \quad (27)$$

$$= 1 + \sum_{k=1}^{\infty} \frac{\zeta^{n+k}}{n+k} + \sum_{k=2}^{\infty} \left[\sum_{l=1}^{k} \frac{1}{2(n+l)(n+k-l)} \right] \zeta^{(2n+k)} + \cdots. \quad (28)$$

From equation (28) we see that except for the contribution of unity at $t=0$, the first contribution is at the $(n+1)$th multiple. The second summation contributes only at the $(2n+k)$th multiple for $k \geq 2$. This relation clearly shows that removing the first n complex-cepstrum contribution eliminates the first n time-domain multiples and reduces the remaining multiples to at most $1/(n+1)$ of their original amplitude. Actual values are given in Table 2 for the first eight multiples, after removal of the first one, two, and three complex-cepstrum contributions. The resulting time-domain traces are shown in Figure 6. Of course, the more complex-cepstrum contributions we remove, the more exactly we eliminate the multiple generator. However, by simply removing two, or at most three, we can sufficiently attenuate the water-column reverberations or bubble-pulse oscillations for most practical purposes.

We do not need to know the exact locations of the multiple generator in the complex cepstrum if we are willing to tolerate the loss of a small amount of trace information. For example, we know that in water 60 m deep the multiple generator of the water column has its first two complex-cepstrum locations at periods less than .2 sec (in fact, they are at .08 and .16 sec). Addi-

Table 2. Multiple elimination and reduction for the first 8 multiples in time obtained by removing the first $p_1(t)$, first two $p_2(t)$, and first three $p_3(t)$ nonzero complex cepstrum contribution of the multiple generator $m(t)$.

Multiple	$m(t)$	$p_1(t)$	$p_2(t)$	$p_3(t)$
0	1	1.0000	1.0000	1.0000
1	$-R$	0.0000	0.0000	0.0000
2	R^2	0.5000 R^2	0.0000	0.0000
3	$-R^3$	-0.3333 R^3	-0.3333 R^3	0.0000
4	R^4	0.3750 R^4	0.2500 R^4	0.2500 R^4
5	$-R^5$	-0.3666 R^5	-0.2000 R^5	-0.2000 R^5
6	R^6	0.3681 R^6	0.2222 R^6	0.1667 R^6
7	$-R^7$	-0.3678 R^7	-0.2262 R^7	-0.1429 R^7
8	R^8	0.3679 R^8	0.2229 R^8	0.1562 R^8

tionally, if we use an air gun whose spectrum peaks at 18 hz, the period of the air gun is .055 sec, so that we have three complex-cepstrum contributions of the bubble pulse at periods less than .2 sec. By removing all complex-cepstrum locations for $T < .2$ sec, we eliminate the first two water-column multiples and reduce the remainder to at most one-third of their original amplitude. Also, we eliminate the first three bubble-pulse oscillations and reduce the remaining oscillations to at most one-fourth of their original amplitudes. In effect, we have reduced the amplitude of the remaining terms to the point where they are negligible for most purposes.

Figure 7a shows the impulse train used previously convolved with the multiple generator and the band-pass wavelet. The trace is weighted with $a = 0.98$, and Figure 7b is the resulting complex cepstrum. We deconvolve by removing all complex-cepstrum contributions for $-18 \leq T \leq 18$. This deconvolution includes only one contribution of the multiple generator. In Figure 7c we see the deconvolved trace and in Figure 7e the

original impulses which we are attempting to recover. A deconvolution which includes the second complex-cepstrum contribution of the multiple generator greatly improves the result, as seen in Figure 7d. The power of this type of deconvolution is that one can improve the trace substantially by simply excluding one or two multiple-generator, complex-cepstrum contributions. In addition, any source convolution products which are maximum-phase can be easily deconvolved by setting the complex cepstrum equal to zero for negative periods. This will have no effect on the reflector series if the trace has been weighted sufficiently to make the reflector series minimum-phase and unaliased.

THE CONTINUOUS RAMP-FREE PHASE SPECTRUM

Given a function of time, its Fourier transform $F(\omega)$ consists of a real part $u(\omega)$ and an imaginary part $v(\omega)$. For properly sampled, time-limited data the complex Fourier coefficients will be samples of the continuous and unaliased $F(\omega)$.

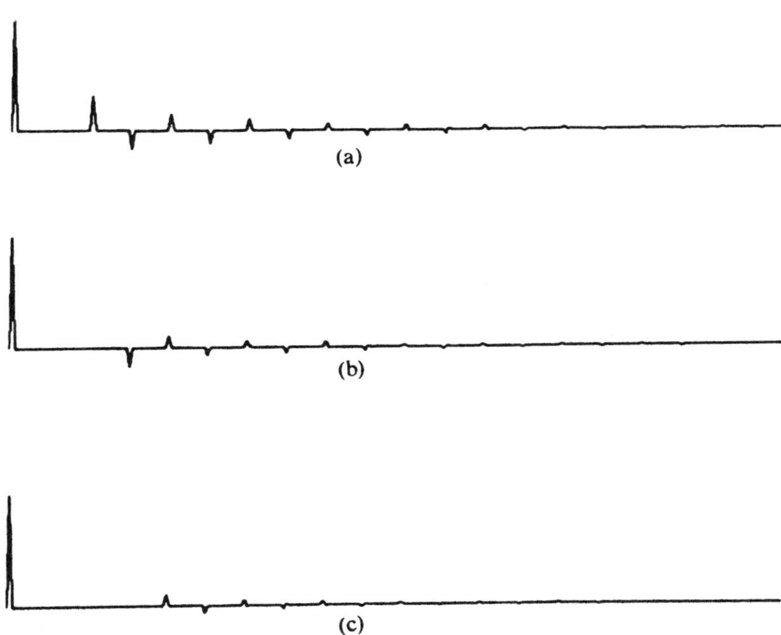

(a)

(b)

(c)

FIG. 6. The deconvolved time series obtained by removing (a) the first one, (b) the first two, and (c) the first three nonzero complex cepstrum contributions of the multiple generator of Figure 2. [Note that eliminating two (or three) complex cepstrum contributions not only eliminates the first two (or three) multiples in time but results in a substantial reduction of the remaining multiples.]

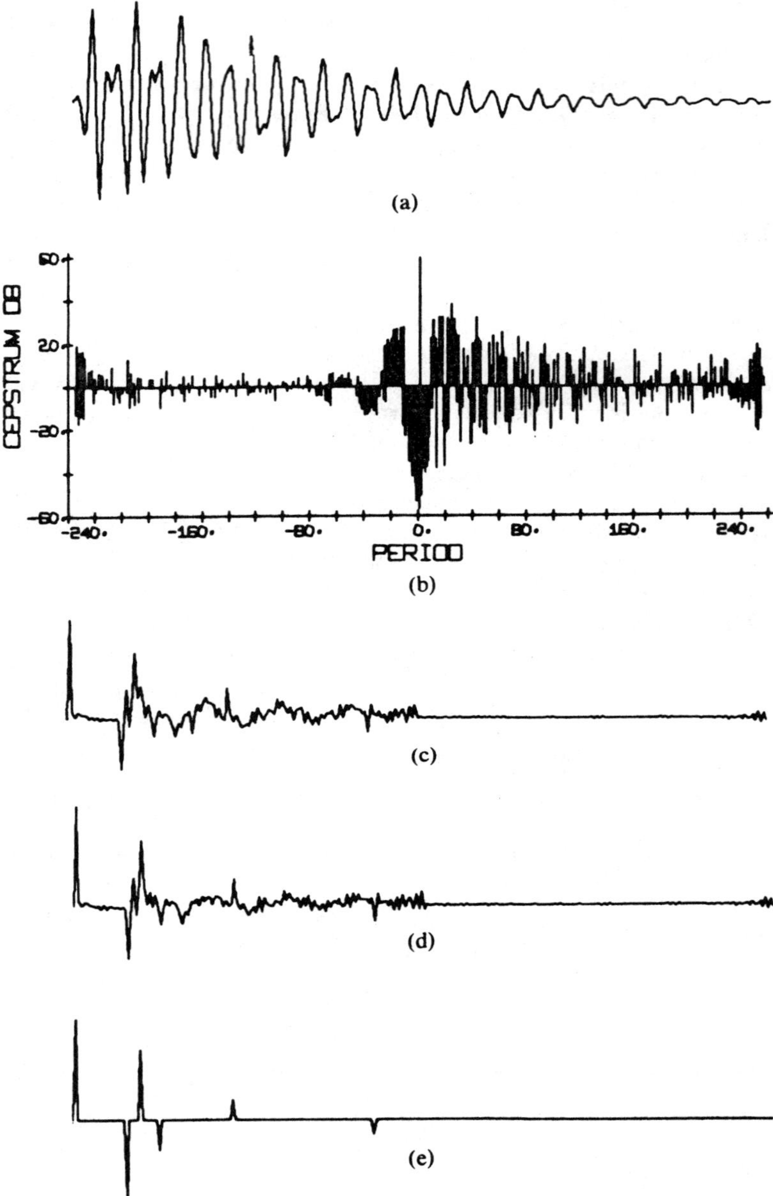

FIG. 7. (a) Time series formed by convolving the impulse train of Figure 5a with the source wavelet of Figure 3a and the multiple generator of Figure 2a. (b) Complex cepstrum of the time series of Figure 5a weighted with $a = 0.98$. A small amount of aliasing is present. (c) Deconvolved trace obtained by defining all complex-cepstrum contributions for $-18 \le T \le 18$ as zero. (d) Deconvolved trace obtained as in Figure 7c but with the second complex-cepstrum contribution of the multiple generator at $T = 26$ also removed. Notice the substantial improvement over Figure 7c. (Note in 7c and d we have only unweighted half the trace in order to eliminate the amplification of the convolutional noise which occurs at the end of the time series as in Figure 5d.) (e) The impulse train we are attempting to recover from the time series of Figure 7a.

The next step in performing the transformation into the complex cepstrum is to take the complex natural logarithm.

For example, $f(t)$ transforms to $F(\omega)$, where $F(\omega) = u(\omega) + iv(\omega)$. Writing this in terms of amplitude and phase, we obtain

$$
F(\omega) = [u^2(\omega) + v^2(\omega)]^{1/2}
$$
$$
\cdot \exp \left\{ i \tan^{-1} \left[\frac{v(\omega)}{u(\omega)} \right] \right\} \qquad (29)
$$
$$
= A(\omega) \exp [i\phi(\omega)].
$$

The complex natural logarithm of this expression is

$$
\hat{F}(\omega) = \log [A(\omega)] + i\phi(\omega). \qquad (30)
$$

The greatest difficulty in computing the complex cepstrum is computing the proper phase curve. While $u(\omega)$ and $v(\omega)$ are continuous, $\tan^{-1} [v(\omega)/u(\omega)]$ is in general not continuous. Since the inverse tangent function is multivalued, we must choose the branches which make it continuous. If the principal value is determined, the resulting phase curve is restricted to $-\pi < \phi \leq \pi$. Schafer (1969) has suggested an algorithm for locating the resulting discontinuities or "jumps" in the phase curve and a procedure for "unwrapping" the phase curve. The success of this approach is limited by the fineness with which the original phase curve is sampled. Usually it is necessary to extend the original time function by appending zeros. This results in increased frequency-domain sampling since padding with zeros corresponds to smoothly interpolating between the original frequency-domain coefficients. However, in the presence of a large ramp (linear phase component), this unwrapping procedure still encounters difficulty. This can be alleviated by using an iterative unwrapping algorithm. We unwrap the principal value of the phase curve as suggested by Schafer (1969) and determine the apparent ramp. Next, remove the apparent ramp from the original principal value of the phase and unwrap the resultant. We continue in this manner until there is no apparent ramp.

We have also used an alternative approach to computing a continuous curve which considers the derivative of the phase with respect to frequency. We note that

$$
\frac{d}{dx} [\tan^{-1} x] = \frac{1}{1 + x^2}. \qquad (31)
$$

If $x = v(\omega)/u(\omega)$, and $\phi(\omega) = \tan^{-1}(x)$, then

$$
\frac{d\phi}{d\omega} = \frac{1}{1 + \dfrac{v^2(\omega)}{u^2(\omega)}} \frac{d}{d\omega}\left[\frac{v(\omega)}{u(\omega)} \right], \qquad (32)
$$

and

$$
\frac{d}{d\omega}\left[\frac{v(\omega)}{u(\omega)} \right]
$$
$$
= \left[u(\omega) \frac{dv(\omega)}{d\omega} - v(\omega) \frac{du(\omega)}{d\omega} \right] \Big/ u^2(\omega), \qquad (33)
$$

so that

$$
\frac{d\phi}{d\omega} = \frac{1}{u^2(\omega) + v^2(\omega)}
$$
$$
\cdot \left[u(\omega) \frac{dv(\omega)}{d\omega} - v(\omega) \frac{du(\omega)}{d\omega} \right]. \qquad (34)
$$

Integration of equation (34) yields a continuous phase curve.

The Fourier coefficients are continuous, well behaved functions, which are properly sampled. The derivatives $dv/d\omega$ and $du/d\omega$ can be found to any accuracy desired by using difference methods or by finding the Fourier transform of $t \cdot f(t)$. The integration can also be performed to the desired accuracy. In addition to computing a continuous phase curve, this derivative approach offers a method of removing the linear phase shift. The mean value of $d\phi/d\omega$ is the linear phase shift. Determining this mean, removing it from $d\phi/d\omega$, and then integrating gives the continuous, ramp-free phase curve which we desire.

SUMMARY

Homomorphic deconvolution promises to be an important method of removing the unwanted convolutional components from the shallow-water, single-channel marine seismic trace. The complex cepstrum, the particular homomorphic system considered, offers several advantages over conventional deconvolution. Deconvolution in the complex cepstrum does not require the usual minimum-phase source assumption. Maximum or mixed-phase source components are deconvolved as easily as minimum-phase components. The relation of the complex cepstrum to the original time function is in terms of the periodicities of the time function. Since subbottom reflectors always arrive later in the trace than the bottom reflection, their complex-cepstrum contributions are at correspondingly greater complex cepstrum periods if the reflector series is minimum-phase. Most shallow-water reflector series are minimum-phase (Robinson, 1966). In cases where they are not, a small amount of weighting is usually sufficient to make them minimum-phase.

The bubble-pulse oscillation of an air gun is of short period, often less than the period of the water-column reverberation. All complex cepstra are of infinite extent. However, for an impulsive, periodic time function such as the bubble oscillation or the water-column reverberation, we have shown that merely removing the first few complex-cepstrum contributions reduces their time-domain contributions considerably. Deconvolution via the complex cepstrum allows one to choose readily the extent of the deconvolution and the expense, if any, of the desired reflector information.

Complex-cepstrum analysis requires the computation of a continuous, ramp-free phase spectrum. Algorithms which detect discontinuities in the principal value of the phase curve encounter difficulty in the presence of a large linear term (ramp). This difficulty is eliminated by the methods outlined in this paper.

ACKNOWLEDGMENTS

This work was supported by the National Science Foundation through the Office for the International Decade of Ocean Exploration under grant no. GX-34410, and through the Division of Environmental Sciences under grant GA-27281; and by the Office of Naval Research under grant no. N00014-67-A-0108-0004.

We are grateful to K. McCamy and P. Richards for critically reviewing the manuscript and offering many useful suggestions.

REFERENCES

Backus, Milo B., 1959, Water reverberations—their nature and elimination: Geophysics, v. 24, no. 2, p. 233–261.

Bogert, B. P., Healey, M. J., and Tukey, J. W., 1963, The quefrency analysis of time series for echoes; cepstrum, pseudo-autocovariance, cross-cepstrum, and saphe cracking: Proc. Symp. on Time Series Analysis, M. Rosenblatt, editor, New York, John Wiley and Sons, p. 209–243.

Morse, P. M., Feshbach, H., 1953, Methods of theoretical physics: New York, McGraw-Hill Book Co., Inc., 1977 p.

Oppenheim, A. V., 1965, Superposition in a class of non-linear systems: Tech. Rep. 432, MIT, Res. Lab. of Electr., 62 p.

Pfleuger, J., 1972, Spectra of water reverberations for primary and multiple reflections: Geophysics, v. 37, p. 788–796.

Rice, R. B., 1962, Inverse convolution filters: Geophysics, v. 27, no. 1, p. 4–18.

Robinson, E. A., 1957, Predictive decomposition of seismic traces: Geophysics, v. 22, no. 4, p. 767–778.

——— 1966, Multichannel z transforms and minimum delay: Geophysics, v. 31, p. 482–500.

Schafer, R. W., 1969, Echo removal by discrete generalized linear filtering: Tech. Rep. 466, MIT, Res. Lab. of Electr.

Schulze-Gatterman, R., 1972, Physical aspects of the "air pulser" as a seismic energy source: Geophys. Prosp., v. 20, p. 155–192.

Ulrych, T. J., 1971, Application of homomorphic deconvolution to seismology: Geophysics, v. 36, no. 4, p. 650–660.

——— 1972, Homomorphic deconvolution of some teleseismic events: Bull. SSA, v. 62, p. 1253–1265.

Ulrych, T. J., and Lasserre, M., 1966, Minimum phase: J. Canadian SEG, v. 2, no. 1, p. 22–32.

Ziolkowski, A., 1970, A method of calculating the output pressure waveform from an air gun: Geophys. J. Roy. Astr. Soc., v. 21, p. 137–161.

12

HOMOMORPHIC FILTERING – THEORY AND PRACTICE*

BY

B. BUTTKUS**

ABSTRACT

BUTTKUS, B., 1975, Homomorphic Filtering — Theory and Practice, Geophysical Prospecting 23, 712-748.

The application of homomorphic filtering in marine seismic reflection work is investigated with the aims to achieve the *estimation of the basic wavelet*, the *wavelet deconvolution* and the *elimination of multiples*. Each of these deconvolution problems can be subdivided into two parts: The first problem is the detection of those parts in the cepstrum which ought to be suppressed in processing. The second part includes the actual filtering process and the problem of minimizing the random noise which generally is enhanced during the homomorphic procedure.

The application of homomorphic filters to synthetic seismograms and air-gun measurements shows the possibilities for the practical application of the method as well as the critical parameters which determine the quality of the results. These parameters are:
a) the signal-to-noise ratio (SNR) of the input data,
b) the window width and the cepstrum components for the separation of the individual parts,
c) the time invariance of the signal in the trace.

In the presence of random noise the power cepstrum is most efficient for the detection of wavelet arrival times. For wavelet estimation, overlapping signals can be detected with the power cepstrum up to a SNR of three. In comparison with this, the detection of long period multiples is much more complicated. While the exact determination of the water reverberation arrival times can be realized with the power cepstrum up to a multiples-to-primaries ratio of three to five, the detection of the internal multiples is generally not possible, since for these multiples this threshold value of detectibility and arrival time determination is generally not realized.

For wavelet estimation, comb filtering of the complex cepstrum is most valuable. The wavelet estimation gives no problems up to a SNR of ten. Even in the presence of larger noise a reasonable estimation can be obtained up to a SNR of five by filtering the phase spectrum during the computation of the complex cepstrum. In contrast to this, the successful application of the method for the multiple reduction is confined to a SNR of ten, since the filtering of the phase spectrum for noise reduction cannot be applied. Even if the threshold results are empirical, they show the limits for the successful application of the method.

---

* Paper read at the Thirty-sixth Meeting of the European Association of Exploration Geophysicists, Madrid, Spain, June 1974.
** Bundesanstalt für Geowissenschaften und Rohstoffe, 3 Hannover, Stilleweg 2, West Germany.

The homomorphic method is superior to other methods for the *wavelet estimation*, as no assumptions are imposed on the delay properties of the wavelets. For instance, predictive deconvolution works best only if the input is minimum delay. It is suggested that the *wavelet deconvolution* should be performed with the least squares method, using homomorphic wavelet estimation.

The *suppression of multiples* can only be successful, if the implicit assumption of time invariance of the signal is satisfied. For the discussed deterministic process of homomorphic filtering this condition is more critical if compared with the statistical least squares method.

1. INTRODUCTION

During the last two years homomorphic filtering has been given more attention as an alternative method for different seismic deconvolution problems. The method of homomorphic filtering described by Oppenheim, Schafer, and Stockham (1968), Schafer (1969) and Ulrych (1971) is primarily developed for the problems of echo detection and echo removal.

The algorithm transforms the convolution process into an additive superposition of its components, with the result that the single parts can be separated more easily. While Ulrych (1971) has demonstrated the application of this method in seismology for the separation of overlapping signals, the practical application of the homomorphic filter process in seismic reflection work has been delayed for a long time and is discussed for the first time by Stoffa, Buhl, and Bryan (1974) and by Buhl, Stoffa, and Bryan (1974) for the suppression of air-gun bubble pulses and water column reverberations. Two reasons for this delayed application can be seen:

(1)   For specific problems in seismic reflection work the method of homomorphic filtering is not optimally adaptable to the actual field data.

(2)   Due to insufficient experience with the method the results usually fall short of the expectation.

The aim of this paper is to show how the three problems of

      (1)   wavelet estimation,
      (2)   wavelet deconvolution, and
      (3)   multiple suppression

can be solved with homomorphic filters. We show which critical factors influence the quality of the results, indicate the difficulties of the practical application of the method and give solutions to these problems. The paper can be seen as an addition to the papers of Stoffa et al (1974) and Buhl et al (1974). In contrast to these papers we are concerned with deep sea problems. Unlike Stoffa and Buhl we extend our investigations to the study of homomorphic deconvolution in the presence of noise.

The results are based on empirical studies with synthetic seismograms and where possible on theoretical considerations as well as on the practical application of the method to air-gun measurements.

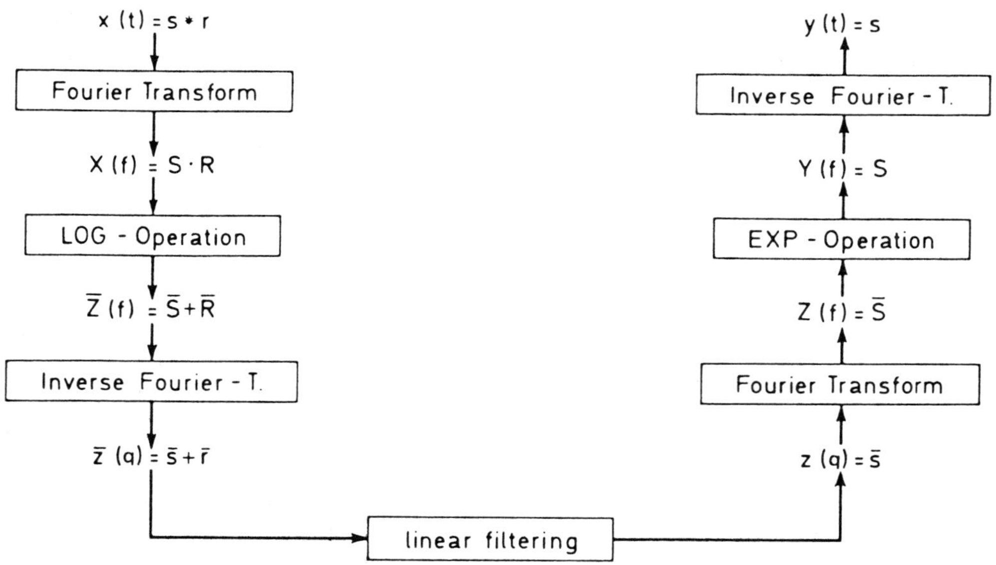

Fig. 1. Realization of homomorphic filtering using the Fourier Transform

2. Algorithm of Homomorphic Filtering

The procedure of homomorphic filtering is shown in figure 1 and includes the following steps:

(1)   The computation of the complex spectrum $X(f)$ of the time series $x(t)$, which is assumed to be given by the following convolution integral

$$x(t) = \int_{-\infty}^{\infty} s(\sigma) \cdot r(t - \sigma)\, d\sigma\,,$$

gives the product superposition of the single components $S(f)$ and $R(f)$ in the frequency domain

$$X(f) = S(f) \cdot R(f) = |X(f)| \cdot e^{i\,\Theta\,(f)}.$$

(2)   The determination of the natural logarithm of the complex spectrum gives the additive superposition of the individual parts:

$$\overline{Z}(f) = \log X(f) = \log S(f) + \log R(f) = \log|X(f)| + i\Theta(f)\,.$$

(3)   The computation of the inverse Fourier transform of the logarithm of the complex spectrum gives the complex cepstrum of the function $x(t)$, which is defined as the inverse Fourier transform of the logarithm of the Fourier transform of $x(t)$:

$$\overline{z}(q) = \mathbf{F}^{-1}\{\overline{Z}(f)\} = \overline{s}(q) + \overline{r}(q)\,.$$

Now one is again in a "time domain", where the additivity of the single parts remains in a simplified form. The periodical parts of the logarithmic complex spectrum are reduced to spikes, as can be seen in the following simple example which treats the superposition of two time delayed signals

$$x(t) = s(t) + a \cdot s(t - t_0) .$$

The logarithmic complex spectrum of this is

$$\overline{Z}(f) = \log X(f) = \log S(f) + \log (1 + a \cdot e^{-i 2 \pi f t_0})$$

with the complex cepstrum

$$\overline{z}(q) = \mathbf{F}^{-1} \{\overline{Z}(f)\} = \mathbf{F}^{-1} \{\log S(f)\} + a\delta(q - t_0) - \frac{a^2}{2} \delta(q - 2t_0) \pm \dots$$

The result of the first three steps is the complex cepstrum, where the undesired parts are to be suppressed. For the homomorphic method this is the filter process. The reverse run through the first three operations as shown on the right hand side of figure 1 provides the final result of the process in form of the filtered trace $y(t)$.

In contrast to the definition of the power cepstrum given by Bogert, Healy, and Tukey (1963), which is the power spectrum of the logarithm of the power spectrum, the complex cepstrum as defined here contains the whole phase information of the trace.

3. Complex Cepstra and Filtering Possibilities in Reflection Seismics

Homomorphic filtering is a deterministic process in the sense that fixed and pre-given parts of the complex cepstrum which are related to the undesired components are eliminated. The success of the method depends primarily on the rate of the separation of the individual components in the complex cepstrum. This means for reflection seismics that—if we model a seismogram as the convolution of the source wavelet with the reflectivity function and a multiple trace—the success depends on the difference of the cepstra of the three single components. Therefore the successful application of the method in seismic reflection work is critically determined by the simplicity and predictibility of the individual components of the seismogram cepstrum.

To elucidate the possibilities and difficulties of homomorphic filtering we show in figure 3 the cepstra of the individual components of a synthetic seismogram. As the cepstra are periodic, the cepstra above q_f can be considered to represent the negative quefrency terms. Considering the fact that maximum- and minimum-delay parts of a function $x(t)$ within the complex cepstrum fall well seperated into the areas $q < 0$ and $q > 0$ (Schafer 1969), our presentation of the complex cepstra shows the maximum- and minimum-delay parts on the right and left side of q_f respectively.

The seismogram and the single convolution components, which are the basic wavelet, the reflectivity function and a multiple trace and an additive random noise, which is superimposed with a signal to noise amplitude ratio of ten, can be seen in figure 2.

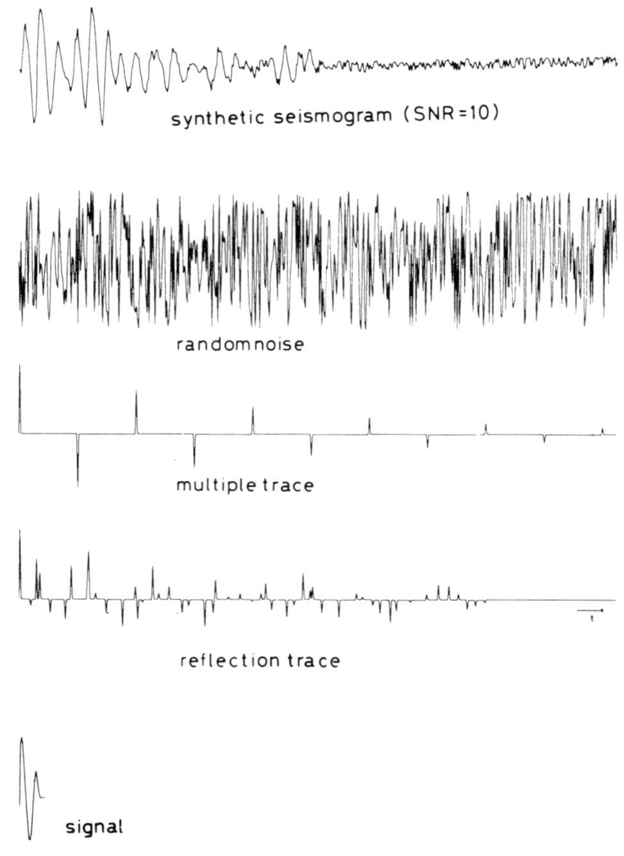

Fig. 2. Model synthetic seismogram.

The most important properties which present themselves for the separation of the single components can be shown with this example:

(1) As it can be seen in figure 3a, the cepstrum of a bandlimited wavelet is concentrated in an interval around the frequency origin of the same order of magnitude as the duration of the wavelet in the time domain. The whitening effect of the log-operation does not critically extend the length of the signal cepstrum.

(2) The complex cepstrum of a nonequally spaced minimum-delay spike series, which is shown in figure 3b, is zero over the region given by the

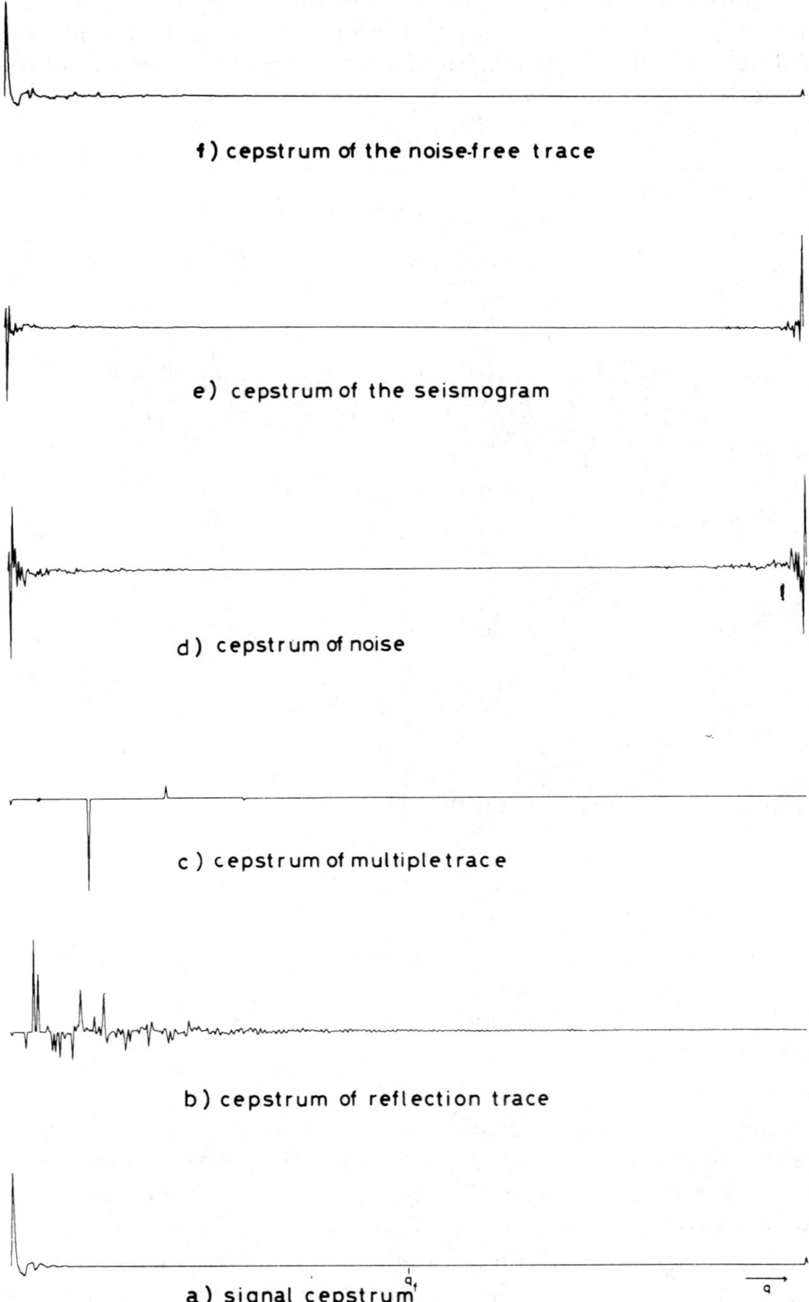

Fig. 3. Cepstrum analysis of the synthetic seismogram from figure 2.

time interval of the first two impulses. The analytical proof of this fact is given by Stoffa et al. (1974). (The minimum-delay condition of the spike train can always be achieved by exponential weighting.) After this region the complex cepstrum will generally show a rather complicated distribution of spikes, which is determined through the initial periods of the impulse function, their multiples and all their linear combinations.

(3) As it is shown in figure 3c, the cepstrum of a periodical impulse train is also an impulse train with the same period. For instance the complex cepstrum of the water column reverberation operator

$$m(t) = \sum_{n=0}^{\infty} (-1)^n R^n \delta(t - nT_w) ,$$

where R is the reflection coefficient at the sea bottom and T_w is the two-way traveltime through the water column, is given by

$$\overline{m}(q) = \sum_{n=1}^{\infty} (-1)^n \frac{R^n}{n} \delta(q - nT_w) \text{ (Stoffa et al. 1974).}$$

Disregarding the noise, the wavelet estimation and wavelet suppression should largely be possible, as the transforms of signals and spike series fall mainly into various parts of the complex cepstrum. The cepstrum of the wavelet is a strongly decreasing function which only exists near the origin, while the cepstrum of an impulse train may extend to very high quefrencies. Thereby, the wavelet estimation can be achieved by separating those cepstrum parts which correspond to the time distance of the first two impulses. As the example shows, the complex cepstrum wavelet component in the noise free case dominates over the contribution of the impulse train. Therefore, the wavelet estimation can be expected to be more accurate than the estimated reflectivity function.

For the suppression of multiples one can make use of the fact that their corresponding cepstrum components are extremly simple and predictable and, therefore, these parts can be eliminated with simple notch filters in the complex cepstrum. Due to

(1) various imaging of the single seismogram components into the complex cepstrum and

(2) simple predictibility of the multiple components

we have, for the problems under discussion, the possibility to separate in the complex cepstrum the individual components with deterministic means.

Figure 3 also indicates the relative large difficulties which already appear in the presence of only little noise. The separation becomes very problematic because the cepstrum parts of the noise, as in our example, can mask the total cepstrum region. The comparison of the three upper traces d, e, f of figure 3 shows this effect, where one can see the cepstra of random noise, of the noisy seismogram, and of the noise free seismogram.

4. Introductory Examples of Homomorphic Filtering

To demonstrate the various possibilities of homomorphic filtering we show the application of the method to three different synthetic cases.

The first example models the *suppression of long period water column reverberations* by homomorphic filtering. Figure 4 shows the chosen synthetic seismogram which can be considered as the convolution of a simple signal, shown in figure 2, with the given reflectivity function and a multiple sequence.

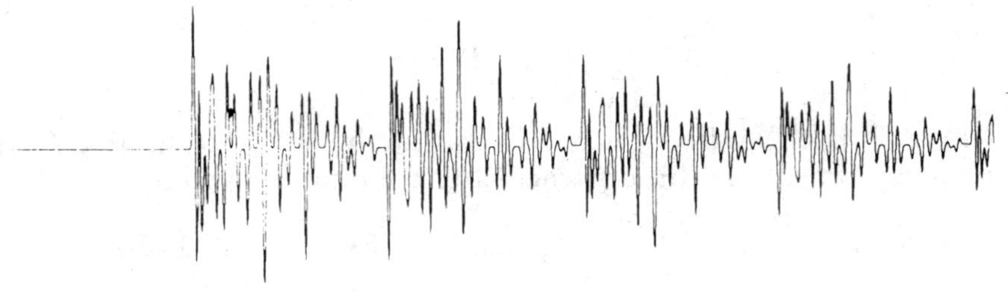

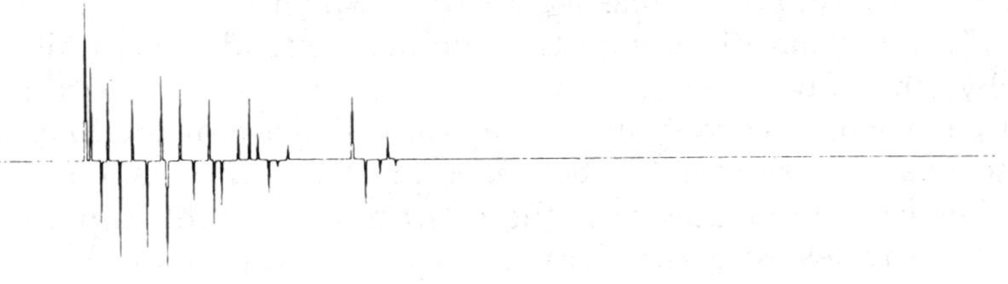

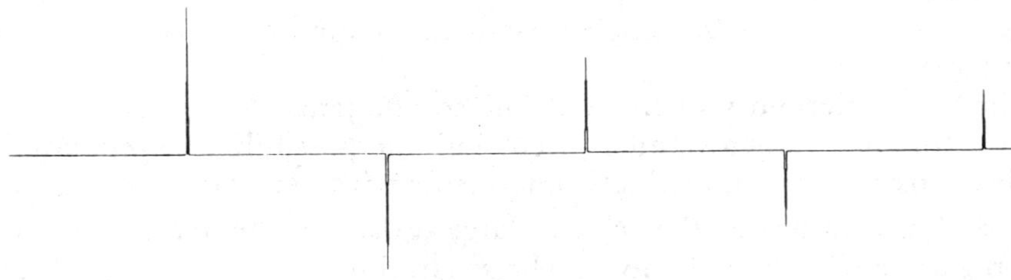

Fig. 4. Synthetic seismogram for the water column reverberation model.

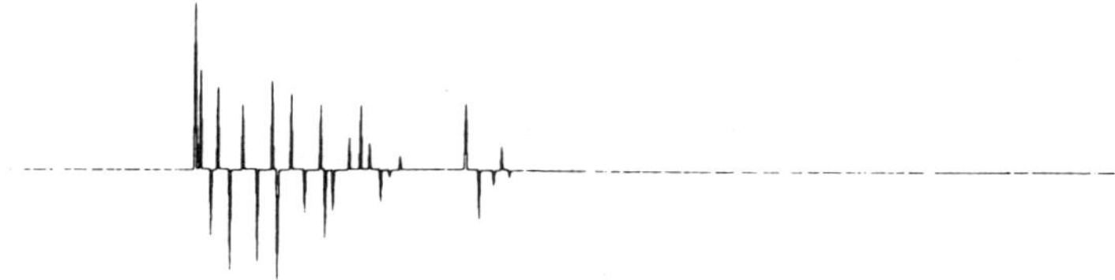

reflection trace

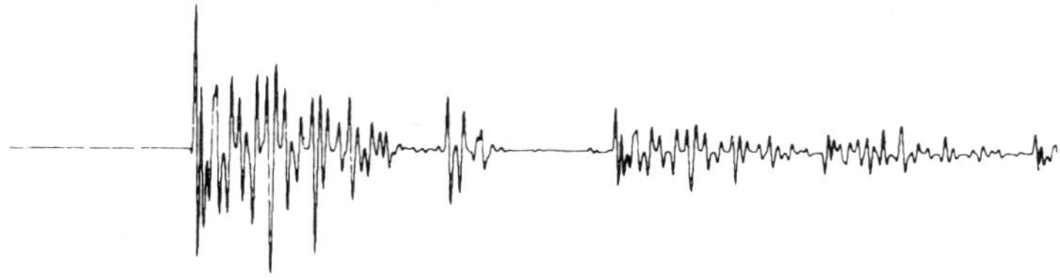

after homomorphic filtering

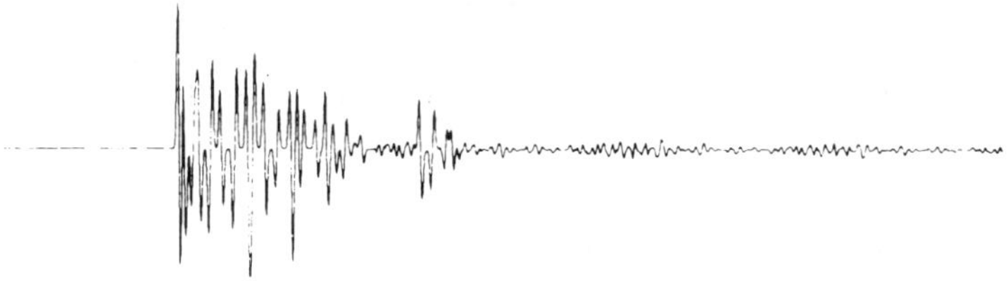

after prediction filtering

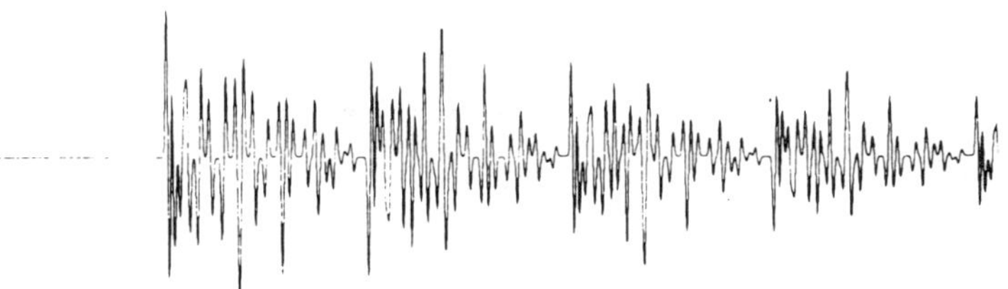

synthetic seismogram

Fig. 5. Application of different de-ringing filters to the synthetic seismogram
of figure 4.

Figure 5 shows the results after the application of different de-ringing
filters. For the homomorphic filtering only the suppression of the first
multiple has been considered, so that the second and third multiples are still
present, but reduced by a factor $1/n$ of their original amplitude, where n is

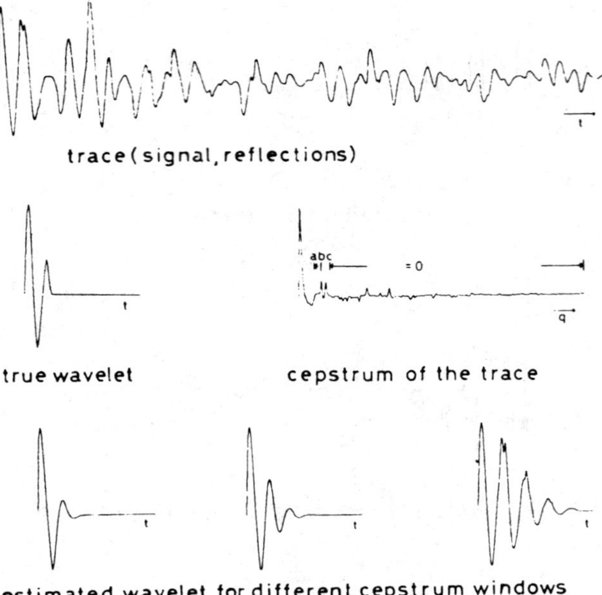

Fig. 6. Wavelet estimation by homomorphic filtering.

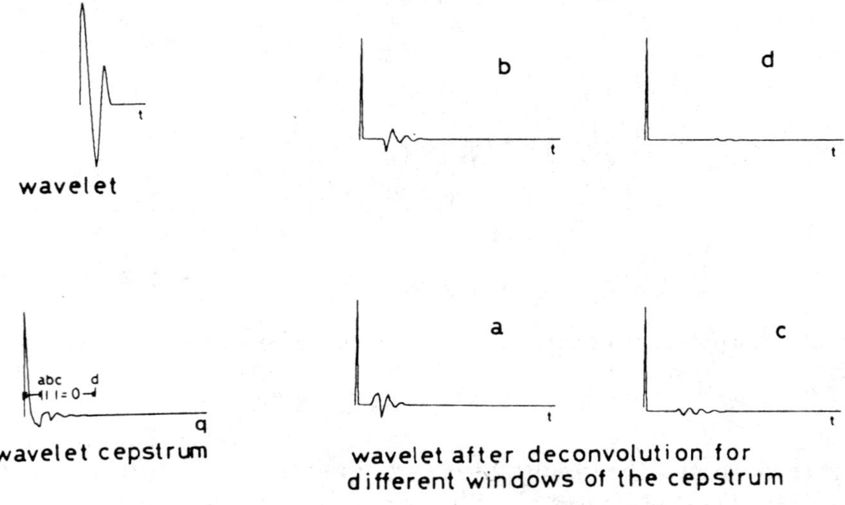

Fig. 7. Wavelet deconvolution of a single wavelet by homomorphic filtering
for different cepstrum windows.

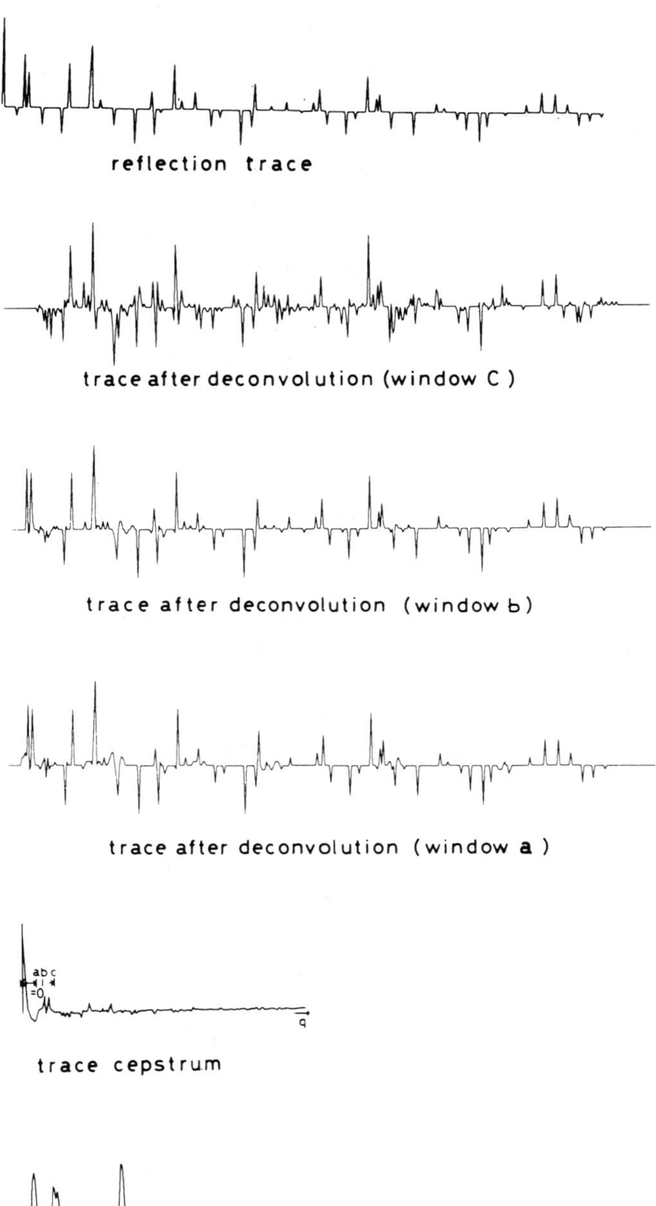

Fig. 8. Wavelet deconvolution of a synthetic seismogram.

the number of the multiple (Stoffa et al 1974). For comparison, in the uppermost trace of figure 5 the reflectivity function is shown and in the third trace the result after predictive deconvolution (Peacock and Treitel 1969). It can be seen that for such a simple noise free example, the deterministic homomorphic method gives the best results, provided the parameters for filtering the complex cepstrum are known.

Example 2 shows the *wavelet estimation* in a synthetic case, which is shown in the uppermost trace in figure 6, as a function of varying length of that window where the cepstrum has been put equal to zero. In the cases *b* and *c* this window (which is shown along with the complex cepstrum of the trace) has been chosen too small, so that here one has a superposition of two and more overlapping wavelets. The main problem in this example, namely how to choose the optimum window, is treated in chapter 7.

In example 3 the *wavelet deconvolution* is demonstrated. Figure 7 shows the deconvolution of a single wavelet as a function of the length of the window over which the cepstrum has been suppressed. The chosen windows are shown in the cepstrum trace. The results demonstrate that for "efficient" wavelet deconvolution the cepstrum window width has to correspond to the length of the wavelet (see chapter 3).

Figure 8 shows the practical application of homomorphic filtering to the wavelet deconvolution on a synthetic seismogram. The comparison of the results with the pre-given reflectivity function in the uppermost trace shows the excellent result which can be achieved by this method under optimum conditions, i.e. without superimposed noise and a sufficiently decreasing signal cepstrum with increasing quefrencies.

5. Critical Parameters in Homomorphic Filtering

The examples given in chapters 3 and 4 shows the possibilities for the practical application of the method in reflection seismics with all the advantages and disadvantages which can be summarized in the following form: The method offers the possibilities of wavelet deconvolution, wavelet estimation and multiple suppression—without "a priori" knowledge of the waveshape or of the number of echoes—by deterministic processing of the complex cepstrum. In addition, the above examples also show which factors generally influence the quality of the results:

(1)   The results depend mainly on the ratio of the signal duration to the two-way traveltime through the uppermost layer, which stronly influences the quality of the wavelet estimation and wavelet deconvolution.

(2)   The amount of random noise is critical.

(3)   The results depend on the choise of predetermined parameters. For the problem under discussion these are, essentially, the window width of the cepstrum over which the influence of the undesired components has to

be eliminated, and the amount of the complex cepstrum which belongs to these components.

(4) The quality of the results depends critically on the following assumptions (generally believed to be satisfied):
   a) the basic wavelet statistics are time invariant,
   b) the seismic trace can be considered as a convolution integral,
   c) a strict periodicity of the multiples exists.

For the practical application of the method the most important parameters are:

(1) the amount of random noise,
(2) the choice of the parameters for the suppression of individual parts in the complex cepstrum.

In the subsequent chapters it is shown how these factors influence the successful application of the homomorphic method and how they can be accounted for in processing.

6. The Influence of Random Noise and its Minimization

As already shown in figure 3, the influence of random noise is quite critical at the computation of the complex cepstrum. In addition to this, figures 9 and 10 indicate that the noise not only masks certain cepstrum parts but that, generally, the noise is pulled up and amplified during the different deconvolution processes. In figure 9 this is shown for the basic wavelet and multiple deconvolution.

The middle trace shows the seismogram before filtering. The trace above displays the result after processing, which should be compared with the corresponding deconvolution result of the noise free (uppermost) trace. In the two lower traces this noise amplification can be seen for the multiple deconvolution process. The comparison of figure 10a (trace before filtering with a SNR = 40) and figure 10d (trace after filtering) shows the noise amplification for the wavelet estimation.

This noise amplification is considerably stronger for the homomorphic filterprocess than for least squares deconvolution processes. Up to now, this is a great disadvantage for the practical application of the method in marine reflection seismics, where usually the random noise parts are not negligibly small. At least three factors of the procedure influence the noise amplification:

(1) The nonlinear logarithmic operation exaggerates small components of the spectrum in the complex cepstrum.
(2) Due to the fact that the noise is additive, it influences the complex cepstrum in a rather complicated way.
(3) The addition of random noise complicates primarily the phase spectrum, which in our experience has often too strong an influence on the complex cepstrum.

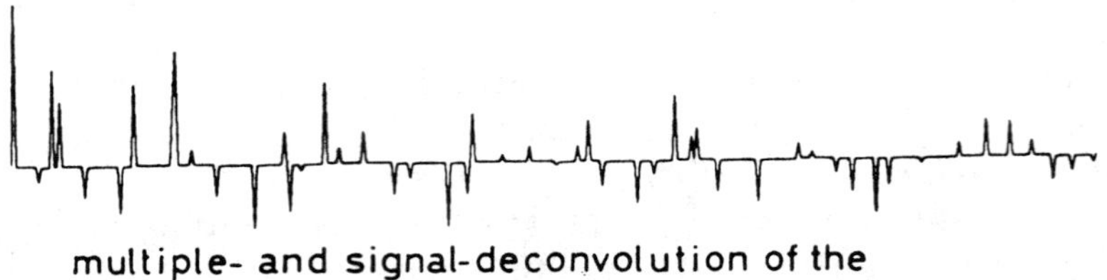

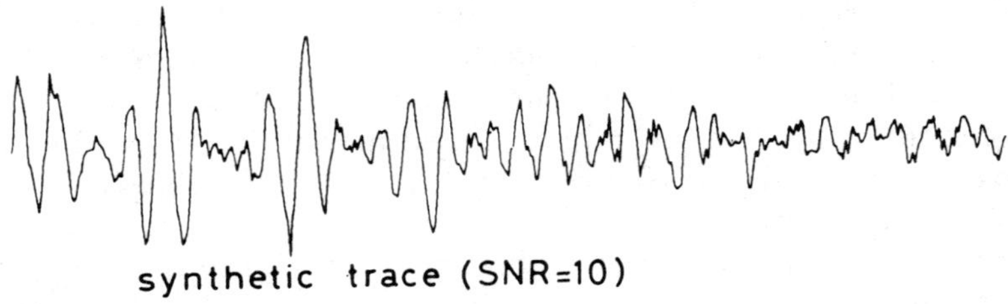

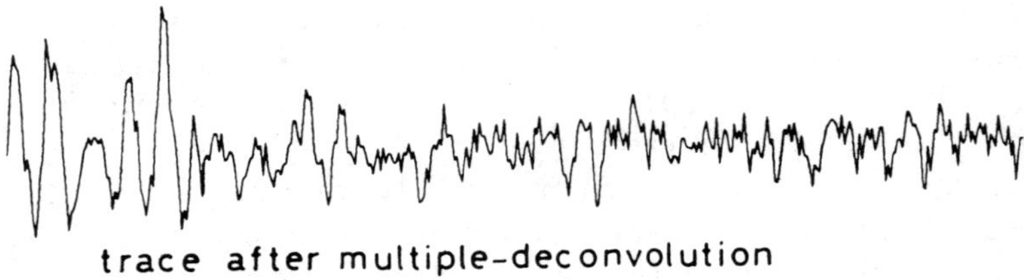

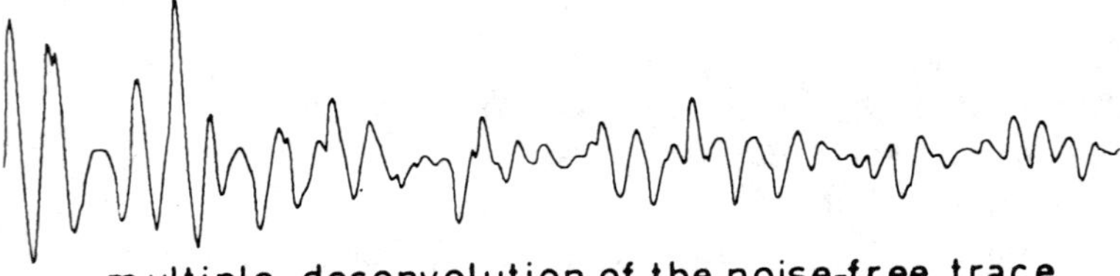

Fig. 9. Homomorphic deconvolution in the presence of random noise.

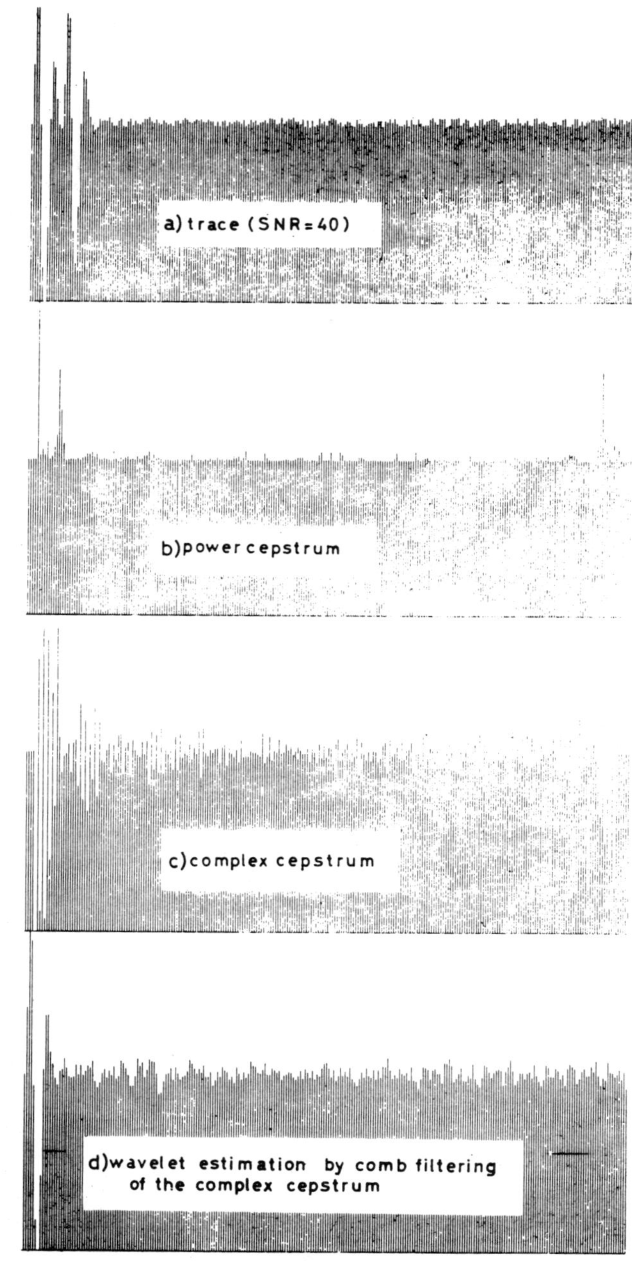

Fig. 10. Wavelet estimation in the presence of noise.

The third fact gives us the possibility to reduce the noise influence on the complex cepstrum and to suppress its amplification during deconvolution processing. The cepstrum procedure (figure 1) requires the determination of the continuous phase spectrum. Our investigations have shown that—especially for long time series, as in our examples—the additive random noise possesses relatively high amplitude/long period continuous phase spectra which mainly influence the cepstrum region occupied by the wavelet (for an example see figure 3). Figure 11 explains the situation using the phase- and

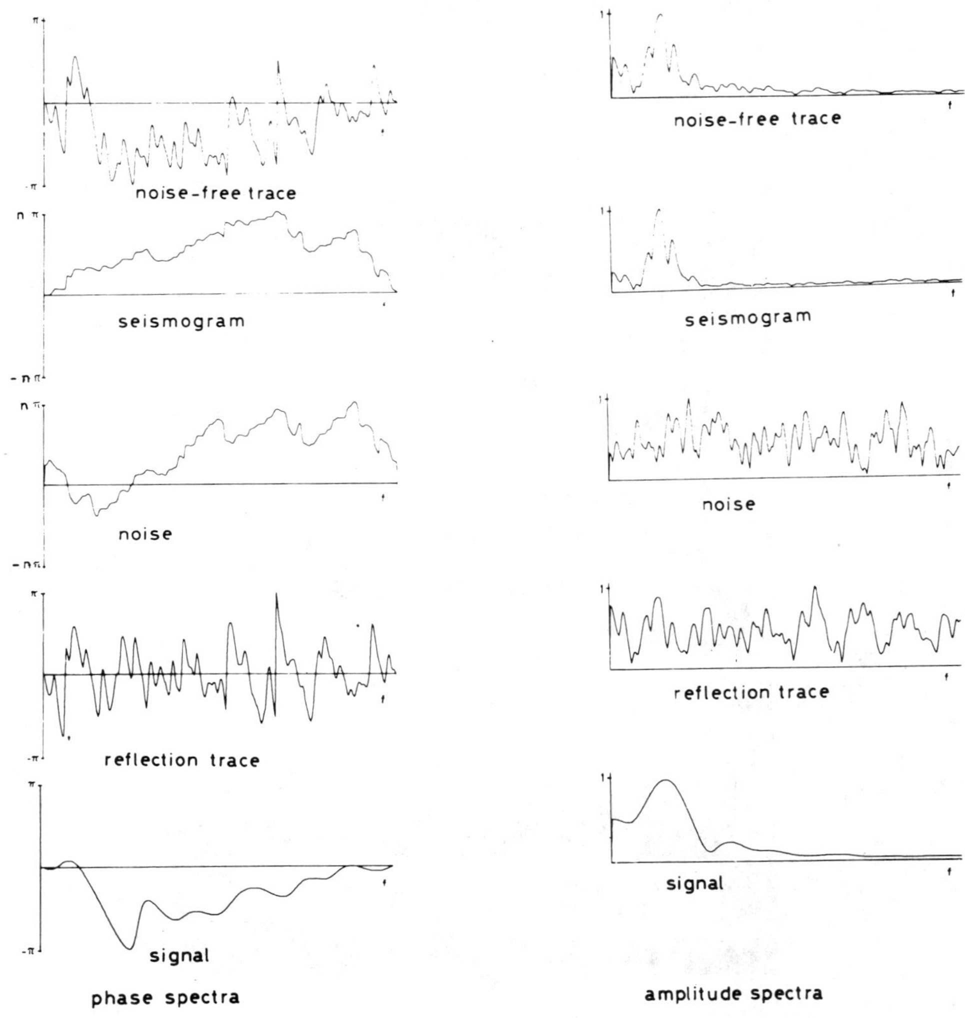

Fig. 11. Amplitude- and phase spectra of the synthetic seismogram of figure 2 and of its single components.

amplitude spectra of the synthetic seismogram from figure 2 and its components. The three upper traces on the left hand side show that noise influences especially the long period parts of the seismogram *phase spectrum* (and more than its amplitude spectrum), with the consequences that

a) the signal region in the complex cepstrum is masked by noise because slowly varying components of the complex spectrum are transformed into the vicinity of the cepstrum origin,

b) the noise is amplified as a consequence of the *continuous* phase spectrum procedure.

In our marine seismic records we have also found such high amplitude/long period continuous phase spectra of the noise which is amplified during processing.

For the practical application of homomorphic filtering in reflection seismics, especially for large time windows where the above effect is more serious, this has the following consequences:

(1)   Care should be taken that the seismic trace is filtered with a signal-matched bandpass filter before homomorphic deconvolution to optimally suppress the noise, as already recommended by Ulrych (1971).

(2)   It is often necessary to high-pass filter the continuous phase spectrum to reduce the strong influence of the noise phase spectrum during the computation of the complex cepstrum. The experimental results confirm this statement, but we do not know whether it is generally true.

For the direct wavelet deconvolution and for the multiple suppression the second recommendation may have a detrimental effect, because this high-pass filtering of the phase spectrum may also effect the phase spectrum of the reflectivity function and thus contribute to a distortion of signal arrivals. For this reason direct wavelet deconvolution of long seismogram sections has to be discarded for cases where the noise is not negligibly small. It follows that for the wavelet deconvolution one has to proceed via homomorphic wavelet estimation to subsequently employ least squares deconvolution in the usual way (Robinson and Treitel 1967). Even though this procedure is more time consuming than each of the combined processes, it offers two advantages:

(1)   The wavelet estimation—and therefore the wavelet deconvolution—is more exact, as no minimum delay assumption is necessary. For instance, predictive deconvolution does not work optimally if the input is not minimum delay.

(2)   The noise influence is less strong.

The influence of the noise becomes a real problem in the suppression of multiples, as phase filtering is not applicable. One has to aim for the suppression of the noise *before* the application of the homomorphic method. Special cases may, however, exist where—due to the delay properties of the noise series—a successful elimination of the noise in the complex cepstrum is possible.

In summary, the situation can be explained as follows: in order to keep the influence of the random noise at the deconvolution procedure reasonably small, one is forced to filter the continuous phase spectrum. This filtering influences not only the noise but also the reflectivity function, and to a lesser degree the signal whose complex cepstrum is revealed by this process. This can be seen in figure 12, where the complex cepstra with and without phase filtering are demonstrated and compared with the complex cepstrum of the noise free trace. As the comparison shows this phase filtering technique has largely reduced the noise influence on the complex cepstrum near the origin. After this operation the main contributions of the random noise fall into the region away from the origin of the complex cepstrum and have only little influence on the wavelet estimation; therefore the effects on the wavelet deconvolution are stronger. This agrees with the results given by Ulrych (1971). In chapter 3 we found that in the noise free case, with regard to the stronger signal cepstrum components, the signal estimation is to be

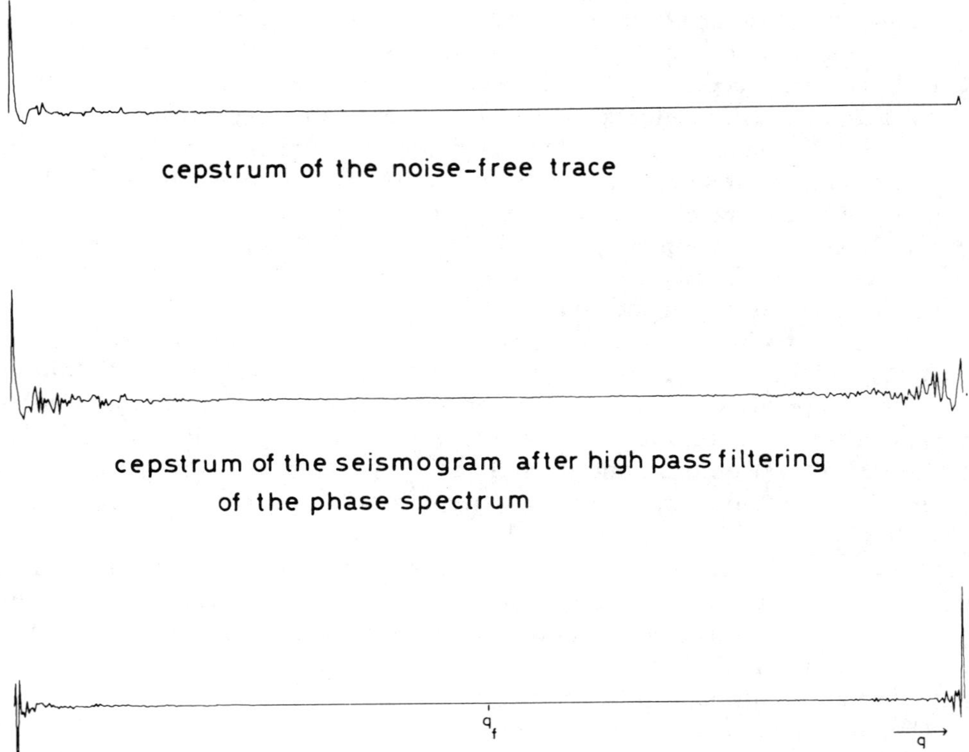

Fig. 12. Reducción of the noise influence on the complex spectrum by means of filtering the phase spectrum during homomorphic processing.

prefered over the wavelet deconvolution. In the presence of noise this becomes still more important, since in contrast to the wavelet deconvolution the noise influence on the wavelet estimation can be strongly reduced by the phase filtering technique which offers the following two advantages:

a) The method reduces the influence of the noise on the results of the deconvolution process.

b) The phase filtering cleans the signal region in the complex cepstrum, which may be a good help in chosing the right parameters for the wavelet estimation.

7. Wavelet Estimation by Homomorphic Filtering

Generally, the main advantage of homomorphic filters for different deconvolution problems over other deterministic filters is the fact that no *a priori* knowledge is necessary. The necessary parameters can be determined during the process itself. In addition, the advantage over least-squares methods, for instance for the wavelet estimation, is that the assumptions about the delay properties of the wavelets are superfluous. But, as shown in the chapters 4, 5, and 6, the success of the wavelet estimation depends largely on the following three factors:

(1)   the signal-to-noise ratio of the input data,

(2)   the degree of overlapping of the cepstrum components of the source wavelet and the reflectivity function,

(3)   the choice of the window for the filtering procedure in the complex cepstrum and the estimated amount of the cepstrum components to be deleted there.

The introductory synthetic examples of chapter 4 show that the separation of the overlapping signal components by a short-pass filter has only limited success. With this kind of filter the error of the signal estimation can become relatively large, especially if later arrivals follow shortly after the first arrival. It is usually not practical to use this form of filtering if distortion free wavelet recovery is desired. Comb filters, which are matched to the signal arrival times and their linear combinations are more appropirate means of wavelet estimation. The advantage of these comb filters over short pass filters for the purpose of signal estimations is shown by the comparison of the results in figure 13. The middle trace shows the influence of the "convolution noise" as a consequence of the truncation of the signal cepstrum. The determination of the matched comb filters necessitates a specific choice of the trace window for the wavelet estimation. The following aspects have to be considered: we have to choose the window with the most simple cepstrum. Subsequently, we should analyze a relatively short interval of the trace. In contrast, we have to choose the interval so large, that the noise belonging to the distorted signals,which is generated at the boundaries of the window, is still relatively small. In addition, we have to keep the trace interval for the wavelet estimation rather short in order to keep the influence of

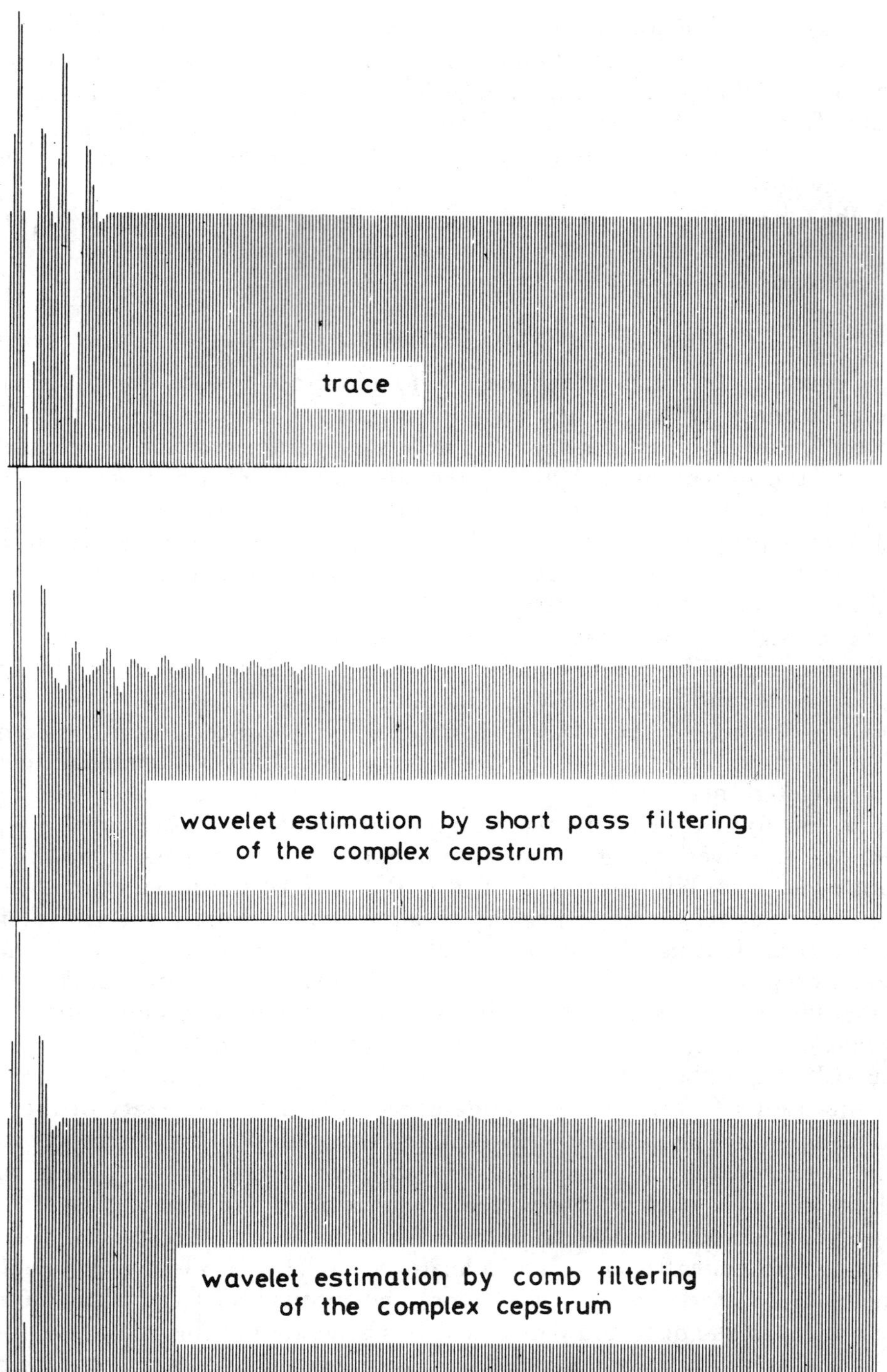

Fig. 13. Wavelet estimation by short pass and comb filtering the complex cepstrum.

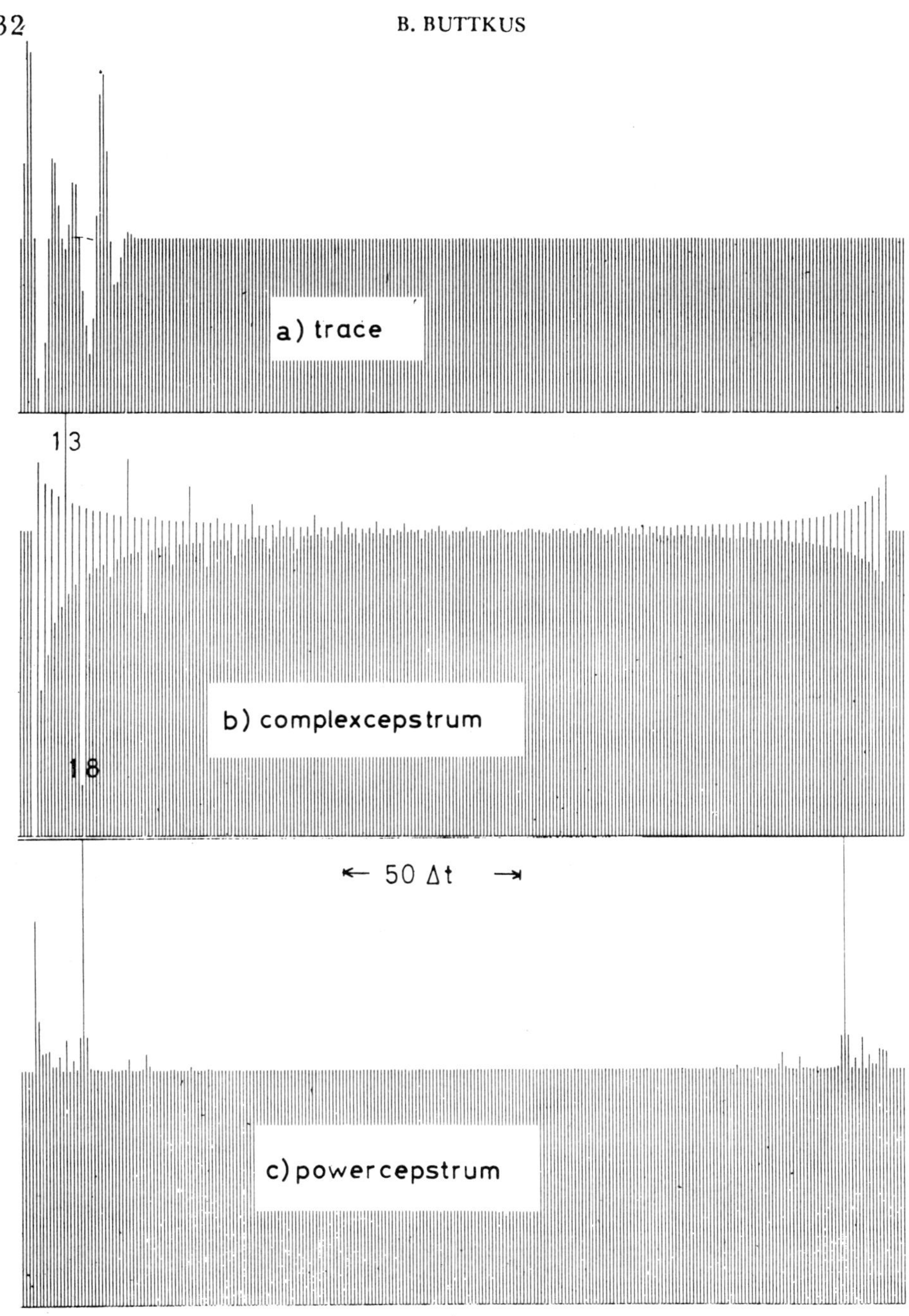

Fig. 14. Detection of signal arrival times by cepstrum analysis in the noise free case;
Signal arrival times: $t_1 = 0$, $t_2 = 13 \cdot \Delta t$, $t_3 = 18 \cdot \Delta t$
cepstrum contributions:
a) of signal 2 at 13, 26, 39 etc.,
b) of signal 3 at 18, 36, 54 etc.,
c) of the linear combinations of signal 2 and 3 at 31, 44, 49 etc. time-units.

the seismogram-noise on the signal within the complex cepstrum and during the processing procedure low.

Studies with synthetic seismograms have shown that an interval of three to five times the signal length seems to be a good compromise between the above conflicting conditions.

The problem of wavelet estimation can be treated in two parts:

(1) the determination of the comb filters matched to the signal arrival times, and

(2) the actual filtering process and the related problem of suppressing the increasing noise. An efficient method to reduce the noise influence is already discussed in chapter 6.

The determination of the signal arrival times, their linear combinations and the estimation of their amplitudes, necessary for the choice of adapted comb filters is a straight-forward detection and parameter estimation problem. In the noise free case the complex cepstrum is itself the best means of determining the signal arrival times and the signal amplitudes, as shown in figure 14. The complex cepstrum shows that there are in fact three overlapping signals with the arrival times $t_1 = 0$, $t_2 = 13 \cdot \Delta t$, $t_3 = 18 \cdot \Delta t$, and the normalized amplitudes 1, 0.4, -0.6, with the complex cepstrum parts at

$$t = \sum_{n=0}^{N} \sum_{m=0}^{M} \delta\left(t - (n\,t_2 + m\,t_3)\right) .$$

It is generally accepted that the success of separating overlapping signals within the cepstrum depends on the fact that the cepstrum envelope is narrower than the distance between successive signals, i.e. the magnitude of the cepstrum must have decayed significantly away from the peak in order to be able to resolve the overlapping cepstrum functions. Practical experience shows, however, that often from the sequence of later cepstrum spikes the signal arrival times and their amplitudes in the vicinity of the cepstrum origin can be predicted. The example in figure 14 demonstrates this possibility of determining the exact comb filter even if the signal cepstrum decreases rather slowly. In figure 14c we also include the power cepstrum to demonstrate that in the noise free case the power cepstrum possesses no advantages over the complex cepstrum.

With increasing noise it becomes more difficult to detect the signal arrivals, to determine the amplitudes and to fix the matched comb filter, which is problematic below a SNR of ten in the complex cepstrum. In agreement with the results obtained by Flinn, Cohen, and Mc Cowan (1973) and Kemerait and Childers (1972), the power cepstrum has been found to be a superior means for the parameter determination in the presence of large noise. As in the complex cepstrum, signal arrivals in the power cepstrum appear as spike sequences with the same period, but only with positive values, so that only the amount of the signal amplitudes (but not their pola-

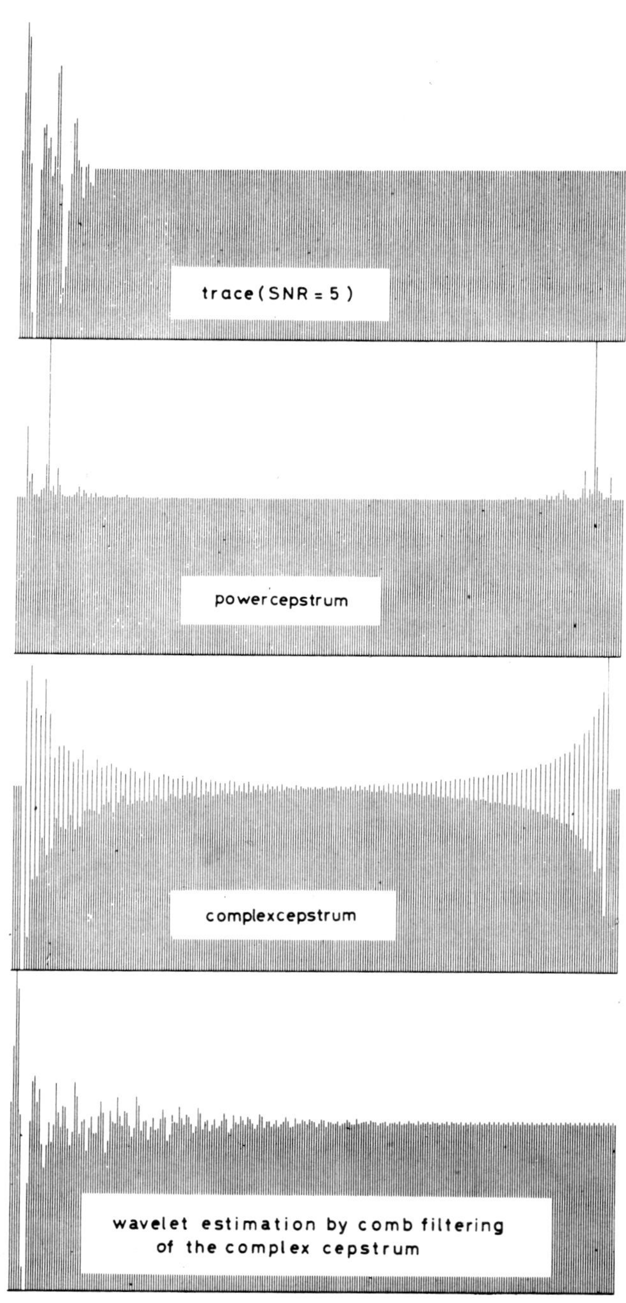

Fig. 15. Detection of signal arrival times and wavelet estimation in the presence of random noise.

rity) can be determined. In comparison to the complex cepstrum, the influence of the random noise is less important and can be eliminated at the computation of the power cepstrum in the logarithmic power spectrum fairly easily by smoothing. Our empirical studies indicate that, in this way, signal detection, in practice, can be reasonably well achieved up to a SNR of three. Figure 15 shows the advantage the power cepstrum offers over the complex cepstrum for the signal detection (an SNR of five is used in the example). While the signal detection gives no problems, the comparison with figure 13 shows the relative large influence of the additive random noise in the signal estimation. The wavelet estimation from the complex cepstrum is only satisfactory up to a SNR of ten. This threshold can be further reduced to a SNR of three to five with the above recommended method of filtering the phase spectrum.

Even though the estimated threshold results are empirical, they show the limits for the successful application of homomorphic wavelet estimation.

The estimation process—based on the phase filtering procedure and the determination of the comb filter—is outlined in the flow diagram in figure 16. The procedure includes the following steps:

(1)   the choice of the trace window,

(2)   the determination of the phase filter for the noise reduction,

(3)   the determination of the matched comb filter by computation of the power cepstrum, picking of the signal arrival times, and application of the phase filter,

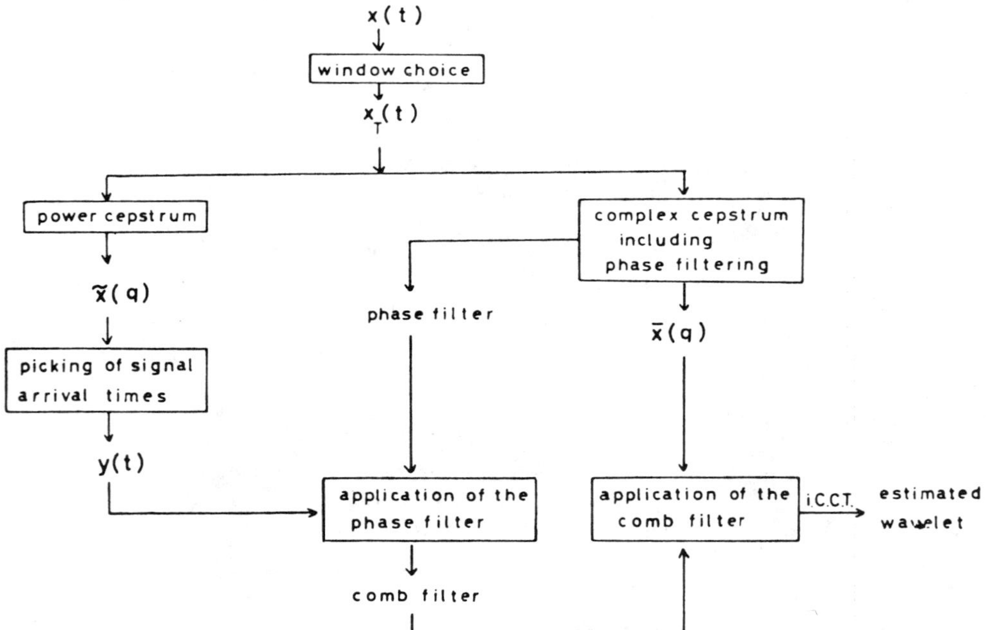

Fig. 16. Flow diagram for signal detection and wavelet estimation.

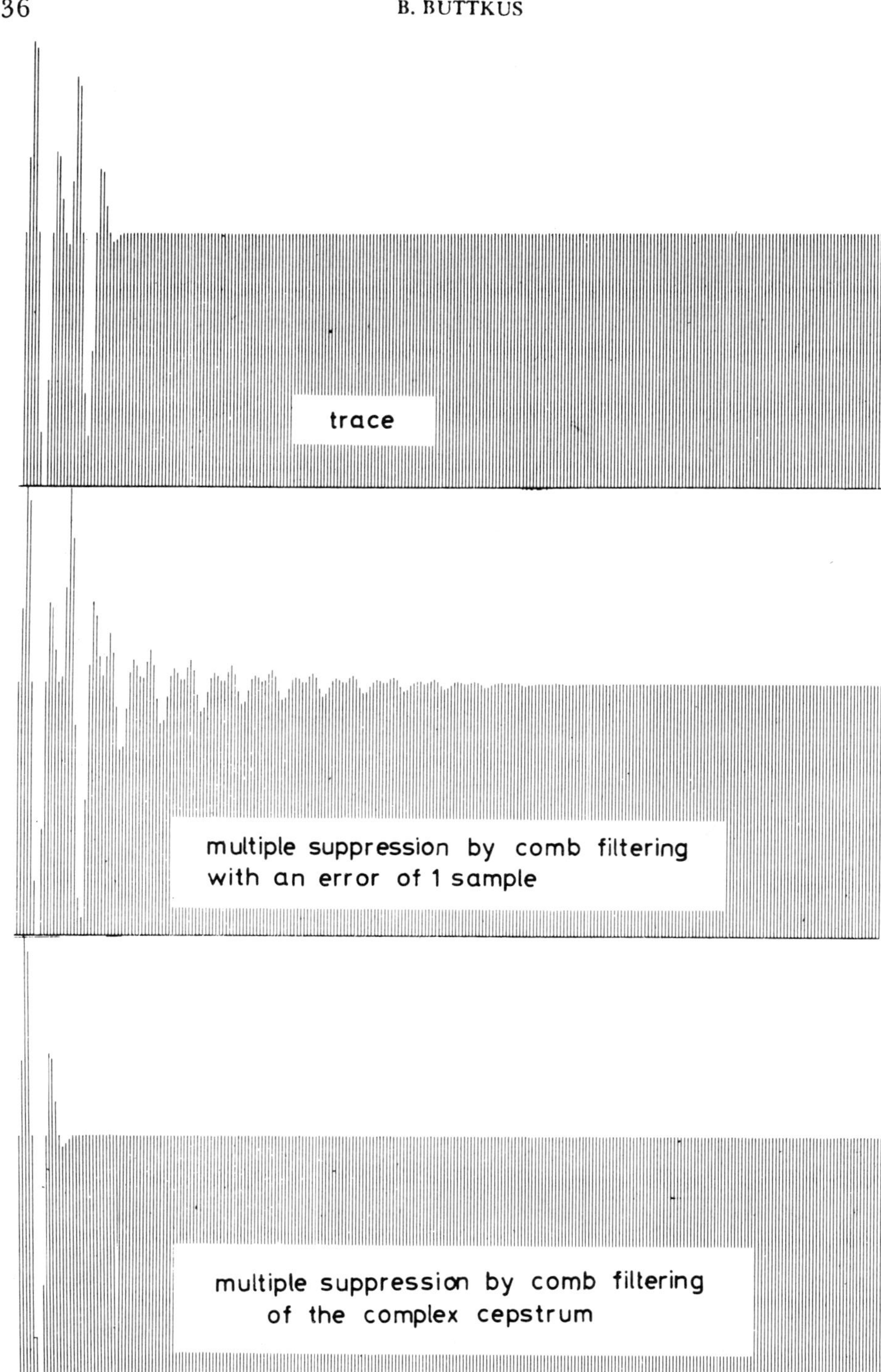

Fig. 17. Influence of the cepstrum window choice for the wavelet estimation.

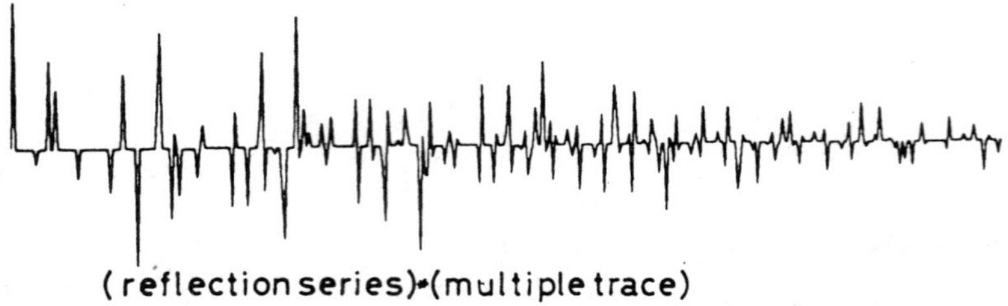

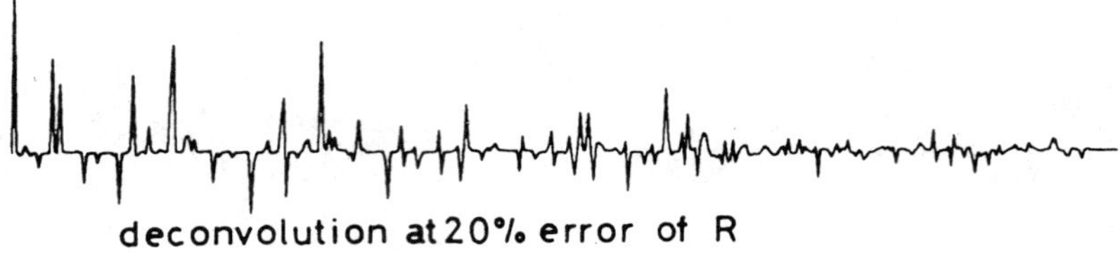

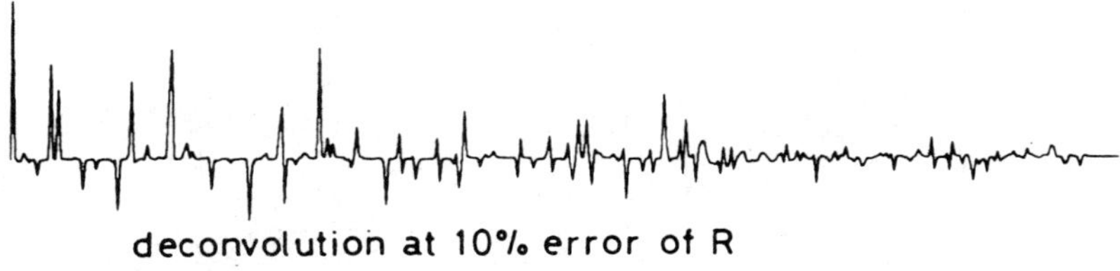

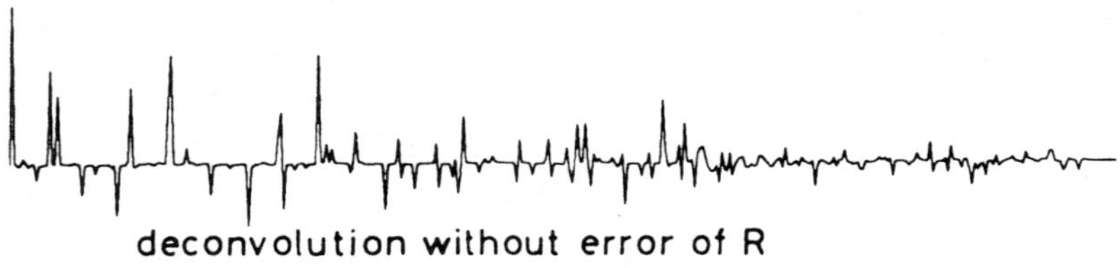

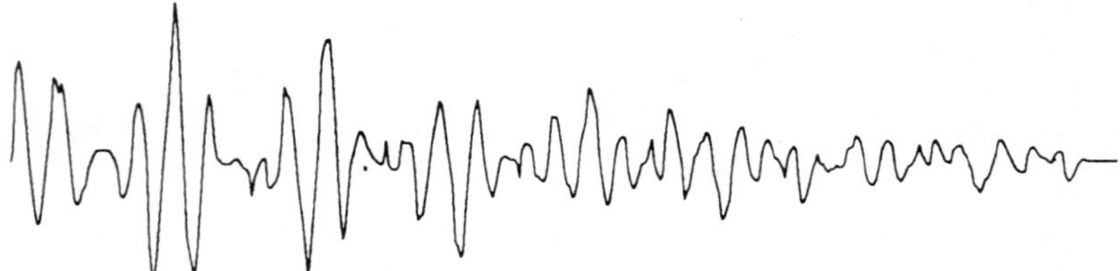

Fig. 18. Influence of the estimated reflection coefficient R at the deconvolution of water column reverberations.

(4)   the computation of the complex cepstrum including phase filtering,
(5)   the application of the comb filter,
(6)   the inverse complex cepstrum transform (i.C.C.T.) which gives the estimated wavelet.

As it can be seen, minimum delay parts as well as maximum delay parts are taken into consideration at the wavelet estimation.

Our investigations have shown that the exact determination of the signal arrivals has to be postulated, as the example in figure 17 shows, where we have demonstrated for the wavelet estimation the influence of a detection error of one sample. The estimation of the amplitudes is less critical. For instance, for the deconvolution of water column reverberations the relative small influence of the estimated reflection coefficient R can be seen in figure 18. The reason for the unequal demands on these parameters is found in the following fact. Uncertainties in determining the signal amplitudes only show up at the locations of the signal arrival times while errors over finite cepstrum intervals reveal themselves everywhere on the trace at later times. For those cases where it becomes difficult to determine the polarity of the signal arrivals it follows that it is allowed to choose a comb filter with zero values at the notch positions without significant signal distortion, if the arrival times are exactly determined.

8. Multiple Suppression by Homomorphic Filtering

Compared with the wavelet estimation the application of homomorphic filters for multiple suppression imposes more problems:
(1)   The recommended procedure for the wavelet estimation of filtering the phase spectrum to reduce the increasing noise cannot be applied. Therefore, the successful application of the method for multiple suppression is confined to a SNR of ten, as already shown in chapter 7.
(2)   Only in certain cases the reflectivity can be approximately expressed as a convolution of a multiple train with the spike series which models the reflection coefficients. In the previously investigated problem of rejecting water column reverberations, this convolution condition is satisfied. Generally, the seismogram—including all multiples—is a more complicated process with the consequence that in the complex cepstrum primaries and multiples are not only additively superimposed.
(3)   The cepstrum separation of multiples and primaries is more difficult than the wavelet estimation.

As for the wavelet estimation, the exact determination of the multiple arrival times has to be postulated while the estimation of the cepstrum parts belonging to the multiples is less critical. As shown in chapter 7, the determination of the comb filter matched to the multiple arrival times can be achieved best by the power cepstrum.

The difference, however, is that the detectibility of multiples is more complicated, because in the cepstrum the multiples are largely reduced (as shown by Schepers 1972) so that their detection can only be realized up to a multiples-to-primaries ratio of three to five. As a consequence of the strong reflection coefficient at the water bottom, the exact determination of the water column reverberation arrival times can be achieved in most cases.

The detection of internal multiples is generally not possible, since for these multiples the threshold value of detectibility and exact arrival times determination is mostly not realized. The practical application of homomorphic filtering, therefore, remains confined to the suppression of water column reverberations.

Schepers (1972) uses the fact of multiple suppression in the cepstrum in order to treat the complex cepstrum as a first approximation of the reflection coefficients. Figure 19 gives an example for the suppression of multiples in the complex cepstrum. This possibility, however, is confined to those cases where the multiple cepstrum parts are small in comparison to the cepstrum parts of later primaries. This depends on the relations of the reflection

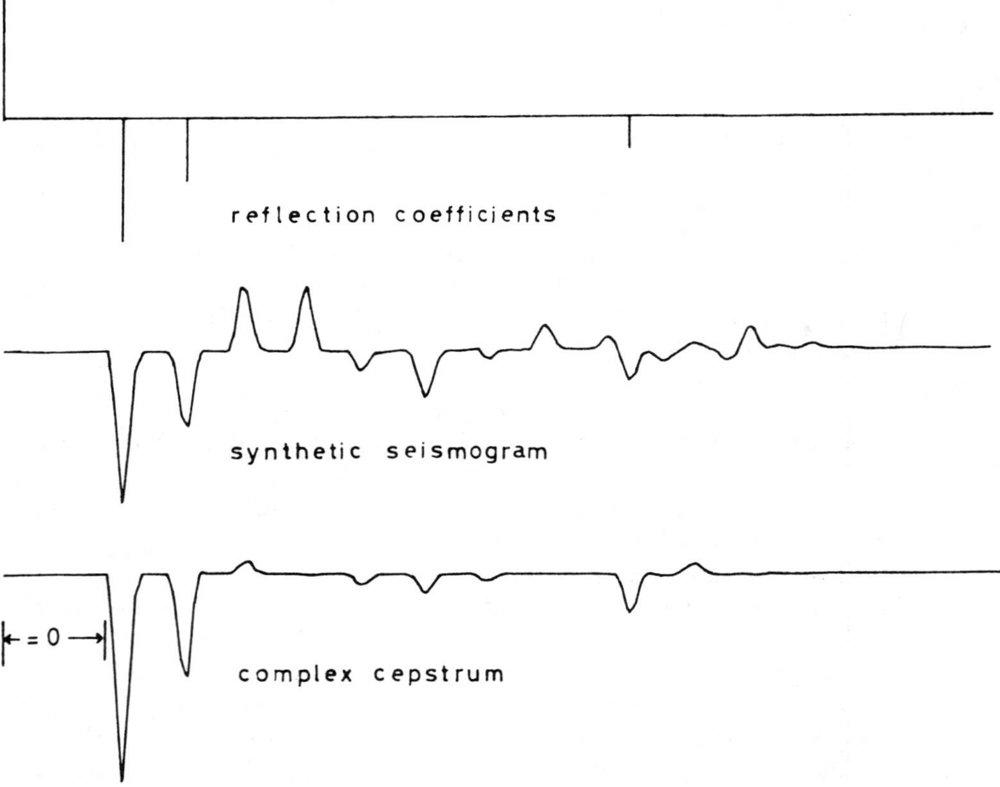

Fig. 19. Suppression of multiples in the complex cepstrum.

740   B. BUTTKUS

coefficients in such a way that the approximation becomes worse if the reflection coefficients decrease rather strongly. Therefore, the multiple suppression in the complex cepstrum is primarily confined to near surface reflectors.

To get further insight into possibilities and restrictions of the method we demonstrate the practical application of homomorphic filtering for long period multiple suppression on a single channel section (figure 20). In the interesting parts of the section the first long period multiple of the sea bottom reflection masks the primary events. Figure 21 shows the section after predictive deconvolution. Although the multiples have been reduced they are still present. Figure 22 shows the section after homomorphic filtering. Here the multiple is suppressed, but at the same time the interesting structure has not become more distinguishable. It has, in fact, been more distorted. Two reasons can be seen for this:

(1) The actual seismic traces do not satisfy the above postulated requirements, as the actual field data are caracterized by the following two properties:

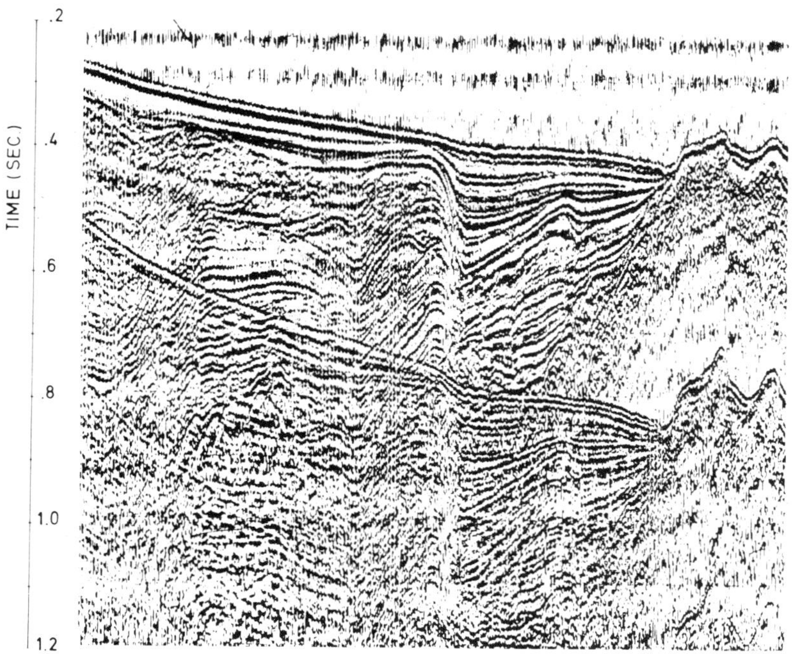

Fig. 20. Single channel air-gun section offshore Korea.

Fig. 21. Korea section after predictive filtering.

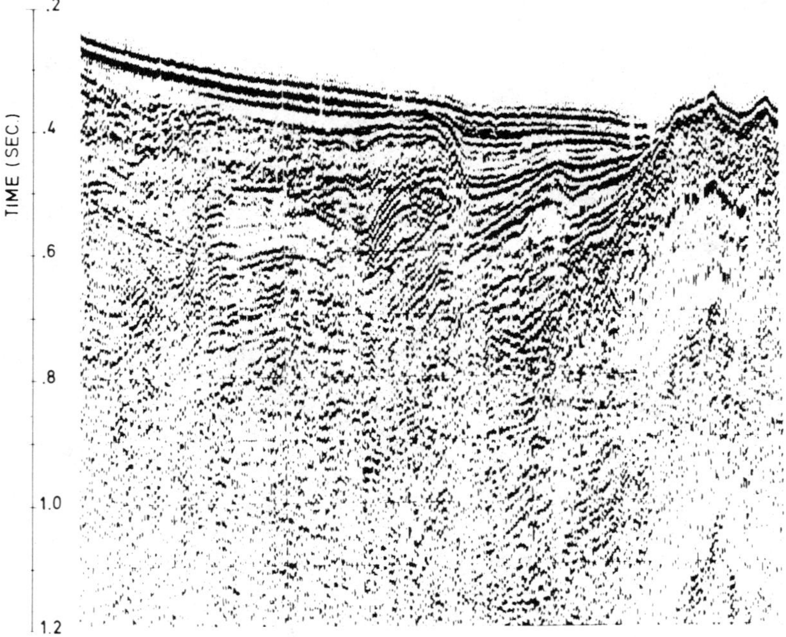

Fig. 22. Korea section after homomorphic filtering.

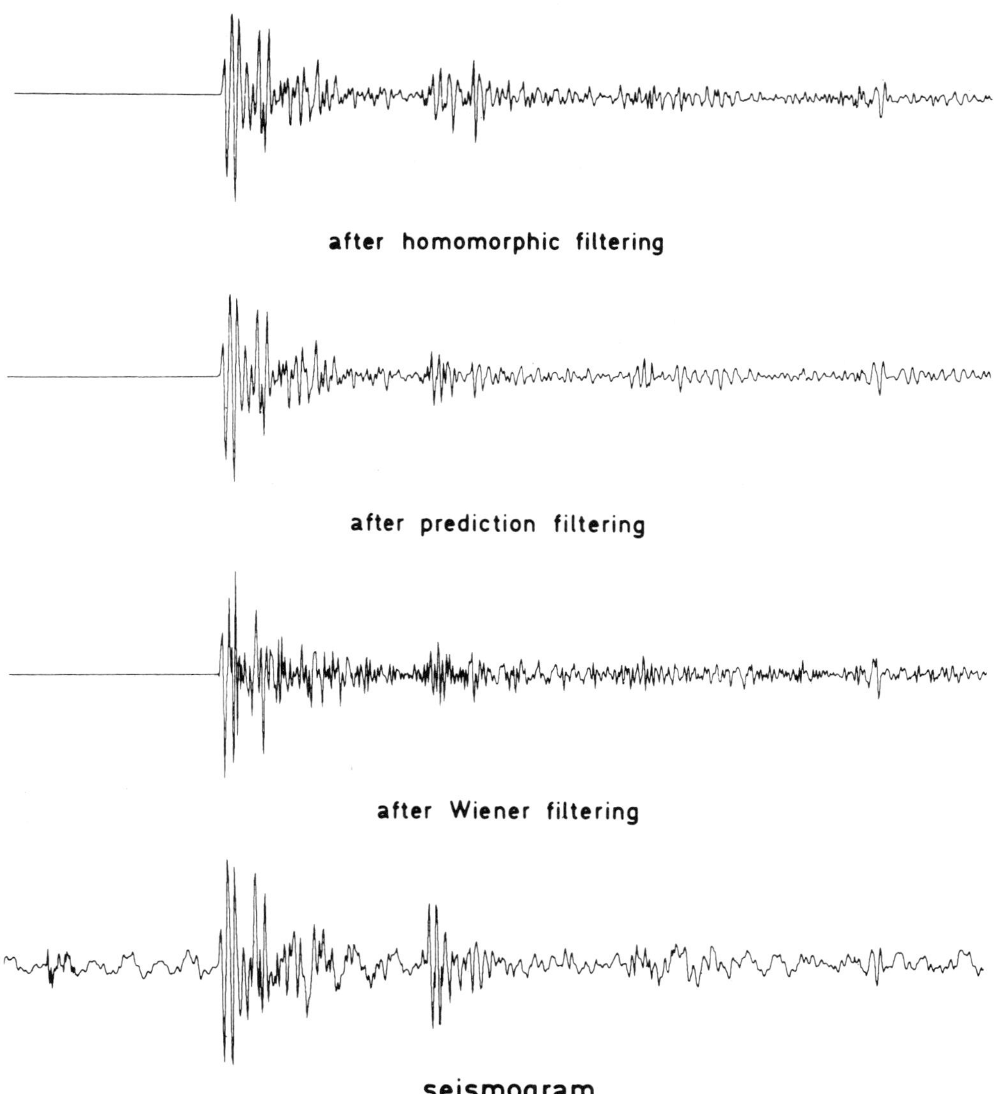

Fig. 23. Application of different deringing filters to a single trace of the Korea section.

a) a relative high noise component with a SNR of five,
b) a time variation of the signal. An example of the filtering effect is
seen in figure 23, where a single trace is featured before and after
filtering. In the uppermost trace one can recognize the blow-up of
the second half of the multiple as affected by homomorphic filtering. The reason for this is the deterministic approach. Deviations
from the assumption of time invariance of the signal are more critical

than for least squares methods, as the comparison with the predictive filtering result shows.

(2) A sufficient handling of the method is not guaranteed: due to the fact that the exact determination of long period multiple arrivals through the power cepstrum is difficult to achieve, we chose as a first approximation a rather broad cepstrum window over which the multiple component is eliminated with the result that the reflections are largely destroyed (as already explained in chapter 7).

For further improvements of the results we tried to determine the exact multiple arrival time in the sequence of power cepstra of succeeding traces. Figure 24 shows the power cepstra of four neighbouring seismograms of the Korea section, shown in figure 25, from which the exact reverberation time can be picked. On the other hand, the determination of the contribution of the multiples in the cepstrum is difficult. The best value for this cepstrum part can be calculated from the auto-covariance of $x(t)$. The multiple energy at the period T_w causes a negative spike in the auto-covariance function at a lag of T_w and the ratio of the auto-covariance values at the lags T_w and zero gives an estimate of the reflection coefficient for which we can compute the complex cepstrum component.

The above picking procedure and cepstrum estimation lead to a 50—70 % multiple reduction (see figure 26). Due to the variation of the signal with

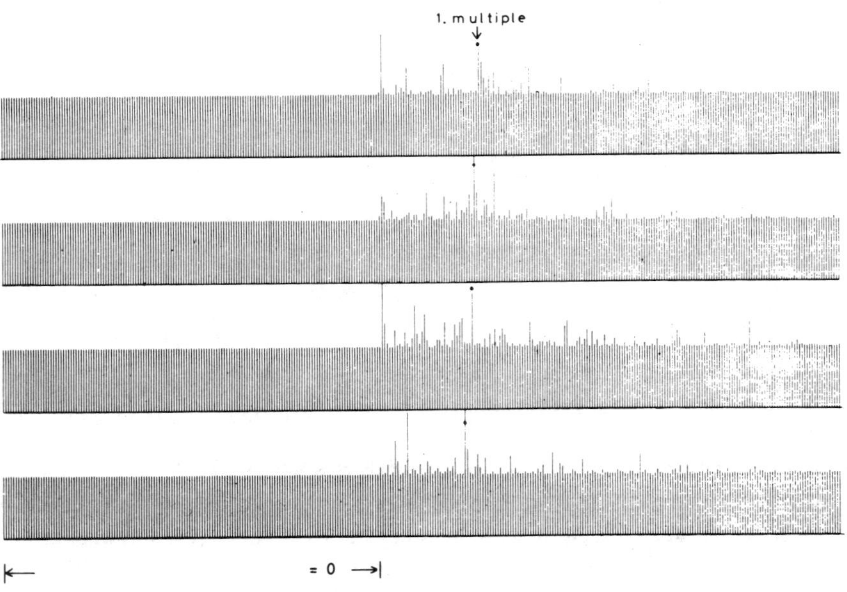

Fig. 24. Power cepstra for the exact determination of the arrival time of the first long period multiple (part of the Korea section).

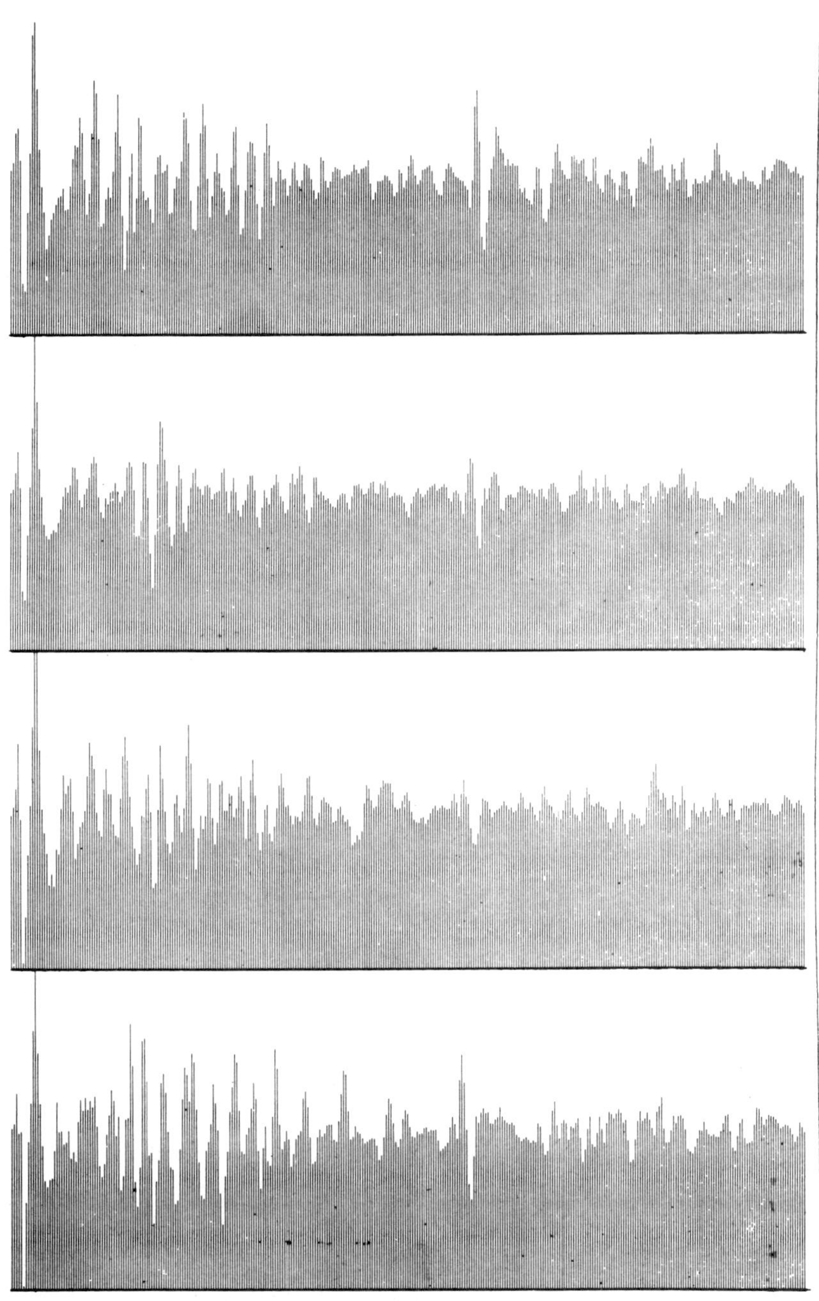

Fig. 25. Single traces of the Korea section.

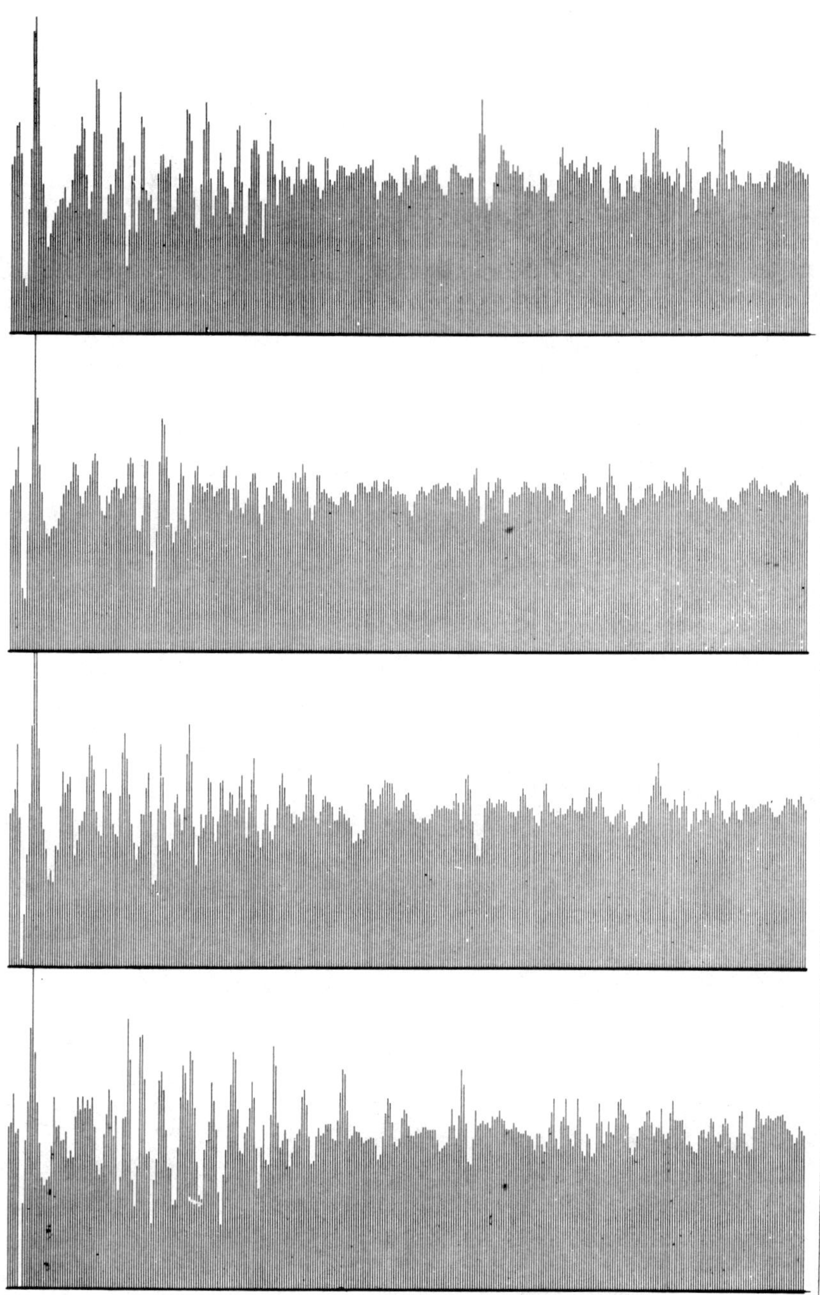

Fig. 26. Seismograms of figure 25 after homomorphic filtering.

time the multiple suppression is restricted in our example to this value even
for an exact parameter estimation.

Nevertheless, with this optimum window choice and the estimation of the
cepstrum component of the multiple the results of homomorphic filtering
are at least of comparable quality to predictive filtering as can be seen in
figure 27. The multiple of the water bottom reflection is largely reduced
while the continuity of the primaries going down into the multiple region is
guaranteed.

The example of the Korea section has been chosen to demonstrate that
the main problem of the successful application of the homomorphic method
is not the exact parameter determination, which for water bottom multiple
can generally be achieved. Rather, it is the implicit assumptions of the homomorphic method concerning the field data which may lead to unsatisfactory results. On the other hand, if the assumption of time invariance is
satisfied, our results with homomorphic filtering to suppress water column
multiples are promising and agree with those obtained by Buhl et al (1974).

9. SUMMARY AND CONCLUSIONS

Deconvolution by homomorphic filtering is an attractive method as it,
firstly, reduces a convolution to an additive superposition of the components and, secondly, often already achieves the separation of the individual
components in the complex cepstrum.

Its application to the seismic reflection method is, however, confined to
the deterministic filter process. This has the disadvantage that the cepstrum
components to be filtered should be known beforehand.

The power cepstrum offers the possibility of determining the parameters
for the filtering process in a fairly exact way in both cases for the wavelet
estimation as well as for suppression of water column reverberations.

The basic advantage of the method is found in the wavelet estimation as
no assumptions about delay properties of the wavelets are necessary. Even in
the presence of noise the phase filtering procedure treated in this paper leads
to an improved wavelet estimation. However, the use of the method for the
suppression of multiples is more problematic. These difficulties are less due
to the parameter determination; more severe are the effects of the deviations
of the implicit assumption of time invariance and the blow-up of random
noise during processing.

Up to now users of the method were primarily concerned with the ex-
ploitation of the additivity property. For the application of the method in
reflection seismics one should, in our opinion, be equally concerned with the
separability of the components. Till now only quefrency differences are used
for the separation of the components. Our present studies are intended to
account for statistical differences in the individual components in order to
exploit the advantages of the complex cepstrum *and* those offered by the

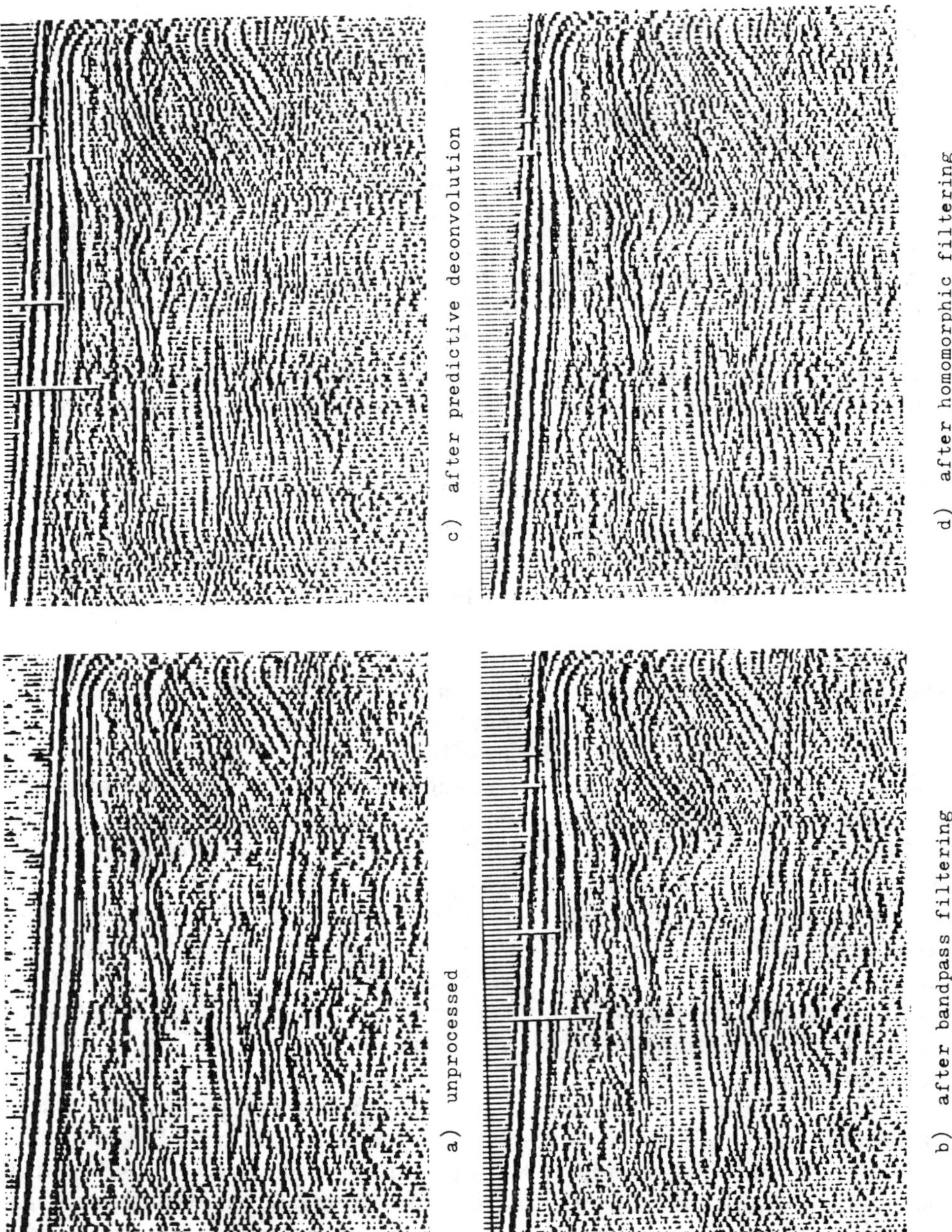

Fig. 27. Application of different de-ringing filters to air-gun measurements
to a part of the Korea section (figure 20).

least squares method. The separation of minimum- and maximum delay parts in the complex cepstrum in connection with statistical assumptions about the reflectivity function is the basis for this possibility.

REFERENCES

BUHL, P., STOFFA, P. L., and BRYAN, G. M., 1974, The Application of Homomorphic deconvolution to shallow-water marine seismology — Part II: Real Data, Geophysics 39, 417-426.

BOGERT, B. P., HEALEY, M. J., and TUKEY, J. W., 1963, The Quefrency Analysis of Time Series for Echoes: Cepstrum, Pseudo-Autocovariance, Cross-Cepstrum and Saphe Cracking, Proc. Symp. on Time Series Analysis, M. Rosenblatt (editor), New York, John Wiley and Sons, 209-243.

FLINN, E. A., COHEN, T. J., and Mc COWAN, D. W., 1973, Detection and Analysis of Multiple Seismic Events, Bull. Seis. Soc. Am. 63, 1921-1936.

KEMERAIT, R. C., and CHILDERS, D. G., 1972, Signal Detection and Extraction by Cepstrum Techniques, IEEE Transactions on Information Theory IT-18 (6), 745-759.

OPPENHEIM, A. V., SCHAFER, R. W., and STOCKHAM jr., T. G., 1968, Nonlinear Filtering of Multiplied and Convolved Signals, Proc. IEEE 56, 1264-1291.

ROBINSON, E. A., and TREITEL, S. 1967, Principles of Digital Wiener Filtering, Geophysical Prosp. 15, 311-333.

SCHAFER, R. W., 1969, Echo Removal by Discrete Generalized Linear Filtering, Technical Report 466, MIT, Res. Lab. of Electr.

SCHEPERS, R., 1972, Bearbeitungsverfahren zur Bestimmung oberflächennaher Strukturen aus Einkanal-Reflexionsseismogrammen bei senkrechtem Einfall, Berichte des Institutes für Geophysik der Ruhr-Universität Bochum, no. 2.

STOFFA, P. L., BUHL, P., and BRYAN, G. M., 1974, The Application of Homomorphic Deconvolution to Shallow-Water Marine Seismology — Part I: Models, Geophysics 39, 401-416.

ULRYCH, T. J., 1971, Application of Homomorphic Deconvolution to Seismology, Geophysics 36, 650-660.

PEACOCK, K. L., and TREITEL, S., 1969, Predictive Deconvolution: theory and practice, Geophysics 34, 155-169.

13

Reprinted from *Geophysics* **42**:1146–1157 (1977)

HOMOMORPHIC DECONVOLUTION
BY LOG SPECTRAL AVERAGING

ROBERT M. OTIS* AND ROBERT B. SMITH‡

A homomorphic system is used to map the convolution of a source function with the impulse response of the earth into the sum of the log spectra of the source function and the earth's response. If the source function is considered stationary and the earth's response spatially nonstationary, by averaging the log spectra of several reflection records, the log spectrum of the source function will be enhanced and the log spectrum of the earth's response will average out. An inverse homomorphic system maps the average log spectrum to an estimate of the source function. The source function is then deconvolved from reflection records. Application of log spectral averaging as a method of deconvolution to a seismic reflection profile obtained in a shallow water-geologically complex area indicates the homomorphic technique to be an effective method for suppressing the air gun bubble pulse oscillation.

INTRODUCTION

The seismic trace is generally assumed to be the convolution of a source wavelet with the impulse response of the earth. Air guns, sparkers, and explosive devices are designed to produce an impulsive signal as a means of increasing signal resolution. However, in most cases the resultant source function is not impulsive, and interpretational difficulties arise from ghosts, reverberations, multiples, and bubble pulses.

A widely used technique to compress wavelets to idealized impulses is unit-predictive deconvolution or least-squares zero-lag Wiener inverse filtering (Robinson and Treitel, 1967; Peacock and Treitel, 1969). This method involves the computation of a filter that is based on the autocorrelation function coefficients of the seismic trace and yields an impulse when convolved with the source function. Two important assumptions are made using unit-predictive deconvolution: (1) the impulse response of the earth is a white series, and (2) the source function is minimum phase. In actual practice, however, these criteria may not be satisfied and, consequently, the application of a unit-predictive filter may not improve the resolution of the seismic records.

Homomorphic deconvolution, which eliminates the computation of the autocorrelation function and thus the necessity of the minimum phase constraint, was introduced by Schafer (1968) and Oppenheim et al (1968) and has been used to deconvolve seismic records (Stoffa et al, 1974; Buhl et al, 1974). Homomorphic deconvolution involves the transformation from a vector space where functions are convolved to a vector space where functions are added. In the additive space the functions may have characteristics which allow their separation, not feasible in the original convolutional space. Thus, by separating the functions in the additive space, eliminating the unwanted functions, and transforming back into the vector space of convolution, functions are deconvolved homomorphically.

The transformation from the convolutional space to the additive space involves three operations: the Fourier transform, the complex natural logarithm, and the inverse Fourier transform. The additive space is referred to as the quefrency domain and has units of time. All aspects of the quefrency domain are named by reversing the letters of the first syllable of their frequency domain analogues. For example,

Presented at the 44th Annual International SEG Meeting, November 12, 1974 in Dallas. Manuscript received by the Editor May 1, 1975; revised manuscript received December 15, 1976.
*Mobil Research and Development Corp., Dallas, TX 75221; formerly, University of Utah, Salt Lake City, UT 84112.
‡University of Utah, Salt Lake City, UT 84112.

frequency becomes quefrency, low-pass filtering becomes short-pass liftering, high-pass filtering becomes long-pass liftering, and so on. The function obtained by transforming a function in the time domain to the quefrency domain is called the complex cepstrum.

Ulrych (1971, 1972) observed from earthquake data that the source function had low-quefrency components in the complex cepstrum while the earth's response had high-quefrency components. By applying a long-pass lifter to the complex cepstrum, he was able to recover the earth's impulse response. Buhl et al (1974) assumed the same characteristics for marine reflection data. However, in shallow water the earth's response had low-quefrency components and consequently when a long-pass lifter was applied to the complex cepstrum, near-surface reflections were removed.

We present a method which uses a homomorphic system to produce an estimate of the source function to be used as a basis for deconvolution. The technique does not assume minimum phase inputs and retains information about near-surface reflections. We observe that if the source function is stationary, and if the earth's response is spatially nonstationary, then the complex cepstrum of each of several sesimic records will be the sum of (1) a function which corresponds to the source function and is nonvariable, and (2) a function which corresponds to the earth's impulse response and varies from record to record. By averaging the complex cepstra of several records, an estimate of the nonvariable function can be obtained since the variable functions will tend to a mean value. Then, by inverse transformation to the time domain, an estimate of the source function is produced which can be used to deconvolve the seismic data.

The complex cepstrum is the inverse Fourier transform of the complex log spectrum, and the Fourier transform is a linear transformation. Thus, the complex log spectra can be averaged rather than the complex cepstra. Therefore, the technique of averaging complex log spectra to obtain an estimate of the source function will be referred to as log spectral averaging.

Log spectral averaging has two important advantages: (1) the homomorphic transformations between the time and quefrency domains preserve phase, so the estimate of the source function will contain phase information concerning the actual source function; and (2) if the estimate of the source function is used to construct a Wiener inverse filter to use as a deconvolution kernel, the assumption

that the earth's response is a white series is eliminated. Anticipation components of the filter can be based upon the phase information preserved in the source estimate.

HOMOMORPHIC SYSTEMS AND LOG SPECTRAL AVERAGING

A homomorphic system G can be represented as the cascade of the three systems shown in canonic form in Figure 1. The first system A is known as the characteristic system and has the property:

$$A[f(t) * g(t)] = A[f(t)] + A[g(t)],$$

i.e., A transforms the functions originally combined by convolution ($*$) to the addition of the functions individually processed by A. A linear system L (Figure 1) separates the added functions and A^{-1} transforms the filtered functions back into the space of convolution.

A commonly used characteristic system A is the cascade of (1) the z-transform of $h(n)$,

$$H(z) = \sum_{-\infty}^{\infty} h(n)z^{-n},$$

(2) the complex logarithm of $H(z)$,

$$\ln H(z) = \ln |H(z)| + j[\arg (H(z))],$$

and (3) the inverse z-transform of $\ln H(z)$,

$$\hat{h}(n) = \{1/(2\pi j)\} \oint \ln H(z)z^{n-1}dz,$$

where $z = \exp (jn\Delta T/N)$ and j is the square root of -1. If the contour of integration lies along the unit circle and $\ln H(z)$ is analytic, then $\hat{h}(n)$ is defined as the complex cepstrum of $\hat{h}(n)$ (Schafer, 1968). The inverse system A^{-1} can be shown to be (1) the z-transform of $\hat{h}(n)$,

$$\ln H(z) = \sum_{-\infty}^{\infty} \hat{h}(n)z^{-n},$$

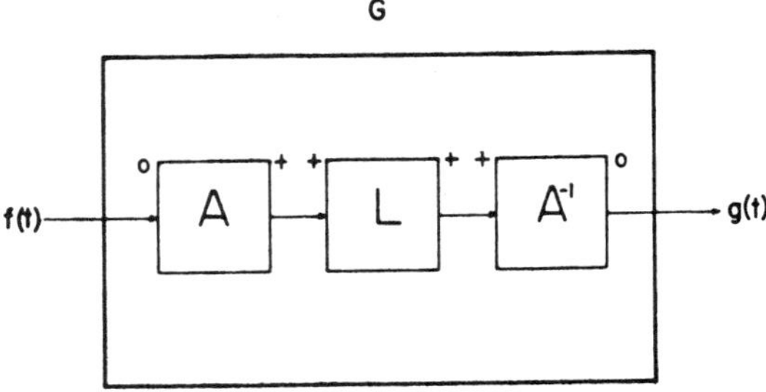

FIG. 1. The canonic representation of a homomorphic system (after Oppenheim et al, 1968).

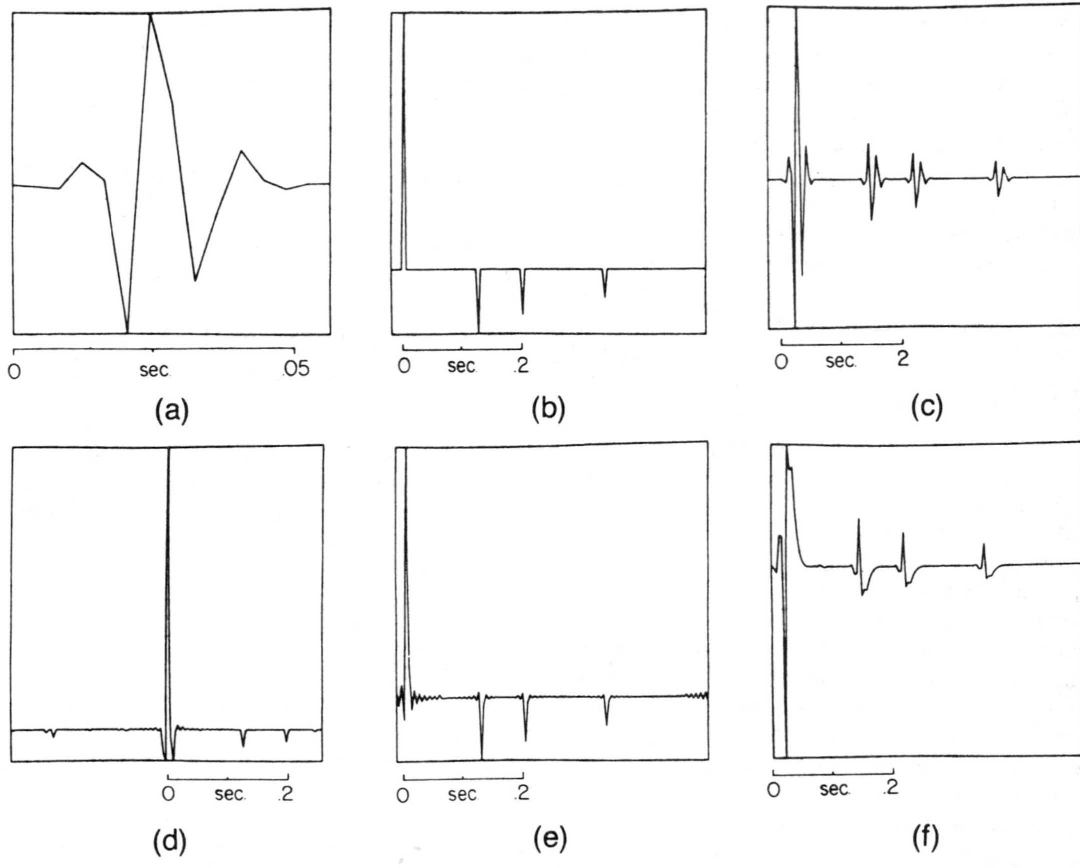

FIG. 2. Example of long-pass liftering: (a) A mixed phase wavelet. (b) The idealized impulse response for a 3-layer earth over an infinite half-space. (c) An idealized seismogram constructed by convolving the mixed phase wavelet with the idealized impulse response. (d) The complex cepstrum of the idealized seismogram. (e) The deconvolved seismogram obtained by long-pass liftering the complex cepstrum. (f) The deconvolved seismogram obtained using a unit-predictive filter.

(2) the complex exponential of $\ln H(z)$,

$$H(z) = \exp(\ln H(z)),$$

and (3) the inverse z-transform of $H(z)$,

$$h(n) = \{1/(2\pi j)\} \oint H(z) z^{n-1} dz,$$

where the contour of integration again is along the unit circle.

Two difficulties are immediately encountered when computing the complex logarithm, $\ln H(z)$: (1) the complex logarithm is not unique and (2) if the phase of $H(z)$ has a linear component, it will not be analytic and the contour integration cannot be done. These problems can be resolved by (1) defining the complex logarithm to be the line integral of $d\alpha/\alpha$, where α is a dummy variable of integration, along a continuous Riemann surface (Oppenheim, 1965), and by (2) removing the linear component of phase before com-

puting the inverse z-transform and replacing it before computing the complex exponential.

In homomorphic deconvolution, if we assume

$$h(n) = g(n) * f(n),$$

the linear system L (Figure 1) is used to separate the two additive functions, $\hat{g}(n)$ and $\hat{f}(n)$. The choice of L depends upon the charactistics of $f(n)$ and $g(n)$ which would allow their separation in the quefrency domain. For example, it has been assumed that in some reflection data the components of the log amplitude and phase spectra of the source function will be slowly varying while the components of the log amplitude and phase spectra of the earth's response will be rapidly varying (Ulrych, 1971, 1972; Stoffa et al, 1974; Buhl et al, 1974). In the complex cepstrum, low-quefrency components will correspond to the source function while high-quefrency components will correspond to the earth's response. A long- or short-pass lifter can separate the two functions in the quefrency domain, thus deconvolving the two functions in the time domain.

To illustrate long-pass liftering, a mixed phase wavelet (Figure 2a) was constructed by specifying the zeroes of its z-transform to lie both inside and outside the unit circle. An impulse response (Figure 2b) representing a simple 3-layered earth model, was convolved with the mixed phase wavelet to generate an idealized seismogram (Figure 2c). The complex cepstrum of the seismogram was computed (Figure 2d) and the linear system L was chosen to be a long-pass lifter, i.e., designed to remove components of low quefrency and retain components of high quefrency. Figure 2e shows the impulse response of the 3-layered earth model obtained by inverting the liftered cepstrum to the time domain. The impulse response of Figure 2b has been accurately recovered, and the mixed phase wavelet has been homomorphically deconvolved.

In comparison, Figure 2f shows the impulse response obtained by applying unit-predictive deconvolution to the idealized seismogram in Figure 2c. It is clear that the homomorphic technique, which makes no assumptions about the phase of the wavelet, recovered the impulse response more accurately than unit-predictive deconvolution technique which assumed a minimum phase source function.

A seismic trace is assumed to be the convolution of a source function with the impulse response of the earth, and therefore liftering the complex cepstrum would eliminate the source function. However, in actual practice, the source function is not necessarily restricted to low quefrencies. For example, a signature of an air gun was windowed with a half cosine bell (Figure 3a), and the complex cepstrum of the windowed signature was computed (Figure 3b). Note that significant information about the source function is contained in high-quefrency components.

If the air gun signature of Figure 3a is convolved with the earth's impulse response, some components of the complex cepstrum of the source function may overlap some components of the complex cepstrum of the earth's impulse response. Since low-quefrency components of the earth's impulse response represent near-surface reflections (Stoffa et al, 1974), long-pass liftering will in general remove components of the source function and the near-surface reflection data. Thus, in this case, liftering is an effective method of deconvolution only if one is interested in deep reflection data (Buhl et al, 1974).

If one is interested in near-surface information

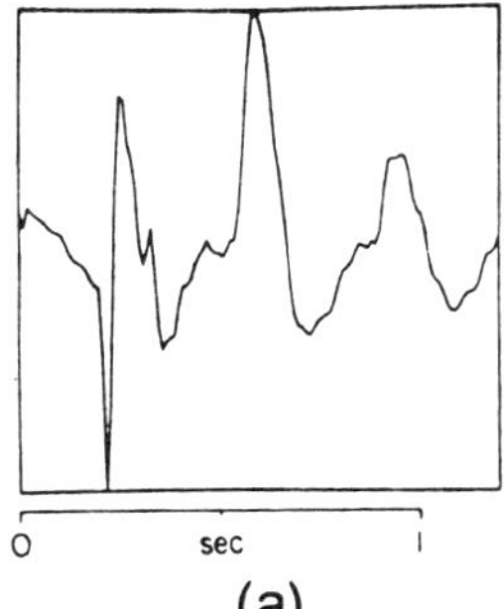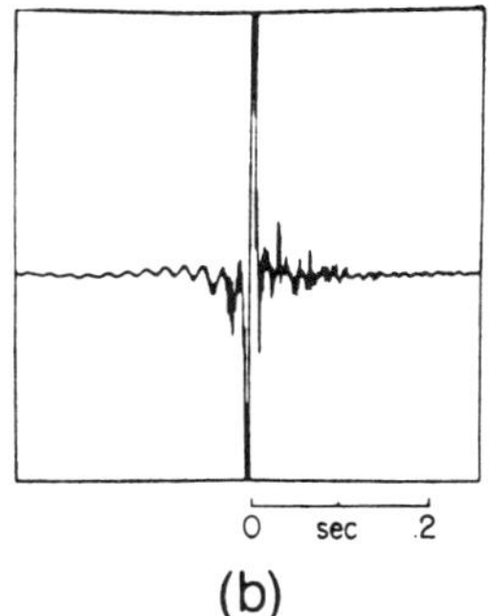

FIG. 3. (a) The windowed signature of a 1 inch^3 air gun. (b) The complex cepstrum of the windowed air gun signature.

where a mixed phase source function has been convolved with a non-white impulse response, we propose an alternate method of homomorphic deconvolution that allows the estimation of the source function and its removal by inverse filtering or direct deconvolution. This technique has been used in image processing (Cole, 1973) and in the restoration of old recordings (Stockham et al, 1975). We assume that one signal which remains constant is convolved with a second signal which changes such that it may be considered stationary over a short time interval, but nonstationary over long time intervals. For many short intervals, the convolved signals are mapped by a homomorphic transformation to the sum of a function which changes from section to section and a function which remains constant. Averaging the complex cepstra of these sections would seem to enhance the unchanging function and suppress the changing signals. An inverse homomorphic transformation of the average complex cepstrum would then produce an estimate of the unchanging signal which may then be deconvolved from each section to obtain the changing signal.

In seismic profiling, if the source function can be considered stationary and the geological structure considered spatially variable, then the complex cepstrum of each trace will be the sum of a constant function corresponding to the source function and a variable function corresponding to the earth's response. If $f(n)$ denotes the source function and $g(n)$ the earth's response, and $h(n)$ the convolution of $f(n)$ and $g(n)$, the simple average of the complex cepstra of selected traces along a profile is expressed as

$$\hat{h}(n) = 1/N \sum_{i=1}^{N} \hat{h}_i(n),$$

where the subscript i designates the ith selected

record. Because the inverse z-transform is a linear transformation, averaging log spectra is equivalent to averaging complex cepstra, and the above equation is expressed as

$$1/N \sum_{i=1}^{N} \ln H_i(z) = 1/N \sum_{i=1}^{N} \ln F_i(z) + 1/N \sum_{i=1}^{N} \ln G_i(z).$$

Because of the stationarity of the source function,

$$1/N \sum_{i=1}^{N} \ln F_i(z) = \ln F(z),$$

and

$$1/N \sum_{i=1}^{N} \ln H_i(z) = \ln F(z) + 1/N \sum_{i=1}^{N} \ln G_i(z).$$

Thus, the successful recovery of the source function depends upon whether the second term on the right side of the above equation tends to some constant value for all z, or in the quefrency domain, whether $g(n)$ tends to an impulse at zero quefrency.

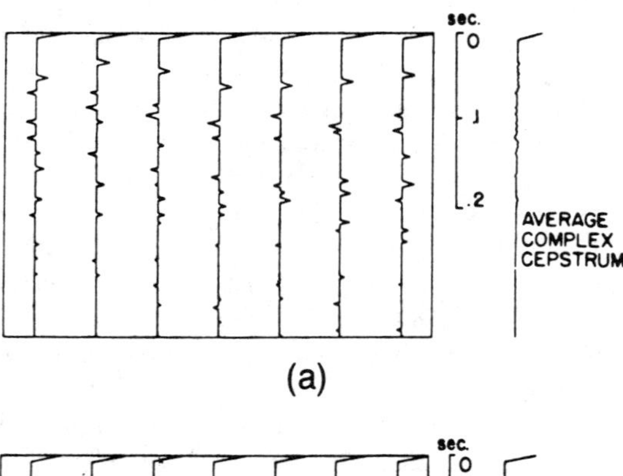

(a)

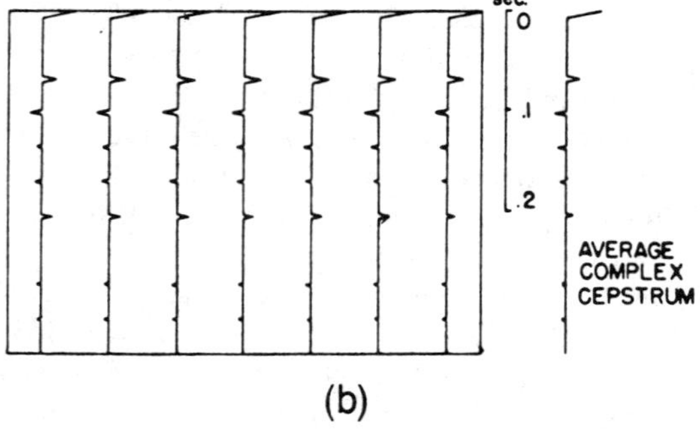

(b)

FIG. 4. (a) Example of how averages of the complex cepstra of idealized impulse responses in a geologically complex area tend to an impulse at zero quefrency, and (b) how the averages in a flat layered environment do not.

If for some z each value of $\ln G_i(z)$ for $i = 1, 2, \ldots, N$ is considered a random variable, and if for each z the random variables are assumed to have identical probability distributions with the same mean and variance, then by the Central Limit Theorem, the sum

$$1/N \sum_{i=1}^{N} \ln G_i(z)$$

will tend to that mean for large N. If these conditions are met, then the above term will tend to a constant value for all z and the average of the log spectra of seismic records will tend to an estimate of the source function.

Experimentation with averaging log spectra has shown that the conditions set by the Central Limit Theorem (CLT) are satisfied when geologic structure changes at each selected shot point. For example, consider averaging the complex cepstra of several synthetic earth impulse responses, which is of course equivalent to averaging their log spectra. If the geology is changing, the average, $\hat{h}(n)$, for some n will be the average of several impulses and many zeroes as shown in Figure 4a. Thus for all n, the function $\hat{h}(n)$ will tend to the desired impulse at zero quefrency. However, if the geologic structure does not change such as in flat-layered structure, the average will not tend to the impulse at zero quefrency. In flat-layered structure it is easily seen (Figure 4b) that the averages for each n do not have the same mean and therefore do not satisfy the conditions of the CLT, and the technique will not enhance the source function. Fortunately, in many cases records may be selected to force the desired variability in geologic structure and subsequently obtain a reasonable estimate of the source function.

Estimates of the source function are generally characterized by high-frequency noise. This noise can be removed by applying a band-pass filter, and then truncating the filtered estimate with a half cosine-bell window $w(n)$,

$$w(n) = \begin{cases} 1/2 + 1/2 \cos[\pi(n-1)/N] & 1 \leq n \leq N, \\ 0 & \text{elsewhere}, \end{cases}$$

where N is the window length, chosen to be sufficiently long to determine the character of the source, but not to introduce oscillations in the frequency response of the estimate. We choose the window length long enough to include two or three oscillations of the bubble pulse and to terminate at a zero crossing of the estimate.

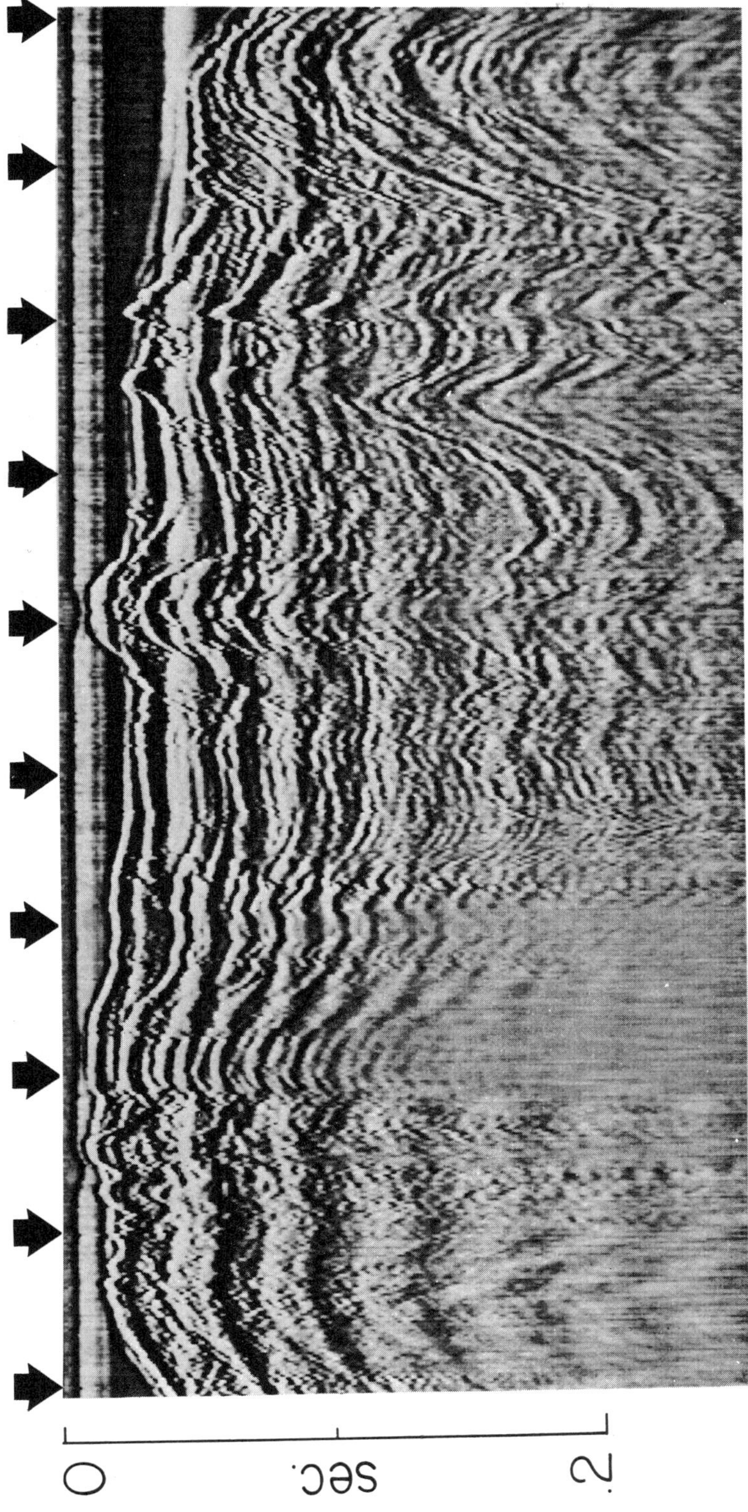

Fig. 6. (a) Unprocessed seismic reflection data obtained using a 1 inch³ air gun and a single channel hydrophone streamer.

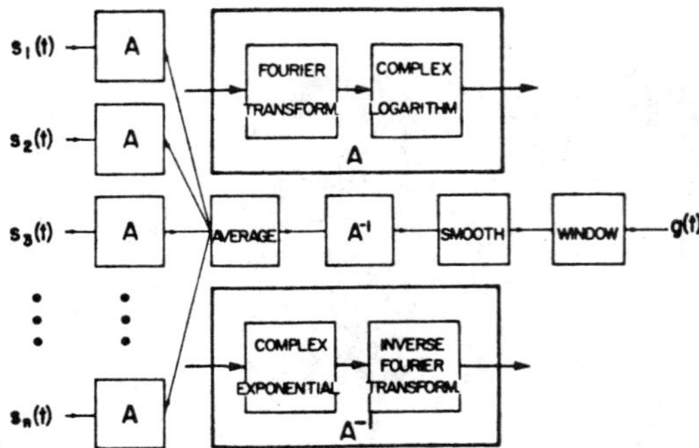

FIG. 5. Flow diagram of method used in log spectral averaging.

The procedure used in log spectral averaging to obtain an estimate of the source function is illustrated by a flow diagram in Figure 5 and is summarized by the following steps: (1) obtain representative records from the seismic profile, (2) compute the complex log spectrum of each record, (3) average the complex log spectra, (4) invert the resultant average to the time domain, and (5) smooth and window.

APPLICATION OF LOG SPECTRAL AVERAGING

A sample profile of unprocessed reflection data obtained with a 1 inch^3 air gun and a single channel hydrophone streamer was digitized at a 1 msec sampling interval (Figure 6a). This profile exemplifies a complex geologic structure in a shallow water environment. The complex log spectra of ten traces (Figure 6b) chosen from the sample profile at equally spaced intervals were computed and averaged. The resultant function was then inverted to the time domain, smoothed with a low-pass 50 Hz filter and windowed with a half cosine bell. The final estimate of the source function is also shown in Figure 6b.

A first attempt to deconvolve the estimate from the records in the sample profile was made by simply dividing the frequency response of each record by the frequency response of the estimate. Examination of the results (Figure 7) shows a monofrequency domination and a ringing of the near-surface reflections at a period equal to the length of the estimate.

Figure 8a shows a function whose components are the reciprocals of the components of the power spectrum of the estimate. This function will be referred to as the inverse power spectrum of the estimate and is the power spectrum of the function that is convolved with the seismic trace to remove effects of the source function. Figure 8a clearly shows a peak at approximately 80 Hz, the frequency that

dominates the profile in Figure 7. Figure 8b shows the estimate's complex cepstrum which is subtracted from the complex cepstrum of the seismic record when the estimate is deconvolved. The impulse denoted by the arrow in Figure 8b results from truncating the estimate by the window applied when constructing the estimate and causes the ringing in the profile. Also note in Figure 8b that most of the energy in the complex cepstrum is at positive quefrencies indicating that the estimate is near-minimum phase.

A second attempt to deconvolve the estimate was made by convolving each record of the sample profile with an eight-point memory-only Wiener inverse of the estimate (Figure 9). Both the monofrequency domination and the ringing are absent, and there is no loss of resolution. Experimentation with other inverse filters up to 100 points in length indicated no appreciable improvement in the resolution of the processed profile or removal of the bubble pulse oscillations.

Figure 10a shows the power spectrum of the eight-point Wiener inverse. Comparison of the spectra in Figures 10a and 8a reveals that a least-squares inverse has a slowly varying power spectrum, which has no dominating component as did the inverse power spectrum of the estimate. Figure 10b shows the complex cepstrum of the Wiener inverse. The absence of an impulse corresponding to the length of the estimate as seen in Figure 8b further indicates that no ringing is introduced.

Because the profile in Figure 9 was deconvolved with a minimum-phase least-squares filter, it would seem that unit-predictive deconvolution would pro-

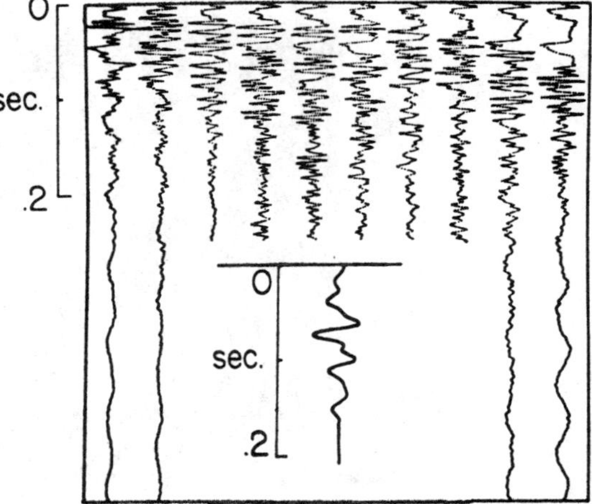

FIG. 6. (b) Ten traces selected from the unprocessed profile, and the estimate of the source function obtained by log spectral averaging. Arrows indicate the selected traces.

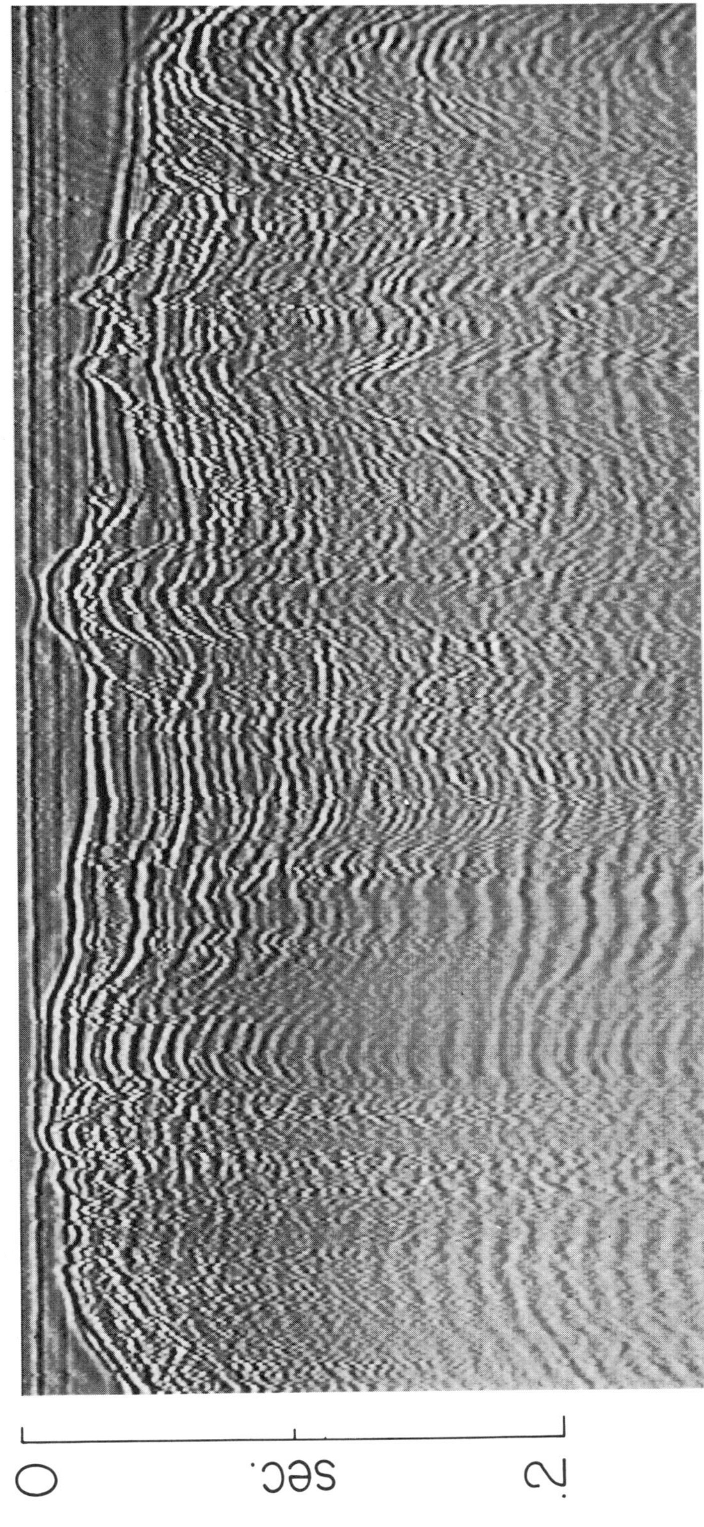

FIG. 7. Deconvolved seismic profile obtained by dividing the frequency response of each trace by the frequency response of the source estimate.

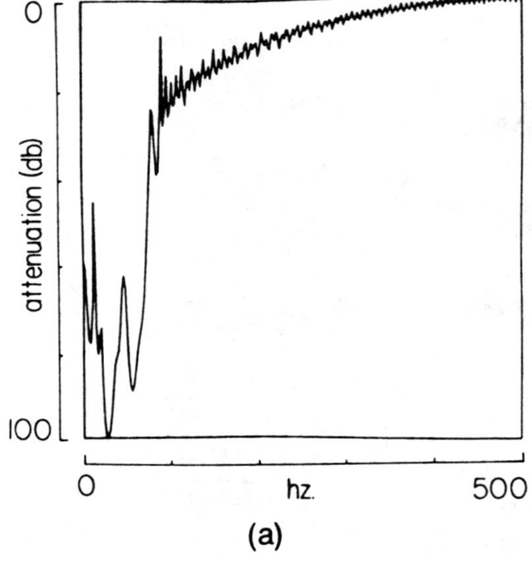

(a)

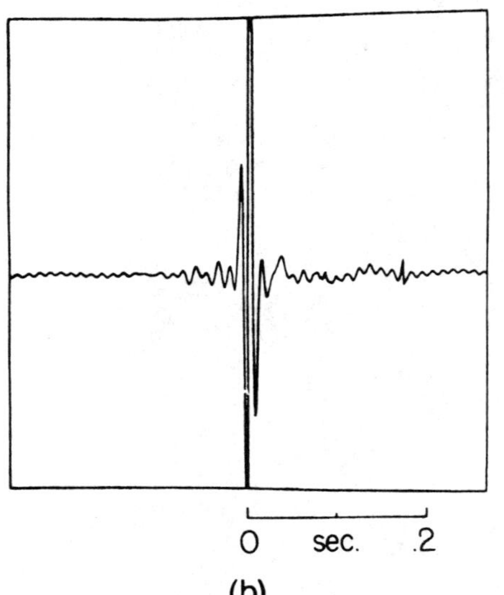

(b)

FIG. 8. (a) Inverse power spectrum of the source estimate. (b) Complex cepstrum of the source estimate.

duce similar results. To compare the results of log spectral averaging with unit-predictive deconvolution, the sample profile was filtered with a 50 Hz high-pass filter, and deconvolved with a ten-point unit-predictive operator (Figure 11a). Note that as many as 4 oscillations of the bubble pulse are still present. We believe that Figure 11 illustrates a case where the assumption that the earth's response is a white series was not valid, thus preventing unit-predictive deconvolution from satisfactorily removing the bubble pulse oscillations.

In addition to the removal of bubble pulse oscillations, log spectral averaging allows the incorporation of both bandpass filtering and deconvolution into one efficient time domain convolution. The filtered and predictively deconvolved profile (Figure 11) required the application of a 64-point operator to remove low-frequency components and a 10-point inverse filter operator for deconvolution. The homomorphically deconvolved profile (Figure 9) required the application of one eight-point operator to both filter and deconvolve because the bandpass filter used to smooth the estimate during design became a band-reject filter when the estimate was used to deconvolve the data. During design, a 50 Hz low-pass filter was applied to the estimate used, and thus, as indicated in Figure 8a, the deconvolved data contains only frequencies above 50 Hz.

CONCLUSIONS

Log spectral averaging can be a useful tool in the deconvolution of the source function from seismic reflection data. If the assumptions of a stationary source function and a spatially variable earth's response are satisfied, then the deconvolution process can be reduced to the convolution of a short operator in the time domain. In addition, bandpass filtering can be included in the deconvolution by using the appropriate filter when constructing the prototype source function. Thus, computation time is saved because filtering and deconvolution are incorporated into one operation.

Other advantages of log spectral averaging are that it is not restricted to minimum-phase inputs and it does not require a random earth's response. Thus, a Wiener inverse based on an estimate from averaging log spectra may be considered more representative of the actual source function than a Wiener inverse based on the autocorrelation of a seismic trace. Finally, application of the homomorphic technique, log spectral averaging, does not remove shallow reflectors, and thus, information about the near surface can be obtained that would normally be obscured by the source function or eliminated by liftering the complex cepstrum.

ACKNOWLEDGMENTS

This research was supported by the National Science Foundation grant no. GA-12870, and the Advanced Research Projects Agency, project number DAHC15-73-C-0363. We wish to thank Dr.

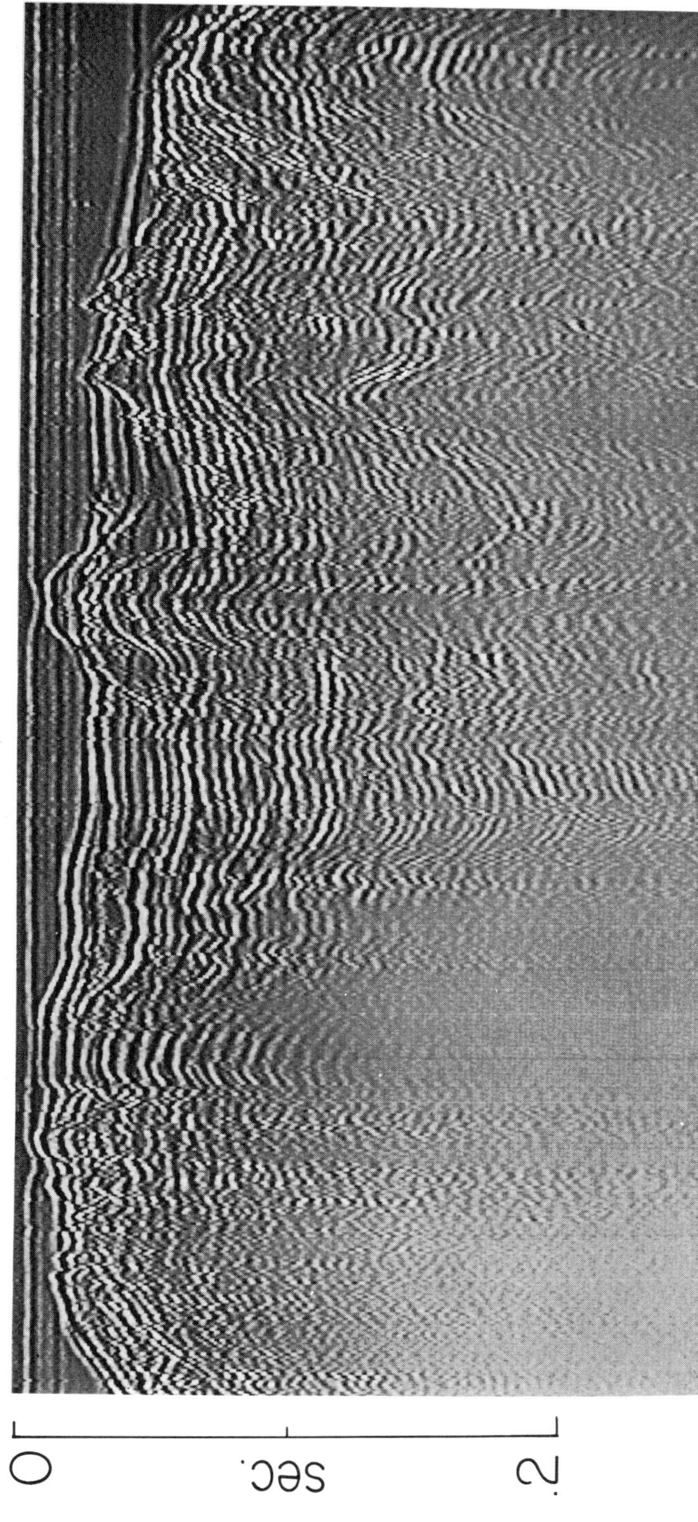

Fig. 9. Deconvolved seismic profile obtained by convolving an 8-point Wiener inverse of the estimate with each trace.

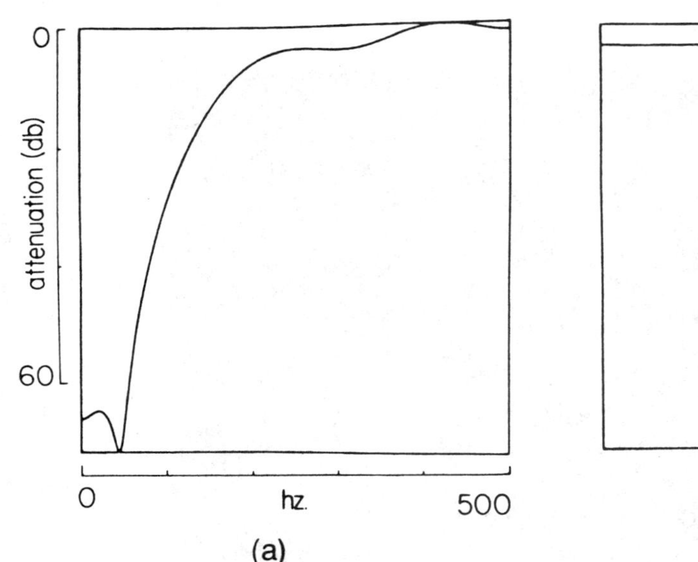
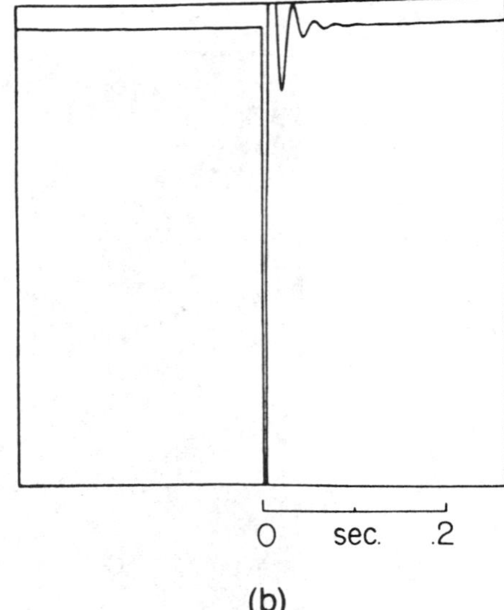

(a)   (b)

FIG. 10. (a) Power spectrum of the Wiener filter. (b) Complex cepstrum of the Wiener filter.

Thomas G. Stockham, Jr., Dr. E. Randall Cole, and Dr. Wayne J. Peeples for their ideas and encouragement, and are grateful to R. B. Warnock, M. Milochik, and many others who have given interest, criticism, and support. Special thanks are due Dr. Lawrence Wood, Dr. Kenneth L. Peacock, and Dr. Sven Treitel of Amoco Production Co. for critically reviewing this manuscript.

REFERENCES

Buhl, P., Stoffa, P. L , and Bryan, G. M., 1974, The application of homomorphic deconvolution to shallow water marine seismology—Part II: Real data: Geophysics, v. 39, p. 417–426.

Cole, E. R., 1973, The removal of unknown image blurs by homomorphic filtering: UTEC-CSs-74-029, Computer Sci. Dept., University of Utah, Salt Lake City.

Oppenheim, A. V., 1965, Superposition in a class of non-linear systems: Tech. rep. 432, MIT, Res. Lab. of Electr., 62 p.

Oppenheim, A. V., Schafer, R. W., and Stockham, T. G., 1968, Nonlinear filtering of multiplied and convolved signals: Proc. IEEE, v. 65, p. 1264–1291.

Peacock, K. L., and Treitel, S., 1969, Predictive deconvolution: theory and practice: Geophysics, v. 34, p. 155–169.

Robinson, E. A., and Treitel, S., 1967, Principles of digital Wiener filtering: Geophys. Prosp., v. 15, p. 311–333.

Schafer, R. W., 1969, Echo removal by discrete generalized linear filtering: Tech. rep. 466, MIT, Res. Lab. of Electr.

Stockham, T. G., Jr., Cannon, T. M., and Ingebretson, R. B., 1975, Blind deconvolution through digital signal processing: IEEE Trans. Audio Electroacoust., v. AU-23.

Stoffa, P. L., Buhl, P., and Bryan, G. M., 1974, The application of homomorphic deconvolution to shallow-water marine seismology—Part I: Models: Geophysics, v. 39, p. 401–416.

Ulrych, T. J., 1971, Application of homomorphic deconvolution to seismology: Geophysics, v. 36, p. 650–660.

——— 1972, Homomorphic deconvolution of some teleseismic events: Bull. SSA, v. 62, p. 1253–1265.

Fig. 11. Filtered and deconvolved seismic profile obtained using a 64-point 50 Hz high-pass filter and a 10-point unit-predictive filter.

Applications of Short-Time Homomorphic Signal Analysis to Seismic Wavelet Estimation

JOSÉ M. TRIBOLET, MEMBER, IEEE

Abstract—The use of homomorphic systems to deconvolve seismic reflection and teleseismic data has been proposed and explored by a number of researchers with varying success. A careful study of the methods employed reveals a number of problems involving both the class of characteristic systems used and their numerical implementation. Furthermore, the analysis strategies used introduce deterministic constraints on the estimation of the seismic wavelet which may lead to serious problems in determining the earth impulse response.

Several novel results are presented in this paper. A class of homomorphic systems matched to the bandpass nature of seismic signals is discussed and improved and more reliable implementation algorithms are developed. The concept of short-time wavelet estimation by homomorphic filtering is introduced. This technique takes into account the specific time-varying characteristics of seismic traces. Strategies for homomorphic wavelet estimation are proposed and illustrated. The recovery of the earth impulse response may then be accomplished by combining homomorphic wavelet estimation with parametric inverse filtering.

I. INTRODUCTION

AN IMPORTANT class of seismic analysis techniques is based on a representation of the seismic signal as a convolution of components, with the basic signal processing task being deconvolution.

There are a number of approaches typically available for carrying out a deconvolution. One of the most common is the use of linear inverse filtering; that is, processing the composite signal through a linear filter whose frequency response is the reciprocal of the Fourier transform of one of the signal components. Obviously, in order to use inverse filtering, such components must be known or estimated. Homomorphic signal processing [1] represents an alternative approach to seismic deconvolution. This class of filtering techniques represents a generalization of linear filtering problems.

The objectives of this paper are first to review the various ways in which homomorphic filtering has been explored for use with seismic data processing; and second, to present a number of new results in homomorphic signal analysis involving not only the theory and implementation of homomorphic systems, but the overall analysis strategy as well. Before doing so, however, we discuss in the next section the basic principles of homomorphic filtering.

Manuscript received October 11, 1977; revised February 24, 1978. This work was supported by the Advanced Research Projects Agency monitored by the Office of Naval Research under Contract N00014-75-C-0951-NR049-308.

The author was with the Department of Electrical Engineering and Computer Science, Research Laboratory of Electronics, Massachusetts Institute of Technology, Cambridge, MA. He is now with Bell Laboratories, Murray Hill, NJ 07974.

II. HOMOMORPHIC SYSTEMS FOR CONVOLUTION

When we are faced with the problem of filtering signals which have been added, we very often use a linear filter. A principal reason is that linear filters are analytically convenient for dealing with signals that have been added. The principal advantage with linear filtering applied to added signals is that if the behavior of the filter for each of the components separately is known, then the behavior for the sum is the sum of the responses, as a consequence of the principle of superposition.

The notion of homomorphic deconvolution is to apply the same basic reasoning to deconvolution. Towards this end we consider a class of systems matched to convolution in the same way that linear systems are matched to addition. Then, the class of systems that we consider have the superposition property illustrated in Fig. 1. It can be shown that this class of systems can be represented by the cascade of three systems as shown in Fig. 2. The first system, with transformation $D_*[\cdot]$ maps a convolution of components to a sum. The system $L[\cdot]$ is a linear system and the system $D_*^{-1}[\cdot]$ is the inverse of the system $D_*[\cdot]$ and consequently maps a sum of components to a convolution. The system $D_*[\cdot]$ and its inverse are specified by the characteristics of the signal space with which we are dealing. Thus, all the flexibility within this class lies in the linear system $L[\cdot]$.

A. Full-Band Homomorphic Systems

The transformation $D_*[\cdot]$ has been defined, more specifically, by the property of the z-transform (or Fourier transform) of its output being equal to the complex logarithm of the z-transform (or Fourier transform) of its input, i.e., if

$$\hat{x}[n] = D_*[x(n)] \tag{1a}$$

then

$$\hat{X}(z) = \log X(z). \tag{1b}$$

Since the z-transform is a complex-valued function, its logarithm must be carefully defined. In particular, since the transformation $D_*[\cdot]$ maps convolution to addition, the transformation of (1b) is required to map multiplication to addition, and this imposes a particular interpretation on the complex logarithm. The input space to $D_*[\cdot]$ is restricted to the set of all *full-band* sequences $x(n)$, that is, all stable sequences with stable inverses. Otherwise, the complex logarithm would not be analytic on the unit circle and the output of $D_*[\cdot]$ would be unstable. Then, in particular, low-pass, high-pass and bandpass filtered sequences are not allowable inputs to a homo-

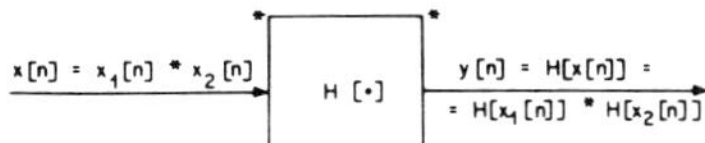

Fig. 1. Convolutional superposition property.

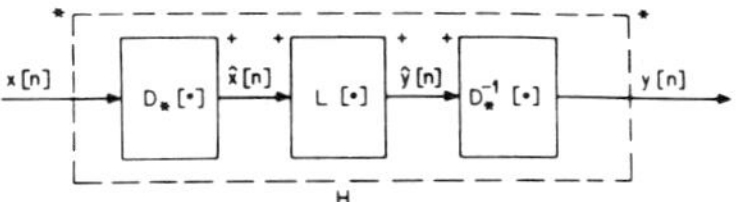

Fig. 2. Canonic representation of homomorphic systems for convolution.

morphic system with a system $D_*[\cdot]$ as in (1). We shall refer to this class of systems as *full-band homomorphic systems*. The output $\hat{x}(n)$ of the system $D_*[\cdot]$ is referred to as the *complex cepstrum* of the input $x[n]$.

B. Properties of the Complex Cepstrum

In carrying out a deconvolution using homomorphic filtering, an appropriate choice for the linear system $L[\cdot]$ must be made. This choice is intimately connected to a number of properties of the complex cepstrum, as summarized below.

Property 1: The complex cepstrum of a convolution of two (or more) signals is the sum of the individual complex cepstra.

Property 2: The complex cepstrum of a minimum-phase sequence is zero for $n < 0$, and the complex cepstrum of a maximum-phase sequence is zero for $n > 0$.

Property 3: The complex cepstrum $\hat{w}[n]$ of a pulse $w[n]$ whose spectrum is smooth tends to be concentrated around low-time values.

Let us denote by $r[n]$ an impulse train of the form

$$r[n] = \sum_{k=1}^{M} r_k \delta[n - n_k] \qquad n_{k+1} > n_k, \quad r_k \neq 0, \quad \forall k$$

where $(r_k, n_k)\ k = 1, \cdots, M$ denote the arrival times and amplitudes associated with each impulse in $r[n]$. Let us define, in general, a *periodic impulse train* with period $T > 1$ as one for which the interarrival times are multiples of T. That is,

$$n_k = n_{k-1} + l_k \cdot T, \qquad k = 2, \cdots, M, \quad \forall l_n > 0.$$

Property 4: The complex cepstrum $r[n]$ of a periodic impulse train $\hat{r}(n)$ is also a periodic impulse train with the same period, that is,

$$\hat{r}[n] = \sum_{n=-\infty}^{\infty} \hat{r}[k]\ \delta[n - kT].$$

Property 5: Consider a minimum-phase impulse train for which the first interarrival time is T_1, so that

$$n_2 - n_1 = T_1, \qquad n_k \text{ arbitrary}, \quad k > 2.$$

Then, the complex cepstrum is zero for $0 > n > T_1$. In general, $\hat{r}[n]$ is nonzero only at times $0, n_2 - n_1, n_3 - n_1, \cdots, n_m - n_1$ as well as at all positive linear combinations of these times. A similar result can be derived for maximum-phase impulse trains.

Property 6: The cepstral structure of a mixed-phase impulse train may be very sensitive to minor changes in the impulse amplitudes. In general, the cepstra of mixed-phase impulse trains exhibit a rather elaborate structure which, within the limits imposed by our present understanding of the cepstral mapping, offer no clues regarding the corresponding time structure and vice versa.

This characteristic should be kept in mind, as it plays a major role in shaping strategies for data analysis by homomorphic filtering, to be described in Section IV.

III. SEISMIC DECONVOLUTION BY TIME-INVARIANT HOMOMORPHIC ANALYSIS

Seismic signals are often represented in time-invariant form as the convolution of a seismic wavelet $w[n]$, thought to represent the combined source and earth attenuation effects with an impulse train $r[n]$, representing the earth impulse response. A third component $p[n]$ may also have to be included to represent the presence of water reverberations or ghosting mechanisms.

The structure of the complex cepstrum $\hat{s}[n]$ is such that the complex cepstrum of the seismic wavelet $\hat{w}[n]$, that of the reflector series $\hat{r}[n]$, and that of the reverberation train $\hat{p}[n]$, may tend to occupy disjoint time intervals. For example, if the source pulse has a relatively smooth spectrum, $\hat{w}[n]$ will tend to be concentrated near the origin. If the reflector series is minimum-phase then it contributes to the complex cepstrum only for times greater than or equal to the time of the first reflection. For marine seismograms, the reverberation train is a periodic, minimum-phase impulse series, and its complex cepstrum will therefore be right-sided and periodic with the same period.

This tendency for different components to occupy different time intervals in the cepstral domain suggests the possibility of separating these components by gating (i.e., windowing) the cepstrum. A window around the time origin, for example, would tend to retain the contribution from the source pulse. This low-time windowing on the complex cepstrum corresponds to smoothing the log spectrum.

Cepstral gating was first explored for seismic data by Ulrych [2], and was directed towards the deconvolution of teleseismic events. Here, the data were modeled as

$$s[n] = w[n] * r[n] \tag{2}$$

where $r[n]$ is in general a mixed-phase impulse train. In order to render the complex cepstral component associated with the reflector series as simple and predictible as possible, the traces were exponentially weighted as

$$s'[n] = s[n]\ \alpha^n = w[n]\ \alpha^n * r[n]\ \alpha^n = w'[n] * r'[n] \tag{3}$$

prior to homomorphic deconvolution. This has the effect of radially scaling the z-transform of a sequence. In particular, any zero of $R(z)$ outside the unit circle in the z-plane can be moved inside the unit circle by exponential weighting, thus rendering the weighted impulse train $r'[n]$, minimum phase. The weighted seismic wavelet $w'[n]$ was then recovered by low-time gating, with a cutoff frequency which was determined by observation of the theoretical impulse response of the earth at the recording site, which was known beforehand.

Ulrych's approach to teleseismic deconvolution was further pursued by Ulrych *et al.* [3], Clayton and Wiggins [4], and Sommerville [5]. These authors were able to estimate very reasonable source pulses for a number of earthquakes and nuclear explosions using the homomorphic filtering technique. However, they also pointed out a number of problems encountered in such processes. For example, they recognized the computational difficulties of defining the complex logarithm in regions of low spectral amplitudes. They also questioned the low-time assumption for the complex cepstrum of the source pulse since in some instances the source pulse had sharp spectral zeros, thus leading to localized but sharp variation in frequency.

The application of homomorphic filtering in seismic reflection studies is discussed first by Buhl, Stoffa and Bryan [6] in the context of shallow-water marine dereverberation and source deconvolution. Here, the data were modeled as

$$s[n] = w[n] * r[n] * p[n] \tag{4}$$

where $p[n]$ denotes a periodic minimum-phase reverberation operator. Exponential weighting was applied to make the reflector series minimum phase. Letting T_1 denote the two-way travel time within the first earth layer, and T_w denote the water-column reverberation, and assuming

$$NT_w \leqslant T_1 < (N+1) T_w,$$

it follows that high-time gating the complex cepstrum with a cutoff time T_c greater than NT_w, but less than T_1 will deconvolve the seismic source, eliminate the first N multiples in the reverberation pattern and reduce the remaining multiples to at most $1/(N+1)$ of their original pulse.

A similar procedure was also investigated for deep-water dereverberation and source deconvolution by Buttkus [7]. Here, of course, the cepstra of the reverberation train will have high-time characteristics. Dereverberation may then be accomplished by notch gating the cepstrum at the multiples of T_w.

Gallemore [8] has attempted to evaluate the use of homomorphic deconvolution to marine seismic data and to compare its performance to the use of an optimal tapped delay line filter. The criteria for evaluation were the percent of multiple energy removed, percent of reflector distortion, and visual improvement of the data when both methods were applied to synthetic seismic data. The indications from the study were that homomorphic deconvolution appears to have greater potential in shallow-water applications while the tapped delay line filter appears to be more efficient for deep-water data. Buttkus [7] has also discussed the limitations of seismic deconvolution by cepstral gating. Using experimental evidence he concluded that the success of wavelet estimates by low-time gating was largely dependent on the signal-to-noise ratio of the input data, the degree of overlapping of the cepstrum components of the source wavelet and of the reflectivity function, and the choice of the low-time window.

He proposed then to estimate the wavelet by analyzing only a windowed segment within the trace. The procedure involved first the estimation of the arrival times by observation of spikes in the power cepstrum (the power cepstrum is the square of the cepstrum of Bogert, Healy and Tukey), followed by notch filtering the segment's complex cepstrum. Implicit in this procedure is again the assumption that the reflector series segment is minimum phase.

Finally, Otis and Smith [9] attempted to extend Stockham's blind deconvolution technique [10] to seismic processing. The technique does not involve cepstral gating. Here, they considered a suite of seismograms with spatially nonstationary reflector series. Then the complex cepstrum of each seismic record will be the sum of the cepstrum of the source pulse, which is nonvariable, plus the cepstrum of the reflector series, which varies from record to record. By averaging the complex cepstra of several records an estimate of the nonvariable functions can be obtained, since the variable functions will tend to a mean value. This procedure has been referred to as log spectral averaging or *cepstral stacking*. The estimated source wavelet is then used to construct a Wiener inverse filter to use as a deconvolution kernel. A similar approach was also used by Clayton and Wiggins [4]. We outlined above the main issues involved in the use of full-band homomorphic systems for the analysis of time-invariant seismic models. In the remainder of this paper we shall outline recent results [11] in homomorphic signal analysis which greatly enhance the potential and the robustness of homomorphic systems in seismic data analysis.

IV. New Results in Homomorphic Signal Analysis

A. Homomorphic Bandpass Systems

Seismic signals are often bandpass filtered prior to sampling and recording, to discriminate against additive noise. As pointed out in Section II, bandpass signals are not allowable inputs to a full-band homomorphic system. Thus, it is necessary to redefine the characteristic system $D_*[\cdot]$ so that its input is matched to the bandpass characteristics of seismic data. Such characteristic systems may be defined as a cascade of two systems, as illustrated in Fig. 3. The first system, BP, corresponds to a frequency scaling operation that shifts and stretches the passband of the signal to occupy the entire band. The second system is then the complex logarithmic system, which is characteristic for full-band homomorphic signal analysis.

Let $x[n]$ be a bandpass signal satisfying

$$X(e^{j\omega}) = \begin{cases} X(e^{j\omega}) \neq 0 & \omega_1 \leqslant |\omega| \leqslant \omega_2 \\ \\ 0 & \text{elsewhere} \end{cases} \tag{5}$$

and let $\tilde{x}[n]$ be the response of BP to the input $x[n]$. Then $\tilde{x}[n]$ satisfies

$$\tilde{X}(e^{j\tilde{\omega}}) = X(e^{j\omega}), \qquad 0 \leqslant |\omega| \leqslant \pi \tag{6}$$

where

$$\tilde{\omega}(\omega) = \pi \, \frac{\omega - \omega_1}{\omega_2 - \omega_1} \qquad \omega_1 \leqslant |\omega| \leqslant \omega_2. \tag{7}$$

This mapping is illustrated in Fig. 4. The system BP is referred to as the bandpass mapping system. Its role in homomorphic

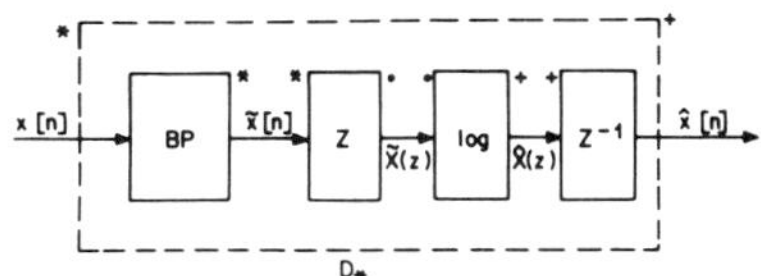

Fig. 3. Characteristic system D_* for homomorphic bandpass filtering.

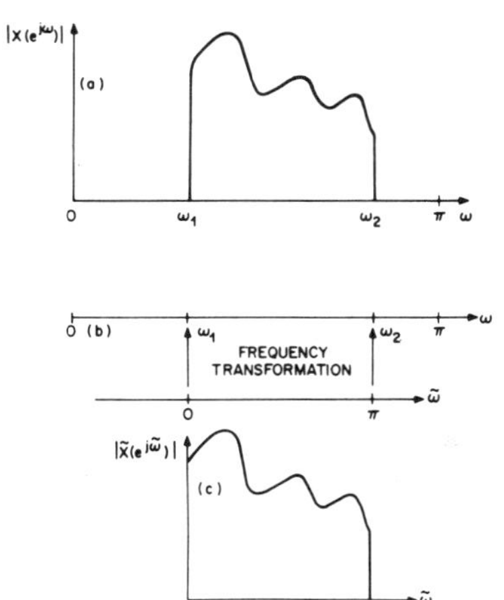

Fig. 4. The bandpass mapping. (a) Magnitude spectrum of bandpass signal $x[n]$. (b) Frequency transformation. (c) Magnitude spectrum of full-band signal $\tilde{x}[n]$.

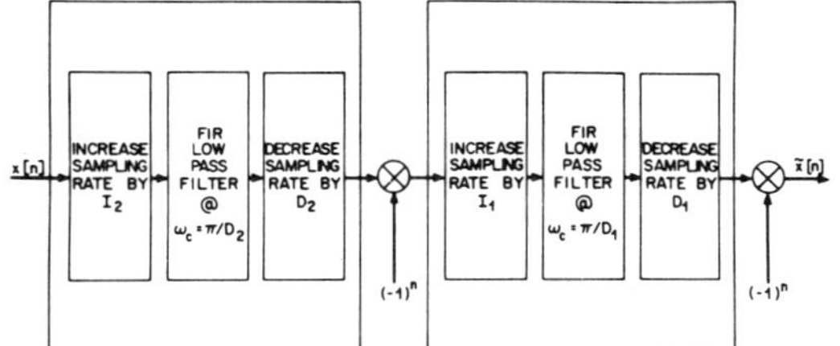

Fig. 5. An implementation of the bandpass mapping BP.

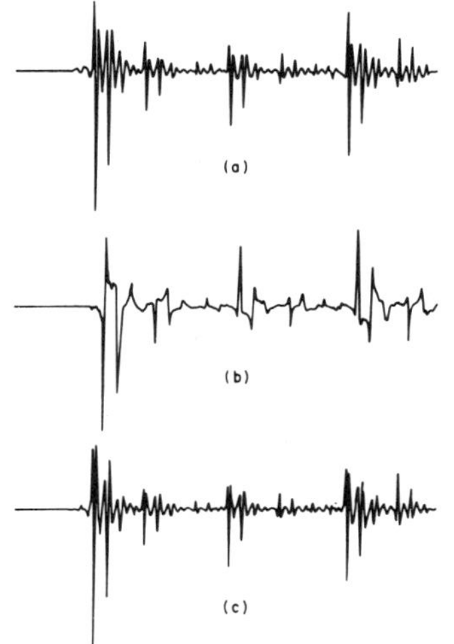

Fig. 6. Effects of bandpass mapping a signal $s[n]$ equal to the convolution of a wavelet $w[n]$ with an impulse train $r[n]$. (a) $s[n]$. (b) $\tilde{s}[n] = BP[s[n]]$. (c) $s^o[n] = BP^{-1}[\tilde{s}[n]]$.

filtering corresponds to the role of a linear bandpass filter in linear filtering in that all out-of-band information is effectively discarded.

The class of homomorphic systems with the characteristic systems of Fig. 4 is referred to as *homomorphic bandpass systems*. The theory and implementation of this class of systems can be found in [11].

Fig. 5 depicts an implementation of the bandpass system BP, based on the use of finite impulse response (FIR) linear-phase digital filters. The cutoff frequencies ω_1 and ω_2 are such that

$$\omega_1 = \omega_2\left(1 - \frac{I_1}{D_1}\right)$$

$$\omega_2 = \frac{I_2}{D_2} \cdot \pi. \tag{8}$$

Let us now illustrate the effects of bandpass mapping the synthetic seismogram $s[n]$ of Fig. 6(a). This signal satisfies the convolutional model

$$s[n] = w[n] * r[n] \tag{9}$$

where $w[n]$ is the bandpass wavelet of Fig. 7(a) and

$$r[n] = \sum_{k=1}^{M} r_k \delta[n - n_k]$$

is the impulse train of Fig. 8(a). The passband cutoff frequencies ω_1 and ω_2 of $s[n]$ and $w[n]$ are given by

$$\omega_2 = \frac{1}{2} \cdot \pi \qquad \omega_1 = \left(1 - \frac{5}{6}\right)\omega_2 \tag{10}$$

which, assuming a sampling interval of 4 ms, corresponds to a passband from 10.4 Hz to 62.5 Hz. Thus, the cutoff frequencies chosen are representative of those we are likely to encounter in seismic reflection studies.

The bandpass mapped components $\tilde{s}[n]$, $\tilde{w}[n]$ and $\tilde{r}[n]$ are illustrated in Figs. 6(b), 7(b), and 8(b), respectively. These signals were computed using the implementation of Fig. 5, with $I_1 = 5$, $D_1 = 6$, $I_2 = 1$ and $D_2 = 2$, where the low-pass filters were approximated by FIR zero-phase optimum Chebyshev filters.

Based on the theoretical analysis of homomorphic bandpass mapping [11], it is possible to show that the bandpass mapped impulse train $\tilde{r}[n]$ may be approximately modeled as

$$\tilde{r}[n] = \tilde{b}[n] * \tilde{p}[n] \tag{11}$$

where $\tilde{p}[n]$ represents an impulse train of the form

$$\tilde{p}[n] = \sum_{k=1}^{M} \tilde{p}_k \delta[n - \tilde{l}_k] \tag{12}$$

and $\tilde{b}[n]$ represents a doublet-like envelope.

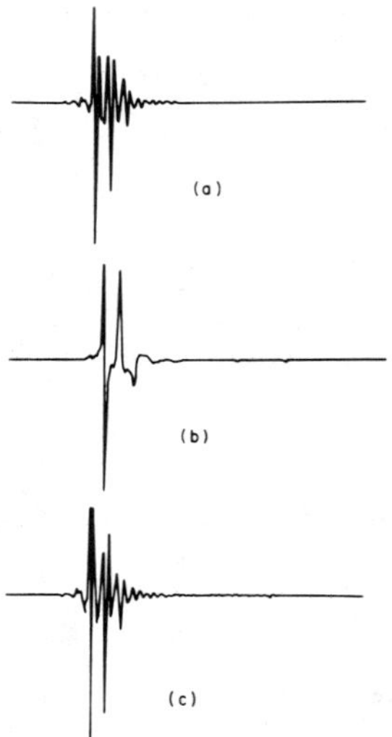

Fig. 7. Effects of bandpass mapping a bandpass filtered wavelet $w[n]$. (a) $w[n]$. (b) $\tilde{w}[n] = BP[w[n]]$. (c) $w°[n] = BP^{-1}[\tilde{w}[n]]$.

Fig. 8. Effects of bandpass mapping a full-band impulse train $r[n]$. (a) $r[n]$. (b) $\tilde{r}[n] = BP[r[n]]$. (c) $r°[n] = BP^{-1}[\tilde{r}[n]]$.

The arrival times and amplitudes $(\tilde{l}_k, \tilde{p}_k)$ are related to the arrival times and amplitudes of $r[n]$ as

$$\tilde{l}_k \cong n_k \cdot \frac{I_1 I_2}{D_1 D_2} \qquad k = 1, \cdots, M \qquad (13)$$

and

$$|r_k| \simeq |\tilde{p}_k| \qquad k = 1, \cdots, M \qquad (14)$$

as it is in fact evident from Fig. 8(a) and 8(b). Note, however, that although the basic amplitude structure of $r[n]$ has been preserved in $\tilde{r}[n]$, its polarity has not, as a consequence of the time-varying nature of the linear transformation BP.

We conclude then that the bandpass mapping essentially preserves the structure of a seismogram $s[n]$, in the sense that $\tilde{s}[n]$ can still be represented as a convolution of a wavelet $(\tilde{w}[n] * \tilde{b}[n])$ with an impulse train $\tilde{p}[n]$, since, using (11)

$$\tilde{s}[n] = \tilde{w}[n] * \tilde{r}[n] \cong (\tilde{w}[n] * \tilde{b}[n]) * \tilde{p}[n]. \qquad (15)$$

The analysis of $\tilde{s}[n]$ in terms of its components enables the recovery of the bandpass filtered reflector series $r°[n]$, defined as

$$r°[n] = BP^{-1}[\tilde{p}[n] * \tilde{b}[n]] \qquad (16)$$

as illustrated in Fig. 8(c). The recovery of the out-of-band spectral components of $r[n]$ *cannot* be handled by means of homomorphic analysis. One alternative might be the use of parametric restoration techniques as suggested by Clayton and Ulrych [12].

B. Phase Unwrapping by Adaptive Integration

The principle computational step in using homomorphic filtering is the computation of the complex cepstrum. The procedure for determination of the complex cepstrum generally relies on an explicit computation of the inverse Fourier transforms of the logarithm of the Fourier transform of the bandpass mapped inputs. This is most easily carried out using the fast Fourier transform algorithm to compute samples of the Fourier transform equally spaced in frequency. Since the Fourier transform is a complex-valued function, its logarithm must be appropriately defined. In particular, the phase of a complex number is only unique to within an integer multiple of 2π, and this ambiguity must be resolved in computing the phase of the Fourier transform. In order that the logarithmic transformation map multiplication to addition as required, it can be shown that the phase must be generated as a continuous function of frequency. Thus, if the phase is computed modulo 2π, as is typically the case, it must be "unwrapped" to obtain a continuous phase curve.

Simple algorithms, based either on examination of the sampled phase to locate jumps of 2π or based on numerical integration of the phase derivative have proven to be generally unrealiable. As a result, a novel adaptive numerical integration phase unwrapping technique has been proposed [13] which has proven to be very reliable.

C. Determinism and Time-Invariant Homomorphic Analysis

An important new result in homomorphic signal analysis deals with the use of exponential weighting as a means of insuring the minimum-phase character of the reflector series. This practice, which has been used by many researchers as discussed in Section III, is valid in noiseless situations but must be questioned otherwise.

As it is shown in [11], *by making the reflector series minimum-phase and low-time gating the cepstra, one is in effect constraining the pulse estimate to be a replica of the first arrival.* Thus, although this strategy may be reasonable in teleseismic analysis, it is often not appropriate in reflection

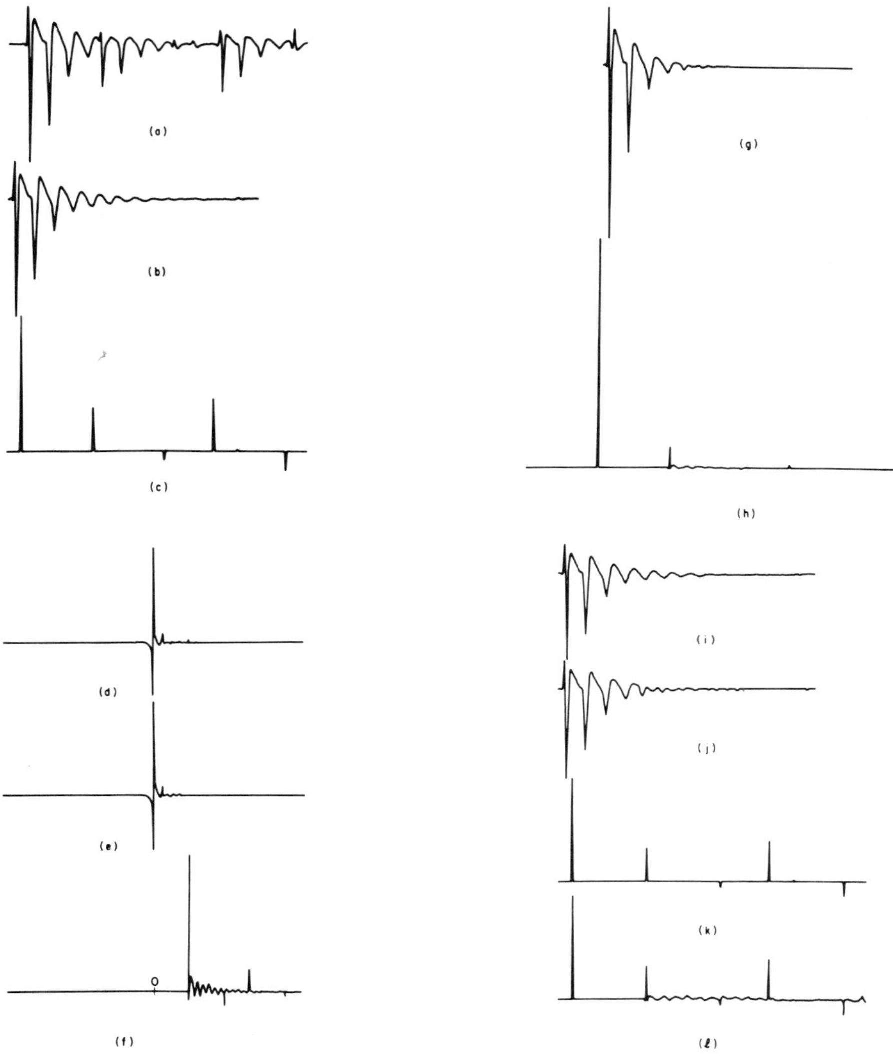

Fig. 9. Time-invariant homomorphic analysis of synthetic seismogram (noise and attenuation effects not included). (a) Synthetic trace $s[n]$. (b) Source pulse $w[n]$. (c) Earth impulse response $r[n]$. (d) Complex cepstrum of exponentially weighted trace $\hat{s}'[n]$. (e) Low-time cepstral component $\hat{s}'^{L}[n]$. (f) High-time cepstral component $\hat{s}'^{H}[n]$. The vertical scale has been magnified ($\times 57$). (g) Low-time component of weighted trace $s'^{L}[n]$. (h) High-time component of weighted trace $s'^{H}[n]$. (i) Source pulse $w[n]$. (j) Deweighted low-time component $s^{L}[n]$. (k) Reflector series $r[n]$. (l) Deweighted high-time component $s^{H}[n]$.

seismology since the seismic wavelet characteristics change in time due to attenuation within the earth. Furthermore, the data often contain significant noise components.

We shall now illustrate the potential and limitations of time-invariant homomorphic analysis. Two types of synthetic seismograms will be analyzed. The first type includes neither attenuation nor additive noise; the second type includes both attenuation and additive noise effects.

In both examples the same earth structure was used, corresponding to a simple three-layer model with a two-way travel time of 300 ms for the top layer and 500 ms for the middle

layer. This earth is illuminated from above by a seismic source, as in marine profiling. Water-layer reverberation effects are not included. The sampling interval was set at 2.44 ms. The corresponding reflector series is illustrated in Fig. 9(c).

In the first example, we shall analyze the synthetic seismogram $s[n]$ of Fig. 9(a), which corresponds to the convolution of the airgun signature $w[n]$ of Fig. 9(b) with the earth impulse response $r[n]$ of Fig. 9(c). This impulse response corresponds to the earth model described above, assuming no loss mechanism.

The seismogram $s[n]$ was exponentially weighted with a

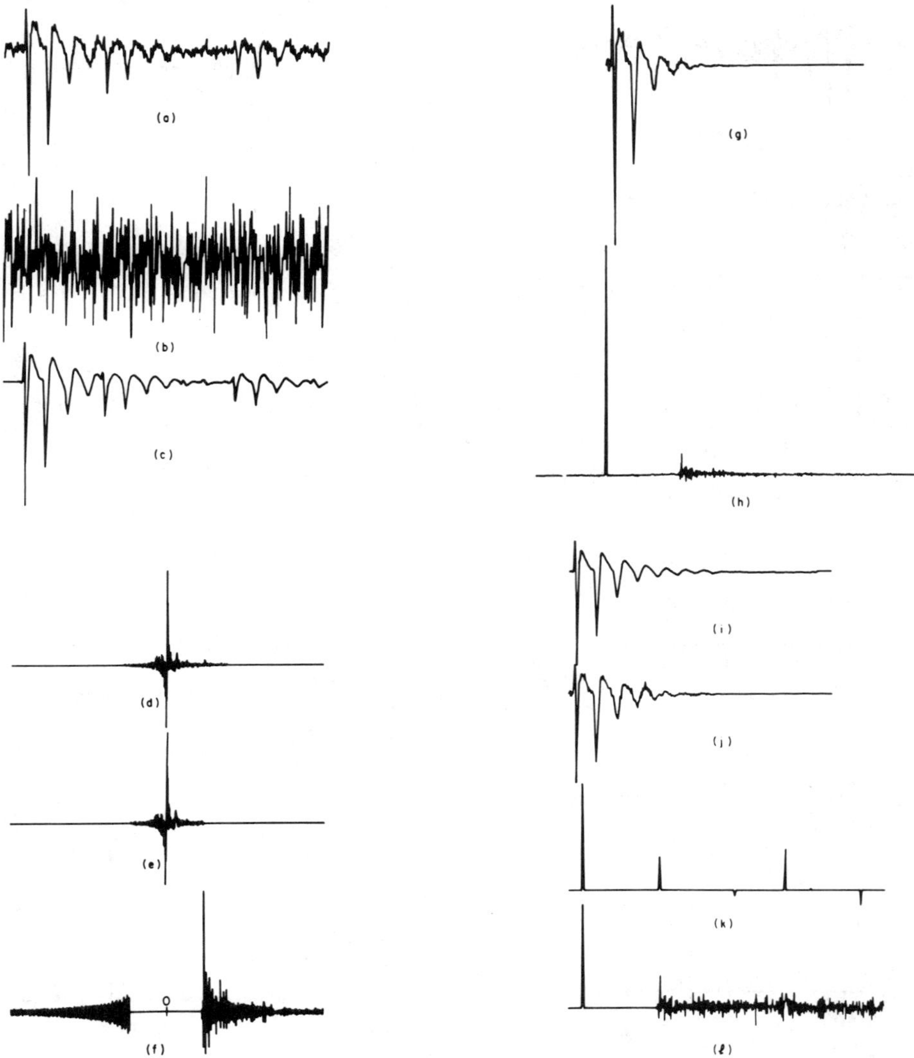

Fig. 10. Time-invariant homomorphic analysis of synthetic seismograms (includes both noise and attenuation effects). (a) Synthetic trace $s[n]$. (b) Noise component $\eta[n]$. The vertical scale has been magnified ($\times 10$). (c) Noiseless synthetic trace (includes attenuation effects). (d) Complex cepstrum of exponentially weighted trace $\hat{s}'[n]$. (e) Low-time cepstral component $\hat{s}'^{L}[n]$. (f) High-time cepstral component $\hat{s}'^{H}[n]$. The vertical scale has been magnified ($\times 57$). (g) Low-time component of weighted trace $s'^{L}[n]$. (h) High-time component of weighted trace $s'^{H}[n]$. (i) Source pulse $w[n]$. (j) Deweighted low-time component of weighted trace $s'^{L}[n]$. (k) Reflection series $r[n]$. (l) Deweighted high-time component $s^{H}[n]$.

weighting equal to 0.99 to yield a weighted seismogram $s'[n]$. Fig. 9(d) depicts the complex cepstrum $\hat{s}'[n]$ of the weighted seismogram, and Fig. 9(e) and 9(f) depict the low-time and high-time components of $\hat{s}'[n]$, denoted by $\hat{s}'^{L}[n]$ and $\hat{s}'^{H}[n]$, respectively. The cepstral cutoff time was set at 293 ms. Fig. 9(g) and 9(h) depict the time sequences associated with such low-time and high-time components, namely $s'^{L}[n]$ and $s'^{H}[n]$. Finally, after exponential deweighting, the results of homomorphic analysis of the seismic trace of Fig. 9(a), namely $s^{L}[n]$ and $s^{H}[n]$, are shown in Fig. 9(j) and 9(l). For com-

parison purposes the airgun pulse $w[n]$ and reflector series $r[n]$ are depicted in Fig. 9(i) and 9(k). Let us consider, in the second example, the seismogram shown in Fig. 10(a), which corresponds to the attenuated noiseless seismogram of Fig. 10(c), corrupted by the additive noise component $\eta[n]$ depicted in Fig. 10(b). The SNR was set at +10 dB. Following the same analysis procedure as in the preceeding example, the noisy seismic trace was analyzed, as illustrated in Fig. 10(d) through 10(l).

The examples above clearly show the matching of the low-time cepstral estimate $s^L[n]$ to the data onset, when the reflector series is made minimum phase. Thus, although leading to very reasonable high-time reflector series estimates in the time-invariant case, it often leads to useless estimates in the noisy, time-varying case.

In summary, the use of homomorphic systems in seismic data processing has been characterized by a purely deterministic approach to signal analysis, in that no account is made for realistic deviation of the data from the idealized time-invariant models. Such philosophy, on which the use of exponential weighting is based, leads typically to results where the first arrival is perfectly resolved as a very sharp impulse—thus indirectly confirming the matching of the wavelet estimate to the first arrival—the remainder being corrupted by noise, with a severity that depends on the particular time-varying nature of the data and the signal-to-noise ratio. Thus, the estimation of the seismic reflector series using homomorphic signal processing requires, in general, a different analysis strategy which takes into account the specific time-varying characteristics of the trace. The development of a framework for such analysis and the investigation of appropriate homomorphic filtering strategies will be the topic of the remainder of this paper.

D. Short-Time Homomorphic Analysis

The homomorphic analysis of signals which approximately follows a convolutional model on a short-time basis is referred to as short-time homomorphic analysis. Here, the basic signal models are of the form

$$s[n] = [w[n] * r[n]] \, v[n] \tag{17}$$

where $v[n]$ represents a short-time window. A first attempt to model short-time windowing effects in the context of homomorphic signal analysis was recently reported [14]. In general, we may model a short-time segment as

$$s[n] = w'[n] * r'[n] \tag{18}$$

where $r'[n]$ represents the windowed reflector segment

$$r'[n] = r[n] \cdot v[n] \tag{19}$$

and $w'[n]$ is defined in the frequency domain as

$$W'(e^{j\omega}) = S'(e^{j\omega})/R'(e^{j\omega}). \tag{20}$$

Note that $w'[n]$ depends not only on $w[n]$ but also on the structure of $r[n]$ and on the window shape, onset and duration. If in carrying out short-time homomorphic analysis, an accurate estimate of the seismic pulse $w[n]$ is desired, then one must choose the window $v[n]$ very carefully. In particular, one should not use windows such that $r'[n]$ becomes minimum phase (or maximum phase) since then $w'[n]$ will match the onset (or offset) of the segment $s[n]$, which will not usually coincide with a wavelet arrival. By choosing windows which have gradual onset and offset transitions, and which are long with respect to the wavelet length and smooth with respect to the reflector series, one minimizes the distortion of the underlying convolutional model by the window-

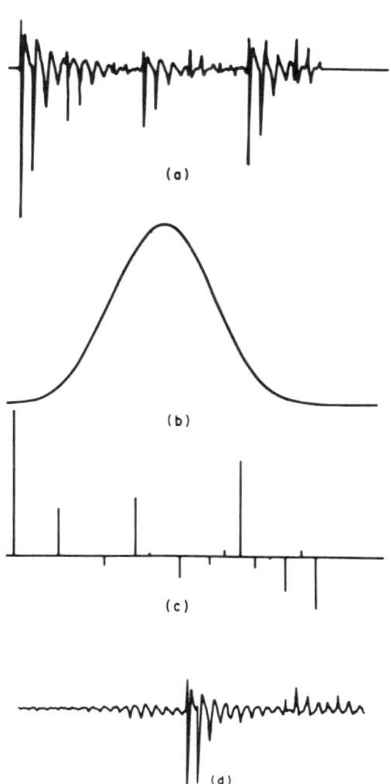

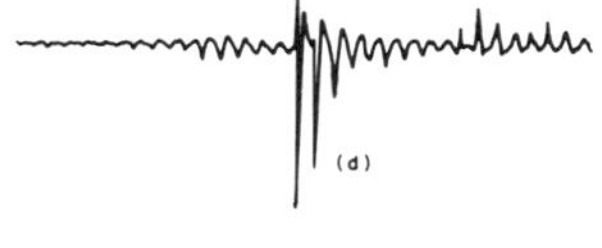

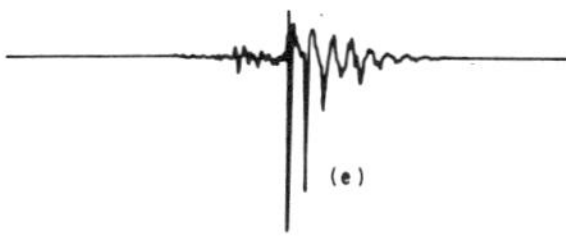

Fig. 11. Short-time homomorphic analysis using a Hamming window squared. (a) Short time segment $s[n]$. (b) Hamming window squared $v[n]$. (c) Seismic reflection segment $r[n]$. (d) Short-time convolutional component $w'[n]$. (e) Low-time cepstral component $w'^L[n]$.

ing operation. In that case, one may approximately model $s[n]$ as

$$s[n] \simeq w[n] * r'[n]. \tag{21}$$

A number of short-time windows have been investigated [11] in the context of short-time homomorphic analysis, namely Hamming, Gaussian, Rayleigh, Linear, and Exponential. It was found that, in general, *the low-time cepstral components of $w'[n]$ and $w[n]$ were essentially identical as long as the windowed reflector series $r'[n]$ was kept mixed phase.*

This is illustrated in Fig. 11. Fig. 11(a) depicts a seismic segment and Fig. 11(c) depicts the corresponding reflector segment $r[n]$. Using a Hamming window squared $v[n]$ as depicted in Fig. 11(b), the sequence $w'[n]$ defined in (18) was computed and is shown in Fig. 11(d). Fig. 11(e) depicts the corresponding low-time cepstral component of $w'[n]$, $w'^L[n]$, which is seen to capture the essential features of the airgun wavelet $w[n]$.

Note that, in general, the low-time component of the seismic segment $s^L[n]$ cannot be assumed to yield an estimate of the seismic wavelet, since the mixed phase windowed reflector segment $r'[n]$ defined in (19) may have significant components in the low-time cepstral region.

V. A Strategy for Short-Time Wavelet Estimation

Based on the above results, a strategy for wavelet estimation by short-time homomorphic analysis was devised which capitalizes on the sensitivity of the cepstral structure of mixed-phase aperiodic impulse trains to time-domain amplitude perturbations. *The procedure consists essentially in using different short-time windows on the same seismic segment.*

Letting $v_k[n]$, $k = 1, \cdots, M$ denote a set of M different short-time windows all defined on the same time-interval, and letting $s_k[n]$ denote the corresponding short-time segments

$$s_k[n] = [w[n] * r[n]] \cdot v_k[n]. \tag{22}$$

It follows, then, that if the windows are chosen as described in Section IV-D,

$$s_k[n] = w[n] * r'_k[n] \tag{23}$$

where $r'_k[n]$ denotes the windowed reflector segment

$$r'_k[n] = r[n] \cdot v_k[n]. \tag{24}$$

The complex cepstra of each segment is then of the form

$$\hat{s}_k[n] = \hat{w}[n] + \hat{r}'_k[n] \tag{25}$$

where it is expected that the cepstral structure of every impulse train $\hat{r}'_k[n]$ is essentially independent from every other. The seismic wavelet might now by estimated using averaging methods, either in the cepstral or in the time domain.

Cepstral Stacking: Here, the wavelet estimate $w_e[n]$ is formed by averaging all the complex cepstra $\hat{s}_k[n]$, as follows:

$$\hat{w}_e[n] = \frac{1}{M} \sum_{k=1}^{M} \hat{s}_k[n] = \hat{w}[n] + \frac{1}{M} \sum_{n=1}^{M} \hat{r}'_k[n]. \tag{26}$$

This procedure will give increasingly better results as $M \to \infty$. Thus, it may be applied in a fully automatic scheme involving multichannel/mutiwindowing and, within the limits of spatial source stationarity, also multishot short-time wavelet estimation, since then the total number of elements averaged may be made quite large.

Time Stacking: Here, each individual cepstrum is low-time windowed, and mapped back to the time domain by $D_*^{-1}(\cdot)$. That is, we compute

$$s_k^L[n] = D_*^{-1}[\hat{s}_k[n] \cdot l[n]] \tag{27}$$

where $l[n]$ denotes a low-time window in the cepstral domain. Since, in general, the cepstrum of each impulse train, $\hat{r}'_k[n]$, will have different contributions in the low-time interval, it follows that

$$s_k^L[n] \simeq w[n] * r'^L_k[n] \tag{28}$$

where we defined $r'^L_k[n]$ to be

$$r'^L_k[n] = D_*^{-1}[\hat{r}'_k[n] \, l[n]]. \tag{29}$$

Each low-time component $r'^L_k[n]$ can be essentially modeled as

$$r'^L_k[n] = \delta[n - m_k] + \eta_k[n]. \tag{30}$$

That is, it consists of an impulse plus a noise component, $\eta_k[n]$. We have observed [11] that the cepstral sensitivity of a mixed-phase impulse train to time-domain amplitude changes is mapped, through the low-time filtering operation, into a similar sensitivity regarding the structure of the noise component $\eta_k[n]$. In other words, it seems reasonable to model the noise component at every instant n as a zero-mean random variable with unknown probability distribution. Thus, the wavelet estimate $w_e[n]$ is formed by stacking all the low-time components, after appropriate time-domain synchronization, so that

$$w_e[n] = \frac{1}{M} \sum_{k=1}^{M} s_k^L[n + m_k]$$

$$= w[n] * [\delta[n] + \frac{1}{M} \sum_{k=1}^{M} \eta_k[n + m_k]]. \tag{31}$$

This technique has been used in the short-time analysis of single-trace noisy synthetic seismograms and has yielded good results. Preliminary evaluation shows that by monitoring this technique the quality of the wavelet estimates is superior to that achieved by cepstral stacking for the same number of components M.

We illustrate now the use of short-time analysis and the stacking strategies discussed above.

Consider the short-time analysis of a synthetic seismogram, with impulse response as in Fig. 12(a), excited by an airgun source. Based on the choice of windows depicted in Fig. 12, the low-time components of the windowed reflector segments are depicted in Fig. 13, after appropriate time synchronization. The time-stacked component is depicted in Fig. 14(a). Fig. 14(b) depicts the cepstrally stacked low-time component.

The short-time analysis of the corresponding synthetic seismogram segments, using the windows of Fig. 12, yields the four low-time components of the windowed seismograms depicted in Fig. 15. The resulting time-stacked wavelet estimate is depicted in Fig. 16(b) and is seen to compare well with the original airgun wavelet shown in Fig. 16(a).

VI. The Role of Homomorphic Filtering in Reflection Seismic Data Analysis

Seismic analysis by homomorphic filtering is a powerful and promising technique in that the seismic wavelet may be recovered without any *a priori* assumptions regarding the structure of the wavelet or of the reflection series. In particular, the seismic wavelet is not assumed minimum phase as it happens with zero-lag predictive deconvolution, nor is the reflector series assumed uncorrelated as it happens with Wiener filtering

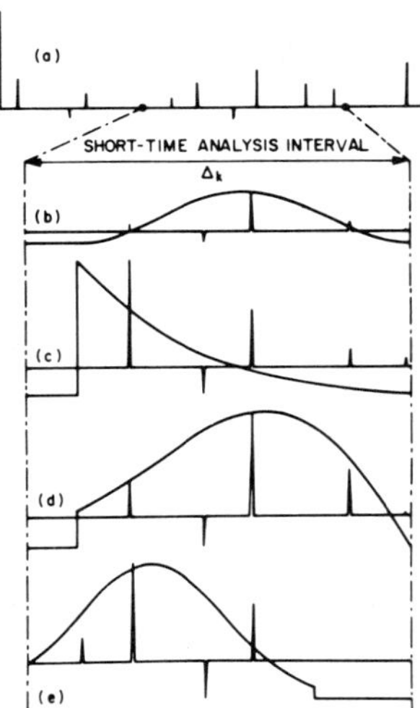

Fig. 12. Analysis of mixed-phase impulse trains. (a) Seismic reflector series $r[n]$. (b) Hamming window $v_1[n]$ and windowed reflector segment $r_1'[n]$. (c) Exponential window $v_2[n]$ and windowed reflector segment $r_2'[n]$. (d) Rayleigh window $v_3[n]$ and windowed reflector segment $r_3'[n]$. (e) Gaussian window $v_4[n]$ and windowed reflector segment $r_4'[n]$.

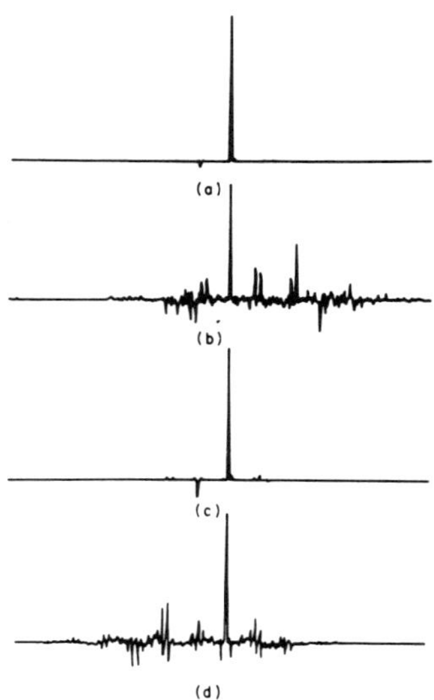

Fig. 13. Low-time homomorphic filtering of windowed reflector segments. (a) Low-time component of Hamming windowed segment $r_1'^L[n + m_1]$. (b) Low-time component of exponentially windowed segment $r_2'^L[n + m_2]$. (c) Low-time component of Rayleigh windowed segment $r_3'^L[n + m_3]$. (d) Low-time component of Gaussian windowed segment $r_4'^L[n + m_4]$.

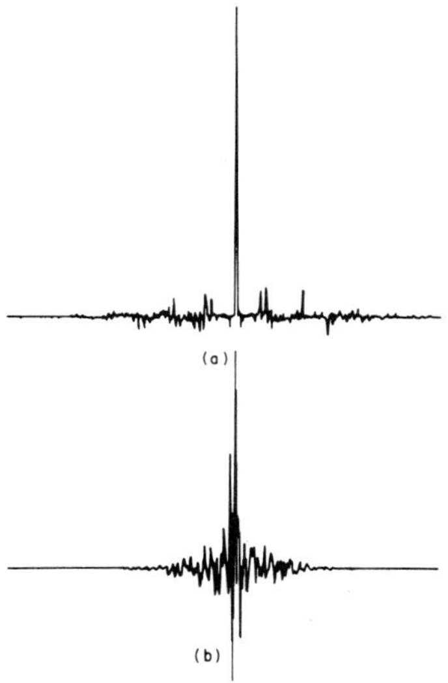

Fig. 14. Stacking of low-time components. (a) Time-stacked low-time component

$$\frac{1}{4} \sum_{k=1}^{4} r_k'^L[n + m_k].$$

(b) Cepstral-stacked low-time component

$$D_*^{-1}\left[\frac{1}{4} \sum_{k=1}^{4} \hat{r}_k'^L[n]\right].$$

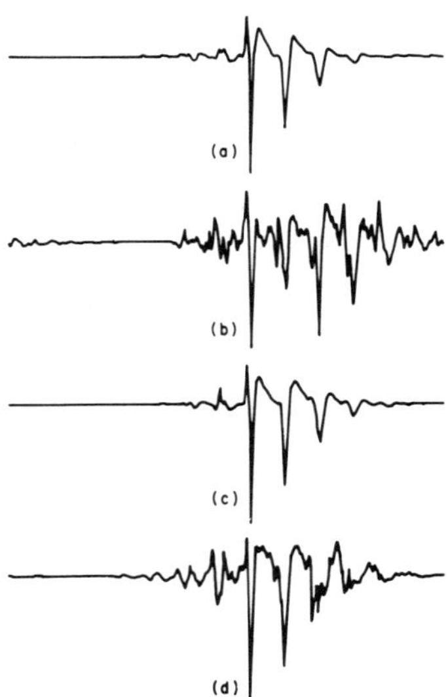

Fig. 15. Short-time analysis of a synthetic seismogram. (a) Low-time component of Hamming windowed section $s_1^L[n + m_1]$. (b) Low-time component of exponentially windowed section $s_2^L[n + m_2]$. (c) Low-time component of Rayleigh windowed section $s_3^L[n + m_3]$. (d) Low-time component of Gaussian windowed section $s_4^L[n + m_4]$.

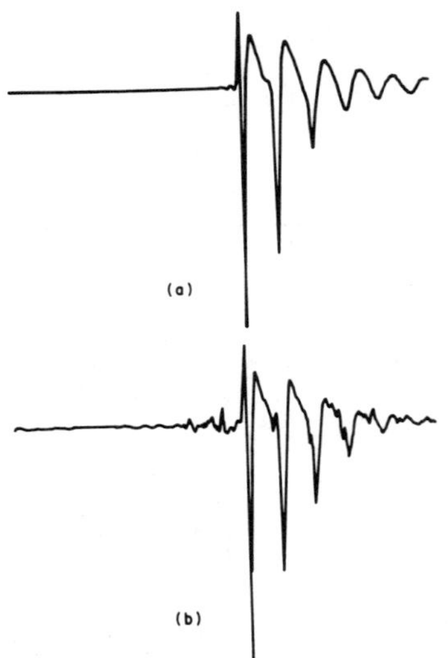

Fig. 16. Wavelet estimation of homomorphic filtering. (a) Original wavelet $w[n]$. (b) Recovered wavelet by time-stacking of low-time components.

$$w_e[n] = \frac{1}{4} \sum_{k=1}^{4} s_k^L[n + m_k].$$

techniques. The homomorphic signal analysis strategy must take into account both the signal models and the analysis goals. In particular, the bandpass nature of seismic data, that is, the fact that the signal's spectrum can be modeled as the product of two components only within a given passband, may be neatly handled by the class of homomorphic bandpass systems discussed in this paper. The use of homomorphic analysis based on time-invariant signal models may be appropriate in many instances, for example, in shallow-water marine dereverberation, and in teleseismic data analysis. Otherwise, the time-varying nature of the data as well as the presence of in-band noise may be handled by analyzing short-time windowed segments and recovering the seismic wavelet which best represents the seismic wavefront for a given suite of reflections. Both cepstral stacking or time stacking can be used for this purpose, either in an automatic scheme or in a monitoring mode.

The recovery of the reflector series within each short-time interval may be handled in various ways. For example, optimum lag Wiener spiking filters [15] may be designed from the seismic wavelet estimate. Alternatively, homomorphic predictive schemes [16], [17], based on the representation of the wavelet estimate by its minimum-phase and maximum-phase components, and by the synthesis of the deconvolution filter by parametric linear predictive modeling of these wavelet components, has proven to be a very versatile technique which often yields improved results relative to optimum lag Wiener filtering [11].

REFERENCES

[1] A. Oppenheim and R. Schafer, *Digital Signal Processing*. Englewood Cliffs, NJ: Prentice-Hall, 1975.
[2] T. Ulrych, "Application of homomorphic deconvolution to seismology," *Geophysics*, vol. 36, no. 4, pp. 650–660, 1971.
[3] T. Ulrych, O. G. Jensen, R. M. Ellis, and P. G. Sommerville, "Homomorphic deconvolution of some teleseismic events," *Bull. Seismological Soc. Amer.*, vol. 62, no. 5, pp. 1253–1265, 1972.
[4] R. W. Clayton, and R. A. Wiggins, "Source shape estimation and deconvolution of teleseismic bodywaves," *Geophys. J. Roy. Astron. Soc.*, vol. 4-F, pp. 151–177, 1977.
[5] P. G. Sommerville, "Time domain determination of earthquake fault parameters from short-period *P*-waves," Ph.D. dissertation, Dept. of Geophysics and Astronomy, The Univ. of British Columbia, 1975.
[6] P. L. Stoffa, P. Buhl, and G. M. Bryan, "The application of homomorphic deconvolution to shallow-water marine seismology-Part I: Models and Part II: Real data," *Geophysics*, vol. 39, no. 4, pp. 401–426, 1974.
[7] B. Buttkus, "Homomorphic filtering, theory and practice," *Geophysical Prospecting*, vol. 23, no. 4, p. 712, 1975.
[8] J. B. Gallemore, "A comparative evaluation of two acoustic signal dereverberation techniques," S.M. and E.E. thesis, Dept. of Elec. Eng. and Comp. Sci., M.I.T., Cambridge, MA, 1976.
[9] R. M. Otis and R. B. Smith, "Homomorphic deconvolution by log spectral averaging," *Geophysics*, 1978, to be published.
[10] T. G. Stockham, Jr., T. M. Cannon, and R. B. Ingebretsen, "Blind deconvolution through digital signal processing," *Proc. IEEE*, vol. 63, pp. 678–692, 1975.
[11] J. M. Tribolet, *Seismic Applications of Homomorphic Signal Processing*. Englewood Cliffs, NJ: Prentice-Hall, 1978.
[12] R. W. Clayton and T. J. Ulrych, "A restoration method for impulsive functions," submitted to *IEEE Trans. Inform. Theory*.
[13] J. M. Tribolet, "A new phase unwrapping algorithm," *IEEE Trans. Acoust., Speech, Signal Processing*, vol. ASSP-25, pp. 170–179, Apr. 1977.
[14] J. M. Tribolet, T. Quatieri, and A. Oppenheim, "Short-time homomorphic analysis," 1977 IEEE Conf. on Acoust., Speech, and Signal Processing, Hartford, CT.
[15] S. Treitel and E. Robinson, "The design of high-resolution digital filters," *IEEE Trans. Geosci. Electron.*, vol. GE-4, no. 1, pp. 25–38, 1966.
[16] J. Tribolet, A. Oppenheim, and G. Kopec, "Deconvolution by homomorphic prediction," 45th Annual International Meeting of the Society of Exploration Geophysicists, 1975.
[17] A. V. Oppenheim, G. E. Kopec, and J. M. Tribolet, "Signal Analysis by Homomorphic Prediction," *IEEE Trans. Acoust., Speech, Signal Processing*, vol. ASSP-24, pp. 327–332, Aug. 1976.

Part IV

KALMAN FILTERING

Editors' Comments
on Papers 15, 16, and 17

Kalman, (1960) and Kalman and Bucy, (1961) developed a new approach to the theory of least-squares estimation. They changed the formulation of the problem by giving not the covariance of the signal process as is necessary in Wiener filtering but instead a model for the signal process as the output of a dynamical linear system driven by white noise. The modeling of the linear system in Kalman's formulation is based on a state space representation of linear time-varying systems (Meditch, 1969) via a first-order vector differential equation instead of a time-varying impulse response. This formalism results in specification of the optimum estimates as the solution to a differential equation whose coefficients are determined by the statistics of the processes involved. Kailath (1974) and Berkhout and Zaanen (1976) give a detailed comparison between Kalman filtering and other least-squares estimation techniques.

Bayless and Brigham (1970) and Crump (1974) discuss a Kalman filtering approach for seismic signal processing. The motivation for this approach is its potential for permitting more flexible modeling assumptions than the Wiener filtering approach. The Wiener filtering method is conventionally applied under the assumptions of a time-invariant system and stationary statistics. The Kalman filtering technique is suitable for handling time-varying models. Bayless and Brigham (1970) present a tutorial on Kalman filtering and illustrate analog computer solutions of Kalman inverse filters for enhancement and deconvolution of continuous signals. Crump (1974) illustrates the use of discrete Kalman filters for deconvolution with synthetic and field-data examples.

Ott and Meder (Paper 15) present an application of Kalman filters for prediction error filtering. They illustrate this with an example for a one-dimensional damped harmonic oscillator that is excited by impulses, at random time distances, of random intensity. From the observed oscillations, they estimate the exciting impulses. This is analogous to deconvolution of seismic traces. Mendel (Paper 16) describes a quantitative evaluation of Ott and Mender's prediction error filter and indicates the reasons for the poor performance observed by Ott and Meder.

Mendel and Kormylo (Paper 17) discuss the application of minimum-variance estimation theory to the problem of estimating the reflection coefficient sequence from a seismic trace. They observe that the problem is equivalent to one of estimating the random disturbance in a state equation and that the minimum-variance estimates of the reflection coefficient sequence are optimal smoothed estimates; an optimal filtered estimate of that sequence does not exist. They develop formulas for single-stage and multistage optimal smoothed estimates and for error variances associated with these estimates. They also present simulation results for synthetic data and compare their results with prediction error filtering and Wiener filtering. Similar results are also given by Mendel and Kormylo (1977) and Mendel (1977). Kormylo and Mendel (1980) also discuss a maximum likelihood deconvolution technique for time-invariant wavelet.

Sims and D'Mello (1978) discuss another Kalman filtering approach, an adaptive estimation and smoothing technique, to accomplish deconvolution of seismic signals under conditions of model uncertainty. McCormack, Verm, and Quay (1978) examine the application of Kalman filtering to deconvolution of nonstationary time series resulting from nonstationary source waveforms in seismic data due to the earth's attenuation. They specifically examine the response of Kalman deconvolution to various estimates of the earth's attenuation factor and the robustness of the deconvolution to errors in the estimate of the attenuation factor. They observe that a small error in the estimate of attenuation factor results in large scaling differences between the actual and the estimated reflection coefficients, whereas the basic character of the true reflection coefficients is preserved.

REFERENCES

Bayless, J. W., and E. O. Brigham, 1970, Application of Kalman Filter to Continuous Signal Restoration, *Geophysics* **35**:2–23.

Berkhout, A. J., and P. R. Zaanen, 1976, A Comparison between Wiener Filtering, Kalman Filtering, and Deterministic Least Squares Estimation, *Geophys. Prospect.* **24**:141–197.

Crump, N. D., 1974, A Kalman Filter Approach to the Deconvolution of Seismic Signals, *Geophysics* **39**:1–13.

Kailath, T., 1974, A View of Three Decades of Linear Filtering Theory, *IEEE Trans. Inf. Theory* **IT-20**:146–181.

Kalman, R. E., 1960, A New Approach to Linear Filtering and Prediction Problems, *ASME Trans. J. Basic Eng.* **82**:35–45.

Kalman, R. E., and R. S. Bucy, 1961, New Results in Linear Filtering and Prediction Theory, *ASME Trans. J. Basic Eng.* **83**:95–108.

Kormylo, J., and J. M. Mendel, 1980, Maximum-Likelihood Seismic Deconvolution, paper presented at the 50th Annual SEG Meeting, Houston.

McCormack, M. D., R. W. Verm, and R. G. Quay, 1978, Deconvolution of Nonstationary Time Series Using a Kalman Filter Approach, paper presented at the 48th SEG Meeting, San Francisco.

Meditch, J. S., 1969, *Stochastic Optimal Linear Estimation and Control,* McGraw-Hill, New York.

Mendel, J. M., 1977, White-Noise Estimators for Siesmic Data Processing in Oil Exploration, *IEEE Trans. Automat. Control* **AC-22**:694–706.

Mendel, J. M., and J. Kormylo, 1977, New Fast Optimal White-Noise Estimators for Deconvolution, *IEEE Trans. Geosci. Electron.* **GE-15**:32–41.

Sims, C. S., and M. R. D'Mello, 1978, Adaptive Deconvolution of Seismic Signals, *IEEE Trans. Geosci. Electron.* **GE-26**:99–103.

15

Reprinted from *Geophys. Prospect.* **20**:549–560 (1972)

THE KALMAN FILTER AS A PREDICTION ERROR FILTER*

BY

N. OTT AND H. G. MEDER**

ABSTRACT

OTT, N. and H. G. MEDER, 1972, The Kalman Filter as a Prediction Error Filter, Geophysical Prospecting 20, 549-560.

In mathematical statistical filtering the deconvolution problem can be solved by two different methods:

1. by inverse filtering,
2. by calculating the prediction error.

Both methods are well known in the theory of WIENER filters.

If, however, the generating process of the signal is known and can be described by a set of linear first order differential equations, then the KALMAN filter can also be used to solve the deconvolution problem. In the case of the inverse filtering method this was shown by BAYLESS and BRIGHAM (1970). But, while their method can only be used if the original signal is a colored random process, this paper shows that in the case of a white process the prediction error filtering method is a more appropriate approach. The method is extremely efficient and simple. This can be demonstrated by an example which may be of special interest for seismic exploration.

1. INTRODUCTION

In seismic exploration deconvolution is a well known technique to remove the "filtering" effect of a medium through which the desired signal must pass. Up to now this medium was considered as a linear dynamical system, and it was described by the classical concept of the impulse response. This concept is the basis of the so-called Wiener filters, which, in their digital forms, are successfully used to solve the deconvolution problem in seismics and other areas (Meder 1969, 1971; Robinson 1967).

However, the concept of describing linear dynamical systems by the impulse response is not the only one possible. In many applications much more a priori knowledge of the internal structure of the considered linear system can be given. In these cases the description of the linear dynamical system by a set of linear first order differential equations is more suitable, because the impulse

---

* Paper read at the Thirty Third Meeting of the European Association of Exploration Geophysicists, Hanover, June 1971.

** IBM Heidelberg Science Center, Heidelberg, German Federal Republic.

response, as a measure of the external system behavior, cannot account for this internal structure. For many purposes, as for example for the examination of certain stability and delay properties, the knowledge of the internal system behavior is of great importance. The main advantage of this description, however, lies in the fact that time varying and even instable systems can just as easily be handled as time invariant and stable systems. This is the reason why Kalman (1960) succeeded in formulating the decisive breakthrough using his filter equations.

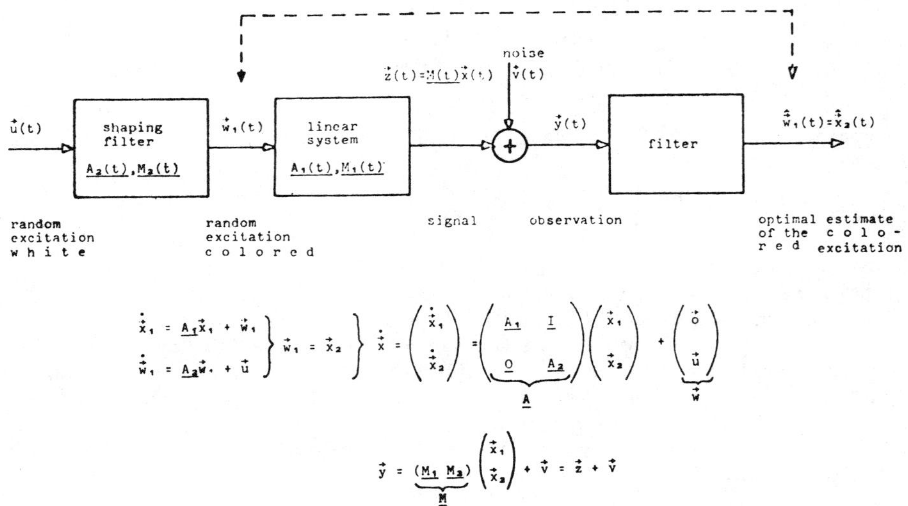

$$\left.\begin{aligned} \dot{\vec{x}}_1 &= A_1\vec{x}_1 + \vec{w}_1 \\ \dot{\vec{w}}_1 &= A_2\vec{w}_1 + \vec{u} \end{aligned}\right\} \quad \left.\vec{w}_1 = \vec{x}_2\right\} \quad \dot{\vec{x}} = \begin{pmatrix} \dot{\vec{x}}_1 \\ \dot{\vec{x}}_2 \end{pmatrix} = \underbrace{\begin{pmatrix} A_1 & I \\ 0 & A_2 \end{pmatrix}}_{A} \begin{pmatrix} \vec{x}_1 \\ \vec{x}_2 \end{pmatrix} + \underbrace{\begin{pmatrix} 0 \\ \vec{u} \end{pmatrix}}_{\vec{w}}$$

$$\vec{y} = \underbrace{(M_1\ M_2)}_{M} \begin{pmatrix} \vec{x}_1 \\ \vec{x}_2 \end{pmatrix} + \vec{v} = \vec{z} + \vec{v}$$

Fig. 1. The KALMAN Filter as an Inverse Filter (Method of BAYLESS and BRIGHAM) (In this method the shaping filter is necessary).

Up to this time the Kalman filter has been used almost exclusively in inertial navigation. The time seems to have come that also in the field of seismic exploration the systems to be considered are no longer "black boxes". The first attempt to solve the deconvolution problem in seismics with the aid of the Kalman filter was made by Bayless and Brigham (1970). They used the well known fact that a *colored* input (excitation) of a linear dynamical system can itself be considered as the output of another linear dynamical system (shaping filter), and, as such, is automatically estimated by the filter as a state vector variable. But deconvolution, as the inverse process of the convolution operation of a linear system, is nothing other than the determination of the input excitation, given the output of the linear system. Thus, inverse filtering with the aid of the Kalman filter can only be done if the random input excitation is a *colored* process. That is the main result pointed out by Bayless and Brigham in their work (see figure 1).

In the theory of Wiener filters it can be proved (Meder 1969; Robinson 1967) that the deconvolution problem can also be solved by calculating the prediction error. In this paper it will be shown that this calculation can also be done with the aid of the Kalman filter. The only additional work to be done is the subtraction of two filter outputs.

It is a result of the white input excitation condition of the Kalman filter that our method to solve the deconvolution problem can only be used if the input excitation has indeed the *white* property. In seismic exploration the reflection coefficients of the various horizons can be assumed to have this property (Bayless and Brigham 1970). Furthermore, since the occurrence times of the impulses are Poisson-distributed, our method seems to be especially appropriate in this application.

In every case this presentation is a supplement to the inverse filtering method by Bayless and Brigham (1970). The principle of functioning and the difference between both methods can be seen in figure 2.

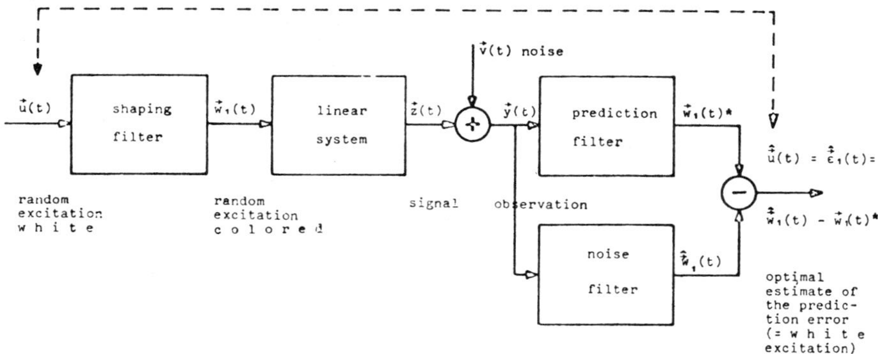

Fig. 2. The KALMAN Filter as a Prediction Error Filter.
If the random excitation $\mathbf{w}_1(t)$ has the white property, the shaping filter is not necessary. In this case the prediction error filter yields: $\hat{\mathbf{w}}_1(t) = \hat{\mathbf{x}}_1(t) — \mathbf{x}_1(t)*$. See figure 1.

2. DESCRIPTION OF THE KALMAN FILTER AND EXTENSION TO THE PREDICTION ERROR FILTER

Since the basic work of Kalman (1960), the Kalman filter equations have been derived and discussed by many authors (Richman and Thau 1966; Sorenson 1966). For this reason a derivation will be excluded from this paper; only the basic relations and the filter equations resulting from these are given, the latter in a form which was derived by one of the authors in another paper (Ott 1970). For our purposes this form is especially appropriate.

In the following, bold capital letters represent matrices, and bold lower case letters column vectors.

The mathematical model of the random signal to be filtered is based on the notion of the state vector, and it is represented by the following vector valued linear first order differential equation:

$$\dot{\mathbf{x}}(t) = \mathbf{A}(t)\,\mathbf{x}(t) + \mathbf{w}(t). \tag{1}$$

This equation is an n-dimensional stochastic differential equation, for the excitation $\mathbf{w}(t)$ represents a stochastic, white random process with a Gaussian distribution function (for a linear filter it is actually unimportant whether $\mathbf{w}(t)$ has a normal distribution or not. But here no details referring to this subject are given; these can be found in the special literature quoted). In contrast to $\mathbf{w}(t)$ the signal $\mathbf{x}(t)$ (state vector) is no longer white, but colored. It has the so-called Markov-property. This results since the linear dynamical system, which is represented by the $n \times n$-dimensional coefficient matrix $\mathbf{A}(t)$ and which generates the signal $\mathbf{x}(t)$ from $\mathbf{w}(t)$, has only a limited band width. For the use of the Kalman filter it is of decisive importance that the matix $\mathbf{A}(t)$ be known. In contrast to other applications—as e.g. inertial navigation, where the dynamical equations of the space- or aircraft are known—in seismic signal exploration the condition that the matrix $\mathbf{A}(t)$ be known cannot be assumed. The problem of determining the pulse shape, and therewith the matrix $\mathbf{A}(t)$, is a modeling and identification problem, which is beyond the scope of this paper.

The Kalman filter is based on current observations of the signal $\mathbf{y}(t)$, which is contaminated by additional noise $\mathbf{v}(t)$. Therefore, what can be observed and fed to the filter as input information is described by the following m-dimensional vector $\mathbf{y}(t)$ (see also figure 1 and 2):

$$\mathbf{y}(t) = \mathbf{M}(t)\,\mathbf{x}(t) + \mathbf{v}(t). \tag{2}$$

Again the $m \times n$-dimensional matrix $\mathbf{M}(t)$ (the so-called measurement or mapping matrix) must be known from the system.

Just as for $\mathbf{w}(t)$, the normal distribution and the property of "being white" is likewise supposed for the superimposed noise $\mathbf{v}(t)$. As already mentioned in the introduction, the supposition of the temporal independence does not mean any sort of restriction, for it is possible in the case of colored noise to interpret this as the output of another linear dynamical system and, extending the dimension of the state vector, to include this secondary system (shaping filter) in the whole system. This is valid for both $\mathbf{w}(t)$ and $\mathbf{v}(t)$, but in the case of $\mathbf{v}(t)$ certain modifications of the equations are necessary (Sorenson and Stubberud 1970, see also figure 1 and 2). For both these random vectors the statistical properties up to the second order (i.e. the mean values and variances) must be known. These expectations are assumed to be as follows:

$$E[\mathbf{w}(t)] = 0, \tag{3a}$$

$$E[\mathbf{v}(t)] = 0, \tag{3b}$$

$$E[\mathbf{w}(t)\,\mathbf{v}(t)^T] = \mathbf{O}, \tag{3c}$$

$$E[\mathbf{w}(t)\,\mathbf{w}(t+\tau)^T] = \mathbf{Q}\,\delta(\tau);\ \mathbf{Q}\ \text{diagonal}, \tag{3d}$$

$$E[\mathbf{v}(t)\,\mathbf{v}(t+\tau)^T] = \mathbf{R}\,\delta(\tau);\ \mathbf{R}\ \text{diagonal}. \tag{3e}$$

Equations (1) and (2) represent continuous processes. To make discrete sequences from these one proceeds by first solving equation (1) with the aid of the so-called transition matrix $\varphi(t, t_0)$ and then by directly transforming the solution into a difference equation. Together with equation (2) the following recursive relationships result:

$$\mathbf{x}_K = \varphi_{K/K-1}\,\mathbf{x}_{K-1} + \mathbf{h}_{K-1} \tag{4}$$

$$\mathbf{y}_K = \mathbf{M}_K\,\mathbf{x}_K + \mathbf{v}_K \tag{5}$$

where the subscript K indicates the discrete time point $t = t_K$ within a time interval subdivided into parts of (not necessarily) equal length and where

$$\mathbf{h}_{K-1} = \int_{t_{K-1}}^{t_K} \varphi(t_K, \tau)\,\mathbf{w}(\tau)\,d\tau. \tag{6}$$

Between the transition matrix $\varphi(t, t_0)$ and the coefficient matrix $\mathbf{A}(t)$ the following relationship exists:

$$\frac{d}{dt}\,\varphi(t, t_0) = \mathbf{A}(t)\,\varphi(t, t_0) \tag{7}$$

This differential equation can only be solved in closed form if the system which generates the signal $\mathbf{x}(t)$ from $\mathbf{w}(t)$ is stationary, i.e. if $\mathbf{A} = const.$

In this case the result is

$$\varphi(t, t_0) \rightarrow \varphi(t - t_0) = e^{\mathbf{A}(t-t_0)}$$
$$= \mathbf{I} + \mathbf{A}\,(t - t_0) + \mathbf{A}^2\,\frac{(t-t_0)^2}{2!} + \dots \tag{8}$$

where $\mathbf{I}$ represents the $n \times n$—dimensional identity matrix.

These are the basic relationships from which the derivation of the Kalman filter equations starts. As criterion for "optimal" the least mean square error of the estimation $\hat{\mathbf{x}}_K$ is chosen, just as with the Wiener filters, namely:

$$E[\mathbf{x}_K - \hat{\mathbf{x}}_K)^T\,(\mathbf{x}_K - \hat{\mathbf{x}}_K)] = Minimum. \tag{9}$$

Under this condition the following set of filter equations can be derived:

$$\mathbf{P}_K^* = E[(\mathbf{x}_K - \mathbf{x}_K^*)(\mathbf{x}_K - \mathbf{x}_K^*)^T] = \varphi_{K/K-1}\,\mathbf{P}_{K-1}\,\varphi_{K/K-1}^T + \mathbf{H}_{K-1}, \tag{10a}$$

$$\mathbf{x}_K^* = \varphi_{K/K-1}\,\hat{\mathbf{x}}_{K-1}, \tag{10b}$$

$$\mathbf{B}_K = \mathbf{P}_K^*\,\mathbf{M}_K^T[\mathbf{M}_K\,\mathbf{P}_K^*\,\mathbf{M}_K^T + \mathbf{R}_K]^{-1}, \tag{11a}$$

$$\hat{\mathbf{x}}_K = \mathbf{x}_K^* + \mathbf{B}_K(\mathbf{y}_K - \mathbf{M}_K\,\mathbf{x}_K^*), \tag{11b}$$

$$\mathbf{P}_K = E[(\mathbf{x}_K - \hat{\mathbf{x}}_K)(\mathbf{x}_K - \hat{\mathbf{x}}_K)^T] = (\mathbf{I} - \mathbf{B}_K\,\mathbf{M}_K)\,\mathbf{P}_K^*, \tag{11c}$$

where

$$\mathbf{H}_{K-1} = E[\mathbf{h}_{K-1}\,\mathbf{h}_{K-1}^T] \simeq \frac{t_K - t_{K-1}}{2}\,(\varphi_{K/K-1}\mathbf{Q}_{K-1}\varphi_{K/K-1}^T + \mathbf{Q}_K). \tag{12}$$

The separation of the filter equations into the sets (10) and (11) indicates that the quantities with asterisks are prediction values for the time point t_K, based upon the observation value $\mathbf{y}_{K-1}$ at time t_{K-1}. However, as soon as the new observation value $\mathbf{y}_K$ at time t_K is available, these prediction values can be updated. The updated quantities, together with the weighting matrix $\mathbf{B}_K$ for obtaining the optimal correction, are summarized in the set (11). (At this point it is unimportant that the matrices $\mathbf{P}_K^*$, $\mathbf{B}_K$ and $\mathbf{P}_K$ are not immediately effected by the correction. Since they do not depend on the observation values $\mathbf{y}_K$ one could calculate and store them even before the measurements are actually made.) $\mathbf{P}_K^*$ and $\mathbf{P}_K$ are the covariance matrices of the respective estimation errors. They are needed since the weighting matrix $\mathbf{B}_K$ depends on them.

We see that the Kalman filter always works so that it first predicts the system state (signal), and then, as soon as the new measurement value is available, updates this prediction.

From understanding this working process it is a very small and simple step to present the Kalman filter as a prediction error filter. All that must be done is the following:

We assume that the superimposed noise $\mathbf{v}_K$ is equal to zero, and that the system under consideration is completely observable and completely controllable. In this case the observation values $\mathbf{y}_K$ are exactly correct and proportional to the signal $\mathbf{x}_K$. But this means that the estimation value $\hat{\mathbf{x}}_K$ passes over to the exact value $\mathbf{x}_K$, and therefore the prediction error $\mathbf{e}_K^*$ can be given as follows:

$$\mathbf{e}_K^* = \mathbf{x}_K - \mathbf{x}_K^* = \hat{\mathbf{x}}_K - \mathbf{x}_K^* \qquad \text{(in the noiseless case, i.e. } \mathbf{v}_K = \mathbf{o}). \tag{13}$$

In the noiseless case the Kalman filter also provides one with the prediction error filter. All that is required in addition is to carry out the subtraction of equation (13).

This fact can now be used in an ideal manner to calculate the excitation of the system $\mathbf{h}_{K-1}$ and $\mathbf{w}(t)$ respectively from the prediction error $\mathbf{e}_K^*$, i.e. to carry out the deconvolution. From the relationship (4), with the aid of the equations (10b) and (13), the following result can be given:

$$\begin{aligned}
\mathbf{h}_{K-1} &= \mathbf{x}_K - \varphi_{K/K-1}\,\mathbf{x}_{K-1} \\
&= \hat{\mathbf{x}}_K - \mathbf{x}_K^* = \mathbf{e}_K^*
\end{aligned}$$

(in the noiseless case, i.e. $\mathbf{v}_K = \mathbf{0}$). (14)

In the noiseless case the random excitation can be exactly determined simply by forming the difference between the two filter outputs. But it is obvious that also in the case of additional measurement noise the difference between these filter outputs can yield good results, for these outputs are the best in the case of noise as well. Of course in the noise case the random excitation cannot exactly be determined; only an optimal estimate of it can be given. This estimate results in:

$$\hat{\mathbf{h}}_{K-1} = \hat{\mathbf{x}}_K - \mathbf{x}_K^*$$

(in the case of additional noise, i.e. $\mathbf{v}_K \neq \mathbf{0}$). (15)

It is a consequence of its method of operation that the prediction error filter is very sensitive to additional noise. This fact is valid also for all other deconvolution filters. However, if the noise level is not too high, very impressive results can be obtained. This can be demonstrated with the following example.

3. EXAMPLE

Let us consider a one-dimensional damped harmonic oscillator which is excited by impulses of random intensity (within certain limits) in random time distances. From the observed oscillations we then try to determine the exciting impulses.

This example is of special interest for seismic exploration since one can approximately identify the exciting impulses with the reflection horizons, and the oscillator movements with the seismic traces. This example has also been used by Bayless and Brigham (1970) in their publication on the inverse Kalman filter, but unlike us they needed a shaping filter to "color" their impulses (see figure 1 and 2).

The impulse response of the oscillator and the exciting impulses may obey the following relationships:

$$g(t) = e^{-dt} \sin \omega_E t, \quad t \geq 0 \tag{16}$$

$$w(t) = \sum_{\nu=1}^{N} a_\nu\, \delta(t - t_\nu), \tag{17}$$

where

d = damping factor of the oscillator
ω_E = "Eigen"-angular frequency of the oscillator
a_ν = amplitudes of the exciting impulses (random numbers)
t_ν = occurring times of the exciting impulses (random numbers)
N = total number of the exciting impulses in the considered time interval (random number)

By convolution of these two relationships one can obtain the random oscillations $x(t)$ of the damped harmonic oscillator. But these oscillations can also be described by a linear second order differential equation (see the appendix):

$$\ddot{x}(t) + 2d \cdot \dot{x}(t) + (d^2 + \omega_E^2)\, x(t) = \omega_E w(t). \tag{18}$$

To bring this relationship into the form of equation (1) one substitutes

$$x = x_1, \tag{19a}$$

$$\dot{x} = x_2, \tag{19b}$$

and obtains the following result:

$$\dot{\mathbf{x}} = \begin{pmatrix} \dot{x}_1 \\ \dot{x}_2 \end{pmatrix} = \underbrace{\begin{pmatrix} 0 & 1 \\ -\beta & -\alpha \end{pmatrix}}_{\mathbf{A}\,=\,const} \begin{pmatrix} x_1 \\ x_2 \end{pmatrix} + \omega_E \underbrace{\begin{pmatrix} 0 \\ w \end{pmatrix}}_{\mathbf{w}(t)}, \tag{20}$$

where

$$\alpha = 2d \tag{21a}$$

$$\beta = d^2 + \omega_E^2. \tag{21b}$$

It is important to observe that equation (20) is not a unique representation of equation (18). This can easily be seen by changing the elements A_{11} and A_{22} of the matrix $\mathbf{A}$ in equation (20) and combining the resulting two equations.

The observed or measured variable is $x = x_1$, therefore:

$$y = \mathbf{M}\mathbf{x} + v = \underbrace{(1 \qquad 0)}_{\mathbf{M}\,=\,const} \begin{pmatrix} x_1 \\ x_2 \end{pmatrix} + v. \tag{22}$$

By transforming the equations (20) and (22) into their corresponding discrete forms one obtains

$$\mathbf{x}_K = \varphi_{K/K-1}\, \mathbf{x}_{K-1} + \mathbf{h}_{K-1}, \tag{23}$$

$$y_K = \mathbf{M}\, \mathbf{x}_K + v_K, \tag{24}$$

where

$$\varphi_{K/K-1} = \varphi = \begin{pmatrix} 1 - \beta\, \dfrac{T_K^2}{2} + \ldots & T_K - \alpha\, \dfrac{T_K^2}{2} + \ldots \\[2ex] -\beta T_K + \alpha\beta\, \dfrac{T_K^2}{2} + \ldots\; ; & 1 - \alpha T_K + (\alpha^2 - \beta)\, \dfrac{T_K^2}{2} + \ldots \end{pmatrix}$$

$$T_K = t_K - t_{K-1} = const \tag{25}$$

and

$$\mathbf{h}_{K-1} = \omega_E \int\limits_{t_{K-1}}^{t_K} \varphi(t_K, \tau)\, \mathbf{w}(\tau)\, d\tau. \tag{26}$$

In pursuance of the relationships (14) and (15) respectively the vector $\mathbf{h}_{K-1}$ can be estimated in an optimal sense. If one wishes to obtain from this estimate the amplitudes a_ν of the impulses, one must observe that in the interval $T_K = t_K - t_{K-1}$, at most one impulse can exist since the Kalman cycle T_K was chosen to be equal to the sampling rate of the original continuous functions. Therefore, within one Kalman cycle the exciting random function can be given as:

$$w(\tau)\; \Big|_{t_{K-1}}^{t_K} = a_{K-1}\, \delta(\tau - t_{K-1}) \qquad \text{if an impulse falls into the cycle } t_K - t_{K-1},$$

$$\text{and} \qquad = 0 \qquad\qquad \text{otherwise.} \tag{27}$$

Substituting this relationship into (26) (with slightly modified limits of the integral), one obtains with (20)

$$\mathbf{h}_{K-1} = \omega_E \int\limits_{t_{K-1}^-}^{t_K^-} \begin{pmatrix} \varphi_{12}(t_K^-, \tau) \\[2ex] \varphi_{22}(t_K^-, \tau) \end{pmatrix} w(\tau)\, d\tau$$

$$= \omega_E \cdot a_{K-1} \cdot \begin{pmatrix} \varphi_{12} \\[1ex] \varphi_{22} \end{pmatrix} \qquad \text{if an impulse falls into the cycle } t_K - t_{K-1},$$

$$\text{and} \qquad = 0 \qquad\qquad \text{otherwise.} \tag{28}$$

With this relation the following final result can be given from equation (15):

$$\hat{a}_{K-1} = \frac{\hat{h}_{1\,K-1}}{\omega_E \cdot \varphi_{12}} \qquad \text{if an impulse falls into the cycle } t_K - t_{K-1},$$

$$\text{and} \qquad = 0 \qquad\qquad \text{otherwise,} \tag{29a}$$

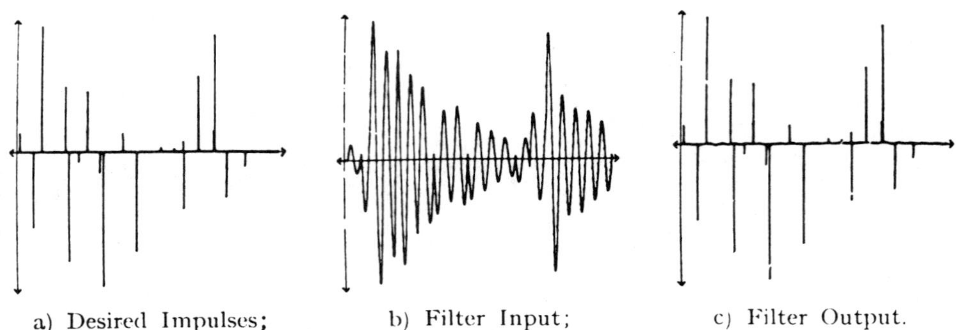

a) Desired Impulses;   b) Filter Input;   c) Filter Output.

Fig. 3. Simulation Result of the Noiseless Case.

The parameters of the one-dimensional damped harmonic oscillator impulse response function were chosen to be: $d = 5$ sec^{-1}, $\omega_E = 2\pi.20$ sec^{-1}.

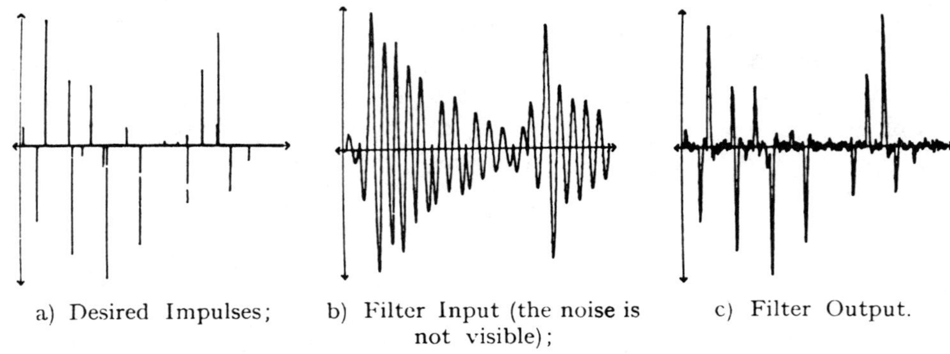

a) Desired Impulses;   b) Filter Input (the noise is not visible);   c) Filter Output.

Fig. 4. Simulation Result of the Case with a Noise Level of 1 % (i.e. $|v|_{max} = 0.01 \cdot |w|_{max}$).

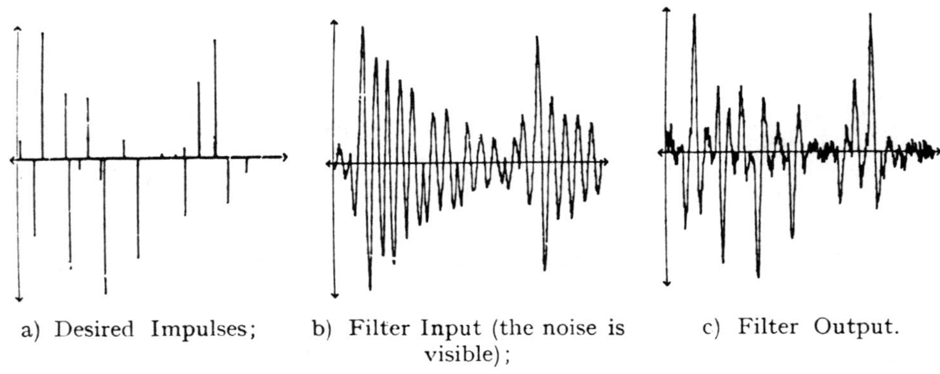

a) Desired Impulses;   b) Filter Input (the noise is visible);   c) Filter Output.

Fig. 5. Simulation Result of the Case with a Noise Level of 5 % (i.e. $|v|_{max} = 0.05 \cdot |w|_{max}$)

270

or

$$\hat{a}_{K-1} = \frac{\hat{h}_{2\,K-1}}{\omega_E \cdot \varphi_{22}} \qquad \text{if an impulse falls into the cycle } t_K - t_{K-1},$$

and $\qquad = 0 \qquad$ otherwise. $\qquad$ (29b)

In this example the noise covariance matrices are given by:

$$Q = \begin{pmatrix} 0 & 0 \\ 0 & \omega_E^2\, \sigma_w^2 \end{pmatrix} \tag{30a}$$

$$R = R = \sigma_v^2 \tag{30b}$$

The simulation results of this example were displayed on the IBM 2250 display-unit and then photographed. At first the noiseless case was displayed, following an example with a maximum noise amplitude of 1% of the maximum value of the exciting function w, and finally an example with a noise level of 5%. The figures 3, 4 and 5 show the results.

APPENDIX

To derive the linear, second order differential equation of the one-dimensional damped harmonic oscillator $x(t)$, one proceeds as follows:

The Fourier-transform of the impulse response function $g(t)$ of the oscillator, equation (16), is given by:

$$G(i\omega) = \mathsf{F}\{g(t)\} = \mathsf{F}\{e^{-d\cdot t} \sin \omega_E t\} \tag{A. 1}$$

$$= \frac{\omega_E}{(i\omega + d)^2 + \omega_E^2} \,.$$

Multiplying both sides of this equation by the denominator of the right-hand side, one obtains:

$$(i\omega)^2 G(i\omega) + 2d \cdot i\omega G(i\omega) + (d^2 + \omega_E^2) G(i\omega) = \omega_E \tag{A. 2}$$

By transforming this expression into the time domain the following equation results (the derivatives of $g(t)$ with respect to time vanish for $t \to \infty$!):

$$\ddot{g}(t) + 2d \cdot \dot{g}(t) + (d + \omega_E^2)\, g(t) = \omega_E\, \delta(t) \tag{A. 3}$$

This relationship describes the time behavior of the impulse response $g(t)$, if the oscillator is excited by the Dirac-δ-function. However, if the damped harmonic oscillator is excited by the random spike function $w(t)$ of equation (17), the response of the system $g(t)$ passes over to the random oscillations $x(t)$. Thus, equation (18) is obtained by a simple substitution of $g(t)$ by $x(t)$ and $\delta(t)$ by $w(t)$ in equation (A. 3).

REFERENCES

BAYLESS, J. W., BRIGHAM, E. O., 1970, Application of the Kalman Filter to Continuous Signal Restoration, Geophysics 35, 2-23.

KALMAN, R. E., 1960, A New Approach to Linear Filtering and Prediction Problems, Transactions of ASME 82D, 35-45.

MEDER, H. G., 1969, Digitale Filter und ihre Anwendungen, IBM-Nachrichten 197, 843-849.

MEDER, H. G., 1971, Digitale Filter, erläutert an geophysikalischen und lagerstättenkundlichen Problemen, Erdöl und Kohle 24, 210-216.

OTT, N., 1970, The KALMAN Filter, IBM Scientific Center Technical Report. No. 70.11.005.

RICHMAN, J., THAU, F., 1966, Elements of the Kalman Filtering Technique, Part I, Theory, Technical News Bulletin of General Precision 9, 12-18.

ROBINSON, E. A., 1967, Multichannel Time Series Analysis with Digital Computer Programs, Holden Day, San Francisco, Cambridge, London, Amsterdam.

SORENSON, H. W., 1966, Kalman Filtering Techniques, *in* "Advances in Control Systems, Theory and Applications", Vol. 3, edited by C. T. Leondes, Academic Press, New York.

SORENSON, H. W., STUBBERUD, A. R., 1970, Linear Estimation Theory, *in* "Theory and Applications of Kalman Filtering", edited by C. T. Leondes, North Atlantic Treaty Organization Advisory Group for Aerospace Research and Development (AGARD).

16

A QUANTITATIVE EVALUATION OF OTT AND MEDER'S PREDICTION ERROR FILTER*

BY

J. M. MENDEL**

ABSTRACT

MENDEL, J. M., 1977, A Quantitative Evaluation of Ott and Meder's Prediction Error Filter, Geophysical Prospecting 25, 692-698.

Ott and Meder's prediction error filter can be rederived so that it correctly handles input noise vectors which are of smaller dimension than the state vector. The poor performance obtained by Ott and Meder for their example can be explained by means of the error covariance matrix for the prediction error filter.

1. INTRODUCTION

Ott and Meder (1972) present a prediction error filter (PEF) which uses a Kalman filter and associated state space representation as its basic ingredients. Some numerical results are presented for a one dimensional damped harmonic oscillator which is excited by impulses of random intensity in random time instances. They demonstrate via simulations that poor performance of the PEF occurs for very modest ratios of | maximum measurement noise | / | maximum input noise | —say, 5%; however, no quantitative reasons are given for this poor performance. In addition, Ott and Meder derive their PEF for a state equation in which the dimension of the input noise vector (their $\mathbf{h}_{k-1}$) is exactly the same as the dimension of their state vector. They then proceed to apply this result to the seismic deconvolution problem in which the white reflection coefficient sequence plays the role of the input noise; but in this case the input noise is a scalar, and they do not handle the reduced dimension of the input noise vector correctly.

The purpose of this paper is to: (1) rederive the prediction error filter so that it correctly handles input noise vectors which are of smaller dimension than the state vector; (2) establish a quantitative measure of performance for the prediction error filter; and (3) explain the poor performance obtained by Ott and Meder for their example.

---

* Received May 1976.

** Department of Electrical Engineering, University of Southern California, Los Angeles, California 90007.

2. Derivation of PEF

We direct our attention at the linear, discrete-time, time-varying system

$$\mathbf{x}(k) = \boldsymbol{\Phi}(k, k-1)\, \mathbf{x}(k-1) + \boldsymbol{\Gamma}(k, k-1)\, \mathbf{w}(k-1) \tag{1}$$

$$\mathbf{z}(k) = \mathbf{H}(k)\, \mathbf{x}(k) + \mathbf{v}(k) \tag{2}$$

where $\mathbf{x} \in R^n$, $\mathbf{w} \in R^q$, $\mathbf{z} \in R^m$, $\mathbf{v} \in R^m$, $\boldsymbol{\Phi}(k, k-1) \in R^{n \times n}$, $\boldsymbol{\Gamma}(k, k-1) \in R^{n \times q}$, $\mathbf{H}(k) \in R^{m \times n}$, k is short for time t_k, and $\mathbf{w}(k)$ and $\mathbf{v}(k)$ are zero mean uncorrelated white noise processes,

$$E\{\mathbf{w}(k)\} = \mathbf{0} \tag{3a}$$

$$E\{\mathbf{v}(k)\} = \mathbf{0} \tag{3b}$$

$$E\{\mathbf{w}(i)\mathbf{w}^T(j)\} = \mathbf{Q}(i)\, \delta_{ij} \tag{3c}$$

$$E\{\mathbf{v}(i)\mathbf{v}^T(j)\} = \mathbf{R}(i)\delta_{ij} \tag{3d}$$

$$E\{\mathbf{w}(i)\mathbf{v}^T(j)\} = \mathbf{0} \tag{3e}$$

matrices $\mathbf{Q}$ and $\mathbf{R}$ do not, as was assumed by Ott and Meder, have to be diagonal matrices.

Our notation in (1) and (2) differs from the notation used in Ott and Meder (1972) to be more consistent with the notation used in the vast literature on linear systems and estimation theory (Meditch 1969).

The Kalman filter equations for the system in eqs. (1) and (2) are (Meditch 1969):

$$\hat{\mathbf{x}}(k|k-1) = \boldsymbol{\Phi}(k, k-1)\, \mathbf{x}(k-1|k-1), \tag{4}$$

$$\mathbf{P}(k|k-1) = \boldsymbol{\Phi}(k, k-1)\, \mathbf{P}(k-1|k-1)\, \boldsymbol{\Phi}^T(k, k-1) + \boldsymbol{\Gamma}(k, k-1)$$
$$\mathbf{Q}(k-1)\, \boldsymbol{\Gamma}^T(k, k-1), \tag{5}$$

$$\mathbf{K}(k) = \mathbf{P}(k|k-1)\, \mathbf{H}^T(k)\, [\mathbf{H}(k)\, \mathbf{P}(k|k-1)\, \mathbf{H}^T(k) + \mathbf{R}(k)]^{-1} \tag{6}$$

$$\tilde{\mathbf{z}}(k|k-1) = \mathbf{z}(k) - \mathbf{H}(k)\, \hat{\mathbf{x}}(k|k-1), \tag{7}$$

$$\hat{\mathbf{x}}(k|k) = \hat{\mathbf{x}}(k|k-1) + \mathbf{K}(k)\, \tilde{\mathbf{z}}(k|k-1), \tag{8}$$

$$\mathbf{P}(k|k) = [\mathbf{I} - \mathbf{K}(k)\mathbf{H}(k)]\, \mathbf{P}(k|k-1). \tag{9}$$

Equations (4) and (5) represent the predictive components of the Kalman filter, whereas eqs. (6) — (9) represent the corrective components of the filter.

Two facts about signal $\tilde{\mathbf{z}}(k|k-1)$ — the *innovations process*—that are important in the derivation and analysis of the PEF are: $\tilde{\mathbf{z}}(k|k-1)$ is a zero mean *white noise process*, and

$$E\{\tilde{\mathbf{z}}(k|k-1)\, \tilde{\mathbf{z}}^T(k|k-1)\} = \mathbf{H}(k)\, \mathbf{P}(k|k-1)\, \mathbf{H}^T(k) + \mathbf{R}(k). \tag{10}$$

These facts about $\mathbf{z}(k|k-1)$ were first proven by Kailath (1968) and have been utilized in many different ways in the estimation theory literature.

In order to derive the PEF, first combine (8) and (4) to show that

$$\hat{\mathbf{x}}(k|k) = \mathbf{\Phi}(k, k-1)\, \hat{\mathbf{x}}(k-1|k-1) + \mathbf{K}(k)\, \tilde{\mathbf{z}}(k|k-1) \tag{11}$$

and then compare the structures of (11) and (1), using the facts that both $\tilde{\mathbf{z}}(k|k-1)$ and $\mathbf{w}(k-1)$ are white noise processes, to obtain the following "correspondence":

$$\mathbf{\Gamma}(k, k-1)\, \hat{\mathbf{w}}_1(k-1) \underline{\Delta} \mathbf{K}(k)\, \tilde{\mathbf{z}}(k|k-1) \tag{12}$$

In this equation, $\hat{\mathbf{w}}_1(k-1)$ denotes the PEF for $\mathbf{w}(k-1)$. Unfortunately, (12) is an overdetermined system of equations (unless $q = n$). We use the pseudo-inverse of $\mathbf{\Gamma}(k, k-1)$, $\mathbf{\Gamma}^\dagger(k, k-1)$ where

$$\mathbf{\Gamma}^\dagger(k, k-1) = [\mathbf{\Gamma}'(k, k-1)\, \mathbf{\Gamma}(k, k-1)]^{-1}\, \mathbf{\Gamma}'(k, k-1), \tag{13}$$

to obtain the following ad-hoc PEF for $\mathbf{w}(k-1)$:

$$\hat{\mathbf{w}}_1(k-1) = \mathbf{\Gamma}^\dagger(k, k-1)\, \mathbf{K}(k)\, \tilde{\mathbf{z}}(k|k-1). \tag{14}$$

Ott and Meder claim to be able to determine $\hat{\mathbf{w}}_1(k-1)$ directly from the overdetermined system (12). In their example, $\mathbf{w}(k-1)$ is a scalar and (12) results in a system of two equations in the one unknown, $\hat{\mathbf{w}}_1(k-1)$. In their eqs. (29a) and (29b), Ott and Meder provide two possible solutions for $\hat{\mathbf{w}}_1(k-1)$; but no rule is given for when to use which solution. This uncertain state of affairs is resolved by means of the pseudo-inverse operation in our equation (14) for $\hat{\mathbf{w}}_1(k-1)$.

Observe that $\hat{\mathbf{w}}_1(k-1)$ enjoys the property of being a white estimator of the white input $\mathbf{w}(k-1)$; however, $\hat{\mathbf{w}}_1(k-1)$ is not an optimal estimator of $\mathbf{w}(k-1)$: It has merely been *defined* as in eq. (12). As such, we refer to the PEF as an *ad hoc estimator*. Mendel (1977) shows that an optimal filtered estimate of $\mathbf{w}(k-1)$, say $\hat{\mathbf{w}}(k-1|k-1)$, does not exist. In fact, optimal (i.e., minimum-variance) estimates of $\mathbf{w}(k-1)$ will be optimal smoothed estimates (i.e., estimates which use future measurements). The quantitative results of Mendel (1977) further support our contention that the PEF is an ad hoc estimator.

3. Performance of PEF

Let $\mathbf{\Psi}_{w1}(k-1)$ denote the error covariance matrix for the PEF $\hat{\mathbf{w}}_1(k-1)$; i.e.,

$$\mathbf{\Psi}_{w1}(k-1) = E\{[\tilde{\mathbf{w}}_1(k-1) - E\{\tilde{\mathbf{w}}_1(k-1)\}]\,[\tilde{\mathbf{w}}(k-1)]^T\} \tag{15}$$

where

$$\tilde{\mathbf{w}}_1(k-1) = \mathbf{w}(k-1) - \hat{\mathbf{w}}_1(k-1). \tag{16}$$

From (14) it follows that $E\{\hat{\mathbf{w}}_1(k-1)\} = \mathbf{O}$; hence, $E\{\tilde{\mathbf{w}}_1(k-1)\} = \mathbf{O}$, so that

$$\boldsymbol{\Psi}_{w1}(k-1) = E\{\tilde{\mathbf{w}}_1(k-1)\,\tilde{\mathbf{w}}'_1(k-1)\}. \tag{17}$$

In the appendix we derive the following expression for $\boldsymbol{\Psi}_{w1}(k-1)$:

$$\begin{aligned}
\boldsymbol{\Psi}_{w1}(k-1) &= \mathbf{Q} + \boldsymbol{\Gamma}^\dagger\mathbf{PH}^T(\mathbf{HPH}^T+\mathbf{R})^{-1}\mathbf{HP}\boldsymbol{\Gamma}^{\dagger\prime} \\
&\quad - \mathbf{Q}\boldsymbol{\Gamma}^T\mathbf{H}^T(\mathbf{HPH}'+\mathbf{R})^{-1}\mathbf{HP}\boldsymbol{\Gamma}^{\dagger T} \\
&\quad - \boldsymbol{\Gamma}^\dagger\mathbf{PH}^T(\mathbf{HPH}^T+\mathbf{R})^{-1}\mathbf{H}\boldsymbol{\Gamma}\mathbf{Q}. \tag{18}
\end{aligned}$$

In this expression $\mathbf{Q} = \mathbf{Q}(k-1)$, $\boldsymbol{\Gamma}^\dagger = \boldsymbol{\Gamma}^\dagger(k,\ k-1)$, $\mathbf{P} = \mathbf{P}(k|k-1)$, $\mathbf{H} = \mathbf{H}(k)$, $\mathbf{R} = \mathbf{R}(k)$, and $\boldsymbol{\Gamma} = \boldsymbol{\Gamma}(k,\ k-1)$.

We propose to evaluate the PEF by means of its error covariance matrix $\boldsymbol{\Psi}_{w1}(k-1)$. In any specific application $\mathbf{Q}(k-1)$, $\mathbf{H}(k)$, $\boldsymbol{\Gamma}(k,\ k-1)$, and $\mathbf{R}(k)$ are known from the state equation description of the system under study, and $\mathbf{P}(k|k-1)$ is computed from the Kalman filter equations (5), (6), and (9), after which $\boldsymbol{\Psi}_{w1}(k-1)$ can be computed via (18).

Because the last two terms in (18) are negative, it would appear to be true that for some systems $\boldsymbol{\Psi}_{w1}(k-1)$ will be smaller than $\mathbf{Q}(k-1)$ [i.e., $\boldsymbol{\Psi}_{w1}(k-1) < \mathbf{Q}(k-1)$]. This is important, since the error covariance matrix of the "zero estimator" of $\mathbf{w}(k-1)$ [i.e., the estimator that estimates $\mathbf{w}(k-1)$ as zero for all k] is $\mathbf{Q}(k-1)$; and, we believe that any useful estimator of $\mathbf{w}(k-1)$ should perform better than the zero estimator.

4. EXAMPLE

Ott and Meder (1972) consider an example which they state "... is of special interest for seismic exploration ...". It is a one dimensional damped harmonic oscillator which is excited by impulses of random intensity in random time instances. For their example

$$\begin{pmatrix} x_1(k+1) \\ x_2(k+1) \end{pmatrix} = \begin{pmatrix} 1 & T \\ -\beta T & 1-\alpha T \end{pmatrix} \begin{pmatrix} x_1(k) \\ x_2(k) \end{pmatrix} + \begin{pmatrix} 0 \\ \omega_E T \end{pmatrix} w(k) \tag{19}$$

$$z(k) = (1 \quad 0) \begin{pmatrix} x_1(k) \\ x_2(k) \end{pmatrix} + v(k) \tag{20}$$

Equation (19) is obtained by discretizing the equation $\ddot{x}(t) + \alpha\dot{x}(t) + \beta x(t) = \omega_E w(t)$, assuming $t_{k+1} - t_k = T$ is small.

When we proceed to calculate $\boldsymbol{\Psi}_{w1}(k-1)$, by means of (18), a very interesting and somewhat unexpected result occurs for this example. Observe from (19) and (20) that $\boldsymbol{\Gamma}(k,\ k-1) = \text{col}\,(0\ \omega_E T)$ and $\mathbf{H}(k) = (1\quad 0)$, so that $\mathbf{H}\boldsymbol{\Gamma} = 0$; and, for this example,

$$\boldsymbol{\Psi}_{w1}(k-1) = \mathbf{Q} + \boldsymbol{\Gamma}^\dagger\mathbf{PH}^T(\mathbf{HPH}^T+\mathbf{R})^{-1}\mathbf{HP}\boldsymbol{\Gamma}^{\dagger T} \tag{21}$$

The second term in (21) is positive semidefinite; hence, we see that $\boldsymbol{\Psi}_{w1}(k-1) \geq \mathbf{Q}$! This means that, for this example, the PEF performs no better and probably worse than the zero estimator of $\mathbf{w}(k-1)$.

These quantitative results explain the poor performance of the PEF (described in section 1) observed by Ott and Meder.

5. Conclusions

We have shown that Ott and Meder's PEF is an ad hoc (i.e., heuristic and suboptimal) estimator of input noise vector $\mathbf{w}(k-1)$. Interestingly enough though, the PEF generates a white estimate of the assumed white input $\mathbf{w}(k-1)$. By means of the PEF error covariance matrix, $\boldsymbol{\Psi}_{w1}(k-1)$, which we have derived herein, we have been able to explain the poor performance obtained by the PEF for Ott and Meder's second-order example.

Finally, we wish to indicate that *optimal* estimators of $\mathbf{w}(k-1)$, which enjoy the same whiteness property of the PEF, have been developed by Mendel (1977).

Appendix

Derivation of $\boldsymbol{\Psi}_{w1}\ (k-1)$

Substitute eq. (16) into eq. (17), to show that

$$\boldsymbol{\Psi}_{w1}(k-1) = \mathrm{E}\{[\mathbf{w}(k-1) - \hat{\mathbf{w}}_1(k-1)]\,[\mathbf{w}(k-1) - \hat{\mathbf{w}}_1(k-1)]^T\}$$

$$\boldsymbol{\Psi}_{w1}(k-1) = \mathbf{Q}(k-1) - \mathrm{E}\{\mathbf{w}(k-1)\,\hat{\mathbf{w}}_1^T(k-1)\} -$$
$$\mathrm{E}\{\hat{\mathbf{w}}_1(k-1)\,\mathbf{w}^T(k-1)\} + \mathrm{E}\{\hat{\mathbf{w}}_1(k-1)\,\hat{\mathbf{w}}_1^T(k-1)\}, \quad (22)$$

where we have used eq. (3c). The last term in eq. (22) is easily developed further by means of eqs. (14), (10), and (6), as follows:

$$\mathrm{E}\{\hat{\mathbf{w}}_1(k-1)\,\hat{\mathbf{w}}_1^T(k-1)\} = \boldsymbol{\Gamma}^\dagger(k,k-1)\,\mathbf{K}(k)\,\mathrm{E}\{\tilde{\mathbf{z}}(k|k-1)\,\tilde{\mathbf{z}}^T(k|k-1)\}$$
$$\mathbf{K}^T(k)\boldsymbol{\Gamma}^{\dagger T}(k,k-1),$$

$$\mathrm{E}\{\hat{\mathbf{w}}_1(k-1)\,\hat{\mathbf{w}}_1'(k-1)\} = \boldsymbol{\Gamma}^\dagger(k,k-1)\,\mathbf{K}(k)\,[\mathbf{H}(k)\,\mathbf{P}(k|k-1)\,\mathbf{H}^T(k) + \mathbf{R}(k)]$$
$$\mathbf{K}^T(k)\,\boldsymbol{\Gamma}^{\dagger T}(k,k-1),$$

$$\mathrm{E}\{\hat{\mathbf{w}}_1(k-1)\,\hat{\mathbf{w}}_1^T(k-1)\} = \boldsymbol{\Gamma}^\dagger(k,k-1)\,\mathbf{P}(k|k-1)\,\mathbf{H}^T(k)\,[\mathbf{H}(k)\,\mathbf{P}(k|k-1)$$
$$\mathbf{H}^T(k) + \mathbf{R}(k)]^{-1}\,\mathbf{H}(k)\,\mathbf{P}(k|k-1)\,\boldsymbol{\Gamma}^{\dagger T}(k,k-1).$$

$$(23)$$

The second term in eq. (22) is developed as follows. Substitute (14) into that term, to show that

$$\mathrm{E}\{\mathbf{w}(k-1)\,\hat{\mathbf{w}}_1^T(k-1)\} = \mathrm{E}\{\mathbf{w}(k-1)\,\tilde{\mathbf{z}}^T(k|k-1)\}\,\mathbf{K}^T(k)\,\boldsymbol{\Gamma}^{\dagger T}(k,k-1).$$

$$(24)$$

Next, we show that

$$E\{\mathbf{w}(k-1)\,\tilde{\mathbf{z}}^T(k|k-1)\} = \mathbf{Q}(k-1)\,\mathbf{\Gamma}^T(k, k-1)\,\mathbf{H}^T(k). \tag{25}$$

To begin, substitute eq. (2) into eq. (7), to show that

$$\begin{aligned}
\tilde{\mathbf{z}}(k|k-1) &= \mathbf{H}(k)\,\mathbf{x}(k) + \mathbf{v}(k) - \mathbf{H}(k)\,\hat{\mathbf{x}}(k|k-1) \\
&= \mathbf{H}(k)\,\tilde{\mathbf{x}}(k|k-1) + \mathbf{v}(k)
\end{aligned} \tag{26}$$

where $\tilde{\mathbf{x}}(k|k-1)$, the single-stage predicted state estimation error equals $\mathbf{x}(k) - \hat{\mathbf{x}}(k|k-1)$. Meditch (1969) derives the following equations for $\tilde{\mathbf{x}}(k|k-1)$ and $\tilde{\mathbf{x}}(k|k)$ [eqs. (5.63) and (5.64), respectively, of Meditch 1969]:

$$\tilde{\mathbf{x}}(k|k-1) = \mathbf{\Phi}(k, k-1)\,\tilde{\mathbf{x}}(k-1|k-1) + \mathbf{\Gamma}(k, k-1)\,\mathbf{w}(k-1) \tag{27}$$

and

$$\begin{aligned}
\tilde{\mathbf{x}}(k|k) &= [\mathbf{I} - \mathbf{K}(k)\,\mathbf{H}(k)]\mathbf{\Phi}(k, k-1)\,\tilde{\mathbf{x}}(k-1|k-1) \\
&\quad + [\mathbf{I} - \mathbf{K}(k)\,\mathbf{H}(k)]\,\mathbf{\Gamma}(k, k-1)\,\mathbf{w}(k-1) \\
&\quad - \mathbf{K}(k)\,\mathbf{v}(k).
\end{aligned} \tag{28}$$

Next, substitute (27) into (26), to show that

$$\begin{aligned}
\tilde{\mathbf{z}}(k|k-1) &= \mathbf{H}(k)\,\mathbf{\Phi}(k, k-1)\,\tilde{\mathbf{x}}(k-1|k-1) + \mathbf{H}(k)\,\mathbf{\Gamma}(k, k-1) \\
&\quad \mathbf{w}(k-1) + \mathbf{v}(k).
\end{aligned} \tag{29}$$

Finally, observe from (28), that $\tilde{\mathbf{x}}(k-1|k-1)$ depends at most on $\mathbf{w}(k-2)$ and $\mathbf{v}(k-1)$; hence,

$$\begin{aligned}
E\{\mathbf{w}(k-1)\,\tilde{\mathbf{z}}^T(k|k-1)\} &= E\{\mathbf{w}(k-1)\,\tilde{\mathbf{x}}^T(k-1|k-1)\}[\mathbf{H}(k)\,\mathbf{\Phi}(k, k-1)]^T \\
&\quad + E\{\mathbf{w}(k-1)\,\mathbf{w}^T(k-1)\}[\mathbf{H}(k)\,\mathbf{\Gamma}(k, k-1)]^T \\
&\quad + E\{\mathbf{w}(k-1)\,\mathbf{v}^T(k)\} \\
&= \mathbf{O} + \mathbf{Q}(k-1)\,\mathbf{\Gamma}^T(k, k-1)\,\mathbf{H}^T(k) + \mathbf{0},
\end{aligned} \tag{30}$$

which is (25).

Now substitute (30) into (24), to show that

$$E\{\mathbf{w}(k-1)\,\hat{\mathbf{w}}_1^T(k-1)\} = \mathbf{Q}(k-1)\,\mathbf{\Gamma}^T(k, k-1)\,\mathbf{H}^T(k)\,\mathbf{K}^T(k)\,\mathbf{\Gamma}^{\dagger T}(k, k-1) \tag{31}$$

In order to obtain the final equation for $\mathbf{\Psi}_{w1}(k-1)$ in (18), substitute (23) and (31), as well as the transpose of (31), into (22), and substitute $\mathbf{K}(k)$ and $\mathbf{K}^T(k)$, from (6), into the resulting expression.

ACKNOWLEDGEMENT

The work reported on in this paper was performed at the University of Southern California, Los Angeles, California, under National Science Foundation Grant NSF ENG 74-02297A01.

REFERENCES

KAILATH, T., 1968, An Innovations Approach to Least-Squares Estimation Part I: Linear Filtering in Additive White Noise, IEEE Transactions on Automatic Control, AC-13, 646-655.

MEDITCH, J. S., 1969, Stochastic Optimal Linear Estimation and Control, McGraw-Hill Book Co., New York.

MENDEL, J. M., 1977, "Single-Channel White Noise Estimators for Deconvolution," Geophysics.

OTT, N. and MEDER, H. G., 1972, "The Kalman Filter as a Prediction Error Filter," Geophysical Prospecting 20, 549-560.

17

SINGLE-CHANNEL WHITE-NOISE ESTIMATORS FOR DECONVOLUTION

JERRY M. MENDEL* AND JOHN KORMYLO*

The Wiener filtering approach to deconvolution is limited by certain modeling assumptions, which may not always be valid. We develop a Kalman filtering approach to deconvolution which permits more flexible modeling assumptions than the Wiener filtering approach. Our approach is applicable to time-varying or time-invariant wavelets as well as to nonstationary or stationary noise processes.

We develop equations herein for minimum-variance estimates of the reflection coefficient sequence, as well as error variances associated with these estimates. Our estimators are compared with an ad hoc "prediction error filter," which has recently been reported on in the geophysics literature. We show that our estimators perform better than the prediction error filter. Simulation results are included, for both time-invariant and time-varying situations, which support our theoretical developments.

INTRODUCTION

In 1954, Enders A. Robinson (1957, 1967) proposed a convolution summation model to describe the signal received by a seismic sensor. We write this model as (Figure 1)

$$z(k) = V_R(k) + n(k), \tag{1}$$

$$V_R(k) = \sum_{j=1}^{k} \mu(j) V^+(k-j), \tag{2}$$

where $V_R(k)$ is the noise free seismic trace; $n(k)$ is "measurement" noise which accounts for physical effects not explained by $V_R(k)$, as well as sensor noise; k is short for time t_k; $V^+(i), i = 0, 1, 2, \ldots, I$, is a sequence associated with the basic seismic wavelet (Robinson, 1967; Anstey, 1970); and $\mu(j), j = 1, 2, \ldots$, is the reflection coefficient sequence. This convolution summation model can be derived from physical principles and some simplifying assumptions, such as: normal incidence (i.e., horizontal layering), each layer is homogeneous and isotropic, small strains, and pressure or velocity (or displacement) satisfy a one-dimensional wave equation. Signal $V_R(k)$, which is recorded as $z(k)$ by a seismic sensor, is a superposition of wavelets which are reflected from the interfaces of subsurface layers. The $\mu(j)$ are related to interface reflection coefficients.

The reflection coefficient sequence often has peaks at subsurface interfaces, and is often assumed to be random and uncorrelated (Robinson, 1967). It can be represented graphically as a sequence of "spikes" which occur at $k = 1, 2, \ldots$ (Figure 2). We would like to learn as much as possible about the peak values of the $\mu(j)$ sequence, including amplitude and time of occurrence, so that information can be assembled about the subsurface geometry and lithology.

If noise were not present, and we had numerical values for sequences $V_R(i)$ and $V^+(i)$, then the reflection coefficient sequence could be reconstructed by passing $V_R(i)$ through an "inverse filter". Let $V_R(z)$, $V^+(z)$, and $\mu(z)$ denote the z-transforms of sequences $V_R(k)$, $V^+(k)$, and $\mu(k)$. Then equation (2) can also be written, in terms of these z-transforms, as $V_R(z) = \mu(z)V^+(z)$, from which we see that $\mu(z) = V_R(z)[1/V^+(z)]$; hence, when $V_R(z)$ is operated upon by the "inverse filter" $1/V^+(z)$, we obtain $\mu(z)$.

In the noisy situation, it is common practice to extract the reflection coefficient sequence using a technique which was developed by Robinson (1957, 1967) known as predictive deconvolution. The major component of this technique is a digital Wiener filter,

Manuscript received by the Editor January 22, 1976; revised manuscript received July 25, 1977.
*University of Southern California, Los Angeles, CA 90007.

280

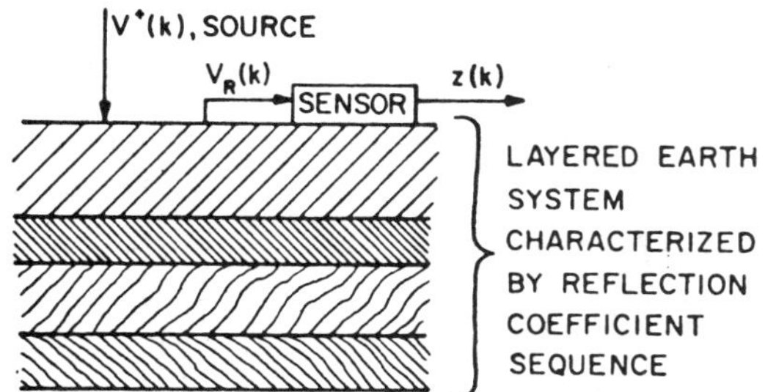

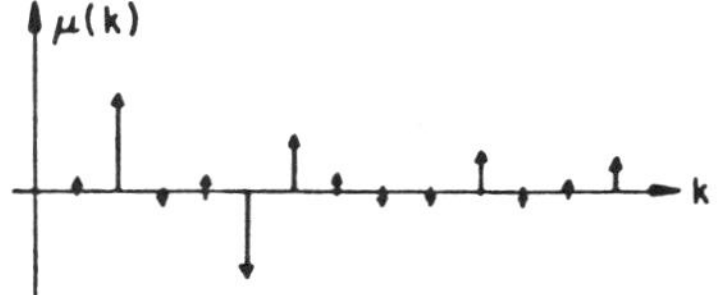

FIG. 1. A layered earth system with source and sensor.

FIG. 2. Example of reflection coefficient sequence $\mu(k)$.

which requires correlation function information. The following modeling assumptions are associated with Robinson's work (Wood and Treitel, 1975): (1) the layered earth is a linear system, (2) the basic seismic wavelet is minimum delay (i.e., minimum phase), and (3) the reflection coefficient sequence is random and uncorrelated.

Because the Wiener filtering approach to predictive deconvolution is limited by these modeling assumptions, which may not always be valid, it is important to look for alternative approaches which will permit more flexible modeling assumptions. Kalman filtering is such an alternative, and has recently begun to receive attention from geophysicists (Bayliss and Brigham, 1970; Crump, 1974, 1975; Ott and Meder, 1972; Berkhout and Zaanen, 1976). *In this paper we shall develop a Kalman filtering approach to obtain optimal estimates of the reflection coefficient sequence.*

Unlike the Wiener filtering approach to predictive deconvolution, which is essentially limited to time-invariant wavelets and stationary noise processes, the Kalman filtering approach to predictive deconvolution is applicable to time-varying or time-invariant wavelets as well as to nonstationary or stationary-noise processes. Additionally, it can be "extended" to nonlinear wavelet models.

From linear system theory (Schwarz and Friedland, 1965, for example), it is well known that the output $y(k)$ of a linear, discrete-time time-invariant, causal system, whose input $r(i)$ is zero prior to time zero, is

$$y(k) = \sum_{j=0}^{k} r(j) h(k - j), \qquad (3)$$

where $h(i)$ is the unit function response of the system. Comparing equations (2) and (3), we are led to the following important system interpretation for the seismic trace model: *signal* $V_R(k)$ *can be thought of as the output of a linear time-invariant system whose unit response is* $V^+(i)$ *and input sequence is the reflection coefficient sequence,* $\mu(i)$ *(Figure 3).*

Whereas the superposition summation model in equation (2) is the starting point for a Wiener filter approach to predictive deconvolution, a different but mathematically equivalent model is the starting point for a Kalman filter approach to predictive deconvolution. Instead of using $V^+(i)$ in its wavelet format, this information is converted into an equivalent state space representation.

We shall assume that the following state space representation of equations (1) and (2) is our starting point for predictive deconvolution:

$$\mathbf{x}(k + 1) = \Phi \mathbf{x}(k) + \gamma \mu(k), \qquad (4)$$

$$z(k) = \mathbf{h}'\mathbf{x}(k) + n(k), \qquad (5)$$

where $\mathbf{x}$, γ, and $\mathbf{h}$ are $n \times 1$ vectors, z is a scalar, Φ is an $n \times n$ matrix, $\mathbf{h}'\mathbf{x}(k) = V_R(k)$, and $\mu(k)$ and $n(k)$ are uncorrelated zero mean white sequences, for which

$$E\{\mu(i)\mu(j)\} = q\delta_{ij}, \qquad (6a)$$

and

$$E\{n(i) n(j)\} = r\delta_{ij}. \qquad (6b)$$

In these equations, symbol δ_{ij} is the Kronecker delta, defined to be unity when $i = j$ and zero otherwise. In equations (4) and (5), $k = 0, 1, 2, \ldots$, and $\mathbf{x}(0)$ initializes equation (4). Additionally, Φ, γ, $\mathbf{h}$, q, and r can be time-varying. For notational simplicity, we do not show the explicit dependence of these quantities on time.

To obtain numerical values for Φ, γ, and $\mathbf{h}$, we

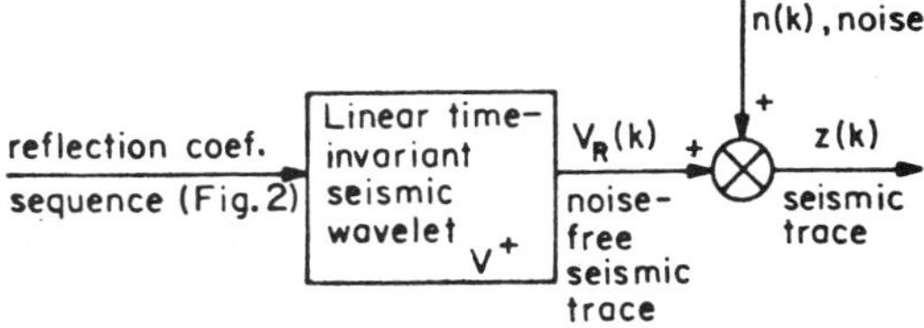

FIG. 3. Linear system interpretation for convolution summation model of a seismic trace. Signal $V_R(k)$ is given by equation (2).

Table 1. Synthetic source signatures and realizations.

$V^+(t)$	A^a	$\underline{b}^a$	$\underline{h}$
.005 SEC $-1360\, te^{-500t}$ (Kramer et al, 1968)	$\begin{pmatrix} 0 & 1 \\ -25 \times 10^4 & -10^3 \end{pmatrix}$	$\begin{pmatrix} 0 \\ 1 \end{pmatrix}$	$\begin{pmatrix} -1360 \\ 0 \end{pmatrix}$
.004 SEC $-1360\, te^{-500t} + 54.4\, te^{-100t}$ (Kramer et al, 1968)	$\left(\begin{array}{cc:cc} 0 & 1 & & 0 \\ -25 \times 10^4 & -10^3 & & \\ \hdashline & 0 & 0 & 1 \\ & & -10^4 & -200 \end{array}\right)$	$\begin{pmatrix} 0 \\ 1 \\ 0 \\ 1 \end{pmatrix}$	$\begin{pmatrix} -1360 \\ 0 \\ 54.4 \\ 0 \end{pmatrix}$
.004 SEC $-1360\, te^{-500t} + 0.5 e^{-15.3t} \sin\!\left(\dfrac{2\pi}{.06}\right) t$ (Kramer et al, 1968)	$\left(\begin{array}{cc:cc} 0 & 1 & & 0 \\ -25 \times 10^4 & -10^3 & & \\ \hdashline & 0 & -15.3 & 104.67 \\ & & -104.67 & -15.3 \end{array}\right)$	$\begin{pmatrix} 0 \\ 1 \\ 0 \\ 1 \end{pmatrix}$	$\begin{pmatrix} -1360 \\ 0 \\ 0.5 \\ 0 \end{pmatrix}$
RICKER WAVELET (Ricker, 1940)[b]	$\left(\begin{array}{cc:cc:cc} -.205 & 787 & & 0 & & 0 \\ -787 & -.205 & & & & \\ \hdashline & 0 & -.205 & 1568 & & 0 \\ & & -1568 & -.205 & & \\ \hdashline & 0 & & 0 & -.205 & 2354 \\ & & & & -2354 & -.205 \end{array}\right)$	$\begin{pmatrix} 0 \\ 1 \\ 0 \\ 1 \\ 0 \\ 1 \end{pmatrix}$	$\begin{pmatrix} -.268 \\ 0 \\ 0.54 \\ 0 \\ -.269 \\ 0 \end{pmatrix}$

a. $\Phi = e^{AT}$ and $\gamma = \int_{t_k}^{t_{k+1}} e^{A(t_{k+1} - \tau)} \mathbf{b}\, d\tau$, where $t_{k+1} - t_k = T$ (Mendel, 1973).

b. The Ricker wavelet realization was obtained by Dr. Keith Glover.

begin with a trace of a basic seismic wavelet (also known as the *source signature*), approximate it, choose a state space representation, and determine Φ, γ, and $\mathbf{h}$ values for that representation either by direct calculations or further approximations. Some realizations of representative source signatures (Kramer et al, 1968; Ricker, 1940) are given in Table 1.

Analytical expressions were available for the first three wavelets; the Ricker wavelet was approximated, as

$$1.0776 \left(\sin \frac{\pi}{6T} t \right) \left(\sin^2 \frac{\pi}{24T} t \right) \exp\left(-\frac{\alpha}{T} t \right),$$

where $T = 10^{-3}/3$ and $\alpha = 6.82 \times 10^{-5}$. In order to obtain Φ, γ, and $\mathbf{h}$ for these wavelets, we (1) compute the Laplace transform of the continuous-time wavelet expressions, separating the transform into a parallel connection of second-order transfer functions; (2) obtain a state space model for each of the second-order transfer functions (see Schwarz and Friedland, 1965, for example); (3) combine the state space models to obtain a continuous-time state space model

for the wavelet; and (4) discretize the continuous-time model, using the formulas in footnote a to Table 1. An example which illustrates this procedure is given in Appendix B. Many other techniques are available for computing Φ, γ, and $\mathbf{h}$ when the trace of the source signature is available, some of which begin by discretizing the wavelet and then proceed directly to obtaining Φ, γ, and $\mathbf{h}$. Further discussions on this important problem[1] are found in Faurre (1976), Chen (1970), and Kalman (1960). For additional discussions on seismic source modeling see Kramer et al (1968), Ricker (1940), Smith (1975), Giles (1967), Mayne and Quay (1970), and Crawford et al (1960).

If a trace of the source signature is not available or cannot be measured (as in land exploration, for example), then parameters in Φ, γ, and $\mathbf{h}$ must be iden-

[1]The state space model in equations (4) and (5) is general enough to permit one to model the wavelet as an ARMA model, or an AR model, or an MA model. Which type of model to choose must be decided upon by the user (see Box and Jenkins, 1970, for example).

tified from the seismogram data $z(k)$. This situation is beyond the scope of the present paper.

Accepting equations (4) and (5) as our starting point for predictive deconvolution, instead of equations (1) and (2), we see that the problem of estimating the white reflection coefficient sequence, $\mu(k)$, is equivalent to the problem of estimating the random disturbance in a state equation. This is a nonstandard problem in estimation theory which has led to some difficulties in trying to do predictive deconvolution via Kalman filtering. The difficulties are due to the fact that (1) we are interested in an optimal estimate of sequence $\mu(k)$, but (2) the Kalman filter obtains optimal estimates of the state vector $\mathbf{x}(k)$ and not $\mu(k)$.

Bayliss and Brigham (1970) and Crump (1974) have attempted to apply Kalman filtering to predictive deconvolution; however, they assume $\mu(k)$ is a colored noise process, which permits them to treat $\mu(k)$ as a state so that they can then estimate the "reflectivity state" by means of a Kalman filter directly. This approach, which is quite different from the one which we shall take, negates the modeling assumption of whiteness for the reflection coefficient sequence. Ott and Meder (1972) estimate $\mu(k)$ by means of a so-called prediction-error-filter; however, as we point out below, their estimator is an ad hoc one, and does not perform as well as ours. We shall obtain optimal estimators[2] of $\mu(k)$ directly from estimation theory.

This paper is limited, for the most part, to single-channel results, i.e., to the situation depicted in Figure 2, as represented by equations (4) and (5) in which both $\mu(k)$ and $z(k)$ are scalars. The more general case, where $\mu(k)$ and/or $z(k)$ are vectors is treated in Mendel (1976a). Optimal *linear* estimators of $\mu(k)$ are developed in the next section. If $\mu(k)$ and $n(k)$ are gaussian, then the next section's estimators are *the* optimal estimators of $\mu(k)$. In the predictive deconvolution application, $\mu(k)$ is often assumed to be white, but is rarely assumed to be gaussian. We use a point process model for $\mu(k)$ in our simulations. We relate our results to the Ott and Meder prediction error filter and simulations which support our theoretical developments are presented.

Before proceeding to the development of optimal linear estimators for $\mu(k)$, let us reiterate the assumptions which are inherent in this paper. We are assuming that the basic seismic wavelet $V^+(i)$ is known a priori and has a given time-invariant or time-varying

state space representation. Process noise[3] $\mu(k)$ and measurement noise $n(k)$ are assumed to be white, with known variances q and r, respectively, either of which may be time-varying. These assumptions are consistent with those often made in the geophysics signal processing literature. Of course, there are many interesting problems that can be studied in which one or more of these assumptions are relaxed. For such problems, our results can still be used to provide baseline data.

OPTIMAL WHITE-DISTURBANCE ESTIMATORS

Introduction

Let us direct our attention to equations (4) and (5). Our objective is to obtain a minimum-variance linear estimator for the white (not necessarily gaussian)-plant-noise sequence $\mu(k)$.

Let

$$Z(j) = \{z(1), z(2), \ldots , z(j)\}, \qquad (7)$$

and $\hat{\mathbf{x}}(k\,|\,j)$ denote the minimum variance estimate of $\mathbf{x}(k)$ which uses all the measurements in $Z(j)$. In terms of the notation $\hat{\mathbf{x}}(k\,|\,j)$, three cases can be distinguished. When $j = k$, so that our estimate of $\mathbf{x}(k)$ is $\hat{\mathbf{x}}(k\,|\,k)$, we say that estimate is the optimal filtered estimate of $\mathbf{x}(k)$. The optimal filtered estimate of $\mathbf{x}(k)$ uses all past and present measurements. When $j < k$, we say that our estimate of $\mathbf{x}(k)$ is an optimal predicted estimate. For example, $\hat{\mathbf{x}}(k\,|\,k-1)$ is the optimal 1-stage (or, single-stage) predicted estimate of $\mathbf{x}(k)$; it is based on past measurements $z(1), \ldots , z(k-1)$. Finally, when $j > k$, we say that our estimate of $\mathbf{x}(k)$ is an optimal smoothed estimate of $\mathbf{x}(k)$. For example, $\hat{\mathbf{x}}(k\,|\,k+2)$ is the optimal 2-stage smoothed estimate of $\mathbf{x}(k)$; it is based on all past and present measurements as well as the two future measurements $z(k+1)$ and $z(k+2)$. Another term for "smoothing" is interpolation.

One of the fundamental results of estimation theory (Meditch, 1969, for example) is that the estimator which minimizes the mean square error $E\{[\mathbf{x}(k) - \hat{\mathbf{x}}(k\,|\,j)]\,'[\mathbf{x}(k) - \hat{\mathbf{x}}(k\,|\,j)]\}$ is the conditional expectation $E\{\mathbf{x}(k)\,|\,Z(j)\}$; i.e.,

$$\hat{\mathbf{x}}(k\,|\,j) = E\{\mathbf{x}(k)\,|\,Z(j)\}. \qquad (8)$$

So, for example, $\hat{\mathbf{x}}(k+1\,|\,k) = E\{\mathbf{x}(k+1)\,|\,Z(k)\}$ and $\hat{\mathbf{x}}(k+1\,|\,k+1) = E\{\hat{\mathbf{x}}(k+1)\,|\,Z(k+1)\}$, and these quantities can be computed from the Kalman

---

[2]An estimator is a formula which uses measurements. When specific values of the measurements are substituted into the formula, the result is an estimate of $\mu(k)$.

[3]In state equation (4), $\mu(k)$ is often referred to as process noise, plant noise, or disturbance. We shall use these terms interchangeably.

filter (KF), which is a digital time-varying filter[4], whose equations are given below.

Prediction equations

$$\hat{\mathbf{x}}(k + 1|k) = \Phi\hat{\mathbf{x}}(k|k), \tag{9}$$

$$P(k + 1|k) = \Phi P(k|k)\Phi' + \gamma q \gamma'. \tag{10}$$

Correction equations

$$\hat{\mathbf{x}}(k + 1|k + 1) = \hat{\mathbf{x}}(k + 1|k) + K(k + 1)$$
$$\cdot [z(k + 1) - \mathbf{h}'\hat{\mathbf{x}}(k + 1|k)], \tag{11}$$

$$K(k + 1) = P(k + 1|k)\mathbf{h}'[\mathbf{h}'P(k + 1|k)\mathbf{h} + r]^{-1}, \tag{12}$$

$$P(k + 1|k + 1) = [I - K(k + 1)\mathbf{h}']P(k + 1|k). \tag{13}$$

The prediction equations use information available from the state equation (i.e., Φ, γ, and q), whereas the correction equations use information available from the new measurement (i.e., $z(k + 1)$, $\mathbf{h}$ and r). Observe that $\hat{\mathbf{x}}(k + 1|k + 1)$ is expressed in equation (11) as a function of $\hat{\mathbf{x}}(k + 1|k)$, and that $\hat{\mathbf{x}}(k + 1|k)$ is expressed in equation (9) as a function of $\hat{\mathbf{x}}(k|k)$. The KF is iterative (sequential) in nature; it is run for $k = 0, 1, 2, \ldots$.

Since $\mathbf{x}(k)$ is $n \times 1$ and $z(k)$ is scalar, $K(k)$ is $n \times 1$, $P(k + 1|k)$ is $n \times n$, and $P(k + 1|k + 1)$ is $n \times n$. Matrix $K(k)$ is often referred to as the Kalman gain matrix (in the multichannel KF, where $z(k)$ is a vector of measurements, $K(k)$ is a full matrix). Matrices $P(k + 1|k)$ and $P(k + 1|k + 1)$ are error covariance matrices of the state vector estimates $\hat{\mathbf{x}}(k + 1|k)$ and $\hat{\mathbf{x}}(k + 1|k + 1)$, respectively; the former is often referred to as the prior error covariance matrix, whereas the latter is often referred to as the posterior error covariance matrix. More specifically, $P(k + 1|k) = E\{[\mathbf{x}(k + 1) - \hat{\mathbf{x}}(k + 1|k][same]'\}$ and $P(k + 1|k + 1) = E\{[\mathbf{x}(k + 1) - \hat{\mathbf{x}}(k + 1|k + 1)][same]'\}$.

A flow chart which summarizes the order in which calculations can be performed in order to implement the predictor/corrector form of the KF is given in Figure 4. The KF is initialized with values of $\hat{\mathbf{x}}(0|0)$ and $P(0|0)$. See Meditch (1969) and Gelb (1974) for discussions on KF initialization. For additional

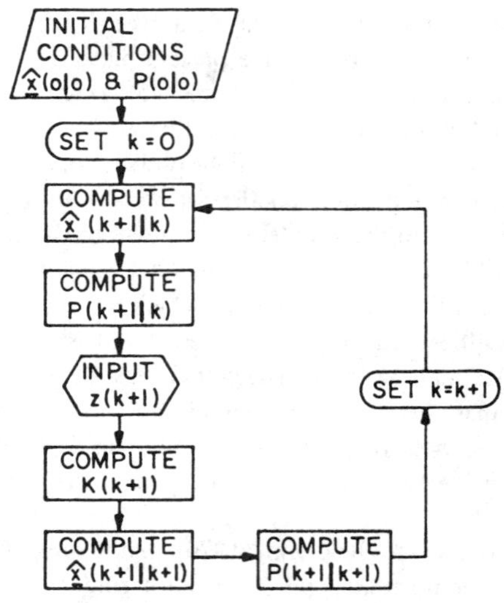

FIG. 4. Flow chart for single-channel predictor/corrector form of Kalman filter equations.

discussions on Kalman filtering which relate the KF to the Wiener filter, see Berkhout and Zaanen (1976) and Mendel and Kormylo (1977).

The single-stage predicted estimate, $\hat{\mathbf{x}}(k + 1|k)$, in equation (9) can be obtained directly from equation (4) by operating on both sides of that equation with the conditional expectation operator $E\{\cdot|Z(k)\}$. The term $E\{\mu(k)|Z(k)\} = E\{\mu(k)\} = 0$, since $Z(k)$ depends only on $\mu(k - 1)$, $\mu(k - 2), \ldots$, and $\mu(1)$ and $\mu(k)$ is a zero mean process.

Let us now direct our attention to obtaining a minimum-variance linear estimator of $\mu(k)$. Operate on both sides of equation (4) with the operator $E\{\cdot|Z(k + 1)\}$:

$$E\{\mathbf{x}(k + 1)|Z(k + 1)\} = \Phi E\{\mathbf{x}(k)|Z(k + 1)\}$$
$$+ \gamma E\{\mu(k)|Z(k + 1)\}. \tag{14}$$

The last term in equation (14) will not be zero because $Z(k + 1)$ depends on $\mu(k)$; hence, we obtain the following equation for an estimator of $\mu(k)$:

$$\gamma\hat{\mu}(k|k + 1) = \hat{\mathbf{x}}(k + 1|k + 1) - \Phi\hat{\mathbf{x}}(k|k + 1). \tag{15}$$

In the following section, we show that γ factors out of the right-hand side of equation (15); hence, equation (15) provides a unique solution for $\hat{\mu}(k|k + 1)$. Observe that $\hat{\mu}(k|k + 1)$ is a single-stage smoothed estimate of $\mu(k)$ which requires not only the optimal filtered estimate of $\mathbf{x}(k + 1)$, but also the optimal single-stage smoothed estimate of $\mathbf{x}(k)$ for its implementation.

---

[4]Many different derivations of the KF appear in the literature; see Kalman (1960), Meditch (1969), Mendel (1973), Sorensen (1966), Kailath (1974), Gelb (1974), Bryson and Ho (1969), Jazwinskii (1970), for example. The predictor-corrector form of the KF is but one of a number of different ways for expressing the KF.

By way of these calculations, we have established the fact that minimum variance linear estimates of $\mu(k)$ will be optimal smoothed estimates.[5] Such estimates make use of future measurements in contrast to optimal filtered estimates which make use only of past and present measurements. In retrospect, this use of future measurements should not be surprising. If, for example, we could solve equation (4) directly for $\mu(k)$, then that solution would depend not only on present state $\mathbf{x}(k)$ but also on future (relative to t_k) state $\mathbf{x}(k + 1)$.

In the remaining paragraphs of this section, we shall explore $\hat{\mu}(k|k + 1)$ and its estimation error in more detail; obtain recursive equations for an optimal l-stage smoothed estimator $\hat{\mu}(k|k + l)$ and its error variance; and discuss some computational aspects of these estimators.

Single-stage optimal smoothed estimates of $\mu(\mathrm{k})$

Here we develop a more useful equation for $\hat{\mu}(k|k + 1)$ than equation (15) and develop an equation for the error variance $\sigma_\mu^2(k|k + 1)$, where

$$\sigma_\mu^2(k|k + 1) = E\{[\tilde{\mu}(k|k + 1) - E\{\tilde{\mu}(k|k + 1)\}]^2\},$$
$$(16)$$

and

$$\tilde{\mu}(k|k + 1) = \mu(k) - \hat{\mu}(k|k + 1). \qquad (17)$$

Theorem 1 —. For the system in equations (4) and (5),

$$\hat{\mu}(k|k + 1) = q\,\boldsymbol{\gamma}'\,P^{-1}(k + 1|k)\,K(k + 1)\,\tilde{z}(k + 1|k),$$
$$(18)$$

and

$$\sigma_\mu^2(k|k + 1) = q - q\,\boldsymbol{\gamma}'\,\mathbf{h}[\mathbf{h}'P(k + 1|k)\mathbf{h} + r]^{-1}\mathbf{h}'\,\boldsymbol{\gamma}\,q. \qquad (19)$$

In equation (18),

$$\tilde{z}(k + 1|k) = z(k + 1) - \mathbf{h}'\hat{\mathbf{x}}(k + 1|k). \qquad (20)$$

The proof of this theorem is given in Appendix A.

We have been able to express $\hat{\mu}(k|k + 1)$ in terms of quantities which can be generated from Kalman filter equations (9) through (13). Signal $\tilde{z}(k + 1|k)$ is known as the innovations process (or prediction error); it possesses the remarkable property (Kailath, 1968) of being a zero-mean white noise sequence. This means that our estimator of the white noise sequence $\mu(k)$, $\hat{\mu}(k|k + 1)$, is itself a white noise sequence. As we shall see later, this property is also shared by Ott and Meder's prediction error filter. We shall also see that this whiteness property is not enjoyed by our l-stage estimators, $\hat{\mu}(k|k + l)$, where $l \geq 2$. Finally, it is straightforward to show, from equation (19), that $\sigma_\mu^2(k|k + 1) \leq q$.

l-stage optimal smoothed estimates of $\mu(k)$

Our estimator of $\mu(k)$, $\hat{\mu}(k|k + 1)$, is based on the set of measurements $Z(k + 1) = \{z(1), z(2), \ldots, z(k), z(k + 1)\}$. In relation to time t_k, $\hat{\mu}(k|k + 1)$ is obtained by looking ahead one sampling time to time t_{k+1} and reflecting "future" measurement $z(k + 1)$ back to time t_k in an optimal manner. In seismic data processing, the entire record is available prior to processing; hence, we should be interested in generalizations of our theorem 1 results to estimators of $\mu(k)$ that look farther ahead into the future than just one sampling point. Suppose, for example, we wanted to base our estimate of $\mu(k)$ on $Z(k + 2) = \{z(1), z(2), \ldots, z(k + 1), z(k + 2)\}$. In this case, we would estimate $\mu(k)$ by $\hat{\mu}(k|k + 2)$. Unfortunately, theorem 1 cannot be used to provide us with this "double-stage" optimal smoothed estimate of $\mu(k)$.

In this paragraph we generalize the results of theorem 1, obtaining $\hat{\mu}(k|k + l)$ and $\sigma_\mu^2(k|k + l)$. Signal $\hat{\mu}(k|k + l)$ is referred to in the sequel as the l-stage optimal smoothed estimate of $\mu(k)$, and $\sigma_\mu^2(k|k + l)$ is the associated l-stage estimation-error variance defined as,

$$\sigma_\mu^2(k|k + l) = E\{[\tilde{\mu}(k|k + l) - E\{\tilde{\mu}(k|k + l)\}]^2\},$$
$$(21)$$

where

$$\tilde{\mu}(k|k + l) = \mu(k) - \hat{\mu}(k|k + l). \qquad (22)$$

Estimator $\tilde{\mu}(k|k + l)$ is obtained by looking l time samples into the future and is based on $Z(k + l) = \{z(1), z(2), \ldots, z(k + l - 1), z(k + l)\}$.

Theorem 2 —. For the system in equations (4) and (5),

$$\hat{\mu}(k|k + l) = \hat{\mu}(k|k + l - 1) + N_\mu(k|k + l)$$
$$\cdot \tilde{z}(k + l|k + l - 1), \qquad (23)$$

where $l = 1, 2, \ldots, \hat{\mu}(k|k) \overset{\Delta}{=} 0$,

$$N_\mu(k|k + 1) = q\,\boldsymbol{\gamma}'\,P^{-1}(k + 1|k)\,K(k + 1),$$
$$(24)$$

$$N_\mu(k|k + l) = q\,\boldsymbol{\gamma}'\,P^{-1}(k + 1)$$
$$\cdot \left[\prod_{i=k+1}^{k+l-1} A(i)\right] K(k + l), \qquad (25)$$

[5]Optimality follows from equation (8).

and[6]

$$A(i) = P(i|i)\,\Phi'P^{-1}(i + 1|i). \qquad (26)$$

Additionally,

$$\sigma^2_\mu(k|k + l) = \sigma^2_\mu(k|k + l - 1) - N^2_\mu(k|k + l)$$
$$\cdot\,[\mathbf{h}'\,P(k + l|k + l - 1)\mathbf{h} + r], \qquad (27)$$

where $\sigma^2_\mu(k|k) \triangleq q$.

A complete proof of this theorem is given in Mendel (1976a) in a multichannel setting. The principal step uses the identity $\gamma\hat{\mu}(k|k + l) = \hat{\mathbf{x}}(k + 1|k + l) - \Phi\hat{\mathbf{x}}(k|k + l)$, where $\hat{\mathbf{x}}(k|k + l)$ and $\hat{\mathbf{x}}(k + 1|k + l)$ are solutions to the minimum-variance smoothing problem. This identity follows from application of the expectation operator $E\{\cdot\,|Z(k + l)\}$ to both sides of equation (4).

Equations (23) and (27) provide us with a means for obtaining $\hat{\mu}(k|k + l)$ and $\sigma^2_\mu(k|k + l)$ in a *recursive* manner. For $l = 1$, equations (23) and (27) reduce to our earlier theorem 1 results.

Suppose we have a complete set of measurements, $z(1)$, $z(2)$, $\ldots$, $z(L)$, available. A number of options are open to us, including using all the data, or using some of the data to estimate the reflection coefficient sequence.

Our estimator equations above are in so-called fixed-point form. A *fixed-point estimator* of $\mu(k)$ is $\hat{\mu}(k|k + l)$ where k is fixed and l is varied. Fixed-point estimates enhance estimates at specific values of k. Suppose that an error-variance ϵ has been established and we wish to estimate $\mu(k)$ for $k = 1, 2, \ldots, L$, such that $\sigma^2_\mu(k|k + l) \leq \epsilon$, all k. Equations (23) and (27) can be used to obtain such estimates in the following manner. Set $k = 1$ in equation (27) and determine the value of l, say l_1, such that $\sigma^2_\mu(1|1 + l_1) \leq \epsilon$; then, evaluate $\hat{\mu}(1|1 + l_1)$. Set $k = 2$ in equation (27) and determine l_2 such that $\sigma^2_\mu(2|2 + l_2) \leq \epsilon$; then evaluate $\hat{\mu}(2|2 + l_2)$. This procedure is repeated for $k = 1, 2, \ldots, L$. Another situation for which a fixed-point estimate is appropriate is one in which we have a preliminary estimate of $\mu(k)$ for all $k = 1, 2, \ldots, L$, obtained by fixing l (we describe this situation below) without regard to achieving a threshold error-variance. From the time-series of values of $\hat{\mu}(k|k + l)$, we detect certain times at which peak values of $\hat{\mu}$ seem to occur. Suppose one of those values occurs at t_{k_1}; i.e., $\hat{\mu}(k_1|k_1 + l)$ is of large amplitude. Further signal enhancement of

our estimate of $\mu(k_1)$ is now possible by means of equation (23). We can increase l to some new value, say l_1, for which $\hat{\mu}(k_1|k_1 + l_1)$ has better resolution than $\hat{\mu}(k_1|k_1 + l)$. This discriminate signal processing procedure can be applied at all values of time for which $\hat{\mu}(k|k + l)$ appears to have prominent peak values.

A *fixed-lag estimator* of $\mu(k)$ is $\hat{\mu}(k|k + l)$, where l is fixed and k is varied. As such, $\hat{\mu}(k|k + l)$ utilizes the following window of future measurements: $z(k + 1)$, $z(k + 2)$, $\ldots$, $z(k + l)$. For example, $\hat{\mu}(1|1 + l)$ utilizes the window of future measurements $z(2)$, $z(3)$, $\ldots$, $z(1 + l)$, whereas $\hat{\mu}(2|2 + l)$ utilizes the window $z(3)$, $z(4)$, $\ldots$, $z(2 + l)$. A fixed-lag estimate of $\mu(k)$ would be useful in those situations (such as the one described above) where we decide to use only a fixed length of future data to obtain the optimal estimate. We may, for example, initially choose a small value for l and obtain a "rough" estimate of $\mu(k)$, after which we can do fixed-point signal enhancement, as described above.

Fixed-point and fixed-lag estimators usually do not make use of all the data. A *fixed interval estimator* of $\mu(k)$ uses all future values of the measurements, $z(k + 1)$, $z(k + 2)$, $\ldots$, $z(L)$. An equation for a fixed-interval estimator of $\mu(k)$ is obtained by iterating equation (23) on l,

$$\hat{\mu}(k|k + l) = \sum_{j=1}^{l} N_\mu(k|k + j)\,\tilde{z}(k + j|k + j - 1), \qquad (28)$$

and by setting $k + l = L$ (and, subsequently, $k = L - l$) in this equation; i.e.,

$$\hat{\mu}(L - l|L) = \sum_{j=1}^{l} N_\mu(L - l|L - l + j)$$
$$\cdot\,\tilde{z}(L - l + j|L - l + j - 1). \qquad (29)$$

This equation holds for $l = 1, 2, \ldots, L - 1$. As such, observe that as we vary l through its range of values, we obtain $\hat{\mu}(L - 1|L)$, $\hat{\mu}(L - 2|L)$, $\ldots$, $\hat{\mu}(1|L)$. Fixed-interval estimates of the reflection coefficient sequence must be computed backwards in time. As such, they often require large amounts of computer storage, since $\hat{\mu}(L - 1|L)$, which is the first computed fixed-interval estimate of μ, requires $\tilde{z}(L|L - 1)$ [to see this, set $l = 1$ in equation (29)], and $\tilde{z}(L|L - 1)$ is computed by running the Kalman filter over the *entire* range of k-values, $k = 1, 2, \ldots, L$.

Earlier we pointed out that $\hat{\mu}(k|k + 1)$ is itself a white process. Consider $\hat{\mu}(k|k + 2)$, where [from equation (28)]

$$^6\ \prod_{i=1}^{l} A(i) = A(1)\,A(2)\ldots A(l - 1)\,A(l).$$

$$\hat{\mu}(k|k + 2) = N_\mu(k|k + 1)\tilde{z}(k + 1|k)$$
$$+ N_\mu(k|k + 2)\tilde{z}(k + 2|k + 1). \quad (30)$$

For $\hat{\mu}(k|k + 2)$ to be a white process, $E\{\hat{\mu}(i|i + 2)$ $\hat{\mu}(j|j + 2)\} = 0$ for all $i \neq j$, and $\neq 0$ for $i = j$. A straightforward calculation[7] of this crosscovariance reveals that $E\{\hat{\mu}(i|i + 2)\hat{\mu}(j|j + 2)\} \neq 0$ for all $i \neq j$; hence, $\hat{\mu}(k|k + 2)$ is not white. This conclusion is true for all $\hat{\mu}(k|k + l)$, $l \geq 2$.

Suppose k is fixed and equation (27) is iterated for $l = 1, 2, \ldots$; then,

$$\sigma_\mu^2(k|k + l) = q - \sum_{j=1}^{l} N_\mu^2(k|k + j)$$
$$\cdot [\mathbf{h}'P(k + j|k + j - 1)\mathbf{h} + r]. \quad (31)$$

It has been our experience that eventually a value of l is reached, say l^*, for which $\sigma_\mu^2(k|k + l^*) \simeq \sigma_\mu^2(k|k + l^* - 1)$; i.e., the error variance reaches a steady state value. From equation (27), we see that when this happens $N_\mu(k|k + l^*) \to 0$; and, subsequently $\hat{\mu}(k|k + l^*) \simeq \hat{\mu}(k|k + l^* - 1)$, which means that our estimate of $\mu(k)$ also reaches a steady-state value.

Equations (24) and (25) are computationally unattractive for two reasons: (1) the inverse of $P(k + 1|k)$ is needed at all values of k, and (2) the product of many $n \times n\, A(i)$ matrices are needed. The totality of multiplications needed to implement equations (24) and (25) in their present form is a (rather large) multiple of n^3, and this is very costly. An alternative is given next.

Our alternative is simply a fast algorithm for computing the smoothing gain $N_\mu(k|k + l)$. [A second alternative is a two-pass algorithm for calculation of $\hat{\mu}(k|k + l)$ and $\sigma_\mu^2(k|k + l)$. This two-pass algorithm gives us a fixed interval estimate of μ, and is a fast algorithm for the calculation of such estimates. The two-pass algorithm is not efficient for calculation of fixed-point estimates of μ; our alternative below should be used for such estimates. Derivations of both alternatives can be found in Mendel and Kormylo (1977).]

Alternative fixed-point algorithm

The following equations can be used to compute $N_\mu(k|k + l)$:

$$N_\mu(k|k + 1) = q\,\boldsymbol{\gamma}'\,\mathbf{h}[\mathbf{h}'P(k + 1|k)\mathbf{h} + r]^{-1}, \quad (32)$$

$$\left. \begin{array}{l} D(k, l) = D(k, l - 1)[I - K(k + l - 1)\mathbf{h}']'\Phi' \\ D(k, 1) \triangleq q\,\boldsymbol{\gamma}' \end{array} \right\}, \quad (33)$$

and

$$N_\mu(k|k + l) = D(k, l)\mathbf{h}[\mathbf{h}'P(k + l|k + l - 1)\mathbf{h} + r]^{-1}, \quad (34)$$

where $l = 2, 3, \ldots$.

A flow chart for the implementation of our alternative fixed-point algorithm is given in Figure 5. Initial values for $\hat{\mu}(k|k)$ and $\sigma_\mu^2(k|k)$ are given in theorem 2, and $D(k, 1)$ is given in equation (33). The following quantities are needed from the KF: $\mathbf{h}'P(k + j|k + j - 1)\mathbf{h} + r$ and its inverse, $K(k + j)$, and $\tilde{z}(k + j|k + j - 1)$. Smoothing gain $N_\mu(k|k + j)$ is computed from equations (32) or (34). Equations for $\hat{\mu}(k|k + j)$ and $\sigma_\mu^2(k|k + j)$ are given in theorem 2. Finally, equation (33) is used to compute $D(k, j + 1)$.

Matrix $D(k, l)$ is $1 \times n$, and the calculations in

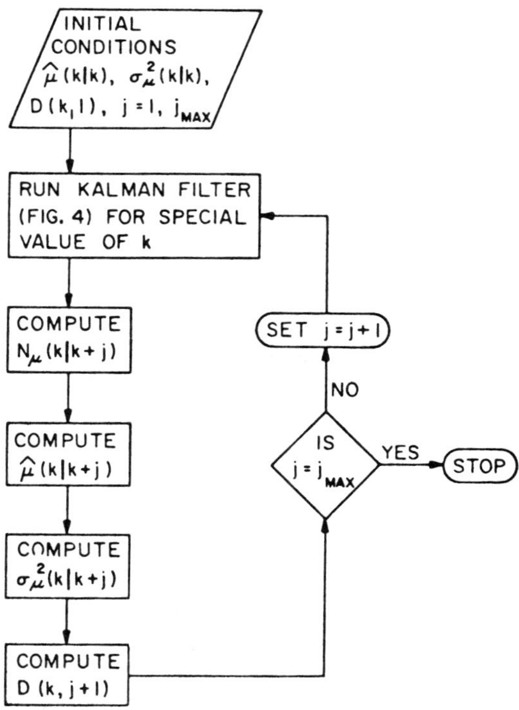

FIG. 5. Flow chart for alternative fixed-point algorithm.

---

[7] $E\{\hat{\mu}(i|i + 2)\,\hat{\mu}(j|j + 2)\}$ contains, as one of its four terms, a term proportional to $E\{\tilde{z}(i + 1|i)\,\tilde{z}(j + 2|j + 1)\}$. If we can find *any* i and j $(i \neq j)$ for which $\tilde{z}(i + 1|i) = \tilde{z}(j + 2|j + 1)$, then this expectation cannot be zero. Clearly, when $i = j + 1$ (e.g., $j = 1$, $i = 2$), $\tilde{z}(i + 1|i) = \tilde{z}(j + 2|j + 1)$.

equations (33) and (34) can be performed without having to multiply any $n \times n$ matrices. Observe, also, that the inverse of $P(k + 1|k)$ is no longer needed. Detailed operation counts are given in Mendel and Kormylo (1977).

Some early work on estimation of states and plant noise is given in Bryson and Ho (1969). They derive $\hat{\mu}(0|1)$ via optimization theory, and then state an extension of this result to multistage processes. Their resulting estimate, $\hat{\mu}(k|N)$, is a fixed-interval smoothed estimate of $\mu(k)$. Bierman (1974) has obtained a square-root version of their algorithm. We have gone much further in this paper than have Bryson and Ho or Bierman. Our results give recursive equations for different types of optimal smoothed estimates of plant noise as well as recursive equations for respective error covariances.

COMPARISON WITH A PREDICTION ERROR FILTER

Ott and Meder (1972) define a "prediction error filter" (PEF) for $\mu(k)$ in terms of the difference between $\hat{x}(k + 1|k + 1)$ and $\hat{x}(k + 1|k)$. In this section we shall examine their estimator and compare its performance with our estimator $\hat{\mu}(k|k + 1)$ [theorem 1].

Combine equations (9) and (11) to show that,

$$\hat{x}(k + 1|k + 1) = \Phi\hat{x}(k|k) + K(k + 1)\,\bar{z}(k + 1|k), \tag{35}$$

in which we recall that $\bar{z}(k + 1|k)$ is the white innovations sequence. Now compare the structures of equation (35) and state equation (4) to obtain the following "correspondence":

$$\gamma\,\hat{\mu}_1(k) \triangleq K(k + 1)\,\bar{z}(k + 1|k). \tag{36}$$

In this equation, $\hat{\mu}_1(k)$ denotes the PEF for $\mu(k)$. Unfortunately, equation (36) is an over-determined system of equations; i.e., it is a system of n equations in one unknown $\hat{\mu}_1(k)$. We use the pseudo-inverse of γ, $\gamma\dagger$, where $\gamma\dagger = (\gamma'\,\gamma)^{-1}\gamma'$, to obtain the following[8] ad hoc PEF for $\mu(k)$:

$$\hat{\mu}_1(k) \triangleq \gamma\dagger\,K(k + 1)\,\bar{z}(k + 1|k). \tag{37}$$

Observe that $\hat{\mu}_1(k)$ enjoys the property of being a white estimator of the reflection coefficient sequence; however, $\hat{\mu}_1(k)$ is not an optimal estimator of $\mu(k)$. It has merely been defined as in equation (37). As such, we refer to the PEF as an ad hoc estimator. Ob-

serve, also, that both $\hat{\mu}_1(k)$ and $\hat{\mu}(k|k + 1)$ look one step into the future; i.e., they both use measurement $z(k + 1)$ to estimate $\mu(k)$.

We wish to compare $\hat{\mu}_1(k)$ and $\hat{\mu}(k|k + 1)$ on the basis of error variances. Let $\sigma^2_{\mu 1}(k)$ denote the error variance for the ad hoc estimator $\hat{\mu}_1(k)$; then, Mendel (1977) has shown that

$$\sigma^2_{\mu 1}(k) = q - 2\gamma\dagger Ph(h'Ph + r)^{-1}h'\gamma q + \gamma\dagger Ph(h'Ph + r)^{-1}h'P\gamma\dagger, \tag{38}$$

in which P is short for $P(k + 1|k)$. It is relatively straightforward to show (Mendel, 1976b) that

$$\sigma^2_{\mu}(k|k + 1) < \sigma^2_{\mu 1}(k), \tag{39}$$

which means, of course, that one will obtain better performance with our single-stage estimator, $\hat{\mu}(k|k + 1)$, than with the PEF, $\hat{\mu}_1(k)$.

Additionally, since we have also shown that we can improve our performance over that obtainable by $\hat{\mu}(k|k + 1)$ by using $\hat{\mu}(k|k + l)$, where $l \geq 2$, it follows that we should be able to achieve much better performance by using $\hat{\mu}(k|k + l)$ instead of $\hat{\mu}_1(k)$. We support these theoretical observations below with simulations.

SIMULATION RESULTS

In this section we present some simulation results which illustrate different aspects of our l-stage smoothers.

Primary reflection data

In this first experiment, we generate a random reflection coefficient sequence; compute the noise-free seismic trace for one of the wavelets in Table 1; add gaussian white measurement noise to the noise-free seismic trace, to obtain the seismic trace $z(k)$ (see Figure 3); and process $z(k)$ by both our l-stage smoothers and Ott and Meder's prediction error filter. Conceptually, we can think of the random reflection coefficient sequence as the primary reflection component of the impulse response of a randomly layered system.

The reflection coefficient sequence $\mu(k)$ must be impulsive in nature and white. A Poisson impulse sequence with zero mean uncorrelated gaussian[9] amplitudes was chosen for $\mu(k)$. In this case,

$$\mu(k) = \sum_i a_i\,\delta(k - m_i),$$

where m_i is a randomly occuring point in time that

---

[8]Ott and Meder do not use the pseudo-inverse of γ. They obtain $\hat{\mu}_1(k)$ directly from equation (36). This leads to n possible estimates; but, no rule is given for which one to choose.

[9]The gaussian nature of the amplitudes was suggested to us by Dr. Roy Pusey of Chevron Oil Field Research Co., La Habra, Calif.

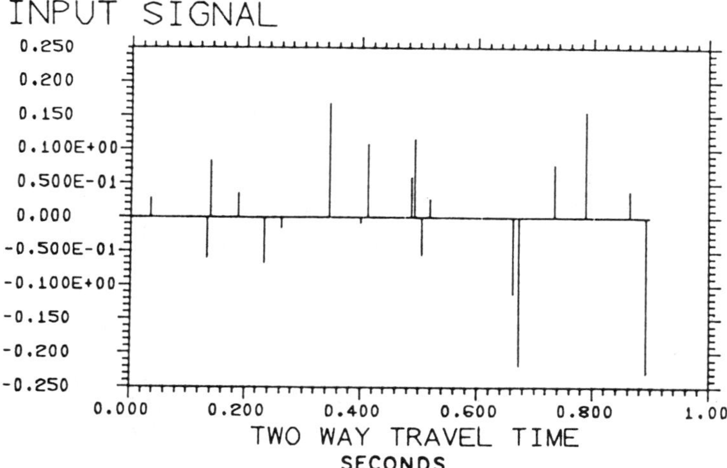

FIG. 6. A reflectivity sequence (Poisson impulse sequence with gaussian uncorrelated amplitudes).

can take on discrete values $\{0, 1, \ldots\}$; the a_i are identically distributed gaussian uncorrelated random variables which are independent of the m_i; and $\delta(j) = 1$ if $j = 0$ and $\delta(j) = 0$ if $j \neq 0$. One can show[10] that $E\{\mu(k)\} = \bar{a}\lambda$ and $E\{\mu(k)\mu(j)\} = \lambda(\sigma_a^2 + \bar{a}^2)\delta_{kj}$, where $\bar{a} = E\{a_i\}$, $\sigma_a^2 = E\{a_i^2\}$, and $\lambda = n/\tau$, where n is the number of random points which occur in the interval $(0, \tau)$. In order to keep the amplitude of $\mu(k)$ below 0.30 [a geological constraint (Anstey, 1970)], we chose $\sigma_a = 0.15$; λ was chosen to be 0.05. In this case, since $\bar{a} = 0$, $q = \lambda\sigma_a^2 = 0.1125 \times 10^{-2}$. A number of realizations of $\mu(k)$ were generated, one of which (Figure 6) was chosen for our simulations because it looked geologically interesting. We attribute no other geological plausibility to this example.

Simulations have been run for both the second- and fourth-order basic seismic wavelets which appear in the first and third rows of Table 1. Results are presented below only for the fourth-order wavelet. A sampling time (i.e., T) of 3 msec was used in all our simulations.

The ideal seismic trace $V_R(k)$ was corrupted by additive gaussian noise $n(k)$, whose variance r was chosen so that signal-to-noise ratio (S/N) is fixed at prespecified levels of 20, 10, 5, 2, or 1. An expression for S/N ratio is obtained from equations (4) and (5), as follows:

$$\frac{S}{N} = \frac{E\{[\mathbf{h}'\mathbf{x}(k)][\mathbf{h}'\mathbf{x}(k)]'\}}{r}, \qquad (40)$$

$$\frac{S}{N} = \frac{\mathbf{h}'[\Phi P_x \Phi' + q\,\gamma\,\gamma']}{r}, \qquad (41)$$

where $\overline{P_x}$ denotes the steady-state solution of the equation (4) covariance equation $P_x(k + 1) = \Phi P_x(k)\Phi' + q\,\gamma\,\gamma'$ (Meditch, 1969). A trace of $z(k)$ for S/N = 10 is depicted in Figure 7.

The alternative l-stage fixed point algorithm, described above, as well as the Ott and Meder prediction error filter, were programmed, and a number of different experiments were performed, the results of which are displayed in Figures 8–13.

Observe, from Figure 8, where we have plotted[11] $\sigma_\mu^2(80|80 + l)$ versus l for the five S/N ratios of interest, that a significant improvement in performance is obtained by choosing $l > 1$. This result is strongly dependent upon the order of the seismic wavelet. We observed that, for the second-order wavelet, there was almost no improvement in performance, for all S/N ratios, when $l > 1$. Observe, from Figure 8, that as S/N ratio gets larger, there is an increase in percentage decrease of error variance (relative to q, shown at $l = 0$) ranging, for $l = 1$ from 19.6 to 66.5 percent and for $l = 8$ from 30.8 to 83.3 percent.

We can expect the biggest payoff for using larger values of l for large S/N ratios. The reason for this is that as S/N $\to 0$, $r \to \infty$, and $K \to 0$; hence, since each $N_\mu(k|k + j)$ depends upon $K(k + j)$, as S/N $\to 0$, all $N_\mu \to 0$, and, therefore [see equation (31)],

---

[10]The calculations are similar to those in Papoulis (1965).

[11]The value $k = 80$ was chosen somewhat arbitrarily; by $k = 80$, the Kalman filter and $P_x(k)$ have reached steady-state values.

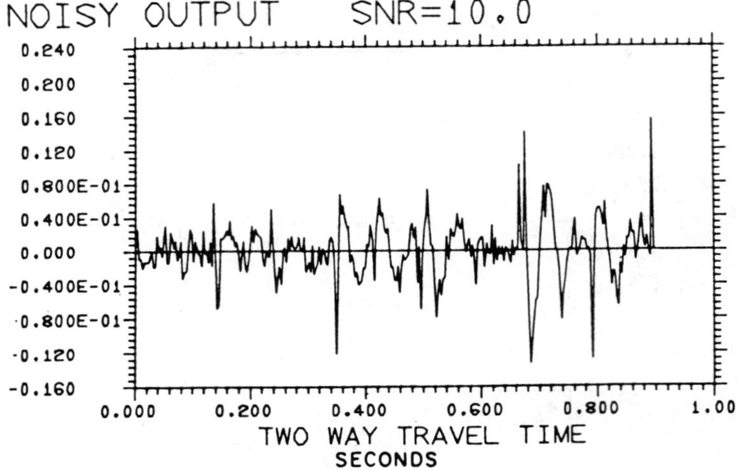

FIG. 7. Seismic trace for S/N = 10.

$\sigma_\mu^2(k|k + l) \to \dot{q}$, regardless of l. This behavior is already evident in Figure 8 for S/N = 1.

We depict $\sigma_{\mu_1}^2(k)$ for the PEF in Figure 9. Upon comparison with respective S/N curves in Figure 8, we see, for example, that for S/N = 10 and $l = 1$ the optimal smoother (OS) is[12] 18 percent better than the PEF, whereas for $l = 6$, the OS is 87 percent better than the PEF. Observe, also, that as S/N $\to 0$, $\sigma_{\mu_1}^2(k) \to q = 0.1125 \times 10^{-2}$. The reason for this can be inferred from equation (38); for, when $r \to \infty$, $\sigma_{\mu_1}^2 \to q$.

Figures 10–13 depict estimates of the reflection

coefficient sequence $\mu(k)$ for a number of different situations. For S/N = 10, we observe from Figures 10–12, that the OS for $l = 1$ does a substantially better job at deconvolving the reflection coefficient sequence than the PEF does, and that visibly significant improvement in deconvolution is obtained using $l = 5$ instead of $l = 1$. Finally, Figure 13 demonstrates that a reasonably good deconvolution of the reflection coefficient sequence is accomplished for a much lower S/N ratio, S/N = 5, and $l = 5$.

Synthetic seismogram data

Our simulation results above have been for a contrived situation where the layered-media system is assumed to be completely represented by the reflec-

[12]These comparisons utilize the steady-state PEF error-covariance values.

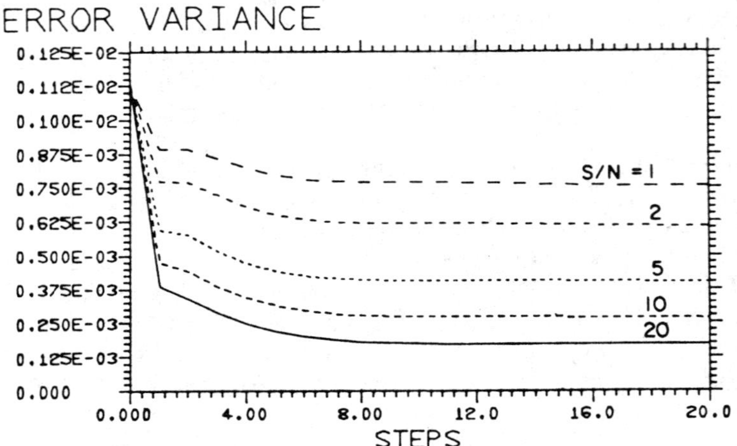

FIG. 8. $\sigma_\mu^2(80|80 + l)$ versus l.

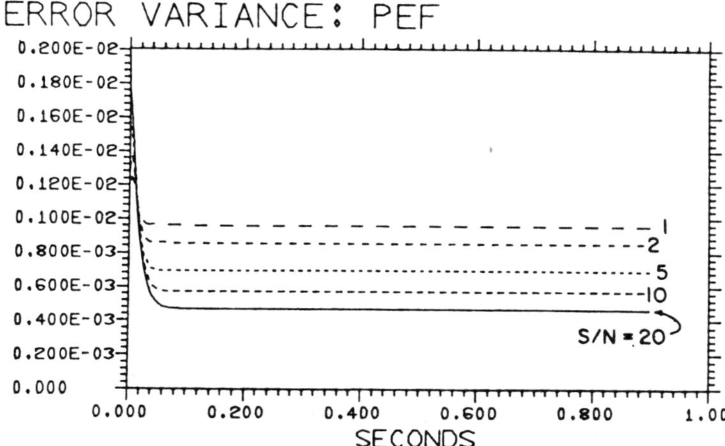

FIG. 9. $\sigma_{\mu_1}^2(k)$ versus $kT(T = 3$ msec$)$.

tion coefficient sequence, which, as we pointed out above, can be associated with the primary reflections. In effect, then, our signal (Figure 3) represents only the primary reflection component of a complete seismogram. It does not include the multiple reflections, which are always present in a real seismogram. In this experiment we associate the 19 spikes in Figure 6 with a 19-layer media system whose reflection coefficients (amplitudes of the spikes) and respective one-way traveltimes are given in Table 2. We then repeated the procedure listed in the first paragraph in "Primary reflection data", using synthetic seismogram data for $V_R(k)$.

Simulations were run for the fourth-order seismic wavelet used above. A trace of $z(k)$ for S/N = 10 is depicted in Figure 14. Figure 15 depicts estimates of $\mu(k)$ for S/N = 10 and $l = 5$. Comparing it with Figure 12, we observe that the OS does almost as well when synthetic seismogram data are used. The multiples which are now in $z(k)$ do not seem to degrade the performance of the OS, at least for this 19-layer system.

Nonnormal incident data were generated by means of the one-way traveltime approximation given in Nahi and Mendel (1976). Results for an end-on spread with sensors spaced 200 ft are depicted in Figures 16 and 17. Due to the shallow nature of our 19-layer media system (approximately 3000 ft) and the re-

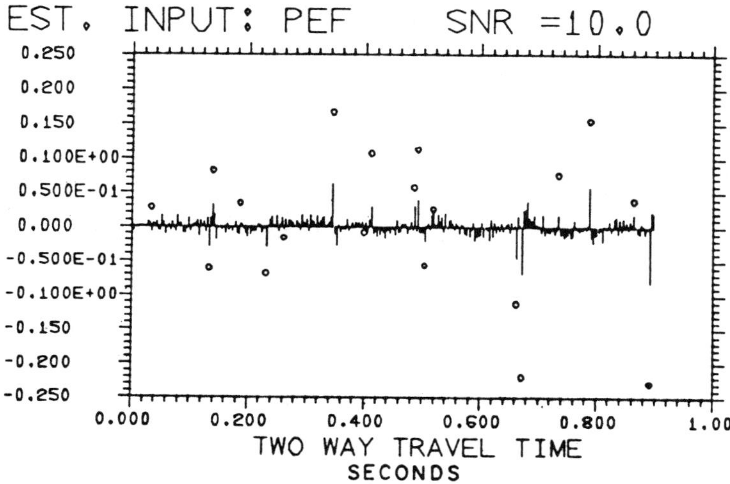

FIG. 10. Output of PEF, $\hat{\mu}_1(k)$ (solid lines) compared with $\mu(k)$ (circles).

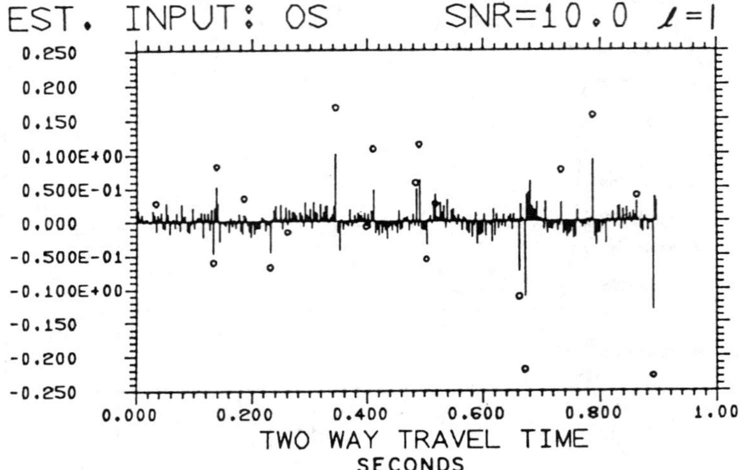

FIG. 11. Output of OS, $\hat{\mu}(k|k + 1)$ (solid lines) compared with $\mu(k)$ (circles).

latively long moveout, the Nahi and Mendel (1976) approximation is very poor. This explains the incorrect steepening slope, with increasing depth, of the hyperbolic-like time-distance curves on these figures. No geologic plausibility should be attributed to our simulated nonnormal incident data. We have included Figures 16 and 17 merely to illustrate the potential utility of our l-stage smoothers on nonnormal incident data. Limited resources prevented us from generating accurate data.

Spherical divergence compensation

Due to spherical divergence, the amplitude of a reflected wavelet is attenuated by a function of depth.

This attenuation, if not compensated, will produce false values for the reflection coefficients. That is to say, there exists some gain function $d(k)$ that relates the observed reflection coefficient $\mu(k)$ to the actual (plane-wave) reflection coefficient $\mu_a(k)$ according to,

$$\mu(k) = \mu_a(k)/d(k), \tag{42}$$

where $\mu(k)$ is defined by equation (2). The form of this gain function is dependent upon the unknown subsurface structure, but is usually approximated by

$$d(k) = \frac{kT\,\overline{V}(kT)}{V_1}, \tag{43}$$

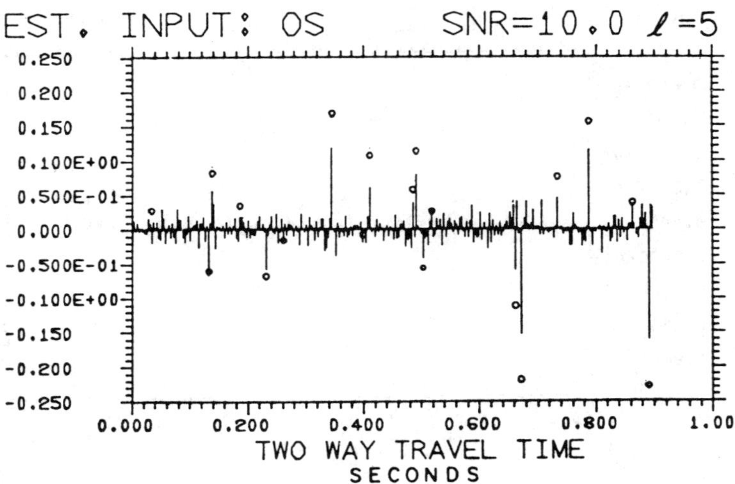

FIG. 12. Output of OS, $\hat{\mu}(k|k + 5)$ (solid lines) compared with $\mu(k)$ (circles).

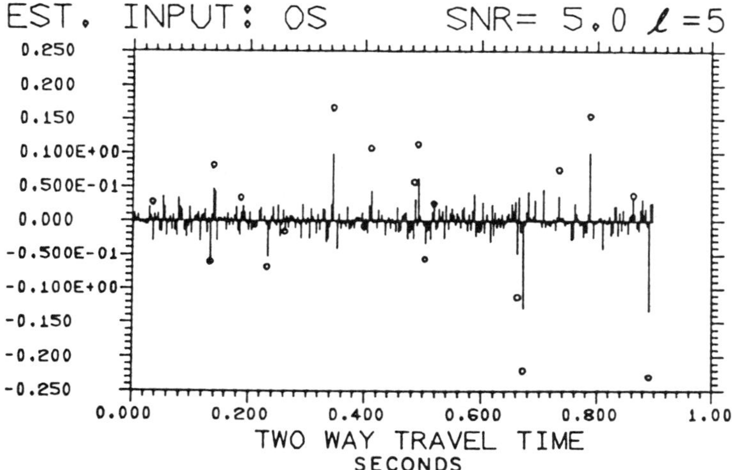

FIG. 13. Output of OS, $\hat{\mu}(k|k + 5)$ (solid lines) for S/N = 5 compared with $\mu(k)$ (circles).

where T is the sampling interval, V_1 is the wave velocity in the first layer, and $\overline{V}(kT)$ is the time-weighted rms velocity as a function of the two-way traveltime (Newman, 1973; O'Doherty and Anstey, 1971).

More importantly, if one assumes that the actual reflection coefficient sequence is a stationary process, then the effective reflection coefficient sequence is nonstationary, violating the assumptions of Wiener filtering. Attempts to overcome this problem using prescaling and time gating (Wang, 1969) are more complex and, as we will show, inferior to the optimal smoother approach.

Here we briefly outline one solution to this problem which uses our l-stage optimal smoother. Additional details are given in Kormylo and Mendel, 1977. Our solution uses a state vector model of the form,

$$\mathbf{x}(k + 1) = \Phi \mathbf{x}(k) + \gamma(k) \mu_a(k), \qquad (44)$$

where here $\gamma(k)$ is a time varying parameter vector. Substituting equation (42) into equation (4), we see that the resulting description is equivalent to equation (44), if

$$\gamma(k) = \gamma \frac{1}{d(k)}, \qquad (45)$$

where γ is chosen to fit the source signature as before. The covariance of the actual reflection coefficient sequence,

$$E\{\mu_a(i) \mu_a(j)\} = q \delta_{ij}, \qquad (46)$$

now replaces equation (6a); and, the estimates $\hat{\mu}_a(k|k + l)$ are found using the same filter equations as before—equations (9)–(13), (23)–(26), substitut-

Table 2. Parameters for 19-layer media system

Layer/ interface	Reflection coefficient	Travel-time	Velocity[1]	Layer/ interface	Reflection coefficient	Travel-time	Velocity
1	.028	.016	5000	11	.114	.003	4742
2	−.061	.050	5711	12	−.057	.006	4992
3	.082	.004	6465	13	.026	.007	8439
4	.034	.023	9251	14	−.112	.072	8056
5	−.068	.022	6362	15	−.220	.005	8710
6	−.016	.015	5229	16	.076	.030	5409
7	.168	.042	5122	17	.156	.027	6549
8	−.008	.028	7690	18	.039	.038	10135
9	.108	.006	9644	19	−.229	.014	5632
10	.058	.038	8183				

[1] Velocity, in ft/sec, was computed from the equation $v_{i+1} = v_i[(1 + r_i)/(1 - r_i)]^{0.80}$, which derives from the equation for r_i, that $r_i = (\rho_{i+1} v_{i+1} - \rho_i v_i)/(\rho_{i+1} v_{i+1} + \rho_i v_i)$, and the quarter-power law for ρ_i (Gardner et al, 1974), that $\rho_i \approx 0.23\, v_i^{0.25}$.

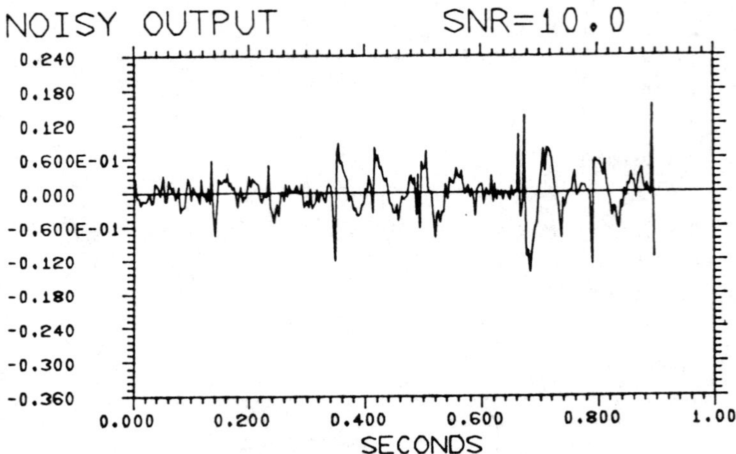

FIG. 14. Synthetic seismogram trace for S/N = 10.

ing $\gamma(k)$ for γ and μ_a for μ. These estimates are optimal estimates of the actual reflection coefficient sequence, $\mu_a(k)$.

Signal-to-noise ratio is a poorly defined quantity for nonstationary data; so, it is difficult to characterize the noise level for these systems. If we use the definition in equation (40), we obtain an SNR which is a function of time; i.e.,

$$\frac{S}{N}(k) = \frac{\mathbf{h}' P_x(k) \mathbf{h}}{r}, \qquad (47)$$

where the covariance matrix, $\mathbf{P}_x(k)$, is generated recursively from

$$P_x(k + 1) = \Phi P_x(k) \Phi' + q\,\gamma(k)\,\gamma'(k), \qquad (48)$$

and $P_x(0) = 0$. The value of this signal-to-noise ratio function may range over several orders of magnitude, so that any sort of average value is somewhat meaningless. *We shall therefore characterize a nonstationary signal by the minimum value of its SNR over the time-domain of interest.*

A simulation was run using the same fourth-order wavelet model and the same reflection coefficient sequence $\mu_a(k)$ as in "primary reflection data," above. Figure 18 depicts the estimates produced when $l = 5$ and r is chosen to produce a minimum SNR of 1. To simplify matters, we used the gain function

$$d(k) = k, \qquad (49)$$

to produce $\gamma(k)$ in equation (45). By comparing

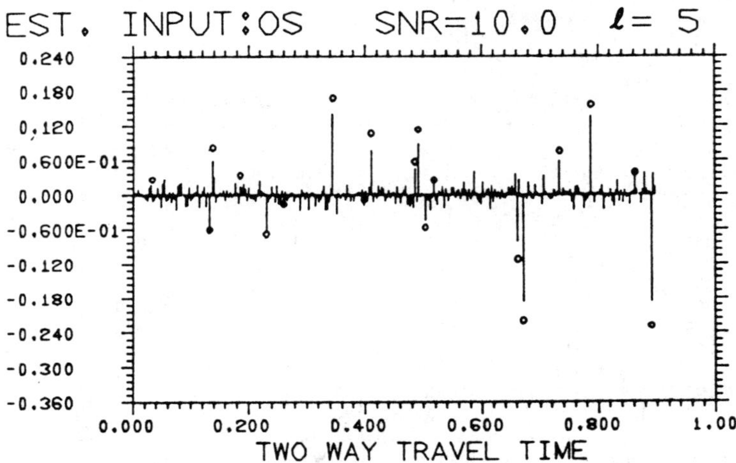

FIG. 15. Output of OS, $\hat{\mu}(k|k + 5)$ (solid lines) compared with $\mu(k)$ (circles), for synthetic seismogram data.

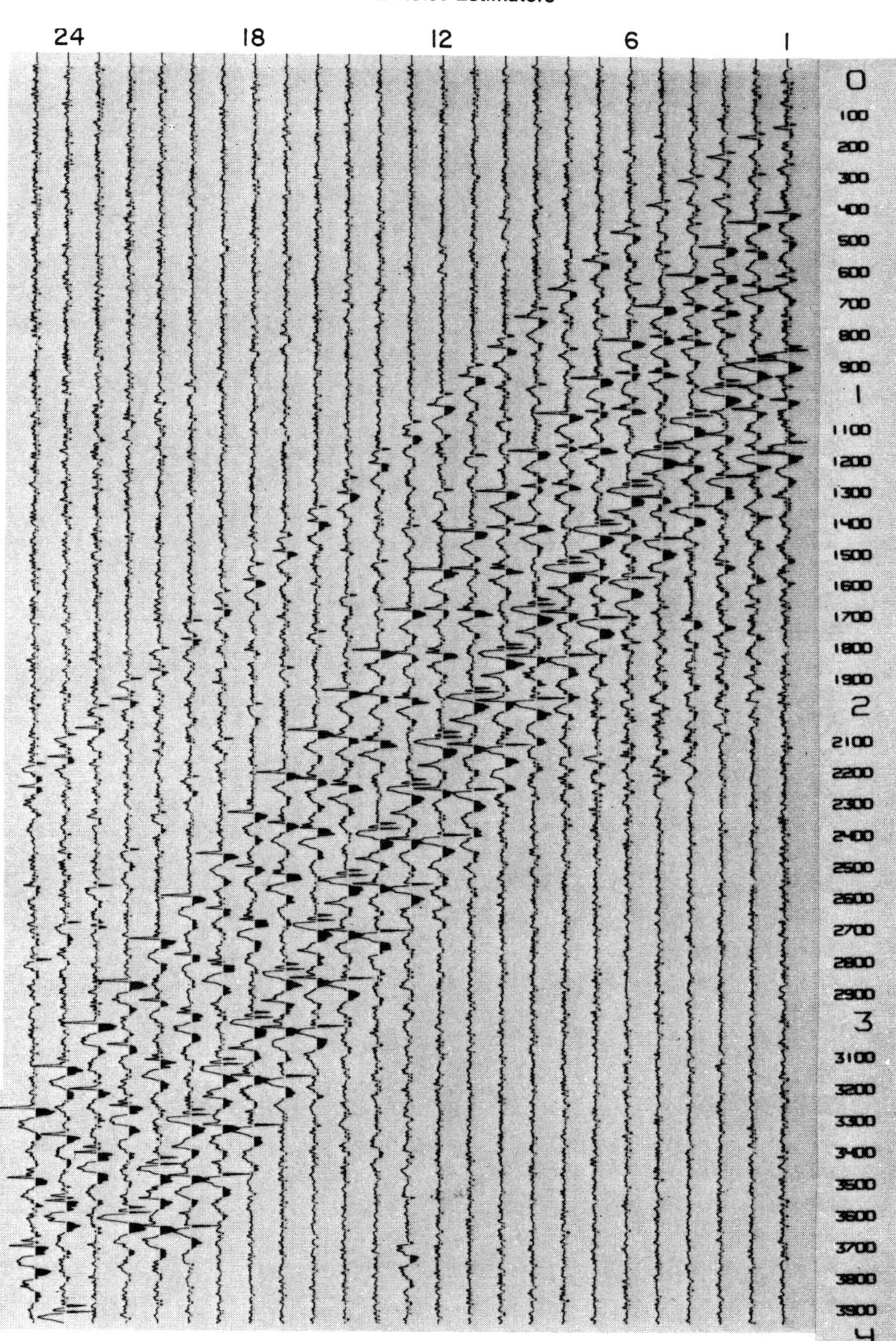

FIG. 16. Approximate nonnormal incident data for S/N = 10.

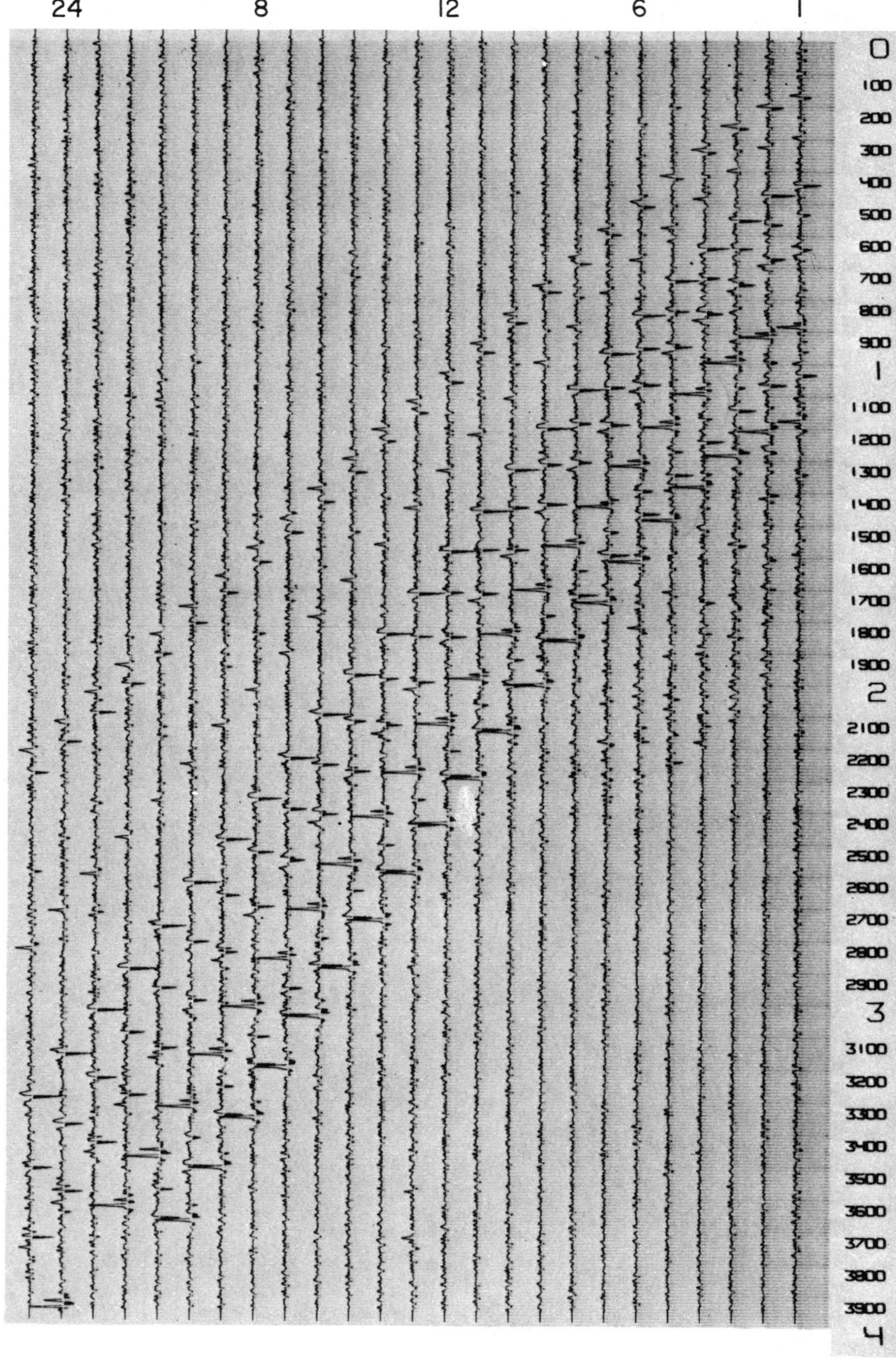

FIG. 17. Optimally smoothed ($l = 5$) approximate nonnormal incident data.

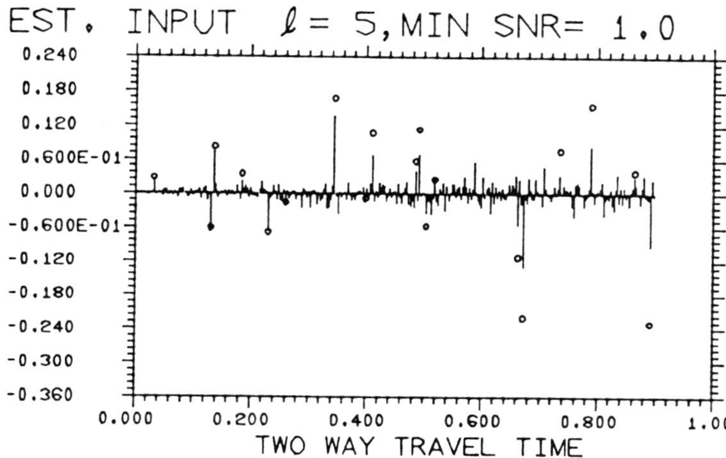

FIG. 18. Output of OS, $\hat{\mu}_a(k|k + 5)$ (solid lines) compared with $\mu_a(k)$ (circles).

Figure 18 to the stationary cases in Figures 12 and 13, we see that very good estimates are obtained for early arrivals, and progressively poorer estimates are obtained for later arrivals, as one would expect since the SNR continuously decreases with traveltime. This is borne out by the estimated error-variance, depicted in Figure 19, which monotonically increases with time. Other runs with lower SNRs showed that the error variance asymptotically approaches $q = 0.1125 \times 10^{-2}$.

For comparison, we also processed the same data using prescaling and a Wiener filter (but no time-gating). To make the comparison as fair as possible, the Wiener filter was designed to use the same data as the optimal smoother. In other words, the filter was an IIR realization with an l-step delay. Furthermore, the filter was designed to be optimal in the steady state for the system parameters at $t = 0.45$ sec ($K = 150$). The formula for the error variance using this filter was derived (Kormylo and Mendel, 1977); values of that error variance are depicted in Figure 19. There, we see that the Wiener filter suffers from a large transient error, that it equals the performance of the Kalman filter only at its design point ($t = 0.45$ sec), and that for $t > 0.45$ the error becomes unmanageable. For this example (early arrivals and a small SNR), the optimal smoother is substantially better for deconvolution than the Wiener filter.

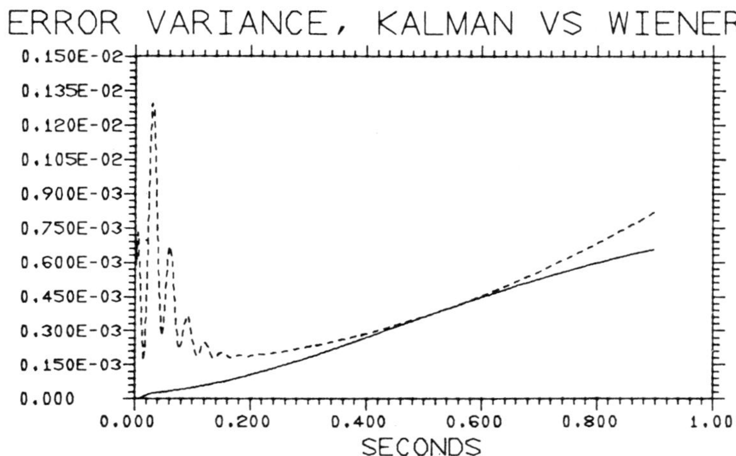

FIG. 19. $\sigma^2_{\mu_a}(k|k + 5)$ versus kT ($T = 3$ msec) for optimal smoother (solid line) and Wiener filter (dashed line).

297

CONCLUSIONS

We have applied minimum-variance estimation theory directly to the problem of estimating the reflection coefficient sequence, and have shown that the problem is equivalent to one of estimating the random disturbance in a state equation. We have also shown that minimum-variance estimates of the reflection coefficient sequence are optimal smoothed estimates; an optimal filtered estimate of that sequence does not exist.

Formulas have been developed for single-stage and l-stage ($l = 1, 2, \ldots$) optimal smoothed estimates of the reflection coefficient sequence, and for error-variances associated with these estimates. The formulas for the l-stage estimators are given in both recursive and nonrecursive forms. From these formulas it is possible to do a variety of different types of signal processing, including optimal fixed-point, fixed-lag, and fixed-interval smoothing.

We have examined Ott and Meder's (1972) PEF and have shown that our optimal smoothed estimates always give better performance than does their PEF; i.e., our optimal smoothed estimates always have lower error variance than the error variance of the PEF.

Simulations which illustrate different aspects of our l-stage smoothers and which contrast their performance against the Ott and Meder PEF have been given. Additionally, we have shown how to compensate for spherical divergence effects, and that it is easy to do this by means of our l-stage estimators.

Total emphasis in this paper has been on the development of optimal linear estimators for white plant noise. We have explained that, in the predictive deconvolution application, the white plant noise is nongaussian and impulsive in nature; hence, it is a point process. Additional work in the study of estimators of point processes, which may be useful for seismic data processing, is underway.

ACKNOWLEDGMENTS

The work reported in this paper was performed at the University of Southern California, Los Angeles, under National Science Foundation grant NSF ENG 74-02297A01; Chevron Oil Field Research Co. contract-76; and Teledyne Exploration Co., contract TEC-76. The authors would like to acknowledge the very helpful discussions they had with some of the staff at Chevron Oil Field Research Co. about different aspects of seismic data processing problems in oil exploration and seismic sources, especially Dr. Roy Pusey and Dr. Dan Stalmach; with Dr. Benjamin Friedlander (formerly with Stanford University) of Systems Control, Inc., about fast implementations of the l-stage optimal smoother algorithms; and with Dr. Keith Glover (formerly of the University of Southern California), of Cambridge University, about approximate state space realizations. They would also like to thank Michael Chan, a graduate student in the Electrical Engineering Department, University of Southern California, for generating the synthetic seismogram data, and Chevron Oil Field Research Co. for use of their seismic section plotter. Finally, the first author would like to thank the reviewers of the paper, especially Dr. Milo Backus, for their very comprehensive reviews and constructive suggestions.

REFERENCES

Anstey, N. A., 1970, Seismic prospecting instruments, vol. 1: Signal characteristics and instrument specifications: Berlin, Gebruder Borntraeger.

Bayliss, J. W., and Brigham, E. O., 1970, Application of the Kalman filter to continuous signal restoration: Geophysics, v. 35, p. 2–23.

Berkhout, A. J., and Zaanen, P. R., 1976, A comparison between Wiener filtering, Kalman filtering, and deterministic least-squares estimation: Geophys. Prosp., v. 24, p. 141–197.

Bierman, G. J., 1974, Sequential square root filtering and smoothing of discrete linear systems: Automatica, v. 10, p. 147–158.

Box, G. E. P., and Jenkins, G. M., 1970, Time series analysis, forecasting, and control: San Francisco, Holden Day, Inc.

Bryson, A. E., Jr., and Ho, Y. C., 1969, Applied optimal control: New York, Blaisdell.

Chen, C. T., 1970, Introduction to linear system theory: New York, Holt-Rinehart and Winston.

Crawford, J. M., Doty, W. E. N., and Lee, M. R., 1960, Continuous signal seismograph: Geophysics, v. 25, p. 95–105.

Crump, N. D., 1974, A Kalman filter approach to the deconvolution of seismic signals: Geophysics, v. 39, p. 1–13.

———— 1975, Techniques for the deconvolution of seismic signals: Proc. IEEE Confer. on decision and control, Houston, p. 2–8.

Faurre, P. L., 1976, Stochastic realization algorithms, *in* System identification: advances and case studies: R. K. Mehra and D. G. Lainiotis, editors, New York, Academic Press, p. 1–25.

Gardner, G. H. F., Gardner, L. W., and Gregory, A. R., 1974, Formation velocity and density: The diagnostic basis for stratigraphic traps: Geophysics, v. 39, p. 770–780.

Gelb, A. (editor), 1974, Applied optimal estimation: Cambridge, MIT Press.

Giles, B. F., 1967, Pneumatic acoustic energy source: Presented at 29th annual meeting of EAEG, Stockholm.

Jazwinskii, 1970, Stochastic processes and filtering theory: New York, Academic Press.

Kailath, T., 1968, An innovations approach to least-squares estimation—Part 1: Linear filtering in additive white noise: IEEE Trans. on automatic control, v. AC-13, p. 646–655.

———— 1974, A view of three decades of linear filtering theory: IEEE Trans. on infor. theory, v. IT-20, p. 146–181.

Kalman, R. E., 1960, A new approach to linear filtering and prediction problems: Trans. ASME, J. Basic Eng., ser. D, v. 82, p. 34–45.

Kormylo, J., and Mendel, J. M., 1977, On the treatment of spherical divergence for Kalman filtering deconvolution: Presented at the 47th Annual International SEG Meeting, Calgary.

Kramer, F. J., Peterson, R. W., and Walter, W. C. (editors), 1968, Seismic energy sources 1968 handbook: Presented at the 38th Annual International SEG Meeting, Denver.

Mayne, W. Harry, and Quay, R. G., 1970, Seismic signatures of large air guns: Presented at the 40th Annual International SEG Meeting, New Orleans.

Meditch, J. S., 1969, Stochastic optimal linear estimation and control: New York, McGraw-Hill Book Co., Inc.

Mendel, J. M., 1973, Discrete techniques of parameter estimation: The equation error formulation: New York, Marcel Dekker, Inc.

———— 1976a, White noise estimators for seismic data processing in oil exploration: Proc. Joint Automatic Control confer., Lafayette, Ind. Purdue Univ., p. 642–651.

———— 1976b, Single-channel white noise estimators for predictive deconvolution: Presented at the 46th Annual International SEG Meeting, Houston.

———— 1977, A quantitative evaluation of Ott and Meder's prediction error filter: Geophys. Prosp. (in press).

Mendel, J. M., and Kormylo, J., 1977, New fast optimal white-noise estimators for deconvolution: Spec. iss., IEEE Trans. on Geosci. Electron, no. GE-15, p. 32–41.

Nahi, N. E., and Mendel, J. M., 1976, A time-domain approach to seismogram synthesis for layered media: Presented at the 46th Annual International SEG Meeting, Houston.

Newman, P., 1973, Divergence effects in a layered earth: Geophysics, v. 38, p. 481–488.

O'Doherty, R. F., and Anstey, N. A., 1971, Reflections on amplitudes: Geophys. Prosp., v. 19, p. 430–458.

Ott, N., and Meder, H. G., 1972, The Kalman filter as a prediction error filter: Geophys. Prosp., v. 20, p. 549–560.

Papoulis, A., 1965, Probability, random variables, and stochastic processes: New York, McGraw-Hill Book Co., Inc.

Ricker, N., 1940, The form and nature of seismic waves and the structure of seismograms: Geophysics, v. 5, p. 348–366.

Robinson, E. A., 1957, Predictive decomposition of seismic traces: Geophysics, v. 22, p. 767–778.

———— 1967, Predictive decomposition of time series with application to seismic exploration: Geophysics, v. 32, p. 414–484 (1954 Ph.D. thesis).

Schwarz, R. J., and Friedland, B., 1965, Linear systems: New York, McGraw-Hill Book Co., Inc.

Smith, S. G., 1975, Measurement of airgun waveforms: Geophys. J. R. Astr. Soc., v. 42, p. 273–280.

Sorenson, H. W., 1966, Kalman filtering techniques, *in* Advances in control systems, theory and applications, vol. 3: C. T. Leondes, editor, New York, Academic Press, p. 219–292.

Wang, J., 1969, The determination of optimal gate lengths for time-varying Wiener filtering: Geophysics, v. 34, p. 683–695.

Wood, L. C., and Treitel, S., 1975, Seismic signal processing: Proc. IEEE, v. 63, p. 649–661.

APPENDIX A.
PROOF OF THEOREM 1

Derivation of equation (18)

Meditch (1969) derives the following equation for $\hat{x}(k|k+1)$:

$$\hat{x}(k|k+1) = \hat{x}(k|k) + A(k) K(k+1) \tilde{z}(k+1|k), \quad (A\text{-}1)$$

where

$$A(k) = P(k|k) \Phi' P^{-1}(k+1|k). \quad (A\text{-}2)$$

Substitute equations (A–1) and (11) into (15), making use of equation (9), to show that

$$
\begin{aligned}
\gamma \hat{\mu}(k|k+1) \\
&= \hat{x}(k+1|k) + K(k+1)\tilde{z}(k+1|k) \\
&\quad - \Phi[\hat{x}(k|k) + A(k)K(k+1)\tilde{z}(k+1|k)], \\
&= \Phi\hat{x}(k|k) + [I - \Phi A(k)]K(k+1)\tilde{z}(k+1|k) \\
&\quad - \Phi\hat{x}(k|k), \\
&= [I - \Phi A(k)] K(k+1) \tilde{z}(k+1|k). \quad (A\text{-}3)
\end{aligned}
$$

Next, premultiply both sides of equation (A–2) by Φ,

$$\Phi A(k) = \Phi P(k|k) \Phi' P^{-1}(k+1|k); \quad (A\text{-}4)$$

but, from equation (10), observe that

$$\Phi P(k|k) \Phi' = P(k+1|k) - \gamma q \gamma'. \quad (A\text{-}5)$$

Substitute equation (A–5) into (A–4) to show that

$$\Phi A(k) = I - \gamma q \gamma' P^{-1}(k+1|k); \quad (A\text{-}6)$$

hence,

$$I - \Phi A(k) = \gamma q \gamma' P^{-1}(k+1|k). \quad (A\text{-}7)$$

Finally, substitute equation (A–7) into (A–3) to obtain

$$\hat{\mu}(k|k+1) = q \gamma' P^{-1}(k+1|k) K(k+1) \tilde{z}(k+1|k), \quad (A\text{-}8)$$

which is equation (18).

Derivation of equation (19)

As mentioned in Single-stage optimal smoothed estimates of $\mu(k)$, $\tilde{z}(k+1|k)$ is a zero-mean white-noise sequence, and (see Kailath, 1968),

$$E\{\tilde{z}^2(k+1|k)\} = h' P(k+1|k)h + r. \quad (A\text{-}9)$$

We shall use this fact in our development of $\sigma_\mu^2(k|k+1)$.

In order to evaluate $\sigma_\mu^2(k|k+1)$ from equation (16), we must first compute $E\{\bar{\mu}(k|k+1)\}$. From equation (17), we see that

$$E\{\bar{\mu}(k|k+1)\} = E\{\mu(k)\}$$
$$- E\{\hat{\mu}(k|k+1)\} = 0, \quad (A-10)$$

since $E\{\mu(k)\} = 0$ by definition, and

$$E\{\hat{\mu}(k|k+1)\} = q\,\boldsymbol{\gamma}'\,P^{-1}(k+1|k)\,K(k+1)\,E\{\tilde{z}(k+1|k)\}; \quad (A-11)$$

but, as pointed out above, $E\{\tilde{z}(k+1|k)\} = 0$. We see, therefore, that

$$\sigma_\mu^2(k|k+1) = E\{\bar{\mu}^2(k|k+1)\}$$
$$= E\{[\mu(k) - \hat{\mu}(k|k+1)]^2\}$$
$$= q - 2E\{\mu(k)\,\hat{\mu}(k|k+1)\}$$
$$+ E\{\hat{\mu}^2(k|k+1)\}. \quad (A-12)$$

The last term in equation (A-12) is easily developed further by means of equations (A-8), (A-9), and (12) as follows:[13]

$$E\{\hat{\mu}^2(k|k+1)\} = q\,\boldsymbol{\gamma}'\,P^{-1}(k+1|k)$$
$$K(k+1)\,E\{\tilde{z}^2(k+1|k)\}$$
$$K'(k+1)\,P^{-1}(k+1|k)\,\boldsymbol{\gamma}q,$$

$$E\{\hat{\mu}^2(k|k+1)\} = q\,\boldsymbol{\gamma}'\,P^{-1}(k+1|k)\,K(k+1)$$
$$[\mathbf{h}'P(k+1|k)\mathbf{h}+r]$$
$$K'(k+1)\,P^{-1}(k+1|k)\,\boldsymbol{\gamma}q$$
$$= q\,\boldsymbol{\gamma}'\mathbf{h}[\mathbf{h}'P(k+1|k)\mathbf{h}$$
$$+ r]^{-1}\mathbf{h}'\,\boldsymbol{\gamma}q. \quad (A-13)$$

Next, we shall develop $E\{\mu(k)\,\hat{\mu}(k|k+1)\}$. From equation (A-8),

$$E\{\mu(k)\,\hat{\mu}(k|k+1)\} = q\,\boldsymbol{\gamma}'\,P^{-1}(k+1|k)$$
$$K(k+1)\,E\{\mu(k)\tilde{z}(k+1|k)\}. \quad (A-14)$$

At the end of Appendix A, we show that

$$E\{\mu(k)\,\tilde{z}(k+1|k)\} = \mathbf{h}'\,\boldsymbol{\gamma}q, \quad (A-15)$$

so that,

---

[13] We use the facts that the transpose of scalar $\hat{\mu}$ is the same as $\hat{\mu}$, and that covariance matrix $P(k+1|k)$ is symmetric; hence, $[P^{-1}(k+1|k)]' = P^{-1}(k+1|k)$.

$$E\{\mu(k)\,\hat{\mu}(k|k+1)\}$$
$$= q\,\boldsymbol{\gamma}'\,P^{-1}(k+1|k)\,K(k+1)\mathbf{h}'\,\boldsymbol{\gamma}q$$
$$= q\,\boldsymbol{\gamma}'\mathbf{h}[\mathbf{h}'P(k+1|k)\mathbf{h}+r]^{-1}\mathbf{h}'\,\boldsymbol{\gamma}q. \quad (A-16)$$

Finally, substitute equations (A-13) and (A-16) into (A-12) to show that,

$$\sigma_\mu^2(k|k+1) = q - q\,\boldsymbol{\gamma}'\mathbf{h}[\mathbf{h}'P(k+1|k)\mathbf{h}+r]^{-1}\mathbf{h}'\,\boldsymbol{\gamma}q, \quad (A-17)$$

which is equation (19).

We return now to the proof of equation (A-15), which is somewhat delicate in nature since it relies on a number of results, proven in detail by Meditch (1969), that we shall not reprove, but shall merely state.

To begin, substitute equation (5) into (20), to show that

$$\tilde{z}(k+1|k) = \mathbf{h}'\mathbf{x}(k+1) + n(k+1) - \mathbf{h}'\hat{\mathbf{x}}(k+1|k)$$
$$= \mathbf{h}'[\mathbf{x}(k+1) - \hat{\mathbf{x}}(k+1|k)] + n(k+1)$$
$$= \mathbf{h}'\tilde{\mathbf{x}}(k+1|k) + n(k+1), \quad (A-18)$$

where $\tilde{\mathbf{x}}(k+1|k)$, the single-stage predicted state estimation error, equals $\mathbf{x}(k+1) - \hat{\mathbf{x}}(k+1|k)$. Meditch (1969) derives the following equations for $\tilde{\mathbf{x}}(k+1|k)$ and $\tilde{\mathbf{x}}(k|k)$ [equations (5.63) and (5.64), respectively]:

$$\tilde{\mathbf{x}}(k+1|k) = \boldsymbol{\Phi}\tilde{\mathbf{x}}(k|k) + \boldsymbol{\gamma}\,\mu(k), \quad (A-19)$$

and

$$\tilde{\mathbf{x}}(k|k) = [I - K(k)\mathbf{h}']\,\boldsymbol{\Phi}\tilde{\mathbf{x}}(k-1|k-1)$$
$$+ [I - K(k)\mathbf{h}']\,\boldsymbol{\gamma}\,\mu(k-1)$$
$$- K(k)\,n(k). \quad (A-20)$$

Substitute equation (A-19) into (A-18), to show that

$$\tilde{z}(k+1|k) = \mathbf{h}'\boldsymbol{\Phi}\tilde{\mathbf{x}}(k|k)$$
$$+ \mathbf{h}'\boldsymbol{\gamma}\mu(k) + n(k+1). \quad (A-21)$$

Observe, from equation (A-20), that $\tilde{\mathbf{x}}(k|k)$ does not depend upon $\mu(k)$; it depends at most upon $\mu(k-1)$; hence,

$$E\{\mu(k)\,\tilde{z}(k+1|k)\} = \mathbf{h}'\,\boldsymbol{\Phi}E\{\mu(k)\,\tilde{\mathbf{x}}(k|k)\} + \mathbf{h}'\,\boldsymbol{\gamma}E\{\mu^2(k)\} + E\{\mu(k)\,n(k+1)\}$$
$$= 0 + \mathbf{h}'\,\boldsymbol{\gamma}q + 0, \quad (A-22)$$

which is equation (A-15). In the development of equation (A-22), we have used the additional facts that $\mu(k)$ and $n(k)$ are uncorrelated, and $\mu(k)$ is a zero-mean sequence.

APPENDIX B
DERIVATION OF A SYNTHETIC SOURCE STATE
SPACE MODEL

Here we obtain the state space model, given in the third row of Table 1, for the synthetic source signature,

$$V^+(t) = -1360t\, e^{-500t} + 0.5\, e^{-15.3t} \sin\left(\frac{2\pi}{.06}t\right),$$
$$(B-1)$$

by means of the first three steps of the first procedure. To begin, we take the Laplace transform of $V^+(t)$:

$$V^+(s) = \frac{-1360}{(s+500)^2}$$
$$+ 0.5\, \frac{(2\pi/.06)}{(s+15.3)^2 + (2\pi/.06)^2}. \qquad (B-2)$$

Observe that $V^+(s)$ is already the sum of two second-order transfer functions. For more complicated signatures, we would have to express their more complicated Laplace transforms as a sum (i.e., a parallel connection) of second-order transfer functions by means of partial fraction expansion techniques. To proceed further, it is convenient to write $V^+(s)$, as

$$V^+(s) = -1360\, F(s) + 0.5\, G(s), \qquad (B-3)$$

where

$$F(s) = \frac{1}{(s+500)^2} = \frac{1}{s^2 + 10^3 s + 25\times 10^4}, \qquad (B-4)$$

and

$$G(s) = \frac{(2\pi/.06)}{(s+15.3)^2 + (2\pi/.06)^2}. \qquad (B-5)$$

Next, we develop second-order state space models for $F(s)$ and $G(s)$. Many different state space models are possible. Two, which are quite useful, and, which are the ones we shall use for $F(s)$ and $G(s)$ are given next.

Let

$$F(s) = \frac{1}{s^2 + a_1 s + a_2}, \qquad (B-6)$$

then $f(t)$ is the solution of the following second-order differential equation:

$$\ddot{f}(t) + a_1 \dot{f}(t) + a_2 f(t) = \delta(t), \qquad (B-7)$$

where $\delta(t)$ is the unit impulse function. Equation (B–7) can be expressed in state space format, as

$$\begin{pmatrix} \dot{x}_1(t) \\ \dot{x}_2(t) \end{pmatrix} = \begin{pmatrix} 0 & 1 \\ -a_2 & -a_1 \end{pmatrix} \begin{pmatrix} x_1(t) \\ x_2(t) \end{pmatrix} + \begin{pmatrix} 0 \\ 1 \end{pmatrix} \delta(t), \qquad (B-8)$$

where $x_1(t) = f(t)$ and $x_2(t) = \dot{f}(t)$. This state equation is valid for all values of a_1 and a_2 [i.e., it is valid for real, multiple, or complex poles of $F(s)$]. When we apply equation (B–8) to (B–4), we obtain

$$\begin{pmatrix} \dot{x}_1(t) \\ \dot{x}_2(t) \end{pmatrix} = \begin{pmatrix} 0 & 1 \\ -25\times 10^4 & -10^3 \end{pmatrix} \begin{pmatrix} x_1(t) \\ x_2(t) \end{pmatrix} + \begin{pmatrix} 0 \\ 1 \end{pmatrix} \delta(t). \qquad (B-9)$$

Next, let

$$G(s) = \frac{\beta}{(s-\alpha)^2 + \beta^2}. \qquad (B-10)$$

Observe that $g(t)$ is the solution of the following second-order differential equation, the eigenvalues of which are always complex:

$$\ddot{g}(t) - 2\alpha \dot{g}(t) + (\alpha^2 + \beta^2)\, g(t) = \beta\, \delta(t). \qquad (B-11)$$

This equation can be expressed in state space format as,

$$\begin{pmatrix} \dot{x}_3(t) \\ \dot{x}_4(t) \end{pmatrix} = \begin{pmatrix} \alpha & \beta \\ -\beta & \alpha \end{pmatrix} \begin{pmatrix} x_3(t) \\ x_4(t) \end{pmatrix} + \begin{pmatrix} 0 \\ 1 \end{pmatrix} \delta(t), \qquad (B-12)$$

where $x_3(t) = g(t)$. To verify equation (B–12), take the Laplace transform of its two first-order differential equations, eliminate $X_4(s)$, and show that $X_3(s)$ is given by the right-hand side of equation (B–10); but, $X_3(s) = G(s)$. When we apply equation (B–12) to (B–5), we obtain

$$\begin{pmatrix} \dot{x}_3(t) \\ \dot{x}_4(t) \end{pmatrix} = \begin{pmatrix} -15.3 & 104.67 \\ -104.67 & -15.3 \end{pmatrix} \begin{pmatrix} x_3(t) \\ x_4(t) \end{pmatrix} + \begin{pmatrix} 0 \\ 1 \end{pmatrix} \delta(t). \qquad (B-13)$$

Finally, to obtain the state space model for $V^+(t)$, we combine equations (B–9) and (B–13), as follows:

$$
\begin{pmatrix} \dot{x}_1(t) \\ \dot{x}_2(t) \\ \dot{x}_3(t) \\ \dot{x}_4(t) \end{pmatrix} = \overbrace{\left(\begin{array}{cc|cc} 0 & 1 & & \\ -25 \times 10^4 & -10^3 & & \raisebox{1.5ex}{0} \\ \hline & & -15.3 & 104.67 \\ 0 & & -104.67 & -15.3 \end{array} \right)}^{A} \begin{pmatrix} x_1(t) \\ x_2(t) \\ x_3(t) \\ x_4(t) \end{pmatrix} + \overbrace{\begin{pmatrix} 0 \\ 1 \\ 0 \\ 1 \end{pmatrix}}^{b} \delta(t). \qquad \text{(B–14)}
$$

Observe also, from equation (B–3), that $V^+(t)$ can be written in terms of states $x_1(t)$ and $x_3(t)$, as

$$V^+(t) = -1360\, f(t) + 0.5\, g(t) = -1360\, x_1(t) + 0.5\, x_3(t),$$

or,

$$
V^+(t) = \overbrace{\begin{pmatrix} -1360 & 0 & 0.5 & 0 \end{pmatrix}}^{h'} \begin{pmatrix} x_1(t) \\ x_2(t) \\ x_3(t) \\ x_4(t) \end{pmatrix} \qquad \text{(B–15)}
$$

Matrices $\mathbf{A}$, $\mathbf{b}$, and $\mathbf{h}$ are indicated in equations (B–14) and (B–15) and are the quantities which appear in the third row of Table 1.

AUTHOR CITATION INDEX

SUBJECT INDEX

About the Editors

VIJAY K. ARYA received the B.E. and M.E. degrees with honors from the University of Rajasthan in 1960 and 1961, respectively, receiving the gold medal awards for merit. He spent the next three years as Assistant Professor of Electrical Engineering at the Birla College of Engineering, Pilani, India. From 1963 to 1967 he was a research assistant and then a research associate at the Communication and Control Systems Laboratory at Ohio State University, Columbus, where he received the Ph.D in 1967.

Dr. Arya joined the Shell Oil Company in 1967 as a research engineer engaged in computer control of petrochemical processes, and in 1972 moved to the geophysics research group of Shell Development Company, Houston. Since 1980 he has been with the Rocky Mountain Division-Exploration, of Shell Oil where he is engaged in various aspects of seismic data processing. He is a member of the Society of Exploration Geophysicists, the European Association of Exploration Geophysicists, the Institute of Electrical and Electronics Engineers, and Sigma Xi; and is also chairman of the Acoustics, Speech and Signal Processing chapter of the Houston section of IEEE.

J. K. AGGARWAL received the B.S. in mathematics and physics from the University of Bombay in 1956, the B.Eng. from the University of Liverpool, England in 1960, and the M.S. and Ph.D. from the University of Illinois, Urbana in 1961 and 1964, respectively. He joined the University of Texas in 1964 as an assistant professor and is now the John J. McKetta Energy Professor of electrical engineering and computer sciences at Austin campus. While still associated with the University of Texas, he was also a visiting assistant professor to Brown University (1968) and visiting associate professor at the University of California at Berkeley (1969–1970).

Dr. Aggarwal is an active member of the Institute of Electrical and Electronics Engineers, Pattern Recognition Society, and Eta Kappa Nu; a former member of the Circuits and Systems Society; and past chairman of the Technical Committee on Signal Processing from 1974 to 1975. An associate editor of *Pattern Recognition* and *IEEE Transactions on Pattern Analysis and Machine Intelligence* and guest editor for a special issue of *Computer Graphics and Image Processing,* he has coedited special issues on digital filtering and image processing (*IEEE Transactions on Circuits and Systems,* March 1975) and on motion and time varying imagery (*IEEE Transactions on Pattern Analysis and Machine Intelligence,* November 1980). He has published numerous technical papers and several books: *Notes on Nonlinear Systems* (1972), *Nonlinear Systems: Stability Analysis* (1977), *Computer Methods in Image Analysis* (1977), *Digital Signal Processing* (1979). His current research interests are image processing and digital filters.